High-temperature Solid Oxide Fuel Cells for the 21st Century

High-temperature Solid Oxide Fuel Cells for the 21st Century

Fundamentals, Design and Applications

Edited by

Kevin Kendall and Michaela Kendall

ELSEVIER

AMSTERDAM • BOSTON • HEIDELBERG • LONDON
NEW YORK • OXFORD • PARIS • SAN DIEGO
SAN FRANCISCO • SINGAPORE • SYDNEY • TOKYO
Academic Press is an imprint of Elsevier

Publisher: Joe Hayton
Acquisition Editor: Raquel Zanol
Editorial Project Manager: Mariana Kühl Leme
Editorial Project Manager Intern: Ana Claudia Abad Garcia
Production Manager: Anusha Sambamoorthy
Marketing Manager: Louise Springthorpe
Cover Designer: Victoria Pearson

Academic Press is an imprint of Elsevier
125 London Wall, London, EC2Y 5AS, UK
525 B Street, Suite 1800, San Diego, CA 92101-4495, USA
225 Wyman Street, Waltham, MA 02451, USA
The Boulevard, Langford Lane, Kidlington, Oxford OX5 1GB, UK

British Library Cataloguing in Publication Data
A catalogue record for this book is available from the British Library

Library of Congress Cataloging-in-Publication Data
A catalog record for this book is available from the Library of Congress

ISBN: 978-0-12-410453-2

Contents

List of contributors

S.B. Adler University of Washington, Seattle, WA, USA

M. Cassidy University of St. Andrews, St. Andrews, UK

P.A. Connor University of St. Andrews, St. Andrews, UK

J. Druce International Institute for Carbon-Neutral Energy Research (I2CNER), Nishi-Ku, Fukuoka, Japan

T. Horita National Institute of Advanced Industrial Science and Technology (AIST) Tsukuba Central 5-2, 1-1-1 Higashi, Tsukuba, Ibaraki, 305-8565 JAPAN

J.T.S. Irvine University of St. Andrews, St. Andrews, UK

T. Ishihara International Institute for Carbon-Neutral Energy Research (I2CNER), Nishi-Ku, Fukuoka, Japan

Z. Jaworski West Pomeranian University of Technology, Szczecin, Poland

T. Kawada Tohoku University, Sendai, Japan

K. Kendall Adelan Ltd, Weekin Works, Harborne, Birmingham, UK; School of Metallurgy and Materials, University of Birmingham, UK

J.A. Kilner International Institute for Carbon-Neutral Energy Research (I2CNER), Nishi-Ku, Fukuoka, Japan

T. Li University of Electronic Science and Technology of China, Chengdu, Sichuan, China

N.Q. Minh University of California, San Diego, CA, USA

L. Niewolak Forschungszentrum Jülich, Jülich, Germany

P. Pianko-Oprych West Pomeranian University of Technology, Szczecin, Poland

W.J. Quadakkers Forschungszentrum Jülich, Jülich, Germany

C.D. Savaniu University of St. Andrews, St. Andrews, UK

S. Sui Shanghai Jiao Tong University, Shanghai, China

F. Tietz Forschungszentrum Jülich, Jülich, Germany

W. Wang Chinese Academy of Sciences, Ningbo, China

W. Winkler Retired Professor and Director of Institute for Energy Systems and Fuel Cell Technology at HAW Hamburg, Germany

G.H. Xiu 23-602 Room, 89 Mudan Rd., Shanghai, China

Preface

The original edition of this book [1] 'High-temperature solid oxide fuel cells: Fundamentals, design and applications' (solid oxide fuel cells, SOFCs) was published in 2003 to celebrate a century of research, development and demonstration of these devices since Nernst's breakthrough in 1897 [2] when he first showed his new incandescent lamp based on zirconia formulations. But since the millennium, a number of very significant advances have been made, illustrating the accelerating pace for advance of SOFC technology. We believe that this justifies the publication of the present book, summarising the older results, then emphasising progress made in the last 15 years. The fact that approximately 4000 copies of the original volume were dispersed and downloaded has encouraged us to update the text.

Progress has been made in several areas which are covered in the new chapters:

- Materials invention covered in Chapters 1–7 explain how new electrolyte compositions, such as scandia-doped zirconia; new cathodes such as LSCF and new interconnects like ferritic stainless steel alloys have shown better performance.
- Stack inventions like m-SOFCs described in Chapters 8–10 have allowed smaller and more rapid starting devices to be proven.
- Application projects covered in Chapters 8–13 have taken new directions such that shipments and installations of SOFCs have risen exponentially, and the best example being the many thousands of SOFC CHP generators now installed in Japan.

Some areas such as thermodynamics have not changed much, but additional recent publications are now covered.

Sad events have also occurred, such as the death of the famous zirconia sensor researcher Hans-Heinrich Mobius who delivered Chapter 2 in the previous book. In addition, a number of the previous authors have retired, and so new younger experts are featured in this volume. We thank Subhash Singhal for instigating this book project and also thank all the SOFC authors who have made a contribution to this work.

Other SOFC books have appeared in the interim [3–8], but these have been essentially conference proceedings or have not given a broad coverage of the wider SOFC field. The present volume is unique in reviewing the big picture in all its complexity.

We thank Elsevier for their support and all the readers who have contributed to make this project successful.

Dr. Kevin Kendall and Dr. Michaela Kendall
Metallurgy and Materials
University of Birmingham
k.kendall@bham.ac.uk

References

[1] S.C. Singhal, K. Kendall, High Temperature Solid Oxide Fuel Cells: Fundamentals, Design and Applications, Elsevier Science, Oxford, 2003.
[2] W. Nernst, DRP 104872, 1897.
[3] J.W. Fergus, R. Hui, X. Li, D.P. Wilkinson, J. Zhang, SOFCs: Materials Properties and Performance, CRC Press, Boca Raton, 2009.
[4] J.T.S. Irvine, P. Connor, SOFCs: Facts and Figures, Springer, London, 2013.
[5] J. Milewski, K. Swirski, M. Santarelli, P. Leone, Advanced Methods of SOFC Modeling, Springer, London, 2011.
[6] T. Ishihara, Perovskite Oxide for SOFC, Springer, London, 2009.
[7] H. Yokokawa (Ed.), Fuel Cell Handbook: Fundamentals Technology and Applications, John Wiley and Sons, New York, 2009.
[8] D. Stolten, B. Emonts (Eds.), In: Fuel Cell Science and Engineering, vol. 2, Wiley-VCH, Weinheim, Germany, 2012, p. 963.

Introduction to SOFCs

1

K. Kendall[1,2]
[1]Adelan Ltd, Weekin Works, Harborne, Birmingham, UK, [2]School of Metallurgy and Materials, University of Birmingham, UK

1.1 Introduction

Solid oxide fuel cells (SOFCs) are the most efficient devices yet invented for conversion of chemical fuels directly into electrical power. Originally the materials' compositions were found by Nernst [1–3] in Gottingen in the 1890s as described in Chapter 2, but considerable improvements in theory and experiment are still being made 120 years later. During the last decade, significant progress has been made to improve the materials, to construct better systems, to focus on simpler fuels and to seek premium applications. This chapter describes these improvements, after summarising the background and situation at 2000 covered in the last book [4]. Firstly, we describe the basic principles and then we review the areas of advance, following with brief descriptions of the history, materials, processing, stack designs, systems, fuels and applications.

1.2 SOFC principles

Figure 1.1 shows an SOFC scheme. It contains a solid oxide electrolyte made from ceramic such as yttria stabilised zirconia (YSZ) which behaves as a conductor of oxygen ions at temperatures from 500 to 1500 °C. This ceramic material allows oxygen atoms to be reduced on its porous cathode surface by electrons brought in through the metal interconnect, thus being converted into oxygen ions, which are then transported through the ceramic oxide electrolyte to a fuel-rich porous anode zone where the oxygen can react, say with hydrogen, giving up electrons to an external circuit as shown in Figure 1.1. Only five components are needed to put such a cell together: electrolyte, anode, cathode and two interconnect wires, although other layers are often added as described later in the chapter.

Of course, hydrogen is ideal for SOFC. But hydrogen is not the preferred fuel because it would more likely be used in a polymer electrolyte membrane fuel cell (PEMFC) which gives extra power at room temperature. Hydrocarbon gases such as methane, or propane derived from pipelines or liquid stores, or biofuels like methanol and formic acid are more attractive for SOFCs for several reasons:

(1) They can potentially be reacted directly on the anode, impossible with PEMFCs at room temperature with existing catalysts.
(2) Hydrocarbons react well with steam or air to provide hydrogen and carbon monoxide which then combine with oxygen ions at the anode/electrolyte interface.
(3) The thermodynamic efficiency of such internal reforming processes can be impressive.

High-temperature Solid Oxide Fuel Cells for the 21st Century. http://dx.doi.org/10.1016/B978-0-12-410453-2.00001-4

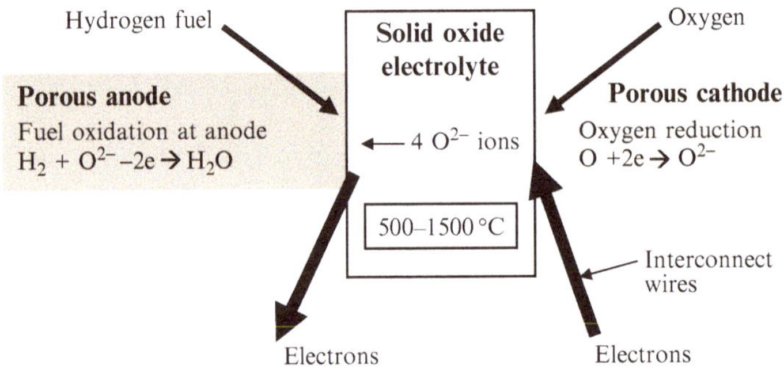

Figure 1.1 Schematic of solid oxide fuel cell (SOFC).

(4) Hydrocarbon fuels give good energy storage density compared to hydrogen, especially when liquid, for example, propane/butane pressurised at room temperature or methane below its critical point of −83 °C.

The anode reactions then become more complicated with a range of possible products shown in the following chemical equations:

$$\text{Direct } CH_4 \text{ oxidation } CH_4 + 4O^{2-} - 8e \rightarrow 2H_2O + CO_2$$

$$\text{Anode CO oxidised } CO + O^{2-} - 2e \rightarrow CO_2$$

$$\text{Steam reforming } CH_4 + H_2O \rightarrow CO + 3H_2$$

$$\text{Partial oxidation } 2CH_4 + O_2 \rightarrow 2CO + 4H_2$$

The key benefits which make the SOFC more attractive than PEMFCs are twofold:

(1) Direct use of liquid fuels which have much superior energy storage capacity (e.g. a factor 6) when compared with compressed hydrogen gas.
(2) Availability of high-quality heat which can be used to give advance in practical applications as described in the following chapters.

Examples of applications demanding this heat are domestic homes which need both power and heat, auxiliary power units (APUs) on vehicles and stationary power generation. In each of these cases, the electrons are extracted from the fuel in the fuel cell at a temperature around 600–1000 °C, as shown in Figure 1.1. A domestic system would then use the heat to produce hot water as currently achieved with simple heat exchangers. In a vehicle, the heat could be used to warm the cab as illustrated in Figure 1.2. A stationary power generation system may use the waste heat to drive a heat engine such as a stirling or gas turbine motor.

It is essential to realise that the heat is inevitably generated in the SOFC by reaction exotherm, ohmic losses, electrode overpotentials, etc., as described in Chapter 3. These losses are present in any fuel cell arrangement and cannot be eliminated completely. Indeed, the heat is necessary to maintain the operating temperature of the cells, replacing the heat lost by thermal conduction and radiation to the

Figure 1.2 Truck with APU running on CNG.

surrounding environment. The benefit of the SOFC over competing fuel cells is the high temperature of the exhaust heat, making its utilisation simple and economic. Thus both electricity and heat are desirable products from the SOFC. The ratio of power to heat can realistically be adjusted from 20% to about 70% depending on the system design, but for basic designs, range from 40% to 60%. These figures compare with about 20–27% for an automobile combustion engine, about 50–60% for the best combined cycle gas/steam turbine power plant and around 40–60% for a diesel combustion generator combined with a steam cycle.

From these basic principles, it is possible to recognise the issues that remain to be solved if SOFCs are to be successful in this century.

1.3 Problems to be resolved

The SOFC process described above is almost magical in its elegance and simplicity, and it is surprising that this process has not yet been commercialised to supplant the inefficient and polluting combustion heat engines which currently dominate our civilisation.

There are several technical reasons why SOFCs have not yet competed successfully with combustion devices, as shown in Figure 1.3, but cost and durability issues also need attention. In the first place, combustion has been accepted for several thousand years, and such established technologies are difficult to supplant. Secondly, combustion does not require any additional materials or interfaces; it can occur freely once

COMBUSTION	SOFC
Established technology	Disruptive technology
Independent of materials	Depends on five materials
Dirty fuels still combust	Clean fuels are necessary for SOFC

Figure 1.3 Difficulties of challenging combustion with SOFC.

ignition of the fuel has occurred. Containment and expansion of the hot gas product is the only requirement for power production. Contrast this situation with the SOFC which needs five materials and four interfaces, each of which can block the power output when contaminated with a single molecular layer of silica. Thirdly, combustion can occur with almost any oxidisable substance so that cheap available complex fuels, for example, petroleum or biomass, may be used. Purity is crucial to fuel cells but has not yet been a barrier to combustion, except where emissions legislation has been imposed.

Consider these three challenging areas in more detail. The historic dominance of combustion is difficult to overcome because the concepts are ingrained in our psyches. From the steam turbine of Hero two millennia ago, to the Watt steam engine two centuries back, to the Diesel engine of 1897, it has been evident that heat engines can utilise the additional hot gas pressure created by combustion to produce power of enormous utility. It is much less obvious that electrochemical potential of fuels can be used in a similar way. Electrochemistry, which can reasonably be dated to Volta's invention of the battery in 1800, has made enormous strides in two centuries but still falls short in competition. A recent example is the push towards battery electric cars which many observers feel should take over soon from combustion engine automobiles. Despite a number of advantages of efficiency and reduced emission, these machines cannot yet beat the consumer preference for combustion vehicles based on price, range, infrastructure and fuel recharge times [5]. Perhaps the main reason for this is the complex nature of automobile technology. A recent paper shows that the use of composite materials in cars is just as important as moving to an electric drive [6]. However, the longer-term effect of emission regulations together with the simplicity and efficiency of electrical drives should eventually lead to fuel cells taking over from combustion devices [7].

The materials problems of SOFCs are profound and dominate this book. Although YSZ has been known for 120 years to provide a suitable electrolyte (Chapter 4), the present anode composition did not appear until the 1960s and the lanthanum strontium cobalt ferrite (LSCF) cathode has only just been optimised in this century as described in Chapters 5 and 6. Moreover, the use of ferritic stainless steel interconnections and the coatings required to inhibit corrosion are of relatively recent origin as defined in Chapter 7. The way in which these materials are best assembled to produce working stacks has been changing continuously since 1990 as illustrated in Chapters 8–10. The nature of the interfaces, especially the catalytic surfaces, has also been critically recognised as detailed in Chapters 11–14.

The effect of fuel purity (Chapter 14) is hugely important because our present infrastructure is based largely on refined petroleum which typically contains hundreds of molecules, including sulphur compounds which corrode SOFC anodes based on nickel in a very short time. Diesel engines and other combustion systems can withstand extraordinary variations in fuel composition, which today differs significantly from town to town and country to country. But fuel cells make use of catalysts which can easily be poisoned by low-level impurities, and there is great difficulty attached to running SOFCs on applications now dominated by diesel technology. Ships, trucks and buses fall into this category. A typical programme to produce APUs for trucks in the United States, the SECA project, was not as successful as it should have been because the processing of diesel fuels before feeding the fuel to the SOFC was extremely costly and time-consuming. Even after reforming a complex fuel like diesel, the lifetime of SOFCs may still be insufficient. However, as diesel emissions become a focus for environmental change, it may be reasonable to predict that fuels will become cleaner and SOFCs may become more durable. This has already happened in the EU where sulphur levels must be controlled to low concentrations. It is also true that diesel fuel from gas-to-liquids refineries is much cleaner than previous formulations. Another promising area of research at present is to use SOFCs to produce the ultimate clean fuel, hydrogen, using electrolysis in the SOEC.

This book attempts to cover these areas more comprehensively than the 2003 volume [4]. In the remainder of this introductory chapter, we examine the main points listed above, starting with a history summary and then moving on to cover the materials, followed by reflections on the component processing, the stacks, the systems, fuels and the new promising applications.

1.4 Historical summary

Whereas combustion was essentially understood in 1772, the understanding of SOFCs has been relatively recent as Chapter 2 describes in detail. Nernst made the breakthrough of observing various types of conduction in stabilised zirconia, that is zirconium oxide doped with several mole per cent of calcia, magnesia, yttria, etc. He found that stabilised zirconia was an insulator at room temperature, conducted ions in red hot conditions, from 500 to 1400 °C and then became an electronic and ionic conductor at white heat, around 1500 °C. He patented an incandescent electric light made from a zirconia filament and sold this invention which he had been using to illuminate his home [1–3]. He praised the simultaneous invention of the telephone because it enabled him to call his wife to switch on the device while he travelled back from the university. The warm-up time was a problem even then [8].

The zirconia lighting filament was not successful in competing with tungsten lamps, and Nernst's invention languished until the late 1930s when a fuel cell concept based on solid zirconium oxide was demonstrated at the laboratory scale by Baur and Preis [9]. They used a tubular crucible made from zirconia stabilised with 15 wt% of yttria as the electrolyte, the so-called Nernst Mass. Iron or carbon was used as the anode and magnetite (Fe_3O_4) as the cathode. Hydrogen or carbon monoxide was the fuel on the inside of the tube, and air was the oxidant on the outside. Eight cells were connected in series to make the first SOFC stack. They obtained power from the device and speculated that this SOFC could compete with batteries. But several improvements were necessary before this would be possible. For example, the electrolyte manufacturing process was too crude and needed optimising, especially to make the electrolyte thinner to reduce the cell resistance from around 2 ohms. In addition, the electrodes were inadequate, especially the cathode Fe_3O_4 which readily oxidised. Also, the power density was small with this stacking arrangement, the

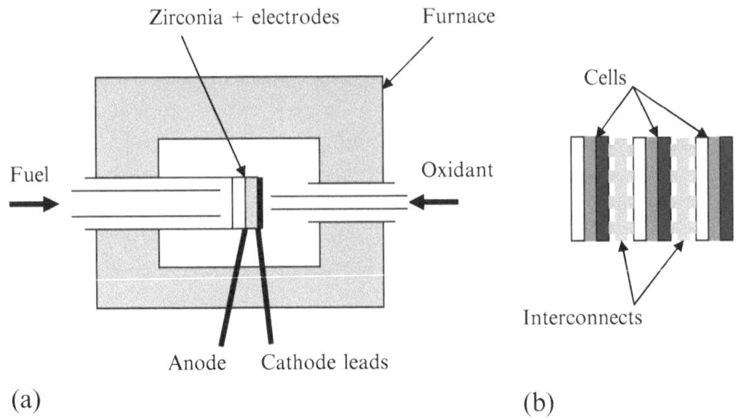

Figure 1.4 (a) The flat button test cell and (b) planar stack of cells and interconnects.

connections between many cells had to be developed and the understanding of fuel reactions and system operation needed much attention.

It was not until the 1950s that experiments began on pressed or tape-cast discs of stabilised zirconia when a straightforward design of test system was developed which is very convenient for pellets and is therefore still in use today, as described in Chapter 12. The essentials of the apparatus are shown in Figure 1.4a. A flat disc of stabilised zirconia is sealed to a ceramic tube and inserted in a furnace held at red heat [10]. A smaller diameter tube is inserted into the ceramic tube to bring fuel to the anode, and another tube brings oxidant gas to the cathode side. Current collector wires and voltage measurement probes are brought out from the electrode surfaces. Once a flat plate of electrolyte had been used, it was easy to see how the flat plate voltaic stack could be built up with interconnecting separator plates to build a realistic electrochemical reactor, as shown in Figure 1.4b. The interconnect plate is essentially made from the anode current collector, and the cathode current collector joined together into one sheet, thus combining the two components. Additionally, the interconnector can contain gas channels which supply fuel to the anode and oxidant to the cathode as well as electrically connecting the anode of one cell to the cathode of the next.

It turned out that there were several problems with flat plate stacks as they were made larger to generate increased power, including sealing around the edges and thermal expansion which caused cracking. However, the configuration of Figure 1.4 has been prominent in zirconia sensors which are now manufactured in large numbers.

1.5 Zirconia sensors for oxygen measurement

An SOFC in reality already exists on every automobile: it is the oxygen sensor device which sits in the exhaust manifold in order to control the oxygen content of the effluent mixture entering the exhaust catalyst. The mixture composition must be controlled to near stoichiometric if the catalyst is to operate at its optimum performance. YSZ is generally used as the electrolyte because this uniquely detects oxygen, and platinum is normally painted on the ceramic surface using proprietary inks to provide the electrodes.

In the original design, which was used from the 1970s as described in Chapter 10, the configuration was similar to that of Baur and Preis [9]. A thimble of YSZ, containing typically 8 wt% yttria, was pressed from powder and fired to 1500 °C to densify the ceramic. Platinum electrodes were applied, and the unit then fixed in a steel plug which could be screwed into the car inlet manifold so that the YSZ cell was protruding into the hot gases. Air was used as the oxygen reference on the cathode side. A wire connection supplied the voltage from the inner electrode to the engine management system, while the other electrode was grounded to the chassis.

Once the exhaust warmed up, above 400 °C, the voltage from the sensor changed with the oxygen concentration in the exhaust gas stream. This voltage varied with the logarithm of oxygen level, giving the characteristic λ-shaped curve of voltage versus oxygen concentration, hence the name 'lambda sensor'. The car computer then used the oxygen sensor signal to manage the engine so that the exhaust composition was optimised for the catalyst. Various improvement have been made to this basic system over the years; for example, a heating element can be built within the thimble in order to obtain a rapid heating sensor as discussed in more detail in Chapter 10.

The major improvement introduced by Robert Bosch GmbH in 1997 was to redesign the zirconia sensor and to manufacture it by a different method. Instead of pressing a thimble from dry powder, a wet mix of zirconia powder with polymer additives was coated and dried like a paint film on a moving belt in a tape-casting machine. The film dried to a thickness of around 100 μm and could be screen printed with the platinum metallisation before pressing three or four sheets together to form a planar sensor array which was fired then diced to size before inserting in the metal boss which screwed into the engine manifold. This process was rather like the ceramic capacitor process developed for the electronic ceramics industry.

There were several benefits of this new device:

(a) There was much less zirconia in it, about 2.8 g.
(b) The thinner ceramic electrolyte gave much faster response.
(c) Heaters and other circuits could readily be printed onto the flat sheets.

An immediate bonus of this technology was the possibility of producing linear response sensors as opposed to the logarithmic response of the thimble type, so as to match the electronic control system more easily. This was achieved by setting the oxygen reference by using one of the sheets as an oxygen pump which could then leak from the cathode compartment through a standard orifice [11].

Oxygen sensors are now widely used in food storage, metal processing and flame controllers, but the main market is automobiles. Zirconia technology for sensors has been so successful in the marketplace, several hundred million dollars in 2010, that it has pushed forward the development of fuel cell materials. The main difference is that the power output of sensors is low so that partially stabilised zirconia can be used. At higher power, fully stabilised zirconia must be used if the electrolyte is to remain stable for long periods. The supply of this material is to be discussed next.

1.6 Zirconia availability and production

The main electrolyte material used in SOFCs at present is YSZ, as described more fully in Chapter 4. Although many other oxide materials conduct oxygen ions, some rather better than zirconia, this material has a number of significant attributes which make it ideal for this application, including abundance, chemical stability, non-toxicity and economics. Against these one can mention several drawbacks, including the high thermal expansion coefficient, plus the problems of joining and sealing the material.

Low-grade stabilised zirconia already commands a large market, especially in refractories, pigment coatings and colours for pottery, but it is only during the last 40 years that technical grade zirconias have been produced for applications such as coatings on gas turbine combustors, hip joint implants and cutting tools. Much of this technology has stemmed from the study of pure zirconia and the effects of small amounts of dopants on the crystal structure and properties. Large effects were seen in the early 1970s, pointing the way to substantial applications of this material [12].

Figure 1.5 shows the trend in world-wide production levels of ionic conductor grade yttria–zirconia over time [13]. It is evident that in 1970 there was very small production at rather high price. However, the introduction of the zirconia lambda sensor to control the emissions of automobiles in the 1970s had a large effect on the production rate, and prices have dropped steadily since that time. The price in 2000 was \$50/kg in 50 kg lots and this changed to \$90/kg in 2013 (no real increase when corrected for inflation). Prices have fluctuated significantly with a spike in 2012. Falls in pure YSZ costs are expected to occur with an estimate of \$40/kg (2000 price) in 2030 as production is predicted to rise to 10,000 te/a. In the year 2013, the sensor application of YSZ was dominant with an estimated world production of 600 metric tonnes, risen from 500 te in 2000. Fuel cell demand had still not taken off with total production of only 15 te in 2013, 40 times less than sensor material.

There is little doubt that large quantities of zirconia will be needed for the SOFC application in the years to come, with annual production rising to more than 1 Mte/year

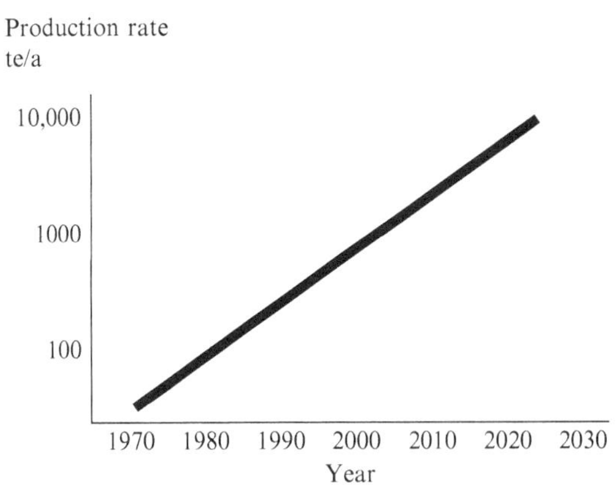

Figure 1.5 Trend in the production of ionic conducting yttria stabilised zirconia powder.

in 2100, rather as titania expanded in the last century for pigment applications. Fortunately, zirconia is one of the most common materials in the earth's crust, being much more available than copper or zinc, for example. Large deposits exist in Australia, Africa, Asia and America, usually as the silicate, zircon $ZrSiO_4$. In terms of cost, the greatest difficulty is purifying this raw material, especially to remove SiO_2 which tends to block the ionic and electron paths in fuel cell systems. A typical zirconia powder for electrolyte application should contain less than 0.1% by weight of silica, and the highest quality YSZ electrolytes contain only 0.005% by weight. Other impurities, like alumina and titania, can be useful in mopping up the damaging silica so that levels of 0.1% by weight are normal. The main impurity, hafnia, is usually present at several wt% but causes no problem because it is an ionic conductor in its own right. Often, zirconia contains small amounts of radioactive α emitter impurities, and this could pose a potential health problem during processing, but otherwise there are no significant toxic hazards known.

Yttria is the principal stabiliser used at present, though both the more expensive scandia and ytterbia give better ionic conductivity. Typically, yttria is added at 13–16% by weight (8–10.5 mol%) to give a fully stabilised cubic material suitable for high current density. Details of these materials are given in Chapter 4. Supply of scarce dopants such as scandia could be a problem in future. However, a more significant issue is the processing of the electrolyte material into a functional device.

1.7 High-quality electrolyte fabrication processes

One of the main issues in slowing down the advance of SOFCs has been the difficulty of making good electrolytes as described in Chapter 4. The electrolyte has to fit several criteria for success:

It must be dense and leak-tight
It has to have the correct composition to give good ionic conduction plus high electronic resistance at the operating temperatures
It must be thin to reduce the ionic resistance
It must be extended in area to maximise the current capacity
It should resist thermal shocks
It must be economically processable

These requirements are not easily reconciled. Industrial ceramic processing has traditionally focused on the pressing of dry powders in metal dies or rubber moulds to make spark plugs, for example. Although zirconia sensors have been made by this technique, and although much academic research has used this method, it is difficult to make thin-walled parts of large area in this way. A stacked tubular design made by powder pressing had been demonstrated in the 1960s, but this had proved to be expensive because of diamond grinding and high resistance due to the 500-μm thick electrolyte [14]. It was far better (Figure 1.6) to move towards the advanced electronic ceramic processes such as chemical vapour deposition, tape casting and extrusion to make the required thin films of electrolyte, as demonstrated in the YSZ sensor market.

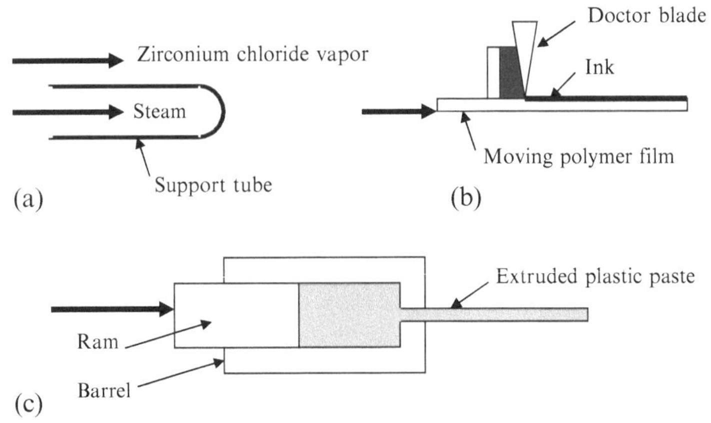

Figure 1.6 Comparison of three electrolyte fabrication processes: (a) electrochemical vapour deposition, (b) tape casting and (c) extrusion.

In the late 1970s, chemical vapour deposition began to be used to make tubular cells at Westinghouse [15]. A porous tubular substrate, around 20 mm diameter, made originally from calcia stabilised zirconia but later from the cathode material, lanthanum strontium manganite (LSM), was placed in a low-pressure furnace chamber and zirconium chloride plus yttrium chloride vapour was passed along the outside of the tube, while water vapour passed down the inside. This deposited a layer of yttria-doped zirconium oxide which blocked the pores and then subsequently grew to about 40 μm in thickness to form the electrolyte layer [15]. The interconnect strip could also be formed from magnesia-doped lanthanum strontium chromite by the same principle. Although these gave good electrochemical performance, the process was lengthy and expensive when compared with tape casting or extrusion. Also, the heavy tubes could not be heated rapidly and required around 6 h start-up time for warming.

Tape casting was originally used to make thin tape materials for electronic applications [16] especially using organic solvents. A slurry of the YSZ powder, with solvent and dispersing agent, for example, methyl ethyl ketone/ethanol mixture with Hypermer KD1 (Croda), was ball milled for 24 h to finely grind the particles and remove agglomerates [17]. Then a polymer and plasticiser mixture was prepared by milling polyethylene oxide and dibutylphthalate with the solvent, then mixing with the particle dispersion, followed by further ball milling. After filtering and vacuum deairing, the slurry was tape cast on a polymer film and dried for 3 h before firing at 1300 °C.

Water-based casting is much more desirable than the solvent system for environmental reasons, and this was developed by Viking Chemicals who prepared their own pure zirconia by solvent extraction techniques [18]. The calcined zirconia powder was bead milled in water with ammonium polyacrylate solution (Darvan 821A, Vanderbilt Minerals) to give a very stable dispersion. To this suspension, a solution of purified ethyl cellulose was added, followed by filtering and deairing. This was tape cast onto polymer film, dried and then fired at 1450 °C. Similar dispersions have been screen printed onto tape-cast anode tapes made by a similar casting procedure to give co-fired supported electrolyte films of reduced thickness which gave enhanced current capacity [19]. Such results were originally reported by Minh and Horne [20] who used the tape calendering method which is similar to tape casting but with a plastic composition [21]. They also corrugated the plates and made monolithic designs by sticking corrugated pieces together in a stacked structure. Of course, the problem with flat plate designs is the thermal shock which prevents rapid heating or cooling. This was a particular problem for monolithic structures which cracked very easily when made more than a few centimetres in length.

To prevent the thermal shock problem, smaller diameter tubes have been produced by extrusion as described in Chapter 10 [22]. Again, these compositions were prepared by mixing zirconia powder with water and polymer, for example, polyvinylalcohol. Extrusion through a tube die gave tubes which could be 2 mm in diameter and 100–200 μm wall thickness, sinterable at 1450 °C. Coextrusion of thin electrolyte was also first demonstrated in 1994 [18]. The recent advance has been the move to such anode-supported cells with 10 μm of YSZ electrolyte and the introduction of LSCF cathodes as described next.

1.8 Anode-supported SOFC materials and reactions

In order to produce lower-resistance electrolyte, thinner YSZ layers are required around 10 μm thick. The nickel cermet anode has therefore been made thicker to provide the mechanical support, while LSCF has been introduced to give improved cathode performance at 700 °C. Figure 1.7 shows a typical cross-section through a cell [23].

Nickel is the main anode material used in SOFC anodes since 1964, largely because of its known performance and economics. Unfortunately, nickel metal does not adhere strongly to YSZ and flakes off unless it is mixed with zirconia. This flaking is driven by the large difference in expansion coefficient between metal and ceramic; YSZ expands at around $11 \times 10^{-6}\,K^{-1}$, whereas nickel expands much more at $14 \times 10^{-6}\,K^{-1}$. By powder mixing 30 vol% nickel oxide with YSZ, followed by firing at 1300 °C to give a porous anode layer by reduction in hydrogen, this mismatch can be reduced. The expansion coefficient of this 'nickel cermet' anode is about $12.5 \times 10^{-6}\,K^{-1}$, allowing

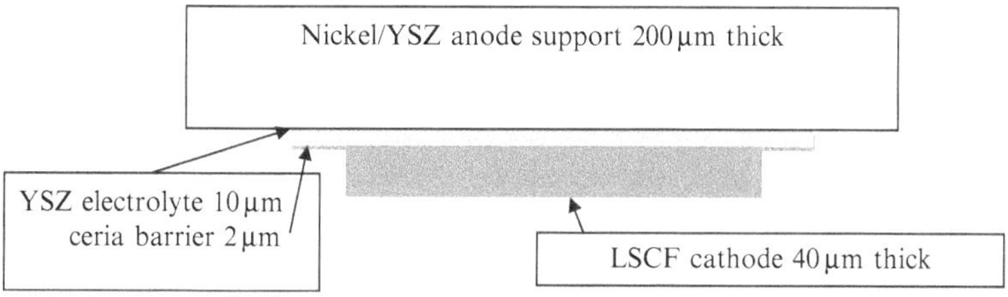

Figure 1.7 Cross-section of typical SOFC with nickel/zirconia anode support, 10 μm YSZ electrolyte, ceria barrier layer and LSCF cathode.

much better adhesion to the electrolyte. The two main demands on the anode are to give rapid, clean reactions with the fuel and to provide good conduction to the interconnect.

The main problem with nickel anodes is their propensity to coke, that is to become coated with a carbon layer on reacting with hydrocarbon fuel. This carbon layer has two deleterious effects: it can disrupt the anode by pushing the nickel particles apart, and it can form a barrier at the nickel surface, preventing gas reactions. Typically, if a hydrocarbon such as methane is fed directly into an SOFC anode, then it will be dead within 30 min as the coking proceeds. Additives to the cermet such as 5% ceria or 1% molybdena can inhibit this process. Alternatively, metals other than nickel can be employed [24].

Cathodes present the main electrode issues in designing and operating SOFCs, as described in Chapter 6. Since these operate in a highly oxidising environment, it is not possible to use base metals, and the use of noble metals is cost prohibitive. Consequently, semiconducting oxides have been the most prominent candidates since 1966 when doped lanthanum cobaltites began to be used, followed in 1973 by manganites [25]. Typically, $La_{0.8}Sr_{0.2}MnO_3$ (LSM) gives a good combination of electronic conductivity and expansion coefficient matching and is available commercially for SOFC applications. Higher conductivity can be obtained at higher dopant levels, but the expansion coefficient then becomes too high. The cobaltite is a much better material from the catalytic and conduction standpoints but is too reactive with zirconia and also expands too much, but has achieved success by using a protective CeO coating, about 2 μm thick. Even the manganite reacts with zirconia above 1400 °C and produces an insulating layer of lanthanum zirconate which increases the resistance enormously. Therefore, firing of the cathode materials on YSZ tends to be kept below 1300 °C, and a minor excess of manganese is used to inhibit the reaction. The manganese can be seen diffusing into the YSZ at high temperatures, a blackened region gradually penetrating the normally white electrolyte.

In order to minimise the resistance at the LSM cathode, especially as the operating temperature of the SOFC is reduced below 900 °C, it has become normal practice to mix the LSM powder with YSZ powder, roughly in 50/50 proportion, to form the first layer of cathode material at the electrolyte surface. This allows a larger 'three-phase boundary' (the line where the gas phase meets both electrolyte and electrode phases) to exist between the oxygen molecules in the gas phase, the LSM particle and the YSZ electrolyte as shown in Figure 1.8. By this means, the cathode contribution to cell resistance can be brought down to about 0.1 ohm for 1 cm^2 of electrode [26]. Alternatively, various doping layers such as ceria or urania can be applied to the YSZ electrolyte before printing on the electrode composition.

The electrode layers have been applied using numerous methods, ranging from vapour deposition and solution coating to plasma spraying and colloidal ink methods such as screen printing and paint spraying, which is perhaps the most economic method. This process is widely used in the traditional ceramic industry to lay down glaze layers from particulate inks to give electrode thicknesses of 50–100 μm. It is advantageous to reduce the number of fabrication steps by adopting composite processes, whereby several layers are produced at one time. One such technique was developed by Minh to tape calender the two electrodes and the electrolyte together

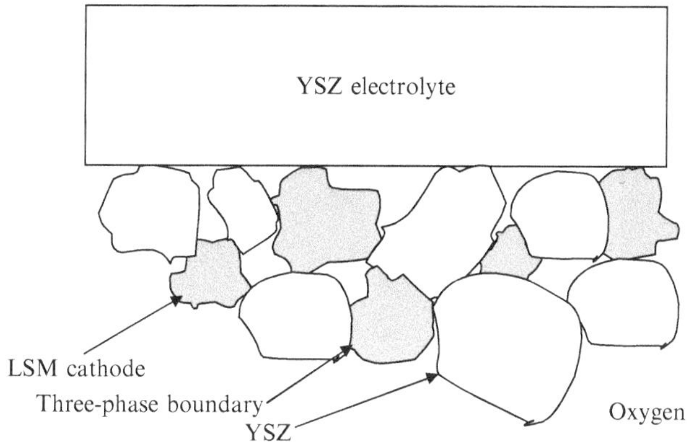

Figure 1.8 Concept of extended three-phase boundary at cathode/electrolyte interface.

as one sheet [27]. Another method that of co-extrusion was developed to make the anode and electrolyte in one step [28]. The problem with low-temperature cathodes such as LSCF, which is described in more detail in Chapters 2, 6, and 10, is that it is sintered at much lower temperatures than nickel cermet, YSZ and even LSM so that it is usually applied after the first firing and needs a second sintering process.

1.9 Interconnection for electrically connecting the cells

The interconnect has always been the most elusive material in electrochemical systems. A perfect interconnect material has not yet been found for even the humble lead/acid battery.

Ideally, an inert and impervious conducting material is needed to withstand both the oxidising potential on the air electrode and the reducing condition at the fuel side. In the SOFC, since 1974, lanthanum chromite has been used to carry out this function for the systems operating near 1000 °C. This material has almost exactly the same thermal expansion coefficient as YSZ, depending on doping. Typically, strontium dopant has been used at 20 mol% to give an expansion coefficient of about $11 \times 10^{-6}\ K^{-1}$. For systems operating at lower temperatures, around 700 °C, metallic alloys like ferritic stainless steel are now being widely used [29] as described in Chapter 7, especially with ceramic coatings to inhibit oxidation on the cathode side.

Magnesium-doped lanthanum chromite was the interconnect material most used in the original Westinghouse tubular design with LSM cathodes, but this has now largely been usurped by stainless steel interconnectors contacting LSCF cathodes in both planar and microtubular designs [30].

For example, Ceramic Fuel Cells Limited (CFCL) used metal interconnects in their small-scale domestic SOFC heat and power unit [31]. The alloy is largely steel with chromium and other additions. This alloy has almost the same expansion coefficient as YSZ and has the benefit of improved processability, strength and toughness compared to ceramic interconnects. But it requires coating to prevent chromium migration.

Another approach is to consider the lead–acid battery and to use wire connections rather than a sheet of interconnect. Wires are brought out from the electrodes and connected externally. This is the approach adopted by Adelan in their microtubular design [32]. Clearly, the design of the cells and how they fit into the overall stack is vitally important in deciding such issues.

1.10 Cell and stack designs

An SOFC is a straightforward five-component entity as described in Figure 1.1. The main problem, which has been exercising engineers for the past 30 years, is that of designing cells which can be stacked to produce significant power output. This power output is directly proportional to the cell area, so the maximum area of YSZ membrane must be packed into the SOFC reactor stack. This is similar to a heat exchanger design exercise. Two plausible solutions are obvious: a stack of flat plates or an array of parallel tubes. The same heat exchanger problems of joining, cracking and leakage are evident in the SOFC stacks because of the complex materials and the high expansion coefficient. Of course the difficulties are greater because of the high temperature of operation, high expansion coefficients and electrical connection short-circuits. Additional difficulties arise because of the low toughness of the ceramic components and the necessity of making electrical connections between all the cells.

Planar cells have the advantage that they can be readily electroded by screen printing, they can be stacked together with narrow channels to achieve high power densities and they can provide short current pathways though the interconnect. If p is the power in Watts per square centimetre of membrane and g is the gap in centimetre between planar electrolyte sheets, then the stack volumetric power density is p/g kW/L, typically 1 kW/L for a Hexis planar stack where p is 0.5 Wcm^{-2} [4]. In a tubular stack packed in a square array, as in the Westinghouse design, the power density depends on the diameter D of cells and the gap g between them according to $\pi Dp/(D+g)^2$ which gives 0.6 kW/L for Westinghouse tubes 2 cm in diameter with a 0.2 cm gap. This is lower than the planar stack because of the relatively large diameter of the tubes. Obviously, high power density depends on having small diameters and less gaps. The microtubular design gives six times better power density at 0.15 cm diameter of electrolyte tube with 0.1 cm spacing as shown in Figure 1.9 [32]. All these figures exclude the volume of thermal insulation and other ancillary parts.

Many companies, including Versa Power, LG (formerly RRFCS) and Hexis, are currently developing planar SOFCs because of the known merits of that design, as explained in Chapters 8 and 9. However, one major problem is still dominant; slow warm-up of existing planar designs, a consequence of the high thermal expansion coefficient and brittleness of YSZ. If the planar stack is heated to 800 °C too rapidly, then it may crack, causing catastrophic failure. Any large YSZ structure will suffer the same problem as thermal gradients are set up through the ceramic. Large tube cells such as those designed by RRFCS [33] required several hours to warm up safely. Thus it is important to use smaller plates or tubes to resist thermal shock. The downside of this is the greater assembly problem for large numbers of small cells.

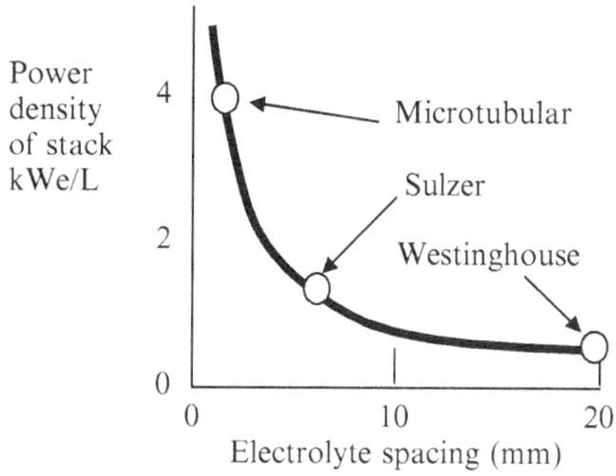

Figure 1.9 Power density of three different stacking geometries.

Large planar cells display two other problems which cause concern: one is the difficulty of making and handling large areas of delicate sheets; the maximum size that has been successfully fabricated is 20 × 20 cm, far smaller than that possible with polymer membranes. The second problem is the gas sealing around the edges of the planar cells; this can be achieved with metal or glass seals, but the required tolerance of around 10 μm in the membrane dimensions causes high cost. Hexis (formerly Sulzer Hexis) avoided this issue by having discs without seals at the outer circumference. Of course, any rigid bonding together of a large ceramic structure also exacerbates the thermal shock issue. Monolithic designs have not been successful for that reason.

The original Westinghouse (later Siemens Westinghouse) tubular design was ingenious because 1.5 m long cells could be manufactured and handled as a result of the inherent strength of the tube structure. A 100 kWe generator could then be built from 1152 cells. Moreover, the sealing problem was eliminated by inserting an air feeder tube down the cell tube. Although the Westinghouse tubular design was large and expensive, it did demonstrate several important features which have lent credence to the SOFC technology picture:

(1) The cells ran for long periods without much deterioration
(2) The efficiency was impressive, around 60%
(3) Methane was used as fuel after desulphurising and pre-reforming
(4) The SOFC exhaust could drive a gas turbine
(5) Emissions were low

In order to understand and predict the performance of such complex stack structures, various mathematical models have been developed, as described in Chapter 13. The most fundamental model starts from the reaction diffusion equations, assuming constant temperature conditions, and calculates the gradient of reactants, products and potentials along a tube or plate of electrolyte [34]. This gives a very sharp reaction front under normal operating conditions if the tube or plate is open ended. The chemical gradient along the SOFC can also be predicted as oxygen ions permeate through the electrolyte [35]. Another important model sets out to calculate temperature and current distributions in a stack of cells [36]. Many such models for different geometries including planar and tubular have been published. The main conclusion is that a 100 mm^2 planar stack with crossflow configuration running on hydrogen at 900 °C will generate a temperature gradient of 140 °C along the plate. The cell will run near 50% efficient if the fuel utilisation is 85% and oxygen 25%. Power density is about 1 kWl^{-1} and 0.6 kW kg^{-1}.

1.11 SOFC reactor systems

Typically, 20% of the volume of a fuel cell system is made up of the cell stack. The rest of the reactor is the balance of plant (BOP) which includes thermal insulation, pipework, pumps, heat exchangers, heat utilisation plant, fuel processors, control system, start-up heater and power conditioning, as described in Chapters 8–10. Arguably, this BOP is the dominant part of the device and should be treated with some concern. One of the major problems of the original Westinghouse design for a 100 kWe cogenerator was its large 16 m^2 footprint and huge weight of 9.3 te. This was not competitive with a standard diesel engine combined heat and power unit.

Originally, SOFCs were designed to compete with large power generation units like central power stations, ships and locomotives, especially to run on coal gas or heavy fuels. During the last 15 years, the realisation has steadily dawned that SOFCs can work well in small, portable, domestic and auxiliary power systems, particularly running on natural gas, propane or biogas, thus reducing risks of large capital investment projects for power stations. Typical examples of such smaller scale developments are those of Hexis [37], Kyocera [38], TOTO [39], AMI [40] and Adelan [41].

A typical schematic for a small SOFC system is given in Figure 1.10. The electrical power output for a mobile power application could be 100 We for communications up to 5 kWe to power a house or to supply air conditioning and auxiliary power in a

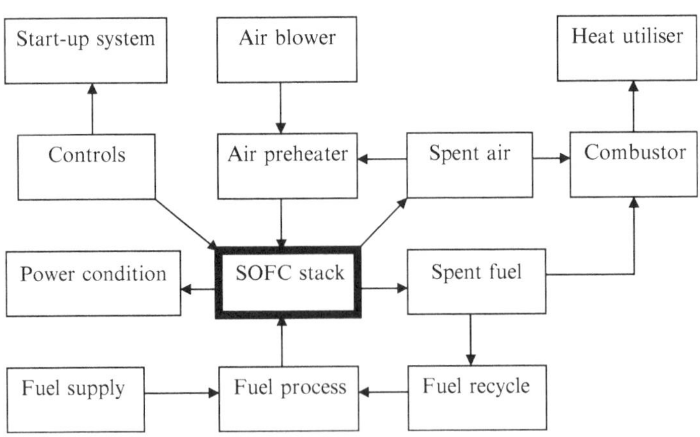

Figure 1.10 Flow sheet showing the BOP surrounding the SOFC stack.

vehicle. Electrical efficiency is not the main performance criterion for some of these applications because the heat can also be used to provide hot water or even chilling.

The main moving part in this plant is the air blower, together with a fuel pump if pressurised fuel is not supplied. All the other parts except valves are solid state and should give the system low maintainance cost and high reliability over a life of many thousands of hours. In a small system, the reliability is the key competitive feature which gives advantage over internal combustion engines. Such heat engines are dominated by moving parts which require oil changes, new spark plugs, rebores, etc. Below 50 kWe, combustion engines are not usually economic because of maintenance costs, so SOFCs have a ready potential market.

1.12 Fuel considerations

One of the great benefits of the SOFC is that it can utilise a wide range of fuels, as described in Chapter 14. The fastest reaction at the nickel anode is that of hydrogen. But other fuels can also react directly on the anode, depending on catalyst composition. For example, carbon monoxide can react on Ni/YSZ but has a higher overpotential than hydrogen [42]. Also, methane can react on the anode but requires ceria or other catalysts to provide suitable sites for direct oxidation.

It was demonstrated in the 1960s that hydrocarbons could be injected directly into SOFCs if steam was supplied [43]. The steam can beneficially be obtained from the spent fuel stream. The main problem with direct use of hydrocarbons is that coke can form to block up and contaminate the anode. There are two damaging reactions which can occur on the nickel:

$$2CO \rightarrow CO_2 + C$$

$$CH_4 \rightarrow 2H_2 + C$$

When carbon formation was investigated in detail, by temperature programmed reaction, three different types of material were discovered on the nickel, as indicated by the temperature required for oxidation [44]. The most stable carbon could not be removed below 1100 K and tended to form when current was flowing through the cell.

The other damaging mechanism of SOFC failure stems from fuel impurities. Sulphur is the most prevalent impurity and can be present up to 1% level in marine diesel fuel. SOFCs cannot operate with this amount of sulphur. More typically, natural gas often has 'odorant' sulphur compounds added to make leaks more easily detectable. The lower levels of such additions, about 10 ppm, are somewhat damaging for SOFC nickel anodes, but the upper limits around 100 ppm cause failure in about 1 h of operation. There are three approaches to solving this problem: firstly, using pure fuel which may be plausible but expensive on an unmanned air vehicle (UAV) which cannot afford to carry a heavy desulphuriser; secondly, adding a sulphur absorber to the fuel processing unit, which is the most common solution and thirdly, using anode metals which are less affected by sulphur.

The second difficulty is the number of additives in conventional fuels which have been formulated for other technologies. For example, regular gasoline contains more than 100 different molecules, some added as lubricants or surfactants. Moreover, the mixture can change with time and place because the standard is dictated by octane number and not composition. Consequently, it is unlikely that SOFCs will be able to run directly on gasoline or diesel although this has been attempted many times over the past 30 years [45]. An objective of research is to formulate a fuel which can operate in an SOFC and a vehicle engine simultaneously, for example, LNG [46].

It is interesting that an SOFC can run more readily on certain molecules. This was tested by injecting pure, dry fuel into a microtubular SOFC while observing the reaction products using mass spectroscopy and the electrochemical performance on a potentiostat [47]. Typical results are shown in Figure 1.11 for small molecules like methanol and formic acid, for comparison with longer molecules in the homologous series.

Methanol on direct injection into an SOFC behaved very well and did not seriously coke up the anode, whereas ethanol gave a serious coking problem, presumably because the methyl group could not oxidise fast enough. In similar fashion, formic acid ran well, albeit at around half the power output of methanol, whereas acetic acid fouled up the anode in less than 30 min. Longer alcohol and acid molecules all gave a significant coking problem unless processed by partial oxidation or steam reforming. Thus, the influence of the fuel is critical in evaluating the competition with conventional combustion engines. Single carbon molecules are preferred.

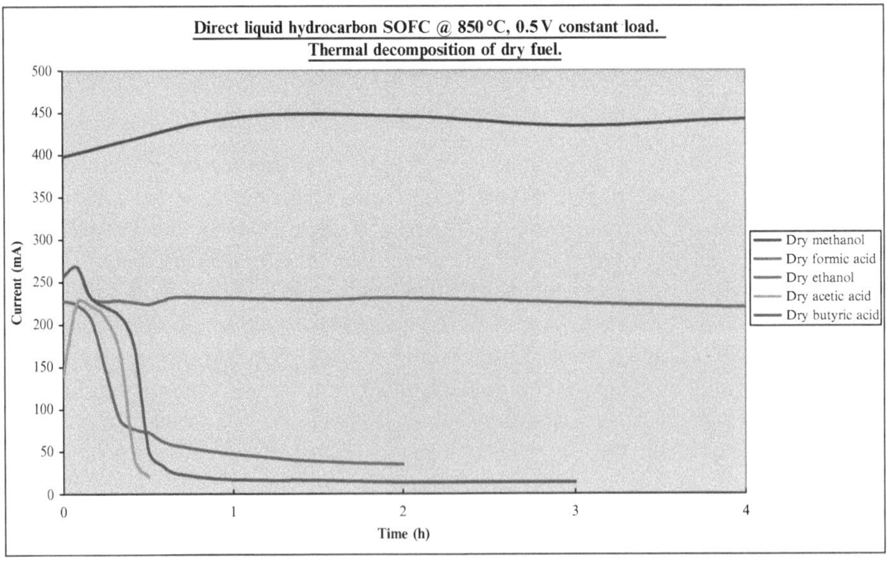

Figure 1.11 Power output of microtubular SOFC on different pure fuels.

1.13 Competition and combination with heat engines in applications

If the SOFC is to be successful commercially, then it must compete with existing heat engines which are currently used to produce electricity from combustion. Such engines operate by burning fuel to heat a compressed volume of gas, followed by expansion of the hot gas in a piston or turbine device driving a dynamo. These cannot maximise the thermodynamic potential of the fuel, as well as being polluting when compared with fuel cells but can be surprisingly economic as a result of a century's development, optimisation and mass production. Ostwald got it famously wrong in 1892 when he said that 'the next century will be one of electrochemical combustion'. Fuel cells are still significantly more costly than conventional engines which can be manufactured for about \$50/kWe. The SOFC advantages of efficiency, modularity, siting and low emissions count for little if they cost \$10,000/kWe. These arguments are considered more fully in Chapters 8–10.

In the 1980s, it was envisaged that SOFCs could compete commercially with other power generation systems, including large centralised power stations and smaller cogeneration units [48,49]. This has not yet happened because costs have remained high despite large injections of government funding for SOFCs in the United States, Asia and Europe. It has been estimated that costs of \$400/kWe could be achieved with mass production using powder methods [50]. Such costs would be competitive with present large power station costs.

There are many possible applications for SOFCs as described in Chapters 8–10. The stationary power market has been most investigated up to this point, since there is a great need for clean and quiet distributed power generation units, for example, in factories, hospitals, hotels and sports facilities located in cities. Traditionally, such demands have been served by 200 kWe diesel engine or gas turbine packages driving dynamos, with the heat supplied to the building for hot water or steam. Other fuel cell systems based on phosphoric acid electrolyte, solid polymer electrolyte and molten carbonate fuel cells have also been developed to fill this niche, but SOFCs can compete because of more useful heat. Combustion engine generators have been difficult to displace because they are economic so that the expected large market for SOFC stationary power has not emerged as predicted. Fuel cell devices will generally not be preferred unless regulations for low emissions are imposed, or unless reliability is necessary, for example, for computer installations in banks. Such regulations have encouraged the installation of hundreds of 'Bloom Boxes', typically 100 kW planar SOFC systems running on pipeline gas, especially in California [51].

The other application which has been much studied is that of domestic cogeneration in homes, requiring about 1 kW of electrical power and several kilowatts of heat. Thousands of these have been installed in Japan but are outnumbered by PEMFCs which may appear simpler but require more difficult fuel processing for pipeline gas [52]. In the EU, demonstrations are occurring on hundreds of SOFC units, competing with PEMFC units. Installing high-efficiency (>50%) SOFCs in every home will cut domestic carbon dioxide emissions by up to 50% [53].

Transport applications have emerged in recent years, especially to compete with PEMFCs which have been rapidly evolving to satisfy the zero emission electric vehicle market by replacing the combustion engine. Several thousands of PEMFC cars are now running across the planet, fuelled at hydrogen filling stations with storage in pressurised hydrogen cylinders at 350 or 700 bar. A typical PEMFC for a vehicle is 50 kWe, runs on pure hydrogen, requires significant quantities of platinum and is still significantly more costly than an internal combustion engine [54]. It does have rapid warm-up and can deliver significant power on a cold-start, but the key question is the renewable source of the hydrogen fuel, which is best obtained from wind turbine surpluses. The SOFC can be useful in vehicles, not to replace the engine, but to supply auxiliary power to supplement or replace part of the existing battery system. Therefore, there is little direct contest with PEMFCs or combustion generators. Typically, 0.1–5 kWe is beneficial, especially in new applications like UAVs, leisure vehicles such as campervans as shown in Figure 1.12 or truck cab generators [55]. The benefits of SOFC for this application are:

(1) runs on hydrocarbon fuel, for example, propane/butane or LNG,
(2) provides useful heat,
(3) runs when the engine is switched off,
(4) is much more efficient than the existing electrical system,
(5) has low emissions.

PEMFCs are also being tested in these applications, and it is not yet clear where the winners will be found.

Smaller, portable SOFC power units which can outperform batteries are also being considered as described in Chapter 10. These can deliver power in the range from 10 mWe to 1 kWe and can run directly on a wide range of fuels from natural gas, to propane, butane, methanol, ammonia and formic acid. The cells are generally small to avoid thermal shock and then the start-up can be quick, in a few seconds [56]. Again, PEMFCs running on hydride stores are competing in this market, for example, in mobile phone chargers.

Figure 1.12 Camper van from Autosleepers with SOFC auxiliary power unit (APU).

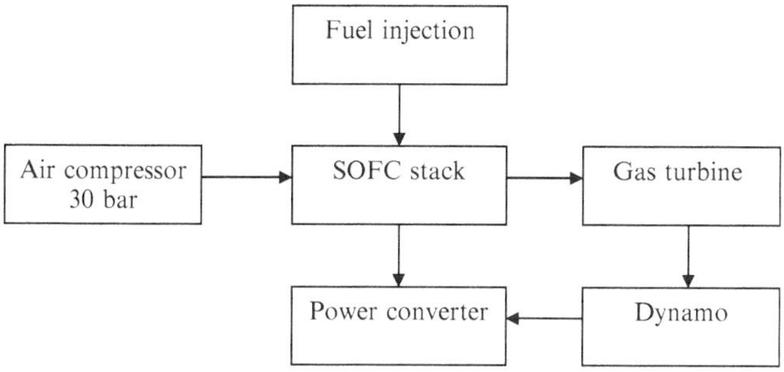

Figure 1.13 Combination of SOFC with gas turbine generator.

The benefits of SOFCs in hot electrolysers have again been recently encouraged. As demand for pure hydrogen has increased for fuel cell electric cars, the improved efficiency of SOFC over PEMFC electrolysers has stimulated substantial R&D [57].

One of the most promising applications of SOFCs for the future is in combination with a gas turbine, as first demonstrated by Westinghouse and further developed by Rolls Royce Fuel Cell Systems (now LG) [58,59]. The flow scheme is shown in Figure 1.13. The SOFC stack forms the combustor unit in a gas turbine system. Air is compressed at up to 30 bar and then fed into the SOFC stack where fuel is injected and electrical power drawn off. Operating near 50% conversion of fuel to electrical power, this SOFC then provides hot gas at 30 bar and 1000 °C to a turbine operating at 35% efficiency. The overall electrical conversion of this system can approach 70%, and this could be further improved by adding a steam turbine [59].

Westinghouse tested this concept using a 100 kWe tubular stack in 2001. The whole system was operated in a pressure chamber at 9 bar abs. A small turbine was used to utilise the 100 kW of hot gas exciting the SOFC system. The overall efficiency of this device was measured at 61%. RRFCS in 2010 demonstrated 50 kW modules which fitted into a 1 MW SOFC design at 5 bar and calculated that the overall efficiency of the pressurised system could readily rise to 75% [59,60]. This is the highest conversion device for fuel to power yet proved. Achieving this with the desired economics and reliability is still difficult largely because fitting 30,000 ceramic parts together in a hot pressure vessel is not trivial.

1.14 SOFC publications

SOFC technology has been advancing at a rapid rate over the decade since the first edition of this volume was published in 2003 [4]. Some significant books have appeared on SOFCs in the meantime [61] but none with comprehensive coverage of the subject. Summarising the complete progress of SOFC up to 2014 in this present volume should help to unify the subject and produce further breakthroughs.

Several conference proceedings are published each year containing material on SOFCs, but these tend to be collections of individual research papers at a particular time rather than a complete compendium of the technology. These include the Electrochemical Society Proceedings series on SOFCs which has been edited by Singhal et al. [62] and the Proceedings of the European SOFC Forums [63].

It is believed that the publication of this volume will provide detailed up-to-date information for the researchers who are about to make SOFCs commercial in the near future, a time when we might see an SOFC in every vehicle, in every home, in more efficient power stations and in novel applications like UAVs and mobile phone chargers.

References

[1] W. Nernst, Uber die elektrolytische Leitung fester Körper bei sehr hohen Temperaturen, Z. Elektrochem. 6 (1899) 41–43.

[2] W. Nernst, DRP 104872, 1897.

[3] W. Nernst, USP 685730, 1899.

[4] S.C. Singhal, K. Kendall, High Temperature Solid Oxide Fuel Cells, Elsevier Science, Oxford, 2003.

[5] K. Kendall, N.J. Shang, I. Staffell, Application of Coulomb's friction law to define energy consumption of new drive trains, in: 4th IET Hybrid and Electric Vehicles Conference HEVC13, London, 6–7 November, 2013, IET Conference Publications, London, 2013.

[6] I. Staffell, J.L. Shang, K. Kendall, Atomic models of strong solids interfaces viewed as composite structures, Appl. Comp. Mater. 21 (1) (2014) 45–55.

[7] K. Kendall, Hopes for a flame-free future, Nature 404 (2000) 233–235.

[8] K. Mendelssohn, The World of Walther Nernst, the Rise and Fall of German Science, Macmillan, London, 1973, pp. 1864–1941.

[9] E. Baur, H. Preis, Über Brennstoff-Ketten mit Festleitern, Z. Elektrochem. 43 (1937) 727–732.

[10] H. Peters, H.H. Möbius, Electrochemical investigation of the formulae CO+. 1. 2 O2 = 2CO, Z. Phys. Chem. 209 (1958) 298–309.

[11] J.H. Lee, Review on zirconia air fuel ratio sensors for automotive applications, J. Mater. Sci. 38 (2003) 4247–4257.

[12] A.H. Heuer, L.W. Hobbs (Eds.), Science and Technology of Zirconia, American Ceramic Society, Columbus, OH, 1981. see also Science and Technology of Zirconia, vols. II–V, American Ceramic Society, Columbus, Ohio, 1984–1993.

[13] Data from several sources.

[14] D.H. Archer, L. Elikan, R.L. Zahradnik, in: B.S. Baker (Ed.), Hydrocarbon Fuel Cell Technologies, Academic Press, New York, 1965, p. 51.

[15] A.O. Isenberg, Energy conversion via solid oxide electrolyte electrochemical cells at high temperatures, Solid State Ion. 3 (1981) 431–437.

[16] C. Bagger, Improved production methods for YSZ electrolyte and Ni-YSZ anode for SOFC, in: 1992 Fuel Cell Seminar, Tucson Arizona, Nov. 29–Dec. 2, 1992.

[17] K. Kendall, Influence of Powder Structure on Processing and Properties of Advanced Ceramics, Powder Technol. 58 (1989) 151–161.

[18] M. Prica, K. Kendall, S.A. Markland, J. Am. Ceram. Soc. 81 (1998) 541–548. K. Kendall, M. Prica, in: U. Bossel (Ed.), First European Fuel Cell Forum, Luzern, 1994.

[19] M. Cassidy, K. Kendall, G. Lindsay, Lower Temperature SOFCs Utilising Screen Printed Electrolytes, in: B. Thorstensen (Ed.), 2nd European SOFC Forum, 6-10th May 1996, Oslo, Norway, 1996, pp. 667–676.

[20] N.Q. Minh, C.R. Horne, in: Proceedings of 14th Risoe Internetional Symposium, Roskilde, Denmark, 1993, p. 337.

[21] K. Kendall, Revolution in ceramic processing, in: F.L. Riley (Ed.), 3rd European Symposium on Engineering Ceramics, Elsevier Applied Science, London, 1989, pp. 97–107.
[22] K. Kendall, Novel ceramic designs for fuel cells, in: International Forum on Fine Ceramics, Nagoya, March 9–11, 1992, pp. 143–148.
[23] S. Yang, T. Chen, Y. Wang, Z. Peng, W.G. Wang, Electrochemical analysis of an anode supported SOFC, Int. J. Electrochem. Sci. 8 (2013) 2330–2344.
[24] S. Park, J.M. Vohs, R.J. Gorte, Direct oxidation of hydrocarbons in solid-oxide fuel cell, Nature 404 (2000) 265–267.
[25] D.D. Button, D.H. Archer, Am. Ceram. Soc. (1966). Washington.
[26] M. Cassidy, C. Bagger, N. Brandon, M. Day, Improved Cathode Performance Using Graded Structures, in: A.J. McEvoy (Ed.), 4th European SOFC Forum, 10–14 July 2000, Luzern, Switzerland, 2000, pp. 637–646.
[27] N.Q. Minh, in: D.J. Viechnicki (Ed.), Proceedings of the 37th Sagamore Army Materials Research Conference, October 1–4, 1990, Plymouth, MA, Army Materials Technology Lab, Watertown, MA, 1990, p. 213.
[28] K. Kendall, E. Wright, A. Golds, Co-formed components for SOFCs, in: M. Dokiya, O. Yamamoto, H. Tagawa, S. Singhal (Eds.), SOFCs IV, The Electrochemical Society Inc., NJ, 1995, pp. 229–233.
[29] Z. Yang, G. Xia, C. Wang, Z. Nie, J. Templeton, J. Stevenson, P. Singh, Investigation of AISI 441 ferritic stainless steel and development of spinel coatings for SOFC interconnect applications, SECA report DE-AC0576RL01830, 2008.
[30] R. Eriksen, S. Gopalan, S. Sampath, Y. Chen, LSM protective coatings on stainless steel as interconnects for solid oxide fuel cells, MRS Proceedings 1644 (2014), http://dx.doi.org/10.1557/opl.2014.330. mrsf13-1644-dd05-03.
[31] US Patent 7981561 B2, Interconnects for solid oxide fuel cells and ferritic stainless steels adapted for use with solid oxide fuel cells, 2011.
[32] K. Kendall, M. Prica, Integrated SOFC tubular system for small-scale cogeneration, in: U. Bossel (Ed.), Proceedings of 1st European SOFC Forum, October 3–7, 1994, Luzern, Switzerland, 1994, pp. 163–170.
[33] T. Ohrn, Z. Liu, Rolls-Royce Coal-Based SECA Program Update (DOE DE-FE0000303), 2011.
[34] K. Kendall, R.C. Copcutt, A.C. King, Reaction-diffusion of fuel with air between planar SOFCs stacked with ceramic felt, Proc. R. Soc. Lond. A452 (1996) 2639–2653.
[35] K. Kendall, J. Billingham, A.C. King, R.C. Copcutt, Analysis of a model for a loaded, planar solid oxide fuel cell, SIAM J. Appl. Math. 60 (2000) 574–601.
[36] S. Ahmed, C. McPheeters, R. Kumar, Thermal–hydraulic model of a monolithic solid oxide fuel cell, J. Electrochem. Soc. 138 (1991) 2712.
[37] A. Mai, B. Iwanschitz, J. Andreas Schuler, R. Denzler, V. Nerlich, A. Schuler, Hexis' SOFC System Galileo 1000 N – lab and field test experiences, ECS Trans. 57 (2013) 73–80.
[38] M. Suzuki, T. Sogi, K. Higaki, T. Ono, N. Takahashi, K. Shimazu, T. Shigehisa, Development of SOFC residential cogeneration system at Osaka Gas and Kyocera, ECS Trans. 7 (2007) 27–30.
[39] K. Kariatsumari, N. Shimizu, TOTO to test 700-W SOFC with 500 °C operating temperature, Nikkei Electron. (July 30) (2007).
[40] A. Crumm, D. Kachman, Integrating micro-fuel cell power systems onto unmanned vehicle platforms, in: AUVSI Unmanned Systems North America Conference 2009, 2009, pp. 259–269.
[41] K. Kendall, Progress in microtubular solid oxide fuel cells, Int. J. Appl. Ceram. Technol. 7 (1) (2010) 1–9.

[42] G.A. Tompsett, C.M. Finnerty, K. Kendall, T. Alston, N.M. Sammes, Novel applications for micro SOFCs, J. Power Sources 86 (2000) 376–382.

[43] R.L. Zahradnik, Diffusional processes in solid electrolyte fuel cell electrodes, J. Electrochem. Soc. 117 (1970) 1443; J. Weissbart, R. Ruka, J. Electrochem. Soc. 109 (1962) 723.

[44] S.J. Livermore, J.W. Cotton, R.M. Ormerod, Fuel reforming and electrical performance studies in intermediate temperature ceria-gadolinia-based SOFCs, J. Power Sources 86 (2000) 411–416; R.M. Ormerod, Stud. Surf. Sci. Catal. 122 (1999) 35.

[45] D. Hennessey, SOFC development at Delphi, in: 2011 Fuel Cell Seminar, 1 November, 2011.

[46] K. Kendall, www.safari-project.eu, first uploaded 2014.

[47] K. Kendall, G.J. Saunders, Reactions of hydrocarbons in small tubular SOFCs, J. Power Sources 106 (2002) 258–263.

[48] P.D. Lilley, E. Erdle, F. Gross, Market potential of SOFC, Report no. EUR 12249 EN, CEC, Luxembourg, 1989.

[49] W.L. Lundberg, SOFC cogeneration system conceptual design, Report no. GRI 89/0162, Gas Research Institute, Chicago, IL, 1989.

[50] I. Staffell, R. Green, The Cost of Fuel Cell Micro-domestic CHP Systems, Imperial College Business School, Discussion Paper, 2012.

[51] G. Bassett, A. England, F. Li, J. Weinberger, A. Wong, The science and economics of the bloom box, ECON 26800, 2010.

[52] I. Staffell, D. Brett, N. Brandon, et al., A review of domestic heat pumps, Energy Environ. Sci. 5 (2012) 9291–9306.

[53] I. Staffell, R. Green, The cost of domestic fuel cell micro-CHP systems, Int. J. Hydrogen Energy 38 (2013) 1088–1102.

[54] Fuel Cell Vehicles, The road ahead, Fuel Cell Today, Royston, 2013.

[55] www.Adelan.co.uk, website (accessed March 2014).

[56] H. Wang, Y. Jiang, L. Philips, Mobile Convergence: Platform, Applications & Services, Parks Associates, Dallas, 2009.

[57] A. Hauch, S.H. Jensen, S. Ramoussen, M. Mogensen, Performance and durability of solid oxide electrolysis cells, J. Electrochem. Soc. 13 (2006) A1741–A1747.

[58] C.J. Steffen, J.E. Freeh, LM. Larosiliere, Solid oxide fuel cell/gas turbine hybrid cycle technology for auxiliary aerospace power, NASA report 213586, 2005.

[59] D.G.D. Agnew, The design and integration of the Rolls Royce fuel cell systems 1MW SOFC, in: Proceedings of the ASME Turbo Expo Conference, Reno, NV, USA, June, 2005.

[60] W. Winkler, P. Nehter, M.C. Williams, D. Tucker, R. Gemmen, General fuel cell hybrid synergies and hybrid system testing status, J. Power Sources 159 (2006) 656–666.

[61] M. Ni, T.S. Zhao (Eds.), Solid Oxide Fuel Cells, From Materials to System Modeling, RSC, 2013; J.W. Fergus, R. Hui, X. Li, D.P. Wilkinson, J. Zhang, SOFCs: Materials Properties and Performance, CRC Press, Boca Raton, 2009; J.T.S. Irvine, P. Connor, SOFCs: Facts and Figures, Springer, London, 2013; J. Milewski, K. Swirski, M. Santarelli, P. Leone, Advanced Methods of SOFC Modeling, Springer, London, 2011; T. Ishihara, Perovskite Oxide for SOFCs, Springer, London, 2009; H. Yokokawa (Ed.), Fuel Cell Handbook: Fundamentals Technology and Applications, John Wiley and Sons, New York, 2009; D. Stolten, B. Emonts (Eds.), Fuel Cell Science and Engineering, vol. 2, Wiley-VCH, Weinheim, Germany, 2012, p. 963.

[62] S. Singhal et al. (Ed.), Proceedings of International Symposium on SOFCs, Electrochemical Soc, Pennington, NJ, 1989, 1991, 1993, 1995, 1997, 1999, 2001, 2003, 2005, 2007, 2009, 2011, 2013, 2015.

[63] U. Bossel, et al. (Ed.), Proceedings of the European SOFC Forum, Baden, Switzerland, 1994,1996,1998,2000,2002,2004,2006, 2008, 2010, 2012, 2014.

History

K. Kendall[1,2]
[1]Adelan Ltd, Weekin Works, Harborne, Birmingham, UK, [2]School of Metallurgy and Materials, University of Birmingham, UK

2

2.1 Introduction

This chapter was written originally by Hans Heinrich Mobius in 2002–2003. After he died in 2011, it became clear that a new chapter was needed, incorporating the early history to the year 2000, but then going on to review events between 2000 and 2013. Therefore, the text has been reworked as a memorial to Mobius's contribution, but giving a further perspective of recent progress in the SOFC field. Therefore, Sections 2.2–2.6 compress the older material, while Sections 2.7–2.12 go on to reflect the key subsequent advances of better ceramic processing, anode-supported cell designs, lower-temperature cathodes especially lanthanum strontium cobalt ferrite (LSCF), ferritic steel interconnects which can exhibit long lifetimes; plus new applications which now range from small systems like sensors, portable SOFCs, auxiliary/remote power to CHP at the MW level to coal gasification plants at the GW scale.

2.2 Before the first solid electrolyte gas cells

Just after Volta's report of the flat-plate battery stack in 1800, Humphry Davy built the biggest battery in the world in London and used it to electrolyse any materials at hand. Experiments using more and more concentrated solutions of alkali hydroxides led to molten flux electrolysis and in 1807 to the discovery of alkali metals [1]. He found that solid alkali compounds were insulating, but the addition of small amounts of water led to high electrical conduction. Faraday followed this up later [2] by considering why the solid alkali salts were insulators, whereas their aqueous solutions were excellent conductors. In his continuing investigations, Faraday introduced the basic terminology, allowing him to classify materials into first and second types of conductors, metallic and electrolytic [3,4]; the first type is now recognised as electronic and the second type as ionic.

As early as 1774, Cavendish [5] had observed an increase in the conductivity of glass on heating. The electrolytic nature of this conduction was explained by Beetz and Buj [6] in 1854. Using mercury, zinc amalgam, various solid metals, carbon and MnO_2 as electrodes, Buj demonstrated galvanic cells and batteries free of water in which glass took over the role of the electrolyte, and he investigated the associated voltage and polarisability. Meanwhile, Gaugain had shown that two glass tubes containing different gases at high temperature produced electricity when in contact [7]. These ideas led to the use of the term 'Solid Electrolyte' towards the end of the

High-temperature Solid Oxide Fuel Cells for the 21st Century. http://dx.doi.org/10.1016/B978-0-12-410453-2.00002-6

nineteenth century, but it was the investigation of electric incandescent lamps by Nernst and his associates which then began to push the subject further [8,9] (Figure 2.1).

After completing his Wurtzburg doctorate in 1887, and his Leipzig habilitation in 1889, Nernst moved to Gottingen in 1890 as a lecturer, wrote a textbook on *Physical Chemistry* in 1893 and was appointed Director of the new PhysicoChemical Institute in 1895. He invented his zirconia incandescent lamp shortly after and then sold his patent to AEG in 1898 for a huge sum, which allowed him to donate 40,000 Marks to his own institute. It is little appreciated that Nernst suggested setting fire to coal seams to raise the Earth's temperature through CO_2 production, having met Arrhenius who studied in Gottingen and who had shown the effect of carbon dioxide creating the greenhouse effect around that time. In 1920, Nernst was awarded the Chemistry Nobel Prize for 'his work in thermochemistry'.

Figure 2.2 shows one of the many designs of the Nernst lamp [10]. When the lamp was switched on, the voltage was applied to the doped zirconia rod, *h*, and to the parallel platinum heating resistor, *i*. Both these components were incorporated in a glass envelope containing air. After sufficient preheating, the current started flowing through the Nernst rod and through the winding *k* of an electromagnet *b*. At a specified electric current, the magnet opened the contacts between *m* and *l* and then the Nernst rod carried all the current and emitted light due to resistive heat generation.

Nernst's insight was to realise that, just as pure water or pure NaCl do not conduct, but mixtures of water and NaCl conduct very well, mixtures of insulating oxides such as zirconia and calcia or yttria could also exhibit strong ionic conduction. In a short

Figure 2.1 Walther Nernst (1864–1941) who discovered the ionic conduction of zirconia doped with other oxides such as calcia, yttria, etc. His incandescent lamps based on these compositions became famous around 1900 when this picture was taken and when 200 of his lamp designs (Figure 2.2a) were used to illuminate the AEG pavilion at the Paris exposition.

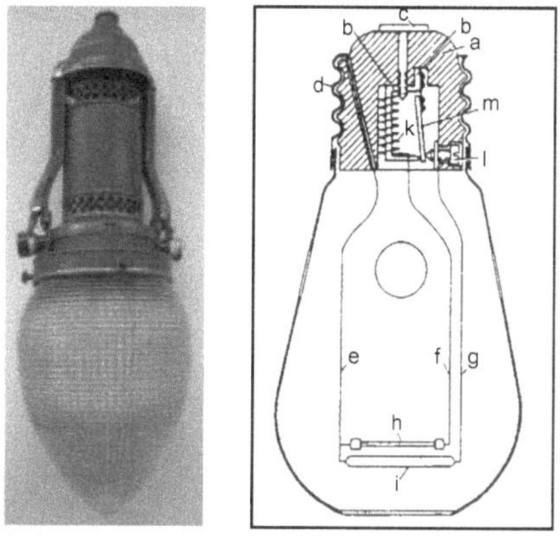

(a) (b)

Figure 2.2 Nernst lamp (a) AEG product and (b) patent drawing where the letters a,b,c, etc. denote components of the lamp assembly.

time, many of the mixed oxides which exhibit high conductivity at elevated temperatures, including the particularly favourable composition 85% zirconia 15% yttria, the so-called Nernst Mass, were recognised [10]. The thesis of Reynolds [11], inspired by Nernst and presented in 1902, expanded this field by measuring the conductivity in the range of 800–1400 °C of numerous binary and ternary systems, among others, formed by ZrO_2 with the oxides of the elements La, Ce, Nd, Sm, Ho, Er, Yb, Y, Sc, Mg, Ca, Th and U, including investigations on the role of composition, concentration, direction of temperature alteration (hysteresis) and other phenomena.

Although the light efficiency of the Nernst lamp exceeded that of the carbon filament lamp by nearly 80%, there were significant problems which we still see today in modern SOFCs. There was a long start-up time before the light came on. Finding a metal interconnector was problematic, and the platinum used in the leads and heater made the lamp expensive. The operating temperature was unstable and had to be controlled with special series resistors. As a result of these complications, the zirconia filament could not compete with tungsten which soon dominated the market.

It was not until 1943 that Wagner [12] (in memory of Walther Nernst who died on 18 November 1941) recognised the existence of vacancies in the anion sublattice of mixed-oxide solid solutions and thus explained the conduction mechanism of the Nernst lamps. We now know that Nernst filaments were oxide ion conductors and the platinum contacts behaved as air electrodes. It follows that Nernst lamps were the first commercially produced solid electrolyte gas cells.

2.3 From solid electrolyte gas cells to solid oxide fuel cells

Even before Nernst had found yttria-stabilised zirconia (YSZ) with great ionic conductivity above 800 °C, electrochemistry had advanced when its connection with thermodynamics was explained by von Helmholtz in 1882 [13]. Then, in 1894, Ostwald

[14] demonstrated that energy from coal could be produced much more efficiently with a galvanic cell than with a steam engine. Of course, the impurities in coal together with the catalyst problems involved in coal gasification had been considered several years earlier by Mond and Langer [15] in their early tests on electrochemical conversion of coal in which they coined the term 'Fuel Cell'.

The agreement between the voltages measured with galvanic solid electrolyte gas cells and calculated thermodynamically was verified by Haber and Moser in 1905 [16]. From 330 to 570 °C, they used glass and from 800 to 1100 °C porcelain as the electrolyte, and partly platinum, partly gold as the material for the electrodes in cells, first with C, CO, CO_2 and O_2, in oxyhydrogen cells and then in hydrogen and oxygen concentration cells. Typical phenomena such as the dependence of the voltage on the gas flux, deviations from zero (asymmetry voltages) and sluggishness in the establishment of constant voltages at low temperatures were observed. Parallel to the publication of these results, Haber filed the first patent on fuel cells with a solid electrolyte [17] as shown in Figure 2.3.

The decomposition equilibria of metal oxides were investigated in 1916 by Treadwell [18] in the region of 1000 °C with quartz and porcelain as solid electrolytes and with a silver/oxygen electrode as the reference system. After these investigations, Baur and Treadwell [19] filed a patent on fuel cells with metal oxide electrodes and a molten salt, held in a porous ceramic, electrolyte. Only after many fruitless experiments with liquid electrolytes of different types did Baur and Brunner [20] come to the conclusion in 1937 that fuel cells have to be made completely solid. But the extensive empirical search by Baur and Preis [21,22] up to the 1950s for suitable solid electrolytes, covering glasses, porcelains, clays and a great variety of oxide mixtures, was unsuccessful.

Zirconia ceramics proved to be the best materials for solid electrolytes and were first used in fuel cells as shown in Figure 2.4 by Baur and Preis during 1937 [21,22]. They wrote on the Degussa tube crucibles used (16–12–190 mm): 'Unsurpassed is the Nernst Mass. But even this mixture is not satisfactory because the current enhances resistance considerably by electrolytic shift (migration away of the cations)'. The problems were possibly caused by the material used for the cathode (Fe_3O_4), which oxidises at 1000 °C in air to form poorly conducting Fe_2O_3. So the wrong conclusion was drawn: 'One has to look for an improvement of the Nernst Mass

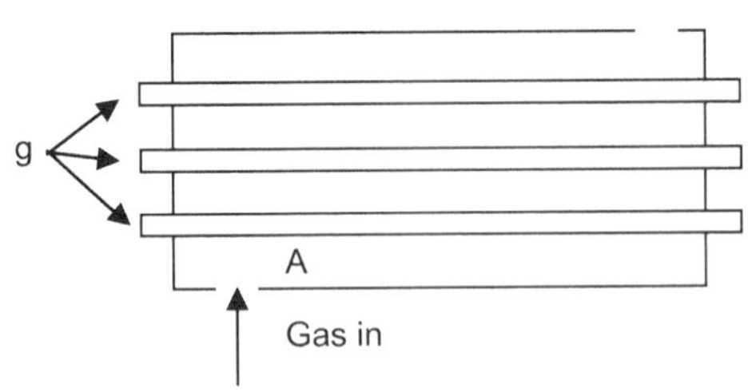

Figure 2.3 Haber's first diagram of cell with solid electrolytes [17]. Generator gas passed through chamber A (440 °C) with parallel glass tubes g covered on both sides with thin layers of noble metal and swept inside by air.

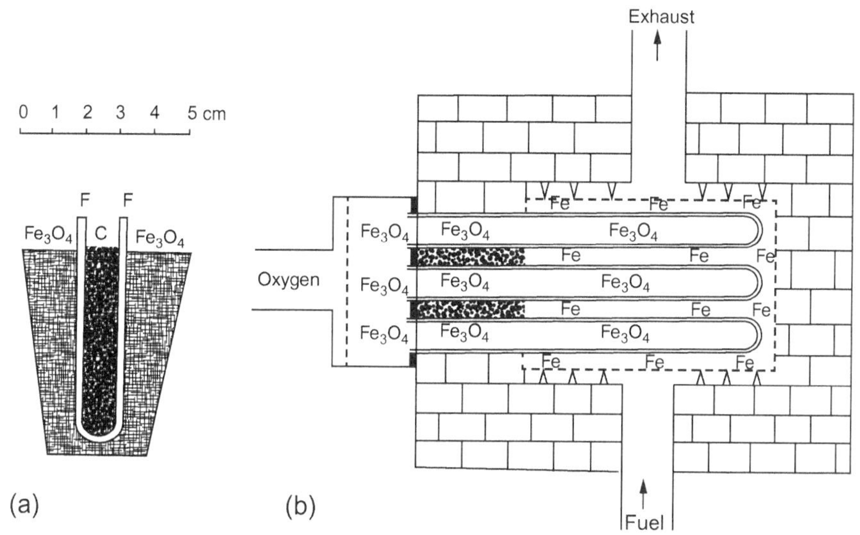

Figure 2.4 Fuel cell arrangement of Baur and Preis (1937): (a) Investigated cells also with Nernst mass as solid electrolyte (F) and (b) proposal for the realisation of stacks of such fuel cells.

or to put alongside it solid conductors of higher value'. The fact is that YSZ is sufficiently conducting if it is made very thin, that is, less than 100 μm. Baur preferred other ceramics containing tungsten oxide and cerium dioxide over the Nernst Mass, and these were also used repeatedly by other investigators. However, as shown for tungstates of Ce, Ca and Zn, in oxidising and reducing conditions, electronic transport through such ceramics is so high that their application in fuel cells cannot be useful except in the temperature range where ionic conduction dominates.

Schottky [23] had shown that the efficiency of solid electrolyte fuel cells decreases to zero if a noticeable part of the conductivity of the electrolyte is electronic. Therefore, the efforts for purification and especially for separation of the polyvalent praseodymium cations from the solid electrolyte material were made. In the case of Th–La mixed oxides, with only 1 mol% $LaO_{1.5}$, the ion transport number 1 was reached, admittedly only with reducing gas on both electrodes (in CO, CO_2 concentration cells); in the oxygen/air cell even at 10 mol% $LaO_{1.5}$, this number was only near 0.8. A perfect disc of $Ce_{0.9}La_{0.1}O_{1.95}$ broke into pieces in CO, CO_2/O_2 between 700 and 840 °C, reaching the mean ionic transport number 0.8. For the available ZrO_2 solid electrolyte (with 50 mol% $YO_{1.5}$) in the oxygen/air cell, the ion transport number was above 0.93.

When Wagner had recognised the mechanism of conduction in the Nernst incandescent lamp, he pointed out in 1943: 'For fuel cells with solid electrolytes anion conductors are to be considered exclusively. From this point of view a systematic investigation of the mixed crystal systems of the type of the Nernst Mass with

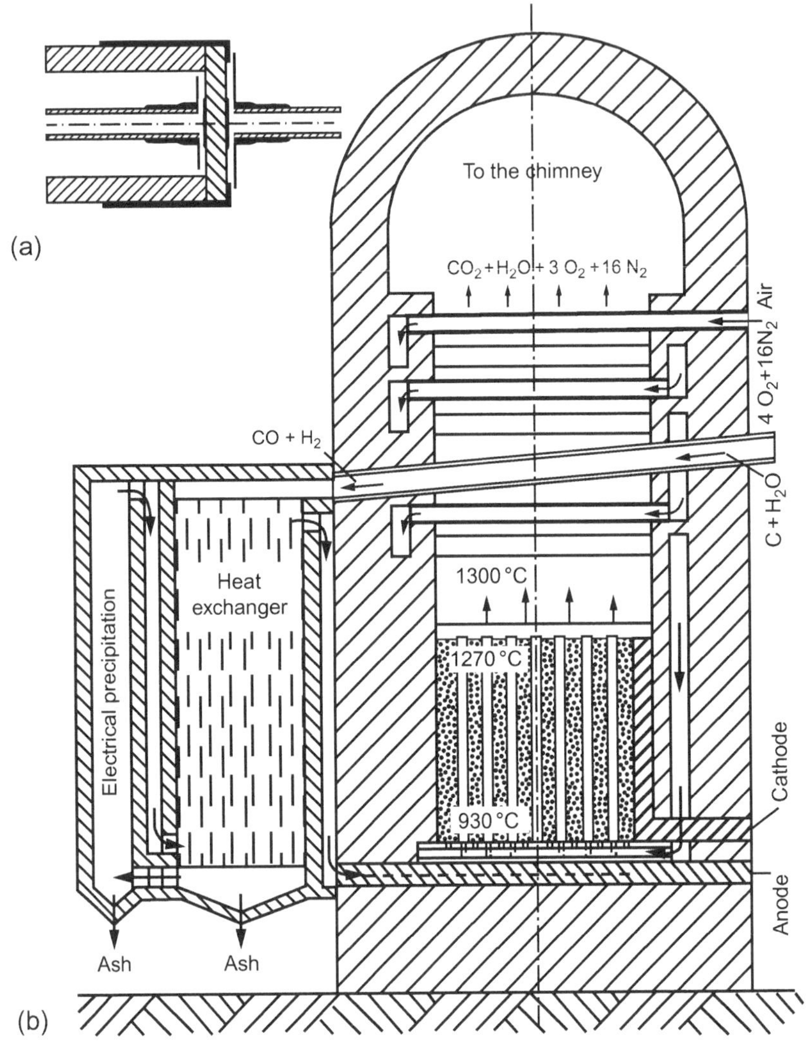

Figure 2.5 Cross-section of galvanic cells used by Mobius and Peters.

roentgenographic and electrical methods seems to be desirable' [12]. This was the start of concentrated research work on solid oxide fuel cells (Figure 2.5).

2.4 First detailed investigations of solid oxide fuel cells

In 1951, Hund confirmed, with X-ray crystallographic and pycnometric investigations, the existence of oxide ion vacancies in the Nernst Mass [24]. Engell [25], having worked from 1936 to 1940 in Darmstadt with Wagner [12], followed his suggestions and left to Peters (in 1951 in Greifswald) the investigation of the lattice structure and the electrical conductivity of some mixed oxides of the type of the Nernst Mass (ZrO_2–Y_2O_3, ThO_2 and CeO_2 with Y_2O_3 and La_2O_3). After Haulle had left for Berlin, Peters finished his thesis in Rostock in 1953 [26] and in 1954 gave Mobius (Figure 2.6) the task of furthering the subject by investigation of galvanic cells using mixed-oxide phases.

The investigations [28], using model fuel cells (Figure 2.5), were started with iron oxides, magnesium ferrite (following Biefeld [29]) and composites of iron and alumina as electrodes and were continued mainly with thin porous layers of platinum, nickel and iron. Very soon it was seen that completely gastight solid electrolyte discs of highly pure substances had to be produced if the measurements were to lead to useful results. Pressure compacting of oxide powders was the original method for making discs, but it was clear that improved methods for making thin ceramic membranes were needed if satisfactory microstructures were to be achieved. The first breakthrough came with the development of chemical vapour deposition later in 1970 as described in Section 2.6. Further breakthroughs in thin ceramic membrane processing are then described in Section 2.7.

Compounds of thorium and cerium were effectively purified as ammonium double nitrates by crystallisation from hot concentrated nitric acid. Pure lanthanum oxide was prepared by fractional precipitation of hydroxides. Then using the well-known experience of fabricating oxide ceramics through high pressure pelleting [30], powders of mixed oxides were pressed to produce gastight discs, 25 mm in diameter and 1–2 mm thick, which were sintered at temperatures up to 1920 °C in a stream of oxygen on a support stack of $Al_2O_3/MgAl_2O_4/ThO_2$ in alumina tubes using a Tammann carbon tube furnace. ZrO_2 for this investigation was available at that time only in the form

Figure 2.6 Hans Heinrich Mobius (1929–2011) whose 1958 patent with Peters [27] was significant for SOFC gas sensors.

of a mixture with Y_2O_3 as a residue from the investigations of Peters because an embargo and the development of the nuclear industry made it difficult to obtain.

In the investigations, carried out from 1955 to 1957, for cells with different composition of the solid electrolyte, the electrode voltages were measured in the temperature range between 300 and 1350 °C and compared with thermodynamically calculated values. On the basis of these results, the Boudouard equilibrium was investigated with $Th_{0.9}La_{0.1}O_{1.95}$ as solid electrolyte in the cell CO, C, Fe/FeO, CO, CO_2, using only the reactive carbon precipitated out of CO; iron in metallic or oxide form in the electrodes supported the establishment of the electrode potential catalytically. And with the ZrO_2 solid electrolyte in a CO, CO_2, Fe_3O_4/Pt, O_2 cell, the CO_2 dissociation equilibrium was investigated [31].

The good agreement between measured and thermodynamically calculated data in these cases led to the most important by-product of SOFC development: Sensors. If solid electrolyte cells, charged with gases of known concentrations, deliver the theoretically expected cell voltages, it also must be possible to calculate unknown gas concentrations backwards from the cell voltages, measured between the cell terminals in gas phases, which can be oxidising or reducing. Accordingly, the potentiometric determination of gas concentrations with solid electrolyte cells and the first designs of probes for the *in situ* analysis of hot gases (reference electrode, e.g. Ni, NiO or a gas with known oxygen partial pressure) were patented from 1958 [27,32–34].

2.5 Progress in the 1960s

The paper of Kiukkola and Wagner [35] stimulated many activities in various parts of the world in the field of solid-state electrochemistry. In this development, zirconia-based solid electrolytes dominated immediately; for example, $Zr_{0.85}Ca_{0.15}O_{1.85}$ was used by Weissbart and Ruka in the first device for the measurement of oxygen concentration in a gas phase using a high-temperature galvanic cell [32]. After 1960, a rapidly increasing number of applications for patents and of papers concerning SOFCs appeared in several countries. Some early contributions are described below.

In the United States, during a short period in 1961–1962, four companies applied for patents on solid oxide fuel cells, partly with series connection [36,37]. The first publication in English about this subject, ‘A solid electrolyte fuel cell’ by Weissbart and Ruka came out in 1962 [38]. From February 1962 to April 1963 in Westinghouse Electric Corporation, a team of 16 people under the leadership of Archer developed solid oxide fuel cell stacks; parallel flat-plate and tubular cells with platinum electrodes were produced and connected in series, using a gold/nickel solder [39]. Flat discs did not lead to satisfactory success; they resulted in bulky stacks and had difficulties with sealing, in contrast to the tubular cells using bell-and-spigot joints (Figure 2.7). The main problem was at the platinum anodes, which did not withstand long-term current loading; they tended to peel off from the electrolyte, probably due to the water vapour formed between the electrolyte and the anode layer. So, tubular designs were prioritised at that time to remove sealing problems.

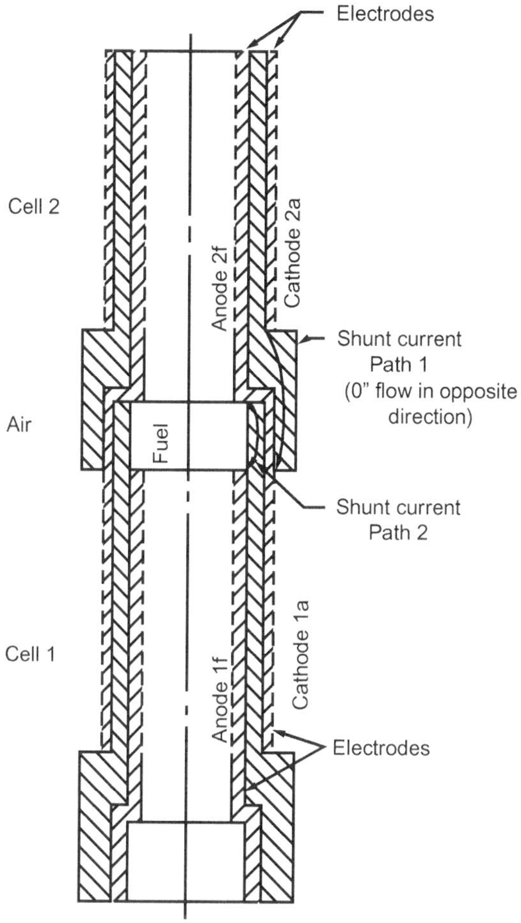

Figure 2.7 Cross-section of a solid electrolyte fuel cell of the bell-and-spigot type with shunt current paths in the seal region developed at Westinghouse (1963).

In Europe, in 1958, Palguyev and Volchenkova published conductivity measurements on $3ZrO_2 2CeO_2$ + 10 wt% CaO and other systems [40]. From 1960 onwards, results of a broadly based research programme on cells with solid oxide electrolytes appeared from the Ural branch of the Academy of Sciences of the USSR [41] under the leadership of Karpachov. Tannenberger et al., starting in 1959 at the Battelle Institute in Geneva, presented a thin film fuel cell concept in a 1962 patent, where a porous ceramic support tube was used as a structural member [42]. From the Battelle Institute in Frankfurt, Sandstede gave in September 1962 the first report on the use of hydrocarbons as a fuel in solid oxide cells, applying a converter containing Ni gauze as catalyst upstream of the cells (discs of $Zr_{0.85}Ca_{0.15}O_{1.85}$, diameter 22 mm, with porous Pt layers), and compared measurements with theoretical calculations [43]. At about the same time, fuel cell work was started in France by Kleitz [44], and in Britain, a patent was filed by Williams in August 1963 [45] to form fuel cells by depositing layers on a porous metallic carrier. In Japan, Takahashi [46], after investigations with alkali carbonate electrolytes, published in 1964 his first results obtained on fuel cells with solid oxide electrolytes. Surveys of these activities were presented at the international fuel cell meetings in 1965, 1967 and 1969 in Brussels. Both Williams and Takahashi were later to produce books on fuel cells, with emphasis on SOFCs.

The mobility of the oxide ions in $Zr_{0.85}Ca_{0.15}O_{1.85}$ was determined using the $^{18}O/^{16}O$ isotope exchange between solid and gas phase by Kingery et al. in 1959 [47] and more precisely by Simpson and Carter in 1965 [48]. In 1962, Schmalzried [49] showed by X-ray intensity measurements that the Zr and Ca cations occupy random sites in the cation sublattice of $Zr_{0.85}Ca_{0.15}O_{1.85}$. In 1963, decrease in conductivity with time was seen as a sign of ageing of the oxide ion conductors, caused by disorder–order transitions, in which the random distribution of the cations and oxide ions in the lattice changed to an ordered state.

In 1965, the fundamentals of gas potentiometry were presented, including the range of free oxygen and of oxygen in equilibria, and the 'neutral' transition range. Calculations and measurements in the case of potentiometric titrations of different gases were in good agreement in all three ranges. The sudden change of the cell voltage of a hydrogen/air cell at the equivalence point when oxygen was fed to the hydrogen had already been shown graphically by Archer et al. [39]. Thus the investigations started for SOFCs led to the development of oxygen sensors (lambda probes) now widely used in automobiles. A zirconia cell working potentiometrically was first proposed by Loos in 1969 as a sensor for O_2 and CO for the regulation of the air/fuel ratio in cars [50]. Another less well-known by-product of SOFC development was the electrochemical thermometry; that is, the determination of elevated temperatures on the thermodynamic scale with CO, CO_2, H_2, H_2O [51] or O_2 concentration cells [52].

The first investigations of polarisation phenomena in solid oxide fuel cells were conducted by the research groups in Sverdlovsk [53], Frankfurt [54], Geneva, Grenoble and Nagoya [55]. In the detailed investigations of fuel cells with cerium–lanthanum mixed oxides by Takahashi et al. [55], the polarisations observed were much smaller at the anode than at the cathode because by partial reduction of the solid electrolyte, a mixed conductor (solid solution of Ce_2O_3 in CeO_2) was formed at the anode giving a depolarising interlayer. Detailed investigations of the polarisation of solid electrolyte cells by determining the complex admittance were first conducted by Bauerle in 1969 [54].

The focal point of work on solid oxide fuel cells during this period was the development of electrode materials. An early problem was poor adhesion of the anode layers, which became obvious in 1963. Spacil as early as 1964 found the now well-known solution of using layers of nickel closely mixed with solid electrolyte material [56].

It was considerably more difficult to find a suitable cathode material. It turned out to be a misconception to think that small concentrations of polyvalent cations in mixed oxides with fluorite structure produce a high electronic conductivity. Layers with a high concentration of praseodymium at the cathode and high concentration of cerium at the anode had to be realised in order to achieve anything near the desired conductivities. Only mixed oxides with uranium proved to be a good material for stable layers with electronic and ionic conductivity, spreading the electrode reactions across the three-phase boundaries of electrode/electrolyte/gas.

This idea was confirmed by the result that thin mixed conducting interlayers between the pure electrolyte and the metallic conductor considerably reduced polarisation phenomena and led to high current densities. An optimised material with polyvalent uranium ions in mixed oxides with fluorite structure suitable for sintering on solid oxide electrolytes without phase boundaries was developed by Tannenberger in 1967 [57]. It proved to be a favourable interlayer in cathodes and anodes. It was about this time that the benefits of solid oxide electrolysis cells were also observed, showing the possibilities for producing hydrogen at half the price of room temperature electrolysis [58].

Indium oxide with different additives was proposed as a cathode material in 1966 [59] and frequently used. However, electronically conducting perovskites soon began to dominate the developments for both cathode and interconnect. The use of $La_{1-x}Sr_xCoO_3$ for the air electrode of solid oxide fuel cells marked the beginning, followed in 1967 by recommendations of $PrCoO_3$ [60] and of mixtures of the oxides of Pr, Cr, Ni and Co [61]. Strontium-doped lanthanum chromite, the most important ceramic interconnection material during the 1980s and 1990s, was proposed by Meadowcroft in 1969 [62].

2.6 On the path to practical solid oxide fuel cells

By 1970, the results of investigations on properties and applications of solid oxide electrolytes were so numerous as to make them very difficult to survey. A comprehensive review by Etsell and Flengas included 674 references [63]. But commercial activity was still confined to potentiometric oxygen sensors, although space applications, high-temperature electrolysis and scale-up for large coal burning power stations had been considered. In the United States, the programme for development of SOFCs ran out in 1970. The greatest obstacles to construction of a working SOFC were ceramic processing, the cathode material, the electronically conducting interconnection material and the selection of suitably profitable application areas.

These issues were intensively studied in the research laboratories of Brown Boveri, where under the leadership of Rohr from 1964, solid electrolyte fuel cells and oxygen sensors were investigated. Between 1969 and 1973, more than 100 oxide substances were synthesised and tested as electrode materials for SOFCs [64]. $LaNiO_3$ doped with Bi_2O_3 and $LaMnO_3$ doped with SrO proved to be particularly suitable, and from 1973 onwards, $La_{0.84}Sr_{0.16}MnO_3$ (LSM) was used widely. Unfortunately, its oxygen conductivity was negligible and the electronic resistance was high below 800 °C. Special investigations were also devoted to the interconnection material. Despite good results with successful tests of single cells at 1000 °C over a period of more than 3 years, plus construction and tests of modules with 25 series-connected cells, the development was not continued after 1975 because the cost for the necessary manufacturing processes did not seem to be sufficiently economic.

A significant step forward was made by Isenberg [65] in 1970 with the development of the electrochemical vapour deposition (EVD) method. Then in 1978, Westinghouse started a new programme, in which the EVD method was used for the perfect closing of the pores in the electrolyte and the interconnecting layers. A new cell design was instrumental in the breakthrough (Figure 2.8); long tubular cells (inside air, outside fuel), electrically interconnected by oxide materials and ductile nickel mesh conductors, were combined together in tube bundles. This design led to the first 5 kW SOFC generator containing 324 cells in 1986 [66] and to the 1152-cell 100 kW SOFC power system which began operation near Arnhem in The Netherlands in January 1998 [67].

Four further SOFC application companies started in the 1990s, and these were to have a major influence on the direction of the technology. Ceramic Fuel Cells Ltd. (CFCL) started in Melbourne, Australia, in 1992 with a vision of planar SOFC plants competing with diesel engines in the 100 kWe range [68]. About the same time, Sulzer Hexis devised a planar stack based on a circular heat exchanger, aimed at the 1 kWe domestic building market [69]. In 1993, Rolls-Royce fuel cell systems (RRFCS) started to exemplify a planar cell design printed on extruded flat-tube ceramic supports (Figure 2.9) in order to assemble a 1 MWe gas turbine unit operating at 5 bar [70]. Meanwhile, Haldor Topsoe took the results achieved by Risoe Laboratory in Roskilde to define a manufacturing plant based on planar, anode-supported square plates. The final application was not well-defined but was based on a manufacturing process built on tape casting of anodes, followed by screen printing of electrodes [71]. This movement, away from thick tubes and complex vapour deposition processes, towards planar thin sheets made by tape-casting and economic screen printing techniques based on electronic ceramic industry methods was a major culture shift which is continuing to the present time. The objective was to bring the costs down while raising both ceramic quality and power density, simultaneously reducing the operating

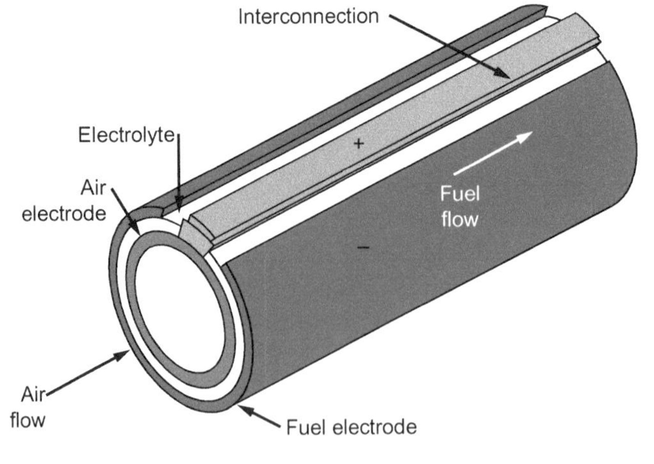

Figure 2.8 Westinghouse tubular design based on chemical vapour deposition of zirconia electrolyte.

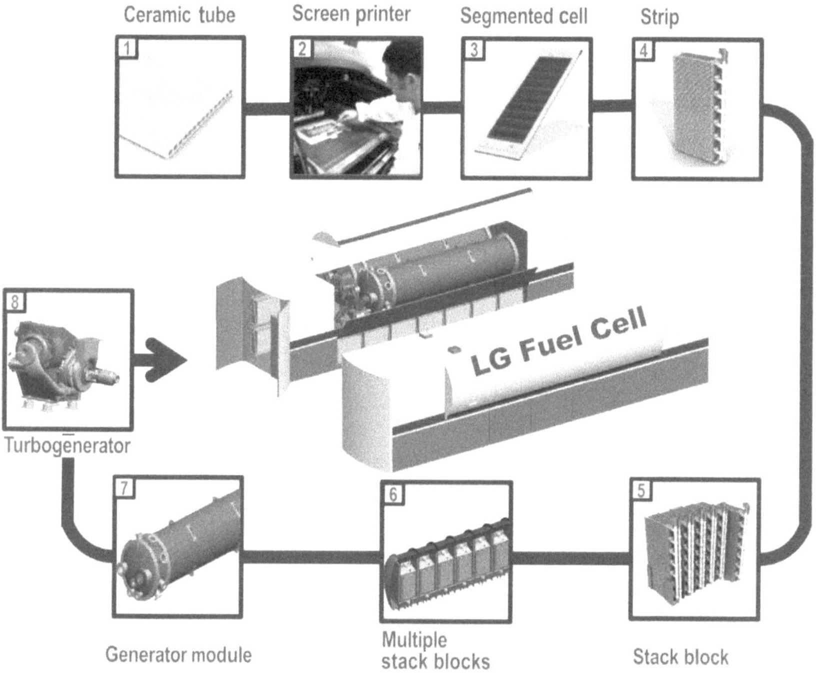

Figure 2.9 Planar design process for RRFCS concept of planar on tubular support.

temperature from 900 to 700 °C. Understanding this large change in direction requires consideration of improved ceramic processing methods, anode support, improved cathodes and cheap interconnects.

2.7 Ceramic processing for high-quality products

Scientific focus on ceramic processing began to pay off in the 1970s onwards as it became clear that the traditional methods of dry powder compaction, plastic clay processing and slip casting (filter moulding of a ceramic particle suspension in water) which had been operated over several millenia in the potteries industry could only offered modest properties when compared with both the theoretical predicted strength and that obtained by melt casting. Sinterability was poor, porosity high and strength miserable, all pointing to the presence of large defects, as indicated in the review of 1986 [72]. A typical comparison between the strengths obtained on the same material by various processes is shown in Figure 2.10 [73]. It was clear that the theoretical strength of an oxide like alumina was near 50 GPa, whereas the best powder-pressed products gave around 0.6 GPa, two orders of magnitude inferior as shown in Figure 2.10. Melt processing to produce sapphire rods could reach 10 GPa and new colloidal processes around 2 GPa, indicating that a factor three improvement was readily achievable by focussing on the powder process issue.

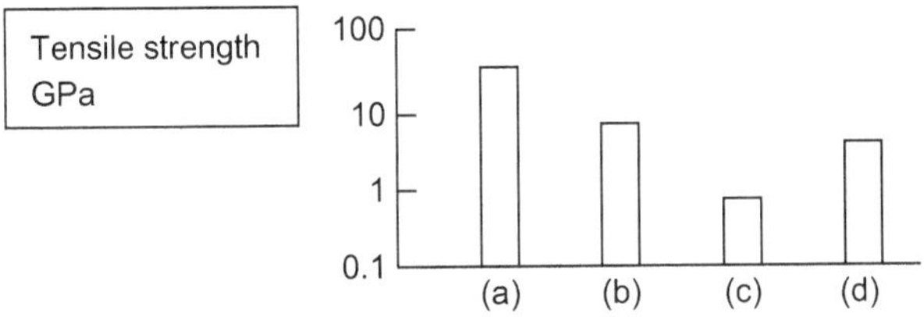

Figure 2.10 Strength of alumina. (a) Theoretical strength, (b) strength of sapphire rod, (c) strength of conventional pressed powder and (d) strength after colloidal processing to remove agglomerates.

This serious problem in powder compaction processes was shown to arise from the aggregation behaviour of fine particles. Whereas particles above 50 μm in diameter tended to flow readily, those below 1 μm stuck together tenaciously because of van der Waals attractions, thereby creating agglomerates which acted as defects in the moulding, and remained in the fired ceramic because the sintering process could not heal such large defects. An additional problem is that, even in a perfectly dispersed powder suspension, flocculation clustering can arise to form large clusters around 60 μm in diameter, especially during the drying process, the clusters then acting as defects in the fired ceramic [74]. This became obvious in the early 1980s when laser scattering instruments such as the Malvern Mastersizer became available for the first time to measure both grain size and aggregate size in ceramic suspensions, illustrating the wide range of agglomerates in the dispersions.

Consequently, the history of successful SOFC demonstrations has largely reflected the new processes and measurement techniques which have been brought to bear in order to solve this agglomerate defect problem [75]. Processes such as vapour deposition, combustion reactions, sol–gel, colloidal processing, injection moulding and polymer extrusion have all had an influence on the commercialisation. For example, the first succesful YSZ sensors were thimbles manufactured by powder pressing, which resulted in thick layers, high resistance and low strength. In the 1990s, these large tubes were replaced by tape-cast sensors similar to electronic capacitor manufacturing by colloidal processing, which gave much smaller devices, thinner layers, low resistance, high strength and lower materials cost (Figure 2.11). Similarly, the already mentioned introduction of the chemical vapour deposition of YSZ by Isenberg in 1970 [65] provided the 60 μm YSZ electrolyte which made the Westinghouse tubular cells so interesting.

During the early 1990s, the adoption of colloidal processing through tape casting and screen printing led to much improved electrolyte and anode structures which were used to produce planar cells of wide application potential. The invention of this process goes back to the early days of electronic ceramic fabrication [76,77] where ceramic powder was milled into a fluid, typically alcohol or methyl ethyl ketone, together with a dispersing agent such as fish oil and a polymer binder to give strength to the dried film. The slurry could be filtered to remove aggregates, then tape cast, dried and fired. Subsequent improvements in surfactants, polymers and solvents have led to substantial rise in product quality over five decades.

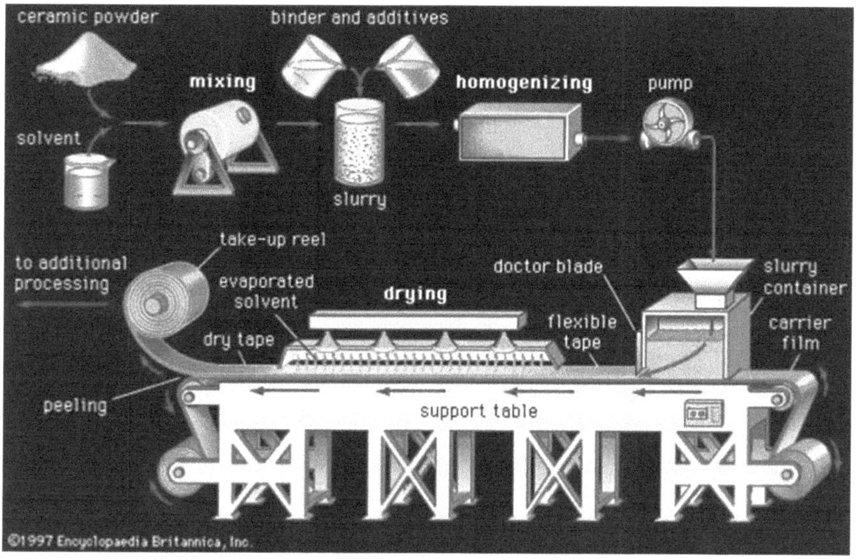

Figure 2.11 Colloidal processing by tape casting (copyright Encyclopaedia Brittanica).

Extrusion of tubes improved from the early porous supports and LSM cathodes of the Westinghouse design ending with the fine microtubes made by AMI and other companies after 2002. The process patented by ICI [78] involved mixing the ceramic powder with a water-soluble polymer such as polyvinylalcohol using sufficient water to obtain a stiff dough which could be rolled or kneaded under high shear stress, followed by extrusion to form microtubes without defects. 1 mm diameter YSZ tubes with 100 mm wall thickness were first extruded in 1991 and after drying, burn-out and firing around 1400 °C, these showed good SOFC behaviour. Coextrusion of anode and electrolyte was first demonstrated in 1994 [79]. However, the electrolyte was still too thick and the process was improved by Crumm and Halloran to give 10 μm YSZ thickness with very high quality [80]. This process gave the benefits of a low-resistance electrolyte coextruded with a highly conducting nickel–YSZ anode. The anode-supported structure, as shown in Figure 2.12, proved to be a winning design.

2.8 Anode support

Figure 2.12 shows an SEM micrograph of an anode-supported SOFC made at NIMTE in Ningbo where several layers were deposited by tape casting. Firstly, a 400 μm support layer of nickel–YSZ cermet was produced. Secondly, a 10 μm thick anode functional layer was tape cast onto the anode tape, followed by a 10 μm thick YSZ electrolyte. Finally, two LSM cathode layers were sprayed on before cosintering [81].

The experimental results were displayed for comparison with the theoretical model in Figure 2.13, showing reasonable agreement at 750 and 800 °C. What is not stated in this paper is the origin of the techniques used to obtain such good performance. The

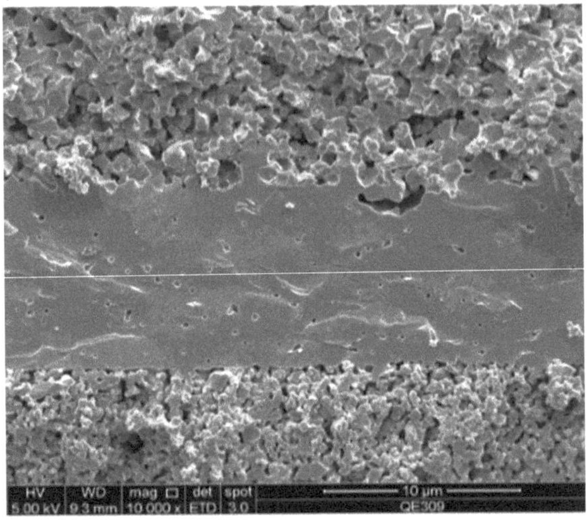

Figure 2.12 Anode active layer on support (top layer) with YSZ electrolyte underneath and LSM at bottom [81].

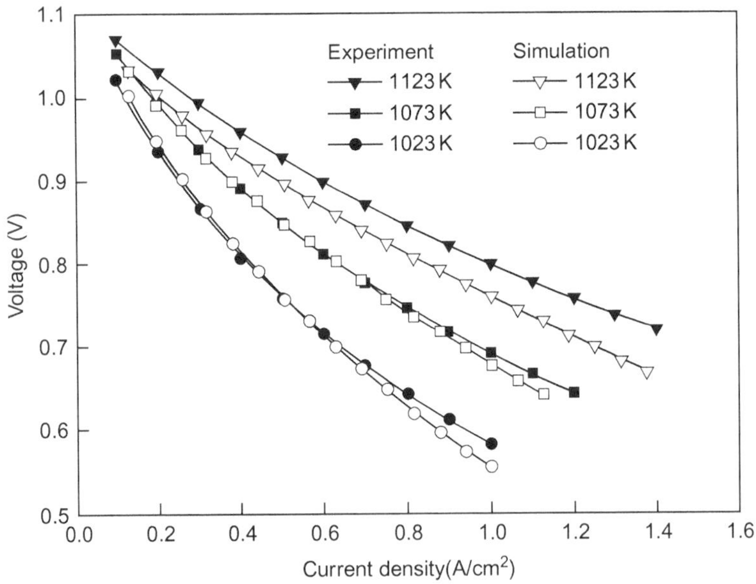

Figure 2.13 Comparison of experimental and model IV curves at different temperatures [81].

co-author Dr. Weiguo Wang joined Risoe Laboratories at Roskilde in Denmark in 2001. At that time, the anode-supported SOFC had been developed for several years by Dr. Mogens Mogensen's group. They had success in 1997 when they found a way to cosinter the anode tape, combined with the anode thin active layer combined with the thin electrolyte layer [82]. The whole planar cell was rather thin, strong and flexible although it was often warped after the sintering stage.

Parallel activities on a thicker anode support structure had also been going on in 1997 at Forschungszentrum Julich by resin casting a 3 mm thick flat-plate of NiO/YSZ mixed powders, followed by printing of electrolyte and cathode layers [83]. This patent was bought by Global Thermoelectric (later Versa Power) in Canada but proved to be too thick and weak so that, ultimately, the tape-casting process became dominant.

Although it is difficult to pin down the precise invention of the thin strong anode-supported planar cell design, it is clear that in the mid-1990s several groups were exploring these structures and two reviews were published [84,85]. The original invention of the wet process came from Dokiya and his colleagues at AIST in Japan published in 1991 [86]. Of course, the cathode properties still remained the most difficult problem since the LSM results shown in Figure 2.13 could not get the operating temperature below 700 °C where the ferritic stainless steel interconnect was lasting for many thousands of hours.

2.9 Better cathodes

It was well known even in the 1960s that the cathode material was the main limitation of SOFC designs once the electrolyte became thinner and the nickel cermet anode was conducting enough to give low specific resistance around 0.1 ohm cm^2. Clearly, if the SOFC was designed to operate near 600 °C, LSM was not satisfactory, though it was used from 1973 for 20 years by most manufacturers to give good results around 900 °C. A new material with good oxygen ion conduction was needed at lower temperatures.

Development of successful alternative cathode compositions started in Japan but accelerated because Brian Steele at Imperial College was convinced that SOFC operation below 700 °C was necessary so that stainless steel interconnects would survive, even going as low as 500 °C to run directly on methanol [87]. The electrolyte which could perform at such low temperatures was cerium gadolinium oxide (CGO). Although this became electronically conducting around 550 °C, giving mixed ionic–electronic conductivity, the prospect of building SOFCs to run much cooler became a priority in London [88], with many studies showing that a suitable cathode was LSCF [89,90]. This had problems in the presence of carbon dioxide and water [91] and was also too reactive with YSZ under ordinary operating temperatures, forming insulating layers of lanthanum zirconate, but its conductivity turned out to be seven times higher than that of LSM [92] as shown in Figure 2.14 with resistance of 0.2 ohm cm^2 at 700 °C.

The Imperial College concept, which was first commercialised in 2001 when the company Ceres Power was formed, proved to be radical because it used a planar configuration, a new electrolyte, a new cathode, new interconnect, a new fuel and new metal support. Perhaps this required too many innovations at once. Thus it deviated immensely from the proven Westinghouse concept which had demonstrated long-term SOFC operation over a 10-year period at 900 °C with methane fuel in a tubular reactor with YSZ electrolyte, LSM cathode and lanthanum chromite–nickel interconnects. The difficulty with the low temperature Imperial College concept was that the domestic building market at which the product was aimed needed to process natural

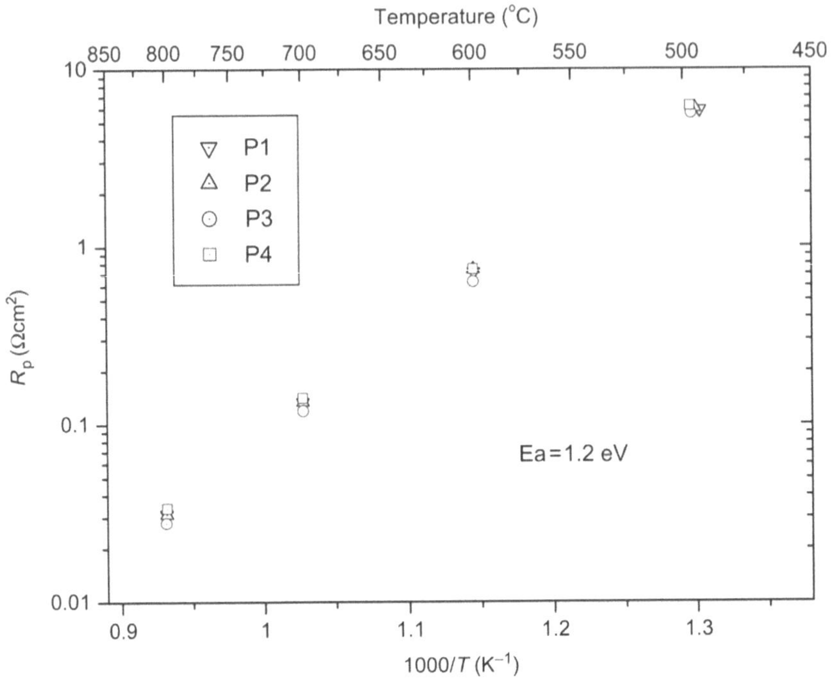

Figure 2.14 Polarisation resistances at different temperatures for LSCF–CGO on YSZ [92].

gas about 700 °C, well above the CGO running temperature. After 10 years with no product sales, the company needed more money in 2012 and the market launch date was put back to 2016.

However, LSCF began to be taken up by other researchers in the YSZ field around 2004 [93,94]. A protective layer of CeO was needed on the YSZ to prevent long-term degradation, some attention was needed to chromium poisoning [95] and then the electrochemical performance was suitable to give 0.4 Wcm^{-2} at 700 °C. The potential for further lowering of operating temperature to 500 °C has been described by Kilner and colleagues [96,97], but since the currently available fuels such as natural gas, propane and butane are processed around 700 °C, there seems to be little incentive to move in that direction at present. The optimum operating temperature around 700 °C must take into account the fuel processing but also the corrosion of economic interconnecting metallic materials as considered next.

2.10 Low-temperature operation with new interconnects

The interconnector material and the way it couples with the stack design has been one of the major obstacles to effective SOFCs. During the period from 1980 to 2000, ceramic interconnection based on the Westinghouse experience of lanthanum strontium chromite appeared to give successful high-temperature tubular operation up to 1000 °C but proved to be less successful in planar stacks [98]. However, when Siemens took over the SOFC development programme in Pittsburgh around the year 2000, they began to adopt their planar Plansee chromium (94% Cr; 5 wt% Fe; 1 wt% Y_2O_3)-based metal interconnect strategy which they had developed in European projects from 1990 onwards. This had been an ambitious idea to develop a completely new metal alloy for corrosion resistance at temperatures ranging from 800 to 1000 °C while maintaining good electrical contact based on a protective layer of semiconducting chromium oxide. The material proved to be expensive and difficult to machine, with the added problem of causing chromium poisoning of the cathode. This meant that surface coatings were needed to limit chromium escape from the surfaces [99], especially spinel or perovskite coatings.

However, it was recognised that some existing ferritic stainless steels such as AISI 430 could withstand these temperatures and be economic [100]. The key question was the growth of insulating oxide layers at the interconnect contact with the cathode. In particular, silicon content in the alloy would oxidise to increase contact resistance over a 1000 h at 800 °C as shown by the weight gain in Figure 2.15, top line. Purification of these alloys to give low silicon stainless steel such as Crofer 22 APU gave improved performance but was costly because of the vacuum processing required. Coating with an oxide material also gave an improvement (Figure 2.15, lower points) [100].

The alternate approach was to use alloy formulations which contained silica but which did not grow the insulating films because the silica became trapped on inclusions. For example, Dulieu et al. [101] found that alloy 1.4509 (German spec, like AISI 441), similar in many ways to 1.4016 (AISI 430), did not form a silica film

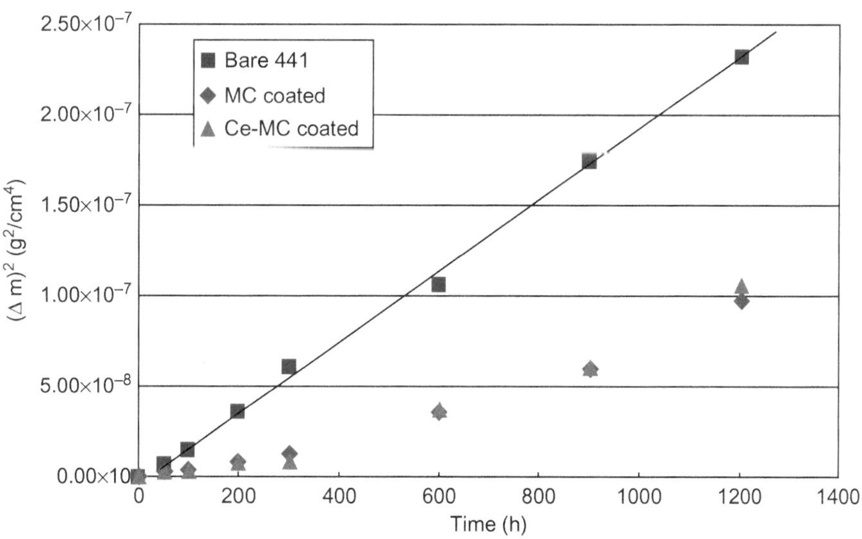

Figure 2.15 Weight gain due to oxidation of ferritic stainless steel 441 showing effect of coatings.

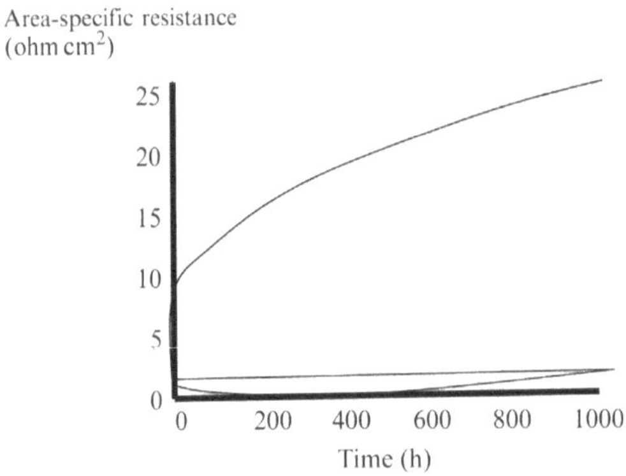

Figure 2.16 Effect of oxidation of AISI 441 in air at 800 °C on area-specific resistance ohm cm^2 [100]; top line, no coating and bottom lines, coated.

at the interface, despite the presence of about 0.5% residual Si in both alloys. The main difference between the two alloys was the presence of traces of Nb and Ti in 1.4509, suggesting that these elements played a role in preventing Si movement to the scale region. Based on this encouraging result, PNNL [100] investigated the metallurgical characteristics, oxidation behaviour and electrical performance of AISI 441. They also applied coatings to show further enhancement of lifetime under SOFC operating conditions. A typical improvement in area-specific resistance with time at 800 °C is shown in Figure 2.16.

Since the first edition of this book in 2003, a large amount of work has gone into advancing this metal interconnect aspect, together with the coating principles, and a recent review gives the wider picture [102].

2.11 Application areas

In the first edition of this book, a range of applications was described as in Figure 2.17, including small generators for military, leisure and vehicle auxiliaries around 100 W to compete with batteries, larger CHP units for domestic homes at 1 kW rising to 100 kW for larger commercial buildings and MW for industry sites and GW for power stations, together with 100–1000 kW for trains, ships, etc. The inserted boxes show where there has been some impact during the past 10 years.

It is clear that sensors are successful and that there have been three substantial energy areas where SOFCs have been put in place:

(1) Bloom energy has installed several hundred generators for back-up on industry sites around 200 kW since 2010
(2) Japanese companies have installed and run several thousand SOFC domestic CHP units about 1 kW since 2007
(3) Many portable SOFCs about 100 W have been sold in the United States for soldier power, back-up power and battery charging since 2008.

The concept of large oxygen generation has not been successful, there are no GW power stations with SOFC technology and there are no trains or ships using SOFCs at present.

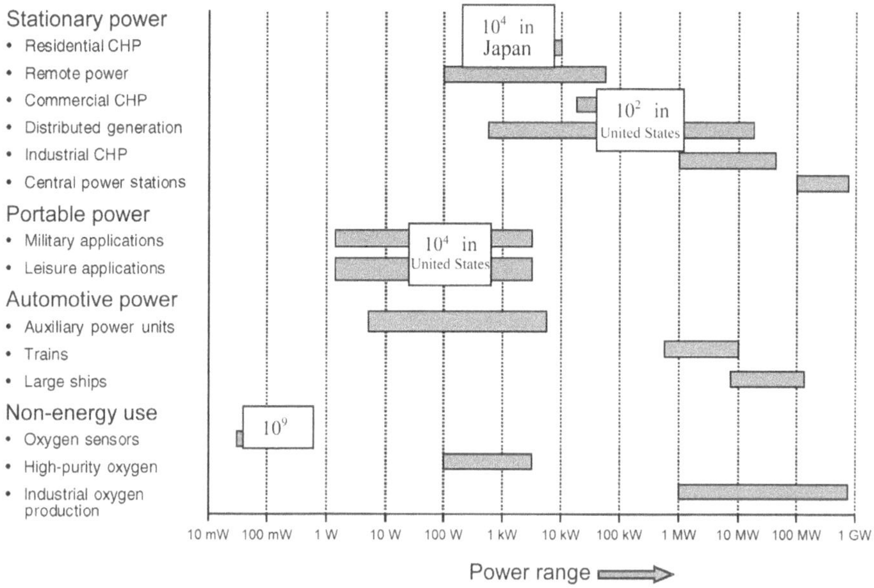

Figure 2.17 Boxes showing applications of SOFCs with systems since 2003.

Perhaps the most intriguing advance is in small portable systems running on butane, propane or methane. These were first demonstrated 20 years ago as micro-tubular SOFCs which could start-up within 10 s [103]. Several simple systems running on butane were described around the year 2000 [104]. Now the most dramatic example is the unmanned air vehicle originating at AMI in 2006 (now Ultra-AMI) [105]. Further application projects on recreational vehicles and trucks are progressing [106]. But wider markets are being predicted for SOFC phone chargers and similar small devices such as Nectarpower from Lilliputian [107], which unfortunately spent 150 M$ and ran out of money in 2014. In addition, the Danish company Topsoe Fuel Cell making planar SOFCs stopped its operations in 2014, while CFCL were bankrupt in 2015.

2.12 Summary

The history of SOFCs is now almost 120 years old and reflects continuing discoveries in electrolyte and electrode materials, advances in ceramic processing to produce thinner and extended area cells, reduction in operating temperature to 700 °C through the use of LSCF cathodes, innovation in interconnecting metal alloys, together with the development of novel commercial applications.

Although the car exhaust sensor has been the largest application since the 1970s, several new niche markets are appearing including cell-phone chargers using butane as fuel, hybrid systems for unmanned air vehicles based on propane, auxiliary power supplies for back-up and leisure use in campervans and trucks and combined heat and power supplies for domestic housing using natural gas. In Japan, many thousands of these domestic cogenerator SOFCs are operating.

It has proved difficult to compete with diesel generation in the 100 kW range, although hundreds of such units now operate in the United States. However, larger systems running on methane or gasified coal look very promising in terms of carbon capture and efficiency as climate change becomes more of a priority in the twenty-first century.

References

[1] H. Davy, On some new phenomena of chemical changes produced by electricity, particularly the decomposition of the fixed alkalis, and the exhibition of the new substances, which constitute their bases, Philos. Trans. Roy. Soc. Lond. 98 (1808) 1–45.
[2] M. Faraday, Experimental Researches in Electricity, Web-Printed Google Books, Taylor and Francis, London, 1839.
[3] M. Faraday, Experimental Researches in Electricity, 1821 (reprinted by Dover Publications, New York, 1965).
[4] M. Faraday, J. Tyndall, Experimental Researches in Electricity, Dent, London, 1922.
[5] H. Cavendish, Franklin Experiments and Observations on Electricity, fifth ed., F. Newbery, London, 1774, p. 411.
[6] W. Beetz, Uber die Leitungsfahigkeit fur Electricitat, welche Isolatoren durch Temperatur-Erhohung annehmen, Ann. Physik u. Chem 92 (1854) 452–466. H. Buj, Uber die elektrische Leitfahigkeit des erhitzten Glases, Ann. Chem. Pharm. 90 (1854) 257–283.
[7] J.M. Gaugain, Note about a new type of gas cell, Paris, Comptes Rendus, XXXVIT., 1853, pp. 584–588.
[8] W. Nernst, Verfahren zur Erzeugung von elektrischem Gluhlicht, DRP 104 872, filed 6.7.1897, also DRP114241,1899.
[9] W. Nernst, Uber die elektrolytische Leitung fester Korper bei sehr hohen Temperaturen, Z. Elektrochem. 6 (1899) 41–43.
[10] W. Nernst, Material for electric-lamp glowers, US Patent 685 730, filed 24.8, 1899.
[11] H. Reynolds, Uber die Leitfahigkeit fester Mischungen bei hohen Temperaturen. Ein spezieller Fall der festen Losungen, Thesis, University of Gottingen, 1902.
[12] C. Wagner, Uber den Mechanismus der elektrischen Stromleitung im Nernststift, Naturwissenschaften 31 (1943) 265–268.
[13] H. von Helmholtz, Die Thermodynamik chemischer Vorgange I, II, III. Sitzungsber. Akad. Wiss. Berlin 2.2.1882, 27.7.1882, 31.5.1883. In Wiss. Abhandlungen, Ambrosius Barth, Leipzig, vol. 2 (1883), pp. 958–978, 979–992; vol. 3 (1895), pp. 92–114.
[14] W. Ostwald, Die wissenschaftliche Elektrochemie der Gegenwart und die technische der Zukunft, Z. Elektrochem. 1 (1894) 81–84, 122–125; Z. Physik. Chem. 15 (1894) 409–421.
[15] L. Mond, C. Langer, A new form of gas battery, Proc. R. Soc. Lond. 46 (1889) 296–304.
[16] F. Haber, A. Moser, Das Generatorgas und das Kohlenelement, Z. Elektrochem. 11 (1905) 593–609.
[17] F. Haber, Verfahren zur Erzeugung von elektrischer Energie aus Kohle und gasformigen Brennstoffen, Austrian Patent 27 743, filed 5.8.1905.
[18] W.D. Treadwell, Uber den Sauerstoffdruck einiger Oxyde and die Kohle-Sauerstoffkette bei hoherer Temperatur, Z. Elektrochem. 22 (1916) 414–421.

[19] E. Baur, W.D. Treadwell, Brennstoffelement, DRP 325 783, filed 20.9. 1916.
[20] E. Baur, R. Brunner, Uber die Eisenoxyd-Kathode in der Kohle-Luft-Kette, Z. Elektrochem. 43 (1937) 725–727.
[21] E. Baur, H. Preis, Uber Brennstoff-Ketten mit Festleitern, Z. Elektrochem. 43 (1937) 727–732.
[22] E. Baur, H. Preis, Brennstoff-Ketten mit Festleitern, Z. Elektrochem. 44 (1938) 695–698.
[23] W. Schottky, Uber stromliefernde Prozesse im Konzentrationsgefalle fester Elektrolyte. Wiss. Vero. Siemens-Werke 14 (1935) no. 2, 1–19 and Foreword.
[24] F. Hund, Anomale Mischkristalle im System ZrO2–Y2O3. Kristallbau der Nernst-Stifte, Z. Elektrochem. Angew. Physik. Chem. 55 (1951) 363–366.
[25] H.-J. Engell, Karl Haulle zum 65. Geburtstag, Ber. Bunsenges. Phys. Chem. 82 (1978) 351–352.
[26] H. Peters, Untersuchungen der Gitterstruktur und der elektrischen Leitfähigkeit an Mischoxyden vom Fluorittyp, Thesis, University of Rostock, 1953.
[27] H. Peters, H.-H. Mobius, Verfahren zur Gasanalyse bei erhohten Temperaturen mit Hilfe galvanischer Fest-elektrolytelemente, DDR-P 21673, filed 20.5.1958.
[28] H.H. Mobius, Theoretische und experimentelle Untersuchungen an Brennstoffelementen mitfesten Elektrolyten, Thesis, University of Rostock, 1958, Mitteilungsbl. Chem. Ges. DDR. 6 (1959) 45–48.
[29] K. Biefeld, Brennstoffbatterie, DE-P 912 106, filed 27.2.1944.
[30] E. Ryschkewitsch, Oxydkeramik der Einstoffsysteme vom Standpunkt der physikalischen Chemie, Julius Springer, Berlin, 1948.
[31] H. Peters, H.-H. Mobius, Elektrochemische Untersuchung der Gleichgewichte CO+0,5 O2=CO2 und C+CO2=2 CO, Z. Physik. Chem. (Leipzig) 209 (1958) 298–309.
[32] J. Weissbart, R. Ruka, Oxygen gauge, Rev. Sci. Instr. 32 (1961) 593–595.
[33] W.E. Tragert, Fuel cell, US Patent 3,138,487, filed 28.2.1961; US Patent 3,138,488, filed 8.3.1961; US Patent 3,296,030, filed 1.11.1962.
[34] W. Oser, Fuelcell, US Patent 3,281,273, filed 5.5.1961.
[35] K. Kiukkola, C. Wagner, Measurements on galvanic cells involving solid electrolytes, J. Electrochem. Soc. 104 (1957) 379–387.
[36] R.J. Ruka, J. Weissbart, Electrochemical method for separating O2 from a gas; generating electricity; measuring O2 partial pressure; and fuel cell, US Patent 3,400,054, filed 24.7.1961.
[37] D.T. Bray, L.D. Lagrange, U. Merten, C.D. Park, Fuel cell having zirconia-containing electrolyte and ceramic electrodes, US Patent 3,300,344, ¢led 2 7.6.1962.
[38] J. Weissbart, R.J. Ruka, A solid electrolyte fuel cell, J. Electrochem. Soc. 109 (1962) 723–726.
[39] D.H. Archer, E.F. Sverdrup, W.A. English, W.G. Carlson, An investigation of solid-electrolyte fuel cells, Tech. Doc. Rept. ASD-TDR-63-448, July 1963.
[40] S.F. Palguyev, Z.S. Volchenkova, Das Problem der festen Elektrolyte fur Brennstoffzellen, Tr. Inst. Khim. Akad. NaukSSSR Ural Filial 2 (1958) 183–200.
[41] S.V. Karpachov, S.F. Palguyev, W.N. Chebotin, A.D. Neuimin, A.T. Filyayev, M.V. Perfilyev, et al., Tr. Inst. Elektrokhim. Akad. Nauk SSSR, Ural Filial, 1 1960 foll.
[42] H. Tannenberger, H. Schachner, W. Simm, Festelektrolytbrennsto¡element, DE-P 1 471 768, filed 22.5.1963; Swiss priority 23.5.1962.
[43] H. Binder, A. Kohling, H. Krupp, K. Richter, G. Sandstede, Elektrochemische Oxydation von Kohlenwasserstoffen in einer Festelektrolyt-Brennstoffzelle bei Temperaturen von 900-1000~C, Electrochim. Acta 8 (1963) 781–793.

[44] J. Besson, C. Deportes, M. Kleitz, 1er Colloque sur les piles a combustible. Bellevue, 6 dec 1962. Contrat de recherches 61 FR 136. 'Etude des e¤ lectrolytes solides pour piles a combustible a haute temperature' and 'Utilisation des electrolytes dans les piles a combustible a haute temperature'. In Les Piles a' Combustible. Editions Technip, Paris, 1965, pp. 87–102, 303–323.
[45] K.R. Williams, J.G. Smith, Fuel cell, GB Patent 1 049 428, filed 15.8. 1963.
[46] T. Takahashi, Research and development of fuel cells in Japan, J. Electrochem. Soc. 34 (1966) 60–69.
[47] W.D. Kingery, J. Pappis, M.E. Doty, D.C. Hill, Oxygen ion mobility in cubic $Zr_{0.85}Ca_{0.15}O_{1.85}$, J Am. Ceram. Soc. 42 (1959) 393–398.
[48] L.A. Simpson, R.E. Carter, Oxygen exchange and diffusion in calcia stabilized zirconia, J. Am. Ceram. Soc. 49 (1966) 139–144. GE report 65-RL-3985M, June 1965.
[49] H. Schmalzried, Uber Zirkondioxyd als Elektrolyt fu r elektrochemische Untersuchungen bei hoheren Temperaturen, Z. Elektrochem. Ber. Bunsenges. Physik. Chem. 66 (1962) 572–576.
[50] C.H. Loos, Vorrichtung zur Regelung des Luft-Brennsto¡-Verha ltnisses in einem Verbrennungsmotor, DE-P 2 010 793, ¢led 6.3.1970, priority NL 22.3.1969.
[51] H.-H. Mobius, Die Messung der thermodynamischen Temperatur mit galvanischen Festelektrolytelementen, Z. Chem. (Leipzig) 1 (1961) 63.
[52] W.T. Lindsay, R.J. Ruka, Electrochemical thermometry: a new approach to the determination of elevated temperatures on the thermodynamic scale, Electrochim. Acta 13 (1968) 1867–1874.
[53] S.V. Karpachov, A.T. Filyayev, S.F. Palguyev, Polarisation of carbon monoxide electrodes on platinum in a solid zirconia-lime electrolyte, Electrochim. Acta 9 (1964) 168–185.
[54] J.E. Bauerle, Study of solid electrolyte polarization by a complex admittance method, J. Phys. Chem. Solids 30 (1969) 2657–2670.
[55] T. Takahashi, K. Ito, H. Iwahara, The efficiency of solid-electrolyte fuel cells, Electrochim. Acta 12 (1967) 21–30.
[56] H.S. Spacil, Electrical device including nickel-containing stabilized zirconia electrode, US Patent 3,503,809, filed 30.10.1964.
[57] H. Tannenberger, Electrode pour pile a electrolyte solide, Swiss Patent, filed 19.7.1967; F-P 1.572.073, filed 18.7.1968.
[58] General Electric, GE process could make cheaper hydrogen, Chem. Ing. News (1968) 48–49; H.S. Spacil, Solid oxygen-ion electrolyte cell for the dissociation of steam, US Patent 3,635,812, filed 5.7.1968; H.S. Spacil, C.S. Tedmon, Electrochemical dissociation of water vapor in solid oxide electrolyte cells, J. Electrochem. Soc. 116 (1969) 1618–1633.
[59] D.D. Button, D.H. Archer, Development of $La_{1\sim x}Sr_xCo_{O3}$ air electrodes for solid electrolyte fuel cells, J. Am. Ceram. Soc, Washington, 1966, Meeting, May.
[60] S.P. Mitoff, High temperature electrical conductor comprising praseodymium cobaltate, US Patent 3,533,849, filed 12.6.1967.
[61] W. Pabst, G. Sandstede, G. Walter, sauerstoffelektrode für galvanische Zellen, insbesondere Kathoden in Brennstoffzellen, DE-P 1 671 721, filed 23.12.1967.
[62] D.B. Meadowcroft, Some properties of strontium-doped lanthanum chromite, Brit. J. Appl. Phys. (1969) 1225–1233 (J. Phys. D, vol. 2).
[63] T.H. Etsell, S.N. Flengas, The electrical properties of solid oxide electrolytes, Chem. Rev. 70 (1970) 339–376.
[64] J. Rohr, High-temperature fuel cells, in: P. Hagenmuller, W. van Gool (Eds.), Solid Electrolytes, Academic Press, New York, 1978, pp. 431–450.

[65] A.O. Isenberg, Recent advancements in solid electrolyte fuel cell technology, in: Nat. Fuel Cell Seminar Newport Beach, CA, Abstr. 14–18 Nov, 1982, pp. 154–156.

[66] P. Reichner, J.M. Makiel, Development status of multi-cell solid oxide fuel cell generetors, in: Fuel Cell Seminar 1986, Tucson, AZ, Courtesy Associates, Washington, 1986, pp. 32–35.

[67] J. Sukkel, Two years experience with the 100 kW SOFC cogeneration unit at Arnhem, in: Proceedings of the 4th European SOFC Forum, Luzern, 2000, pp. 159–166.

[68] K. Foger, B. Godfrey, System demonstration program at Ceramic Fuel Cells Ltd in Australia, in: Proceedings of the 4th European SOFC Forum, Luzern, 2000, pp. 167–173.

[69] R. Diethelm, J. Brun, Th. Gamper, et al., Status of the Sulzer Hexis SOFC system development, in: U. Stimming, S.C. Singhal, H. Tagawa, W. Lehnert (Eds.), SOFC V, 97–40, The Electrochemical Society, Pennington, NJ, 1997, pp. 79–87.

[70] N.T. Hart, M.J. Day, M.N. Pashley, Economic evaluation of 200kWe SOFC system, in: A.J. McEvoy (Ed.), Fourth European Fuel Cell Forum, Switzerland, 2000, pp. 449–458.

[71] S. Linderoth, Low cost fabrication and improved performance of SOFC stack components, in: A.J. McEvoy (Ed.), Fourth European Fuel Cell Forum, Switzerland, 2000, pp. 19–28.

[72] N.McN. Alford, J.D. Birchall, K. Kendall, Engineering ceramics – the process problem, Mater. Sci. Technol. 2 (1986) 329–336.

[73] K. Kendall, Influence of powder structure on processing and properties of advanced ceramics, Powder Technol. 58 (1989) 151–161.

[74] K. Kendall, N. Alford, J.D. Birchall, W.J. Clegg, Flocculation clustering and weakness of ceramics, Nature 339 (1989) 130–132.

[75] K. Kendall, Revolution in ceramic processing, in: F. Riley (Ed.), Third European Symposium on Engineering Ceramics, Elsevier, London, 1989, pp. 97–107.

[76] D.J. Shanefield, R.E. Mistler, Am. Ceram. Soc. Bull. 53 (1974) 416.

[77] L.L. Hench, D.R. Ulrich (Eds.), Ultrastructure Processing of Ceramics, Glasses and Composites, Wiley, NY, 1984.

[78] N. Alford, J.D. Birchall, K. Kendall, European Patent Publication 0288208, 1988.

[79] K. Kendall, M. Prica, Integrated SOFC tubular system for small-scale cogeneration, in: U. Bossel (Ed.), 1st European SOFC Forum, Luzern, 1994, pp. 163–170.

[80] A.T. Crumm, J.W. Halloran, Method for preparation of solid state electrochemical device, US Patent 6749799B2.

[81] S. Yang, T. Chen, Y. Wang, Z. Peng, W.G. Wang, Electrochemical analysis of an anode supported SOFC, Int. J. Electrochem. Sci. 8 (2013) 2330–2344.

[82] Mogens Mogensen, private, communication 2013.

[83] US Patent 599 8056, Anode substrate for a high temperature fuel cell, Forschungzentrum Julich, filed 1997.

[84] H.P. Buchkremer, et al., Advances in the anode supported planar SOFC technology, Electrochem. Proc. 160–170 (1997).

[85] S. Primdahl, et al., Thin anode supported SOFC, Electrochem. Soc. Proc. 99–19 (1999) 793–802.

[86] M. Dokiya, et al., Wet process for planar SOFC, in: F. Grosz, P. Zegers, S.C. Singhal, O. Yamamoto (Eds.), 2nd International Symposium on Solid Oxide Fuel Cells, Athens, Greece, July 2–5, 1991, EUR (Series) 13564 EN, Commission of the European Communities, Brussels, 1991.

[87] B.C.H. Steele, S. Carter, J. Kajda, et al., Optimisation of fuel cell components, in: Proceedings of Second International Symposium on SOFCs, 1991, pp. 517–525.

[88] M. Sahibzada, R.A. Rudkin, B.C.H. Steele, et al., Evaluation of PEN structures incorporating thick film CDG electrolytes, Electrochem. Soc. Proc. 97 (1997) 244–253.
[89] M. Sahibzada, S.J. Benson, R.A. Rudkin, et al., Pd promoted LSCF cathodes, Solid State Ionics 113 (1998) 285–290.
[90] L.A. Lane, S.J. Benson, D. Waller, et al., Oxygen transport in LSCF, Solid State Ionics 121 (1999) 201–208.
[91] S.J. Benson, D. Waller, J.A. Kilner, Degradation of LSCF in carbon dioxide and water atmospheres, J. Electrochem. Soc. 146 (1999) 1305–1309.
[92] W.G. Wang, M. Mogensen, High performance lanthanum ferrite based cathode for SOFC, Solid State Ionics 176 (2005) 457–462.
[93] A. Esquirol, J.A. Kilner, N. Brandon, Oxygen transport in LSCF/GDC composite cathode for IT-SOFCs, Solid State Ionics 175 (2004) 63–67.
[94] A. Esquirol, N. Brandon, J.A. Kilner, et al., Electrochemical characterization of LSCF cathodes for IT-SOFCs, J. Electrochem. Soc. 151 (2004) A1847–A1855.
[95] S.N. Lee, A. Atkinson, J.A. Kilner, Effect of chromium on LSCF SOFC cathodes, J. Electrochem. Soc. 160 (2013) F629–F635.
[96] J.A. Kilner, Accelerated R&D of SOFCs with lowering operation temperature vision, Electrochemistry 77 (2009) 113.
[97] A. Aguadero, L. Fawcett, S. Taub, et al., Materials development for IT-SOFC electrochemical devices, J. Mater. Sci. 47 (2012) 3925–3941.
[98] S.C. Singhal, K. Kendall, High Temperature SOFCs, Elsevier Science, Oxford, 2003.
[99] Y. Larring, T. Norby, Spinel and perovskite functional layers between Plansee metallic interconnect (Cr-5 wt % Fe-1 wt % Y 2 O 3) and ceramic (La0.85Sr0.15) 0.91MnO3 cathode materials for solid oxide fuel cells, J. Electrochem. Soc. 147 (2000) 3251–3256.
[100] Z. Yang, G. Xia, C. Wang, Z. Nie, J. Templeton, J. Stevenson, P. Singh, Investigation of AISI 441 ferritic stainless steel and development of spinel coatings for SOFC interconnect applications, SECA report DE-AC0576RL01830, 2008.
[101] D. Dulieu, J. Cotton, H. Greiner, K. Honegger, A. Scholten, T. Seguelong, in: P. Stevens (Ed.), Procedings of 3rd European SOFC Forum, European Solid Oxide Fuel Cell Forum, Switzerland, 1998, p. 447.
[102] N.J. Magdefrau, Evaluation of solid oxide fuel cell interconnect coatings: reaction layer microstructure, chemistry and formation mechanisms, PhD Thesis, University of Connecticut, 2013.
[103] K. Kendall, Solid oxide fuel cell structures, assigned to Keele University Patent Publication No. WO 94/22178, 1994.
[104] K. Kendall, M. Palin, A small SOFC demonstrator for microelectronic applications, J. Power Sources 71 (1998) 268–270.
[105] A.T. Crumm, T. LaBreche, Interconnect member for fuel cell, assigned to Adaptive Materials Inc., US Patent Pub. No. 2011/0189587A1, 2011.US 2005/0147857 on solid oxide fuel cell with improved current collection by A.T. Crumm, C.J. Reilly, J.R. Rice, 2005.
[106] M. Kendall, A.D. Meadowcroft, K. Kendall, Microtubular solid oxide fuel cells (mSOFCs), in: SOFCxiii Okinawa, October, The Electrochemical Society, Pennington, NJ, 2013.
[107] www.nectarpower.com; from Lilliputian Systems Inc., December 2013 (defunct in 2014).

Thermodynamics

3

W. Winkler
Retired Professor and Director of Institute for Energy Systems and Fuel Cell Technology at HAW Hamburg, Germany

3.1 Introduction

A solid oxide fuel cell is an electrochemical device that converts chemical energy of a fuel and an oxidant gas (air) directly into electricity without combustion. It can be treated thermodynamically in terms of the free enthalpy of the reaction of the fuel with oxidant. Hydrogen and oxygen are used to illustrate the simplest case in the early part (Section 3.2) of this chapter. This treatment allows the calculation of the reversible work at equilibrium for the reversible reaction. Heat must also be transferred reversibly to the surrounding environment in this instance.

The basic solution of any SOFC operation is given by the reduction of the thermodynamic potential by utilising fuel and oxidant. It is quite similar to the cooling of a hot gas flow by extracting heat by a Carnot type engine. This basic solution is principally reversible and thus no loss but the utilisation of a thermodynamic potential to produce reversible work. The irreversible losses, for example, ohmic losses, are caused by materials issues which could, in principle, be solved by selecting improved components.

Any SOFC produces reversible or irreversible heat depending on fuel used, so it is necessary to consider an SOFC system as a heat generator as well as an electricity source. In effect, the whole SOFC system can be treated as a power-generating burner, as in Section 3.4.

In a real engineering device, heat is exchanged within the SOFC in several ways including fuel processing, air preheating, flue gas cooling, etc. Excess air is normally required to prevent overheating, while the conversion of hydrocarbons into hydrogen and carbon monoxide often absorbs heat. The complex heat pathways are described in Section 3.5.

Ultimately, the heat output from the SOFC can be used to drive a heat engine such as a piston engine or gas turbine. These combined SOFC/heat engine cycles are analysed in Section 3.6. Although this is a popular field for modelling studies, there has been little advance in recent years.

First, it is essential to list the symbols and concepts used in this chapter.

Symbols

$\dot{H}$ enthalpy flow
$\dot{m}$ mass flow
$\dot{n}^*$ constant molar flow at the fuel cells anode

High-temperature Solid Oxide Fuel Cells for the 21st Century. http://dx.doi.org/10.1016/B978-0-12-410453-2.00003-8

$\dot{n}_{el}$	molar flow of electrons
$\dot{n}$	molar flow
C_P	temperature-dependent heat capacity
e	elementary charge
e_i	specific exergy of the component i
F	Faraday constant
H	enthalpy
h^*	specific enthalpy related to the standard state
I	electrical current
K	equilibrium constant
LHV	lower heating value
n	molar quantity
N_A	Avogadro constant
n^{el}	quantity of released electrons related on the utilised fuel
n_F	molar quantity of the supplied fuel
p	total pressure
p_0	standard pressure
P_{el}	electrical power
p_{el}	specific electrical power
p_i	partial pressure of the component i
P_{loss}	power loss
P_{rev}	reversible electrical power
Q	Heat
q	specific heat
R	electric or ohmic resistance
R_m	universal gas constant
S	total entropy
s^*	specific entropy related to the standard state
T	absolute temperature
U_f	fuel utilisation
V	voltage or potential
v	*specific volume*
V_N	Nernst voltage
$-W_t$	technical work
$-w_t$	specific technical work
y	molar concentration
β	ballast ratio
$\Delta^r G$	Gibbs enthalpy or free enthalpy of the reaction
$\Delta^r H$	reaction enthalpy
$\Delta^r S$	reaction entropy
Δs	entropy production
ΔV	voltage loss
ζ	exergetic efficiency
η	efficiency
ϑ	celsius temperature
λ	excess air value
μ	fuel-related specific mass
ν	fuel-related quantity
π	pressure ratio

Indices and abbreviations

0	stoichiometric value
A	air
aB	after burner outlet
AH	air heater
AFC	air at thermodynamic state of the fuel cell
An	anode
Ca	cathode
CC	Carnot cycle
CHP	combined heat and power generation
ECO	economizer
EXCO	external cooling
F	fuel
FC	fuel cell
FFC	fuel at thermodynamic state of the fuel cell
FGC	flue gas cooler
FH	fuel heater
G	flue gas
GT	gas turbine
$\mathbf{H_2}$	hydrogen
$\mathbf{H_2O}$	water
HEG	reversible heat engine (flue gas)
HEX	heat exchangers
HP	high pressure
HPA	heat pump (air)
HPF	heat pump (fuel)
I	inlet
i, j	components
INEX	intermediate expansion
irr	irreversible
LP	low pressure
O	outlet
$\mathbf{O_2}$	oxygen
PH	product heater
ref	reformer
rev	reversible
RG	reaction product gas
RH	reheater
SH	superheater
ST	steam turbine
Syst	System

3.2 The ideal reversible SOFC

The use of the first and the second laws of thermodynamics allows a simple description of a reversible fuel cell. The fuel and the air enter the fuel cell as non-mixed flows of the different components, and the flue gas leaves the fuel cell as a non-mixed flow as well if we assume a reversible operating fuel cell. The non-mixed reactants deliver the total enthalpy $\sum n_i H_i$ to the fuel cell, and the total enthalpy $\sum n_j H_j$ leaves the cell with the non-mixed products. Furthermore, the heat Q_{FCrev} must be extracted reversibly from the fuel cell and transported reversibly to the environment. This can be done, for example, if the fuel cell and the environment have the same thermodynamic state. Q_{FCrev} is defined as a positive number if it is transported to the fuel cell. The reversible work $-W_{\mathrm{tFCrev}}$ is delivered by the fuel cell. An idealised description of this model is given in Figure 3.1.

Usually, specific mass- or mol-related figures are used, and the fuel quantity is the reference. The first law of thermodynamics gives, with Figure 3.1

$$q_{\mathrm{FCrev}} + w_{\mathrm{tFCrev}} = \Delta^{\mathrm{r}} H \tag{3.1}$$

The reaction enthalpy $\Delta^{\mathrm{r}} H$ of the oxidation covers the production of the reversible work and heat. The second law of thermodynamics gives

$$\oint \mathrm{d}S = 0 \tag{3.2}$$

The reaction entropy $\Delta^{\mathrm{r}} S$ is a result of the reaction itself and must be compensated by the transport of the reversible heat q_{FCrev} to the environment and Equation (3.2) gives

$$\Delta^{\mathrm{r}} S - \frac{q_{\mathrm{FCrev}}}{T_{\mathrm{FC}}} = 0 \tag{3.3}$$

Equations (3.1) and (3.3) give the reversible work w_{tFCrev}

$$w_{\mathrm{tFCrev}} = \Delta^{\mathrm{r}} H - T_{\mathrm{FC}} \cdot \Delta^{\mathrm{r}} S \tag{3.4}$$

The reversible work w_{tFCrev} of the reaction is equal to the free or Gibbs enthalpy $\Delta^{\mathrm{r}} G$ of the reaction

$$w_{\mathrm{tFCrev}} = \Delta^{\mathrm{r}} G = \Delta^{\mathrm{r}} H - T_{\mathrm{FC}} \cdot \Delta^{\mathrm{r}} S \tag{3.5}$$

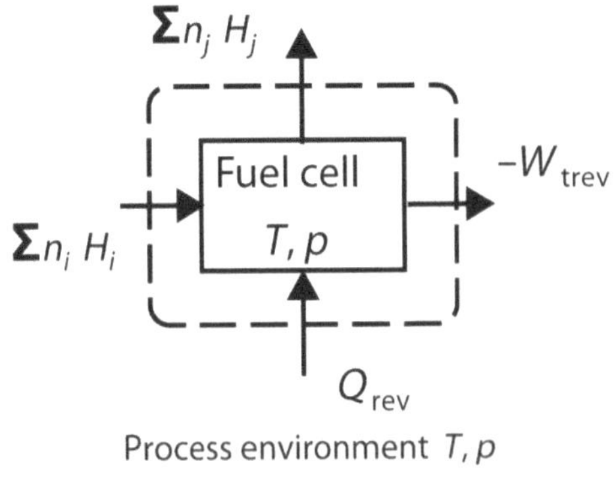

Figure 3.1 The reversible fuel cell, its energy balance and its system boundary.

The reversible efficiency η_{FCrev} of the fuel cell is the ratio of the Gibbs enthalpy $\Delta^{\mathrm{r}}G$ and the reaction enthalpy $\Delta^{\mathrm{r}}H$ at the thermodynamic state of the fuel cell, and Equation (3.5) gives

$$\eta_{\mathrm{FCrev}} = \frac{\Delta^{\mathrm{r}}G}{\Delta^{\mathrm{r}}H} = \frac{\Delta^{\mathrm{r}}H - T_{\mathrm{FC}} \cdot \Delta^{\mathrm{r}}S}{\Delta^{\mathrm{r}}H} \tag{3.6}$$

The process environment of an SOFC cannot exist near the ambient state, and it is thus an artificial model only. A general reversible SOFC system operation is only possible within a system connecting reversibly the process environment with the ambient state. This is the approach in Section 3.5.

An SOFC can also be described as an electrical device where the electrical effects are explained by thermodynamics. Figure 3.2 shows the transport processes within an SOFC connecting the thermodynamic and the electrical effects using the example of hydrogen oxidation.

The oxidation of hydrogen follows the equation

$$\mathrm{H_2} + \frac{1}{2}\mathrm{O_2} \rightarrow \mathrm{H_2O} \tag{3.7}$$

This equation is independent of the process itself. The reaction path in an SOFC depends on the anode and the cathode reactions. Hydrogen is adsorbed at the anode, ionised and the electrons are removed by the connection to the electrical load where the electrical work is used. Oxygen is adsorbed at the cathode connected with the load and ionised by the arriving electrons. The oxide ion is conducted by the electrolyte to the anode. The hydrogen ions (protons) and the oxide ion form a molecule of water. The first reaction **1** on the anode is

$$\mathrm{H_2} \rightarrow 2\mathrm{H^+} + 2\mathrm{e^-} \tag{3.8}$$

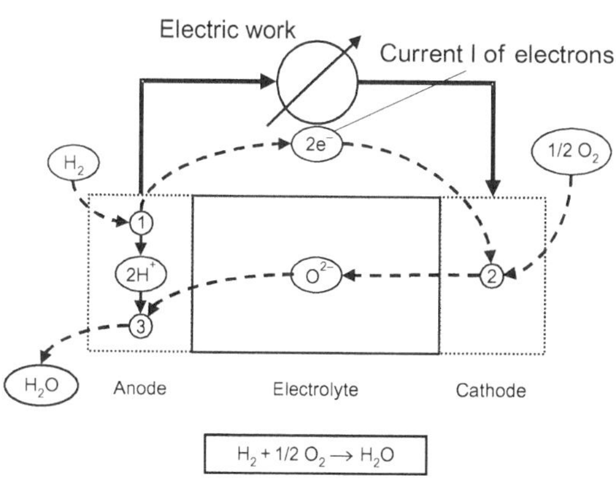

Figure 3.2 Transport processes within an SOFC.

The reaction **2** on the cathode is

$$\frac{1}{2}O_2 + 2e^- \rightarrow O^{2-} \tag{3.9}$$

The oxide ion O^{2-} is conducted through the electrolyte and arrives at the anode. At the anode, water forms **3** according to the reaction

$$2H^+ + O^{2-} \rightarrow H_2O \tag{3.10}$$

As Figure 3.2 shows, the product H_2O is mixed with the anode gas and its concentration increases with increasing fuel utilisation U_f. The fuel utilisation U_f is the ratio of the spent fuel flow and the inlet fuel flow and is defined by:

$$U_f = 1 - \frac{\dot{m}_{FAnO}}{\dot{m}_{FI}} \tag{3.11}$$

where $\dot{m}_{FI}$ is the fuel mass flow at the inlet of the anode and $\dot{m}_{FAnO}$ is the fuel mass flow at its outlet. A similar definition can be made with the molar flow. Because any real process produces irreversible entropy, a reversible SOFC operation is only possible as the limiting process of the real process with $U_f \rightarrow 0$. Equation (3.8) shows that the molar flow of the electrons is twice that of the molar flow of hydrogen, thus

$$\dot{n}_{el} = 2\dot{n}_{H_2} \tag{3.12}$$

The electric current I is a linear function of the molar flow of the electrons or the molar flow of the spent fuel – in this example the molar flow of the spent hydrogen

$$I = \dot{n}_{el} \cdot (-e) \cdot N_A = -\dot{n}_{el} \cdot F = -2\dot{n}_{H_2} \cdot F \tag{3.13}$$

In Equation (3.13), we introduced the elementary charge e

$$e = (1.60217733 \pm 0.00000049) \times 10^{-19} C \tag{3.14}$$

and the Faraday constant F

$$F = e \cdot N_A = (96485.309 \pm 0.029)\mathrm{C/mol} \tag{3.15}$$

as the product of the elementary charge and the Avogadro constant N_A. Equations (3.13) and (3.15) show that the electric current I is a measure of the fuel spent. Thus a current measurement is a very simple method to measure the fuel spent. The matching between thermodynamic and electrical quantities can be done by the power but not by the work. The reversible power can be written as a product of the reversible voltage V_{FCrev} and the current I as well as a product of the molar flow of the fuel $\dot{n}_{H_2}$ and the free enthalpy $\Delta^r G$ of the oxidation reaction

$$P_{\mathrm{FCrev}} = V_{\mathrm{FCrev}} \cdot I = \dot{n}_{\mathrm{H_2}} \cdot w_{\mathrm{tFCrev}} = \dot{n}_{\mathrm{H_2}} \cdot \Delta^{\mathrm{r}} G \tag{3.16}$$

The reversible voltage V_{FCrev} results from Equations (3.13) and (3.16)

$$V_{\mathrm{FCrev}} = \frac{-\dot{n}_{\mathrm{H_2}} \cdot \Delta^{\mathrm{r}} G}{\dot{n}_{\mathrm{el}} \cdot F} \tag{3.17}$$

Equation (3.12) shows the ratio between the molar flow of the electrons and the spent hydrogen as 2. This can be generalised, and n^{el} is the number of the electrons that are released during the ionisation process of one utilised fuel molecule, related on the molar flows we get with Equation (3.11)

$$n^{\mathrm{el}} = \frac{\dot{n}_{\mathrm{el}}}{U_{\mathrm{f}} \cdot \dot{n}_{\mathrm{FI}}} \tag{3.18}$$

Finally, for the reversible voltage V_{FCrev} of the oxidation of any fuel gas,

$$V_{\mathrm{FCrev}} = \frac{-\Delta^{\mathrm{r}} G}{n^{\mathrm{el}} \cdot F} \tag{3.19}$$

It has already been mentioned that the reduction of the thermodynamic potential of the still available reactants during SOFC operation depends on the fuel utilisation. These effects and the related voltage reduction can be described by the change of the partial pressures of the components [1]. We can write Equation (3.4) more precisely as

$$\Delta^{\mathrm{r}} G(T, p) = \Delta^{\mathrm{r}} H(T, p) - T \cdot \Delta^{\mathrm{r}} S(T, p) \tag{3.20}$$

Using the common assumption of the ideal gas, we get

$$\Delta^{\mathrm{r}} G(T, p) = \Delta^{\mathrm{r}} H(T) - T \cdot \Delta^{\mathrm{r}} S(T, p) \tag{3.21}$$

and we can write, with $\mathrm{d}S = (\mathrm{d}H - v.\mathrm{d}p)/T$, for the entropy S_j of any component j

$$S_j(T, p) = S_j^0 + \int_{T_0}^{T} \frac{C_{\mathrm{P}j}(t)}{t} \mathrm{d}t - R_{\mathrm{m}} \cdot \ln\left(\frac{p_j}{p_0}\right) \tag{3.22}$$

where $C_{\mathrm{P}j}$ is the temperature-dependent heat capacity of the component j. The pressure dependence of $C_{\mathrm{P}j}$ can be neglected in this common assumption, Equation (3.21). Using Equation (3.22), we can write for the reaction entropy $\Delta^{\mathrm{r}} S\ (T,p)$

$$\Delta^{\mathrm{r}} S(T, p) = \Delta^{\mathrm{r}} S(T) - R_{\mathrm{m}} \cdot \ln(K) \tag{3.23}$$

with the equilibrium constant K, see, for example, Ref. [2]

$$K = \prod_j \left(\frac{p_j}{p_0}\right)^{\nu j} \tag{3.24}$$

νj is the fuel-related quantity of the component j in the equation of the oxidation reaction. p_0 is the standard pressure (1 bar):

$$p_0 = 1\,\text{bar} \tag{3.25}$$

Using Equations (3.21)–(3.24), we get

$$\Delta^{\text{r}}G(T,p) = \Delta^{\text{r}}G(T) + T \cdot R_{\text{m}} \cdot \ln(K) \tag{3.26}$$

This use of the assumption of an ideal gas allows to express the Nernst potential or the Nernst voltage V_{N} by using Eqs (3.18), (3.19) and (3.26) as

$$V_{\text{N}} = \frac{-\Delta^{\text{r}}G(T)}{n^{\text{el}} \cdot F} - \frac{R_{\text{m}} \cdot T \cdot \ln(K)}{n^{\text{el}} \cdot F} \tag{3.27}$$

The following reversible oxidation of hydrogen H_2, carbon monoxide CO, and methane CH_4 can be analysed as examples by using Equation (3.27):

$$H_2 + \frac{1}{2}O_2 \rightarrow H_2O \tag{3.7}$$

$$CO + \frac{1}{2}O_2 \rightarrow CO_2 \tag{3.28}$$

$$CH_4 + 2O_2 \rightarrow 2H_2O + CO_2 \tag{3.29}$$

The equations (3.7), (3.28), (3.29) determine the reaction enthalpy, the reaction entropy, and thus the free enthalpy and the voltage of the reversible oxidation as formulated in Equations (3.5) and (3.19) with the thermodynamic data of the reactions at the standard conditions 0 (25 °C, 1 bar) as collected in, for example, Ref. [2,3]. A variation of the thermodynamic state of the environment of the reversible cell provides a first idea of the behaviour of real cells under changing operation conditions. It can be assumed that the reaction enthalpy and the reaction entropy depend only slightly on the temperature as a first assumption. Thus the free enthalpy of the reaction at a higher temperature can be approximated with the values of the reaction enthalpy and the reaction entropy at standard conditions, and it depends on the temperature linearly and yields the reversible cell voltage, Equations (3.26) and (3.27). The values of the reversible cell voltage are calculated for the standard state 0, at 1000 °C/1 bar and for 25 °C and 1000 °C at 0.1 and 10 bar. These linearised values of the free enthalpy and the reversible cell voltage at different thermodynamic states are written as $\Delta^{\text{r}}G^*$ and V^*. The water is always assumed to be in the gaseous phase because an

SOFC operates at high temperatures. The values can be found in Table 3.1 and Figure 3.3.

The Equations (3.7) and (3.28) indicate that the volumes of the products of the oxidation reactions of hydrogen H_2 and carbon monoxide CO are smaller than the cumulative volume of the reactants. But the cumulative volume of the reactants and the products of the oxidation reaction of methane CH_4 is the same (Equation 3.29). This is also apparent by the values of the activity K ($\ln(1)=0$). Thus there is theoretically no change of the entropy in the last case, and the real change of

Table 3.1 The reversible oxidation of hydrogen, carbon monoxide and methane

Fuel	H_2	CO	CH_4
$\Delta^r H^0$ in kJ/mol	−241.82	−282.99	−802.31
$\Delta^r S^0$ in J/(mol K)	−44.37	−86.41	−5.13
$\Delta^r G^0$ in kJ/mol	−228.59	−257.23	−800.68
$\Delta^r G^*$ at 1000 °C, 1 bar in kJ/mol	−185.33	−172.98	−795.68
n^{el}	2	2	8
V^0 in V	1.185	1.333	1.037
V^* at 1000 °C, 1 bar in V	0.960	0.896	1.031
$\ln(K)$ at 0.1 bar	1.1513	1.1513	0
$\ln(K)$ at 10 bar	−1.1513	−1.1513	0
V^* at 25 °C, 0.1 bar in V	1.170	1.318	1.037
V^* at 1000 °C, 0.1 bar in V	0.897	0.833	1.031
V^* at 25 °C, 10 bar in V	1.199	1.348	1.037
V^* at 1000 °C, 10 bar in V	1.024	0.960	1.031

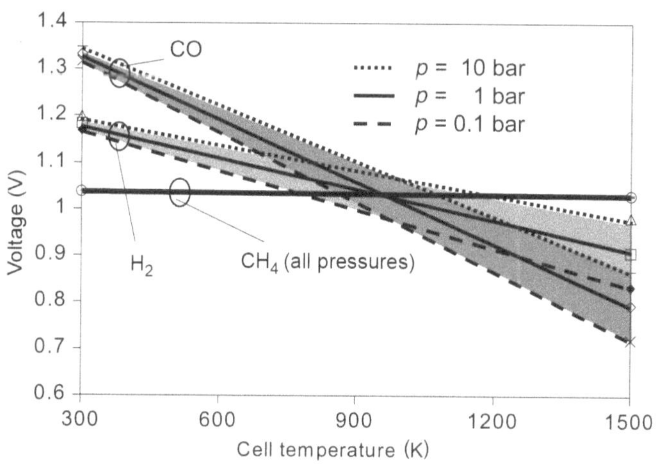

Figure 3.3 The reversible cell voltage of different fuels at different states (p,T) of the environment (linearised model and assumption of ideal gas).

the measured values at the standard state 0 is very limited as indicated in Table 3.1. This is the reason for the very small dependence on temperature of the free enthalpy or the reversible cell voltage of the oxidation reaction of methane, the idealised pressure dependence of the entropy yields no change in the cell voltage caused by the system pressure. The reversible cell voltage of the hydrogen and the carbon monoxide oxidation decreases with increasing system temperature and increases with increasing system pressure.

3.3 Ohmic losses and voltage dependence on fuel utilisation

The thermodynamic relevance of the voltage can be understood by considering Equations (3.13) and (3.16). The voltage is a measure of the exergetic quality similar to the work of a thermodynamic process. The exergy is defined as the potential of the reversible work of a system related to the ambient state 0 [2]. Thus it is clear that the voltage loss ΔV due to the electric resistance R is connected with an additional production of irreversible entropy. We get for the voltage loss

$$\Delta V = I \cdot R \tag{3.30}$$

and for the power loss

$$P_{\text{loss}} = \Delta V \cdot I = I^2 \cdot R = T_{\text{SOFC}} \cdot U_{\text{f}} \cdot \dot{n}_{\text{FI}} \cdot \Delta s_{\text{irr}} \tag{3.31}$$

Equation (3.31) shows that the irreversible entropy production of an ohmic loss P_{loss} in an SOFC is smaller than that in another fuel cell operating at a lower temperature, and a generalised form of Equation (3.13) yields the irreversible entropy production

$$\Delta s_{\text{irr}} = \frac{U_{\text{f}} \cdot \dot{n}_{\text{FI}} \cdot R \cdot \left(n^{\text{el}} \cdot F\right)^2}{T_{\text{SOFC}}} \tag{3.32}$$

Equation (3.13) was generalised here as

$$I = -n^{\text{el}} \cdot U_{\text{f}} \cdot \dot{n}_{\text{FI}} \cdot F \tag{3.33}$$

The influence of the fuel utilisation on the voltage reduction by reducing the thermodynamic potential by producing reversible work can be easily calculated if we consider the fuel utilisation as a function of the changing partial pressure of the components [1].

It should be mentioned that any reactant must have the same thermodynamic state as the reversible cell at its entrance and that the cell operates isothermally. The system approach to guarantee the needed process environment is described in Section 3.5.

Again the oxidation of hydrogen is a good example to illustrate that approach. Using p_i as the partial pressure of the component i we get

$$p_i = y_i \cdot p \tag{3.34}$$

writing y_i for the molar concentration of the component i and p for the total pressure of the system. Using Equation (3.11), we can write

$$U_\mathrm{f} = \frac{y_\mathrm{FI} \cdot \dot{n}_\mathrm{AnI} - y_\mathrm{FO} \cdot \dot{n}_\mathrm{AnO}}{y_\mathrm{FI} \cdot \dot{n}_\mathrm{AnI}} \tag{3.35}$$

if we consider the molar flow of the fuel F as the product of the molar concentration y and the total molar flow at the inlet I and the outlet O of the anode side An. U_f will be used as a variable, thus the outlet O can be interpreted as a space variable along the axis of the parallel flowing fuel and air defined by a certain U_f to be obtained. The local Nernst voltage $V_\mathrm{N}(U_\mathrm{f})$ depends on the local gas concentration. The molar flow on the anode side is constant in our example of the hydrogen oxidation and we get

$$\dot{n}_\mathrm{AnI} = \dot{n}_\mathrm{AnO} = \dot{n}^* \tag{3.36}$$

and Equation (3.35) yields

$$U_{\mathrm{fH}_2} = 1 - \frac{y_{\mathrm{H}_2,\mathrm{O}}}{y_{\mathrm{H}_2,\mathrm{I}}} \tag{3.37}$$

The equation of the reaction (3.7) shows that the molar $\dot{n}_{\mathrm{H}_2,\mathrm{U}}$ flow of the utilised fuel is equal to the molar flow $\dot{n}_{\mathrm{H}_2\mathrm{O},\mathrm{O}}$ of the produced water at the outlet O

$$\dot{n}_{\mathrm{H}_2,\mathrm{U}} = \dot{n}_{\mathrm{H}_2\mathrm{O},\mathrm{O}} \tag{3.38}$$

if the used hydrogen is dry ($y_{\mathrm{H}_2,\mathrm{I}} = 1$). This yields

$$U_{\mathrm{fH}_2} = \frac{\dot{n}_{\mathrm{H}_2,\mathrm{U}}}{\dot{n}^*} = \frac{\dot{n}_{\mathrm{H}_2\mathrm{O},\mathrm{O}}}{\dot{n}^*} = y_{\mathrm{H}_2\mathrm{O},\mathrm{O}} \tag{3.39}$$

Following Equation (3.7) we can write for the cathode side

$$\dot{n}_{\mathrm{O}_2,\mathrm{U}} = \frac{1}{2} \cdot \dot{n}_{\mathrm{H}_2,\mathrm{U}} \tag{3.40}$$

Practical SOFC systems operate with air instead of oxygen and with an excess air $\lambda > 1$. The incoming air flow is defined by the inlet flow $\dot{n}_\mathrm{CaI}$ of the cathode

$$\dot{n}_\mathrm{CaI} = \frac{1}{2} \cdot \lambda \cdot \frac{\dot{n}^*}{0.21} \tag{3.41}$$

The outlet flow of the cathode can be calculated by

$$\dot{n}_{CaO} = \frac{1}{2} \cdot \lambda \cdot \frac{\dot{n}^*}{0.21} - \frac{1}{2} \cdot \dot{n}_{H_2,U} \tag{3.42}$$

The molar oxygen flow at the inlet is

$$\dot{n}_{O_2,I} = \frac{1}{2} \cdot \lambda \cdot \dot{n}^* \tag{3.43}$$

and the molar oxygen flow at the outlet is

$$\dot{n}_{O_2,O} = \frac{1}{2} \cdot \left(\lambda \cdot \dot{n}^* - \dot{n}_{H_2,U}\right) \tag{3.44}$$

Equations (3.43) and (3.44) can be written as a function of U_f

$$\dot{n}_{CaO} = \frac{1}{2} \cdot \dot{n}^* \cdot \left(\frac{\lambda}{0.21} - \frac{\dot{n}_{H_2,U}}{\dot{n}^*}\right) = \frac{1}{2} \cdot \dot{n}^* \cdot \left(\frac{\lambda}{0.21} - U_{fH_2}\right) \tag{3.45}$$

and

$$\dot{n}_{O_2,O} = \frac{1}{2} \cdot \dot{n}^* \cdot \left(\lambda - \frac{\dot{n}_{H_2,U}}{\dot{n}^*}\right) = \frac{1}{2} \cdot \dot{n}^* \cdot (\lambda - U_{fH_2}) \tag{3.46}$$

Now we can express y_i as a function of the fuel utilisation U_f

$$y_{H_2,O} = 1 - U_{fH_2} \tag{3.47}$$

$$y_{H_2O,O} = U_{fH_2} \tag{3.48}$$

and

$$y_{O_2,O} = \frac{\lambda - U_{fH_2}}{\lambda/0.21 - U_{fH_2}} = \frac{\dot{n}_{O_2,O}}{\dot{n}_{CaO}} \tag{3.49}$$

Using Equations (3.24), (3.27) and (3.34), we get for the equilibrium constant K

$$K = \frac{U_{fH_2} \cdot \left(\lambda/0.21 - U_{fH_2}\right)^{1/2}}{(1 - U_{fH_2}) \cdot \left[(\lambda - U_{fH_2}) \cdot p\right]^{1/2}} \tag{3.50}$$

We can use Equations (3.50) and (3.27) to calculate the ideal Nernst voltage V_N as a function of the fuel utilisation U_f. Using (3.33), we see that the current I is proportional to the fuel utilisation U_f

$$U_f = \frac{I}{n^{el} \cdot \dot{n}_{FI} \cdot F} \tag{3.51}$$

The number of electrons n^{el} depends on the fuel (2 for H_2), the Faraday constant F is a constant value and the fuel inlet flow $\dot{n}_{FI}$ is the only variable that influences the relation between fuel utilisation U_f and current I. Any curve of the voltage V depending on the current I can be expressed by any curve of V depending on the fuel utilisation U_f at the same fuel inlet flow $\dot{n}_{FI}$.

Considering Equations (3.50) and (3.27), we see that $V \to +\infty$ for $U_f \to 0$ and $V \to -\infty$ for $U_f \to 1$, respectively. Anyway an integration of V_N as given in (3.27) and by using (3.50) between the integration borders 0 and 1 for U_f delivers a finite result because of the symmetry of the positive and negative contribution at these integration borders. The resulting difference between the reversible solution of the pure substances and the solution of an air supplied cell can be identified as the needed reversible separation work to extract the needed oxygen from the air. The explanation of this difference as an irreversible mixing effect as given in the first edition must be thus corrected. But the model of the ideal gas gives a good approximation for $0 < U_f < 1$ in the regime of the real SOFC operation. This model allows to evaluate the principal influences of the different parameters, system pressure p, SOFC temperature ϑ_{SOFC}, excess air λ and fuel utilisation U_f, on the Nernst voltage V_N.

Figure 3.4 shows the Nernst voltage V_N as a function of the fuel utilisation U_f in an SOFC with H_2 fuel and with the system pressure p as a parameter. The excess air and the SOFC temperature are fixed. The interesting area between $U_f = 0.1$ and $U_f = 0.9$ can be well approximated with the model of the ideal gas. The dotted line shows the adaption of the model. The reduction of the thermodynamic potential within the SOFC reduces V_N between $U_f = 0.1$ and $U_f = 0.9$ by about more than 200 mV. An increase of the system pressure from 1 to 10 bar increases V_N by about 70 mV. An increasing SOFC temperature decreases V_N as shown by Equations (3.27) and (3.50).

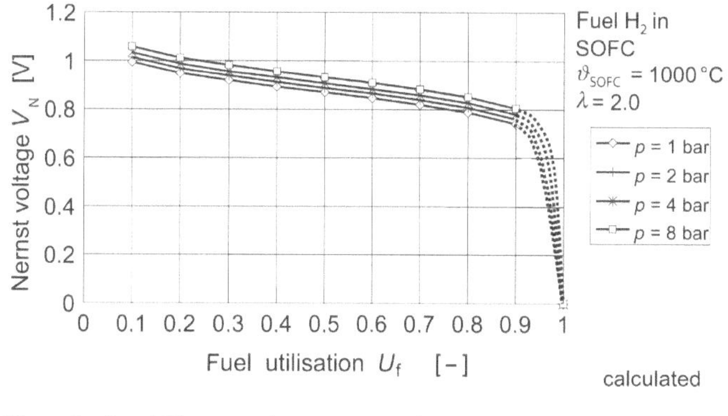

Figure 3.4 The calculated Nernst voltage V_N as a function of the fuel utilisation U_f.

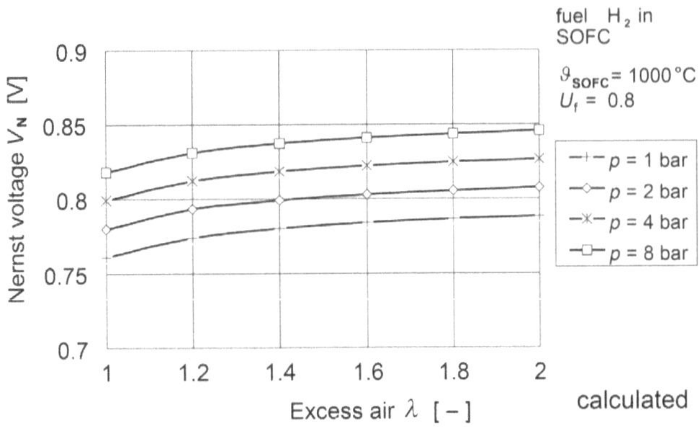

Figure 3.5 The calculated Nernst voltage V_N as a function of the excess air λ.

The excess air λ is a very important process parameter for the design of the total system as shown later. Figure 3.5 shows the Nernst voltage V_N as a function of the excess air λ and the system pressure p as a parameter.

An increasing excess air λ increases V_N slightly. But this influence of the excess air λ on V_N decreases with an increasing excess air. An increase of the excess air λ at values >2 does not really influence V_N anymore. Between $1<\lambda<2$, the voltage increase is ~30 mV. The calculation shows that a certain fuel utilisation U_f leads to a certain V_N. The maximum power P_{elmax} of one cell is determined by the Nernst voltage V_{NO} and the corresponding current I_O depending on the fuel utilisation U_{fO} at the outlet O according to Equation (3.51) and can be written as

$$P_{\mathrm{elmax}} = V_{NO} \cdot I_O \tag{3.52}$$

Figure 3.6(left) illustrates this. It shows the power from a cell as given by the curve of the Nernst voltage V_N as a function of fuel utilisation U_f. The current I is proportional to U_f, see (3.51), for a given substance flow and I is thus marked into brackets in Figure 3.6. An electrical serial connection of a number of cells allows an integration of the curve of V_N as shown at the right side of Figure 3.6. This integration leads to

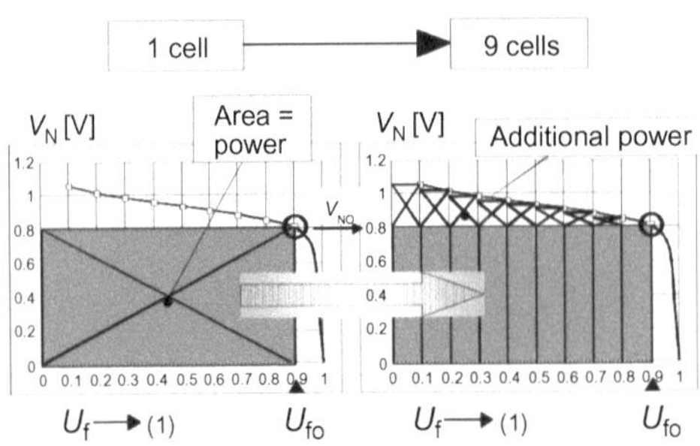

Figure 3.6 The increase in efficiency by cascading single cells.

different voltages in every cell, but the current or the fuel utilisation in every cell must be equal. This cascading of cells allows an increase in power and efficiency compared with a single cell.

The methodology used above for an air supplied oxygen ion conducting SOFC can also be expanded to proton conducting electrolytes and to operation with pure oxygen. Further investigated cases consider the utilisation of H_2 with ballast gases as, for example, after gas processing and most important the utilisation of general hydrocarbons C_nH_m as described in Refs. [4,5]. By using (3.24), (3.27) and (3.34), we can write

$$V_N = \frac{\Delta^r G(p,T)}{n^{el} \cdot F} = \frac{\Delta^r G(T)}{n^{el} \cdot F} + R_m \cdot T \cdot \frac{\ln(\Pi \pi^{\nu_j})}{n^{el} \cdot F} + R_m \cdot T \cdot \frac{\ln\left(\Pi y_j^{\nu_j}\right)}{n^{el} \cdot F} \tag{3.53}$$

Hereby, we use the pressure ratio π:

$$\pi = \frac{p}{p_0} \tag{3.54}$$

for relating the cell pressure p on the ambient pressure p_0. This form allows considering only the concentration term $\Pi y_j^{\nu_j}$ for all investigations regarding reaction progress and ion transport in the electrolyte membrane. For this purpose, we can define now a concentration constant $K_y^{/}$ as

$$K_y' = \Pi y_j^{\nu_j} \tag{3.55}$$

With this definition, we can easily express the reversible utilisation of H_2 in fuel cells for all practical cases in a form as given in Table 3.2.

In some cases, H_2 is delivered in an inert gas mixture and the molar flow at the anode inlet can be found as

$$\dot{n}_{AnI} = \dot{n}_{H_2I} + \dot{n}_{ballast} = \dot{n}_{H_2I} \cdot (1+\beta) \tag{3.56}$$

Table 3.2 Concentration constant for different ways of reversible utilisation of H_2

	Proton conduction	**Oxygen ion conduction**
Oxygen supplied	$K_y' = \frac{2.U_f}{\sqrt{\lambda^2 - U_f^2}}$	$K_y' = \frac{U_f}{1-U_f}$
Air supplied	$K_y' = \frac{2U_f}{(\lambda - U_f)^{1/2}\left(\frac{\lambda}{0.21} + U_f\right)^{1/2}}$	$K_y' = \frac{U_f\left(\frac{\lambda}{0.21} - U_f\right)^{1/2}}{(1-U_f)(\lambda - U_f)^{1/2}}$

and β is the ballast ratio. We finally get for K_y' of any hydrogen-rich ballast fuels reacting with air:

$$K_y' = \frac{2 \cdot U_f \cdot (1+\beta - U_f)}{(1-U_f)(\lambda/0.21 + U_f)^{1/2} \cdot (\lambda - U_f)^{1/2}} \tag{3.57}$$

Any hydrocarbon utilisation is described by:

$$C_nH_m + \left(n + \frac{m}{4}\right)O_2 \rightarrow nCO_2 + \frac{m}{2}H_2O \tag{3.58}$$

The number of electrons released at the combustion of 1 mol of any hydrocarbons C_nH_m can be found as:

$$n^{el} = 4 \cdot n + m \tag{3.59}$$

We finally get for the concentration constant K_y':

$$K_y' = \frac{y_{CO_2}{}^{n} \cdot y_{H_2O}{}^{m/2}}{y_{CnHm} \cdot y_{O_2}{}^{(n+m/4)}} \tag{3.60}$$

and for the pressure constant K_p':

$$K_p' = \pi^{(m-4)/4} \tag{3.61}$$

The pressure influence is shown in Figure 3.7. It can be seen that for $m=2$, the value K_p' is negative, thus a higher pressure increases the Nernst voltage; for $m=4$, there is no pressure influence and for all values of $m>4$, the Nernst voltage is decreasing caused by an increasing positive value K_p'.

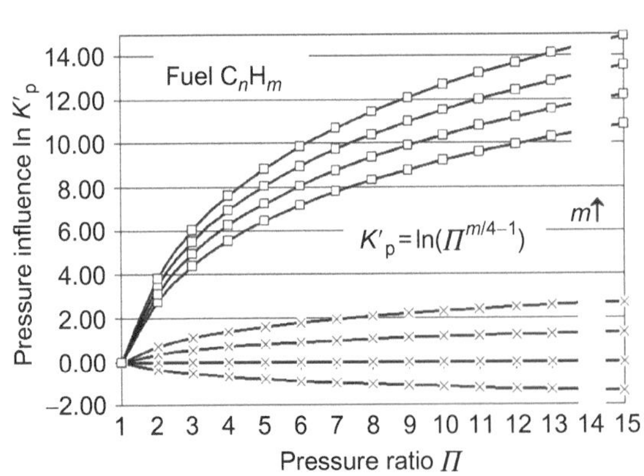

Figure 3.7 Pressure influence of the hydrogen content on Nernst voltage.

3.4 Thermodynamic definition of a fuel cell producing electricity and heat

In designing practical SOFC systems with associated components such as fans, heat exchangers, etc., modelling of an SOFC as a power-generating burner is very helpful (see Figure 3.8). The system is defined as a module consisting of SOFC cells connected in electrical parallel into stacks supplying a common burner with the depleted fuel. The energy balance of the stack provides the necessary requirements for cooling and excess air simultaneously [1].

There are two possible descriptions. The most simple approach is a balance border around the complete module including all stacks and the joint burner from the inlet I of the fuel F and the air A to the outlet aB of the flue gas G after the burner. The more detailed approach is a balance border surrounding all stacks from the inlet I to the outlets O of the anode side AnO and of the cathode side CaO. The calculation of this "power-generating burner" is very similar to the calculation of a combustor of a gas turbine or of a furnace of a boiler. The calculation of the mass flows of the module does not even differ from any calculation of a conventional oxidation. The energy balance of this simpler approach (from *I* to *aB*) delivers

$$\dot{H}_{\mathrm{FI}} + \dot{H}_{\mathrm{AI}} = \dot{Q}_{\mathrm{FC}} + P_{\mathrm{el}} + \dot{H}_{\mathrm{GaB}} \tag{3.62}$$

The total enthalpy flow $\dot{H}_{\mathrm{FI}}$ of the fuel includes the reaction enthalpy (or in technical terms the LHV) as well. The enthalpy flow of the incoming air is $\dot{H}_{\mathrm{AI}}$. Both these enthalpy flows have to cover the energy output of the SOFC module consisting of the produced power P_{el}, the generated heat $\dot{Q}_{\mathrm{FC}}$ and the enthalpy flow of the flue gas $\dot{H}_{\mathrm{GaB}}$. We get with Equation (3.62), the respective mass flows $\dot{m}$ and the respective related enthalpies h^*

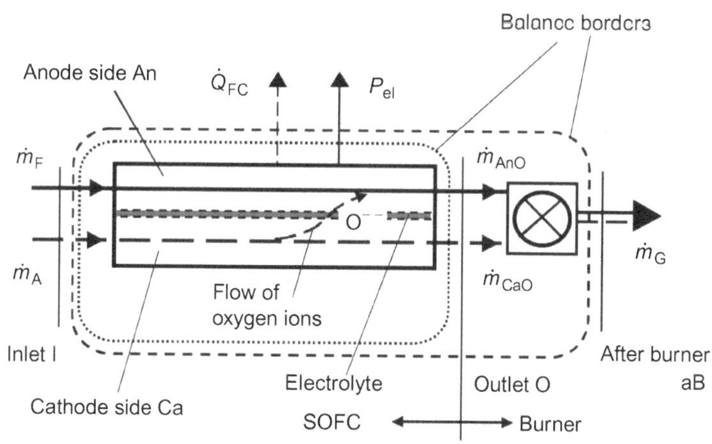

Figure 3.8 The power-generating burner model of an SOFC module.

$$\dot{m}_{\mathrm{FI}} \cdot \left(\mathrm{LHV} + h^*_{\mathrm{FI}}\right) + \dot{m}_{\mathrm{AI}} \cdot h^*_{\mathrm{AI}} = \dot{Q}_{\mathrm{FC}} + P_{\mathrm{el}} + \dot{m}_{\mathrm{GaB}} \cdot h^*_{\mathrm{GaB}} \tag{3.63}$$

The use of the related enthalpies is necessary to match all enthalpies with the LHV related on the chemical standard state (1 bar, 25 °C). The related enthalpy is defined by

$$h^* = h(p, \vartheta) - h_0(1\,\mathrm{bar}, 25\,°\mathrm{C}) \tag{3.64}$$

These equations are sufficient to calculate the necessary excess air λ for a given heat extraction $-\dot{Q}_{\mathrm{FC}}$ of the module, or vice versa to calculate the necessary SOFC cooling by the heat extraction $-\dot{Q}_{\mathrm{FC}}$ for a defined excess air λ as shown below.

The consideration of the stacks only allows a more detailed modelling, and the energy balance delivers

$$\dot{H}_{\mathrm{FI}} + \dot{H}_{\mathrm{AI}} = \dot{Q}_{\mathrm{FC}} + P_{\mathrm{el}} + \dot{H}_{\mathrm{AnO}} + \dot{H}_{\mathrm{CaO}} \tag{3.65}$$

This equation is identical with Equation (3.62) if all the fuel is used in the stacks. The enthalpy flow of the incoming fuel is

$$\dot{H}_{\mathrm{FI}} = \dot{m}_{\mathrm{FI}} \cdot \left(\mathrm{LHV} + h^*_{\mathrm{FI}}\right) \tag{3.66}$$

and the enthalpy flow of the incoming air is

$$\dot{H}_{\mathrm{AI}} = \dot{m}_{\mathrm{AI}} \cdot h^*_{\mathrm{AI}} = \dot{m}_{\mathrm{FI}} \cdot \lambda \cdot \mu_{\mathrm{A0}} \cdot h^*_{\mathrm{AI}} \tag{3.67}$$

The stoichiometric-specific air demand μ_{A0} is defined by the relation of the stoichiometric air mass flow and the corresponding fuel mass flow. The relation of all terms of the energy balance on the mass flow $\dot{m}_{\mathrm{FI}}$ of the incoming fuel allows a generalised consideration. The generated heat is

$$\dot{Q}_{\mathrm{FC}} = \dot{m}_{\mathrm{FI}} \cdot q_{\mathrm{FC}} \tag{3.68}$$

The produced power is

$$P_{\mathrm{el}} = \dot{m}_{\mathrm{FI}} \cdot p_{\mathrm{el}} \tag{3.69}$$

With the fuel utilisation U_{f}

$$U_{\mathrm{f}} = 1 - \frac{\dot{m}_{\mathrm{FAnO}}}{\dot{m}_{\mathrm{FI}}} \tag{3.11}$$

and the enthalpy flow at the anode outlet is

$$\dot{H}_{\mathrm{AnO}} = \dot{m}_{\mathrm{FI}} \cdot (1 - U_{\mathrm{f}}) \cdot \left(\mathrm{LHV} + h^*_{\mathrm{FAnO}}\right) + \dot{m}_{\mathrm{RG}} \cdot h^*_{\mathrm{RGAnO}} \tag{3.70}$$

The flow of the reaction product gas RG is

$$\dot{m}_{\mathrm{RG}} = \dot{m}_{\mathrm{FI}} \cdot U_{\mathrm{f}} + \dot{m}_{\mathrm{O_2}} = \dot{m}_{\mathrm{FI}} \cdot U_{\mathrm{f}} \cdot (1 + \mu_{\mathrm{O20}}) \tag{3.71}$$

with

$$\dot{m}_{\mathrm{O_2}} = U_{\mathrm{f}} \cdot \dot{m}_{\mathrm{FI}} \cdot \mu_{\mathrm{O20}} \tag{3.72}$$

The mass flow at the outlet of the anode side consists of the not utilised fuel and of RG. RG consists of the reaction products CO_2 and H_2O. Its mass flow is equal to the mass flow of the utilised fuel and of the transferred oxygen by the ion conduction through the electrolyte. The stoichiometric demand on oxygen related on the inlet fuel mass flow is given by the figure μ_{O20}. We finally get for the enthalpy flow at the anode outlet

$$\dot{H}_{\mathrm{AnO}} = \dot{m}_{\mathrm{FI}} \cdot \left[(1 - U_{\mathrm{f}}) \cdot \left(\mathrm{LHV} + h^{*}_{\mathrm{FAnO}}\right) + U_{\mathrm{f}} \cdot (1 + \mu_{\mathrm{O20}}) \cdot h^{*}_{\mathrm{RGAnO}}\right] \tag{3.73}$$

The enthalpy flow at the cathode outlet can be calculated by the difference of the enthalpy flow of the non depleted air and the enthalpy flow of the oxygen being transferred to the anode both with the thermodynamic state at the cathode outlet

$$\dot{H}_{\mathrm{CaO}} = \dot{m}_{\mathrm{AI}} \cdot h^{*}_{\mathrm{ACaO}} - \dot{m}_{\mathrm{O_2}} \cdot h^{*}_{\mathrm{O_2CaO}} \tag{3.74}$$

Equations (3.67) and (3.72) yield with Equation (3.74)

$$\dot{H}_{\mathrm{CaO}} = \dot{m}_{\mathrm{FI}} \cdot \left[\lambda \cdot \mu_{\mathrm{A0}} \cdot h^{*}_{\mathrm{ACaO}} - U_{\mathrm{f}} \cdot \mu_{\mathrm{O20}} \cdot h^{*}_{\mathrm{O_2CaO}}\right] \tag{3.75}$$

The specific generated heat q_{FC} results from Equations (3.65)–(3.75) as

$$\begin{aligned} q_{\mathrm{FC}} &= U_{\mathrm{f}} \cdot \mathrm{LHV} + h^{*}_{\mathrm{FI}} - (1 - U_{\mathrm{f}}) \cdot h^{*}_{\mathrm{FAnO}} - p_{\mathrm{el}} \\ &\quad - U_{\mathrm{f}} \cdot \left[\mu_{\mathrm{O20}} \cdot \left(h^{*}_{\mathrm{RGAnO}} - h^{*}_{\mathrm{O_2CaO}}\right) + h^{*}_{\mathrm{RGAnO}}\right] + \mu_{L0} \cdot \left[\lambda \cdot h^{*}_{\mathrm{AI}} - \lambda \cdot h^{*}_{\mathrm{ACaO}}\right] \end{aligned} \tag{3.76}$$

The use of Equation (3.76) depends on a constant excess air λ as probably regulated by an oxidation control by an O_2 measurement after the burner. The necessary excess air λ to cool the cell for a fixed heat extraction q_{FC} can be calculated be rearranging Equation (3.76) as

$$\begin{aligned} \lambda &= \frac{U_{\mathrm{f}} \cdot \left[\mathrm{LHV} - \mu_{\mathrm{O20}} \cdot \left(h^{*}_{\mathrm{RGAnO}} - h^{*}_{\mathrm{O_2CaO}}\right) - h^{*}_{\mathrm{RGAnO}}\right] + h^{*}_{\mathrm{FI}}}{\mu_{\mathrm{L0}} \cdot \left(h^{*}_{\mathrm{ACO}} - h^{*}_{\mathrm{AI}}\right)} \\ &\quad + \frac{-q_{\mathrm{FC}} - p_{\mathrm{el}} - (1 - U_{\mathrm{f}}) \cdot h^{*}_{\mathrm{FAnO}}}{\mu_{\mathrm{L0}} \cdot \left(h^{*}_{\mathrm{ACaO}} - h^{*}_{\mathrm{AI}}\right)} \end{aligned} \tag{3.77}$$

3.5 Thermodynamic theory of hybrid SOFC systems

The process environment of the cell model in Figure 3.1 must be related reversibly to the ambient state to define the reversible system. As mentioned above, we assume $U_f \rightarrow 0$ and the flows consist of unmixed components to assure a reversible process. Figure 3.9 shows the reversible fuel cell–heat engine system that fulfils these requirements. The reactants air and fuel in the ambient state T_0,p_0 are brought to the thermodynamic state of the cell T,p by the reversible heat pumps HPA (for air) and HPF (for fuel). The necessary heat consists of the energy from the environment and of the exergy being supplied by the reversible working heat pumps. The fuel cell FC delivers the flue gas, the work and the heat as given in Equation (3.5). The unmixed product flue gas is brought from the state T,p of FC to the ambient state T_0,p_0 by the reversible heat engine HEG. The reversible work delivered by HEG is the exergy of the flue gas with the state T,p. Finally, we use the Carnot cycle CC for the heat management between FC and the environment. FC can exchange heat reversibly with CC as with the process environment before. CC operates as a heat engine for $\Delta^r S < 0$.

The fuel cell FC delivers the reversible work w_{tFCrev}

$$w_{tFCrev} = \Delta^r G = \Delta^r H - T_{FC} \cdot \Delta^r S \tag{3.5}$$

and the reversible heat q_{FCrev}

$$q_{FCrev} = T_{FC} \cdot \Delta^r S \tag{3.78}$$

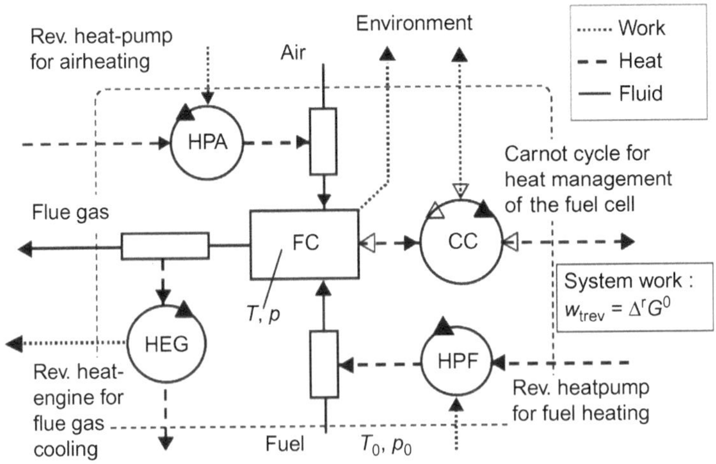

Figure 3.9 The reversible fuel cell–heat engine hybrid system.

FC is the heat source of the Carnot cycle CC and delivers reversible heat q_{FCrev}. The reversible work w_{tCCrev} of CC is defined by

$$w_{\mathrm{tCCrev}} = q_{\mathrm{FCrev}} \cdot \left(1 - \frac{T_0}{T_{\mathrm{FC}}}\right) = T_{\mathrm{FC}} \cdot \Delta^{\mathrm{r}} S \cdot \left(1 - \frac{T_0}{T_{\mathrm{FC}}}\right) \tag{3.79}$$

and the reversible heat q_{FCrev}

$$q_{\mathrm{CCrev}} = q_{\mathrm{FCrev}} \cdot \frac{T_0}{T_{\mathrm{FC}}} = T_0 \cdot \Delta^{\mathrm{r}} S \tag{3.80}$$

The heat pump HPF has to deliver the reversible heat q_{HPFrev} to heat the fuel

$$q_{\mathrm{HFrev}} = h^*_{\mathrm{FFC}} = w_{\mathrm{tHPFrev}} + q_{\mathrm{HPFrev}} \tag{3.81}$$

HPF has to be supplied reversibly with the work w_{tHPFrev} that is equal to the exergy e_{FFC} of the fuel with the thermodynamic state of the fuel cell and the heat q_{HPFrev} from the environment

$$w_{\mathrm{tHPFrev}} = e_{\mathrm{FFC}} = h^*_{\mathrm{FFC}} - T_0 \cdot s^*_{\mathrm{FFC}} = (h_{\mathrm{FFC}} - h_{\mathrm{F0}}) - T_0 \cdot (s_{\mathrm{FFC}} - s_{\mathrm{F0}}) \tag{3.82}$$

$$q_{\mathrm{HPFrev}} = T_0 \cdot s^*_{\mathrm{FFC}} \tag{3.83}$$

In the right part of Equation (3.82), the definition of the exergy of the fuel with the thermodynamic state of the fuel cell is now worked out in more detail. Similar processes are used for the reversible heating of the air and the reversible cooling of the flue gas. The reversible air heating needs the work w_{tHPA}

$$w_{\mathrm{tHPArev}} = \mu_{\mathrm{A}} \cdot e_{\mathrm{AFC}} = \mu_{\mathrm{A}} \cdot \left(h^*_{\mathrm{AFC}} - T_0 \cdot s^*_{\mathrm{AFC}}\right) \tag{3.84}$$

and the reversible heat engine HEG for the cooling of the flue gas G delivers the work w_{tHEGrev}

$$w_{\mathrm{tHEGrev}} = -(\mu_{\mathrm{A}} + 1) \cdot e_{\mathrm{GFC}} = -(\mu_{\mathrm{A}} + 1) \cdot \left(h^*_{\mathrm{GFC}} - T_0 \cdot s^*_{\mathrm{GFC}}\right) \tag{3.85}$$

The total work of the reversible fuel cell–heat engine system is defined by

$$w_{\mathrm{tsystrev}} = w_{\mathrm{tFCrev}} + w_{\mathrm{tCCrev}} + w_{\mathrm{tHPFrev}} + w_{\mathrm{tHPArev}} + w_{\mathrm{tHEGrev}} \tag{3.86}$$

Using (3.5), (3.79), (3.82), (3.84), (3.85) and (3.86), we get

$$w_{\mathrm{tsystrev}} = \Delta^{\mathrm{r}} H^0 - T_0 \cdot \Delta^{\mathrm{r}} S^0 = \Delta^{\mathrm{r}} G^0 \tag{3.87}$$

The reversible work $w_{tsystrev}$ of any fuel cell–heat engine system is independent of the state of the cell and is equal to the free enthalpy of the reaction $\Delta^r G^0$ at the ambient state [6]. The standard condition is assumed to be the ambient state to keep the argument simple. It can be shown that this reversible process model can be expanded to a number of process descriptions, and it is an expansion of the reaction chamber of van't Hoff because it allows the inclusion of work and delivers an overall reversible connection to the ambient state [7].

It is useful to define a simplified process for further analysis because the three reversible heat engines HPA, HPF and HEG do really nothing else than to heat fuel and air by cooling the flue gas – their total reversible work is negligible. Thus the simplified process uses a heat exchanger system for the heat recovery instead of HPA, HPF and HEG, as shown in Figure 3.10. But this simplified reference cycle is generally not reversible [8]. This is caused by the changes of the specific heat capacities of the different substances with the reaction temperature T that change the reaction enthalpy $\Delta^r H\ (T,p)$. A (small) amount of the waste heat of FC must be used to heat the reactants completely. We lose this amount of heat for further reversible use in the Carnot cycle CC.

The simplified fuel cell–heat engine hybrid cycle as a reference cycle fits the reversible system well; the deviation at 1000 °C is −0.76% only for the hydrogen oxidation. Figure 3.11 shows the applications of this cycle. The left side of Figure 3.11 shows the efficiency of the reference cycle. The system efficiency η_{syst} is defined as

$$\eta_{syst} = \frac{\sum w_t}{\mathrm{LHV}} \tag{3.88}$$

The cell temperature T_{FC} is equal to the temperature T of the process environment but more illustrative for further discussion. The work w_{tCC} produced by the Carnot cycle CC increases with increasing T_{FC}, and the work w_{tFCrev} produced by FC decreases with increasing T_{FC} as expected. There is a compensation of both effects and the work w_{tsyst} of the system is independent of T_{FC} (or nearly independent in the case of the simplified process). FC operates reversibly in both cases, but CC does not operate completely reversible in the simplified process due to the fact that a small part of the waste heat of FC is used to heat air and fuel. The practical use of the simplified combined fuel

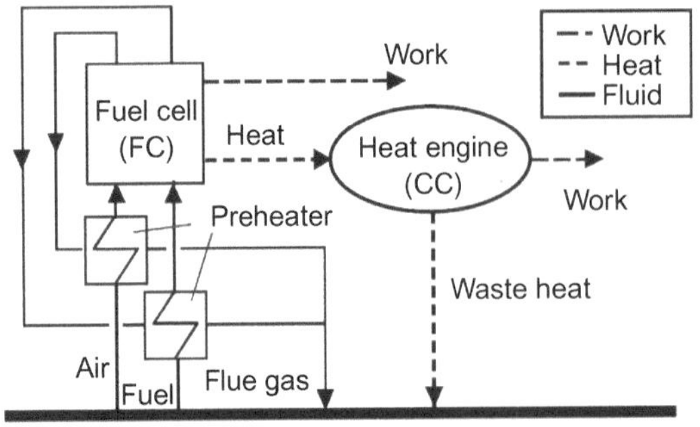

Figure 3.10 Simplified fuel cell–heat engine hybrid system as a reference cycle.

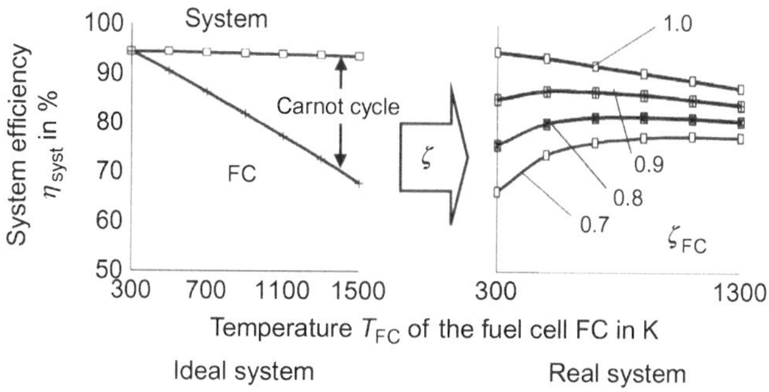

Figure 3.11 The system efficiency of the ideal and the real fuel cell–heat engine hybrid system with an exergetic efficiency $\zeta_{HE}=0.7$ and the oxidation of hydrogen.

cell–heat cycle is the opportunity to use exergetic efficiencies to describe the operational conditions of real cycles by using this very simple model. The exergetic efficiency ζ is defined as

$$\zeta = \frac{w_{treal}}{w_{trev}} \tag{3.89}$$

It is well known from heat engines that ζ is between about 0.7 and 0.8. All types of real cells have efficiencies between 55% and 65%, but there is no significant difference caused by the cell temperature T_{FC} as we would expect from the thermodynamic considerations [1]. Thus it is obvious that real cells operating at lower temperatures do not use their potential work w_{tFCrev} properly. Real high-temperature fuel cells use their potential work reasonably well. The exergetic efficiency of fuel cells is described in Ref. [8] for the H_2 fuel. It can be shown easily that the SOFC has the best exergetic efficiency here. The exergetic efficiency ζ_{FC} (3.89) can be related to the total fuel feed if we assume that the non-utilised fuel can be burnt in an isothermal combustor. We can define the fuel cell and the isothermal combustor as one unit in that case (see Figure 3.8).

The system efficiency η_{syst} of any real hydrogen-fuelled combined fuel cell–heat cycle is plotted over the cell temperature T_{FC} in Figure 3.11 on the right side. The exergetic efficiency of the fuel cell ζ_{FC} is the parameter (0.7; 0.8; 0.9; 1.0), and the exergetic efficiency of the heat engine ζ_{HE} is kept constant at 0.7.

The system efficiency η_{syst} increases with an increasing cell temperature T_{FC} for all exergetic efficiencies $\zeta_{FC} < 1.0$ until a maximum is reached. The location of the maximum of η_{syst} shifts with increasing exergetic efficiencies ζ_{FC} to decreasing T_{FC}. But there are no big changes in the region of the maximum if we change T_{FC} at constant ζ_{FC}. The influence of the Carnot cycle dominates at lower temperatures T_{FC} and lower exergetic efficiencies ζ_{FC}.

The main result of these considerations is the possibility of designing hybrid fuel cell–heat cycles with efficiencies of about 80% [8]. This efficiency value is a target of the U.S. Department of Energy since 1999 [9,10]. It appears useful to operate the cell at the lowest possible cell temperature T_{FC} in the region of the maximum of η_{syst} for reducing material costs of the heat engine and the heat exchangers. An increase in T_{FC} leads to only a negligible increase in the system efficiency η_{syst}.

The use of natural gas or other hydrocarbons changes the system design because of the processing of the fuel before its use in the SOFC. The following investigations will be done for methane as the main component of natural gas to keep the calculations simple. A very common fuel processing for hydrocarbons is the endothermic steam reforming process as shown for methane in (3.90) with the heat demand (3.91)

$$CH_4 + H_2O \rightarrow 3H_2 + CO \tag{3.90}$$

$$\Delta^r H(750\,°C)_{ref} = +14065.1\,kJ/kg_{CH_4} \tag{3.91}$$

Heat is also needed to evaporate the feed water. It is useful to use the waste heat of the cell for these purposes. A general model of a methane fired combined SOFC cycle based on the reference cycle of Figure 3.10 is shown in Figure 3.12 to describe the thermodynamic influences on the system's behaviour as simply as possible [1,11,12].

The SOFC can be modelled as one unit of two parallel operating SOFCs fuelled with hydrogen and carbon monoxide. All irreversible effects including mixing are to be described by $\zeta_{FC} < 1$. The reasons for these irreversibilities of the SOFC and other components are not important for understanding the system's behaviour if they are considered properly in the system. The important relation here is the relation between work and heat within the single components and the temperatures of the heat sources and the heat sinks. The SOFC is the heat source of the fuel processing in reforming and evaporation. If we look at the necessary temperature levels, we find generally

$$T_{FC} \geq T_{ref} > T_{evap} \tag{3.92}$$

This can be assumed for any SOFC system.

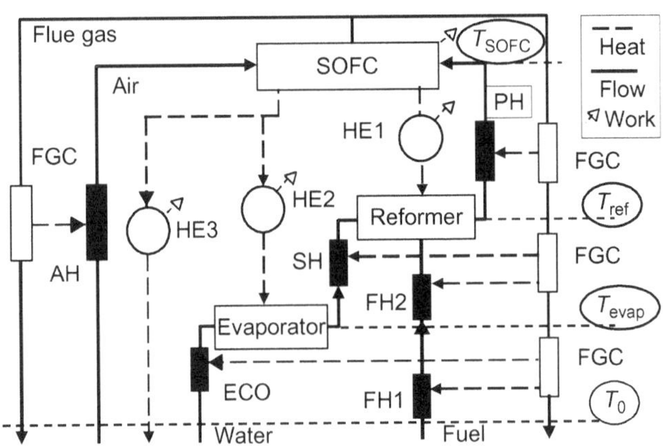

Figure 3.12 Process model for integrated reforming in SOFC systems.

A reversible heat transport between these different temperature levels is possible for an SOFC as the heat source of the two Carnot cycles and for the reformer and the evaporator as their heat sinks. The real engines are the heat engine HE1, operating between the SOFC and the reformer, and the heat engine HE2, operating between the SOFC and the evaporator. The exergetic efficiencies ζ_{HE} (Equation 3.80) describe the irreversibilities in the principal reversible processes. Finally, a third heat engine HE3 must operate between the SOFC and the ambient state if the waste heat of the SOFC is not used completely in the system. The flue gas flow is divided into two streams. The air is heated in the air heater AH by cooling of the major stream of the flue gas (FGC). The reactant (of reforming) feed water is heated in the economiser from the ambient temperature T_0 to the evaporator temperature T_{evap}, and the saturated steam is superheated from T_{evap} to the reformer temperature T_{ref}. The reactant (of reforming) methane is heated in the fuel heaters FH1 and FH2 from T_0 to T_{ref}, and finally, the products (of reforming) hydrogen and carbon monoxide (+steam) are heated from T_{ref} to the T_{SOFC} in the product heater PH. The required heat is supplied by the cooling of the second pass of the flue gas (FGC) from T_{SOFC} to a waste gas temperature $>T_0$, by the SOFC directly (for PH), by the waste heat of HE1 (for $T<T_{ref}$) and by the waste heat of HE2 (for $T<T_{evap}$). An auxiliary burner must be used for the reforming process if the waste heat of the SOFC cannot cover the heat requirement of the reformer and the evaporator. This auxiliary burner is not shown in Figure 3.12. It is possible to use efficiencies η

$$\eta = \frac{\dot{Q}_{used}}{\dot{Q}_{supplied}} \tag{3.93}$$

to describe any real heating process (heat exchanger, burner). The internal reforming in the SOFC is included in this modelling of the system ($T_{SOFC}=T_{ref}$). The external reforming is included in this model as well if the heat engine HE1 is replaced by a burner. The parameters listed in Table 3.3 have been used for analysis of both the systems. The water surplus was fixed at 2 to avoid coke formation [13].

Table 3.3 Standard parameters for analysis of SOFC–heat engine hybrid cycles

SOFC temperature, T_{SOFC}	900 °C
Reformer temperature, T_{ref}	750 °C
Evaporator temperature, T_{evap}	200 °C
Ambient temperature, T_0	25 °C
Excess air, λ	2
Water surplus, n_W	2
Exergetic efficiency SOFC, ζ_{SOFC}	0.60
Exergetic efficiency heat engine, ζ_{HE}	0.70
Efficiency of air heater, η_{AH}	0.90
Efficiency of heat exchangers, η_{HEX}	0.98

The system efficiencies η_{syst} of combined SOFC cycles with integrated and external reforming have been calculated and compared [11,12]. The possible system efficiencies η_{syst} of systems with external reforming are about 5–7% lower than that of a system with integrated reforming. The differences between the processes with external or integrated reforming are caused by the utilisation of the waste heat of the SOFC within the system. External reforming systems use an external burner with an additional entropy production. This increases the usable heat of the heat engine HE3 operating with the heat source SOFC and the heat sink environment, and thus the waste heat of the system increases. An external reformer cannot be used as a heat sink of the waste heat of the SOFC, and the entropy cannot be recycled as within an integrated reforming. An internal reforming in the SOFC has no temperature difference available for a power generation during the heat transport. This leads to a slight decrease of the system efficiency η_{syst} of the internal reforming compared with the integrated reforming.

An important influence on the performance of combined SOFC cycles with an integrated reforming is caused by the operation of the air heater under real conditions. The excess air λ and the efficiency η_{AH} of the air heater (Equation 3.83) are the main operational and design parameters. The system efficiency η_{syst} is plotted against excess air λ in Figure 3.13 with η_{AH} as a parameter. The basic chemical thermodynamics shows that $\Delta^r G$ – and therefore w_{tFCrev} (Equation 3.5) and $w_{tsysrev}$ (Equation 3.78) – is independent of the excess air λ. This can be used to prove the model because η_{syst} is independent of λ for $\eta_{AH} = 1$ as expected. η_{syst} decreases with increasing λ for all $\eta_{AH} < 1$. The influence of η_{AH} increases with increasing λ. The behaviour of the system for $\eta_{AH} = 0.85$ is shown Figure 3.13.

Firstly, η_{syst} decreases slightly with an increasing λ because the work of the heat engine HE3 decreases by compensating increasing heat losses. The other heat engines are operating at full load. The decrease of η_{syst} becomes sharper for an excess air $\lambda \approx 3$ because the heat engine HE3 goes out of service by a lack of available heat. The total waste heat of the SOFC must be used to supply the heat engines HE1 and HE2 which operate between the SOFC and the reformer and the evaporator respectively and to

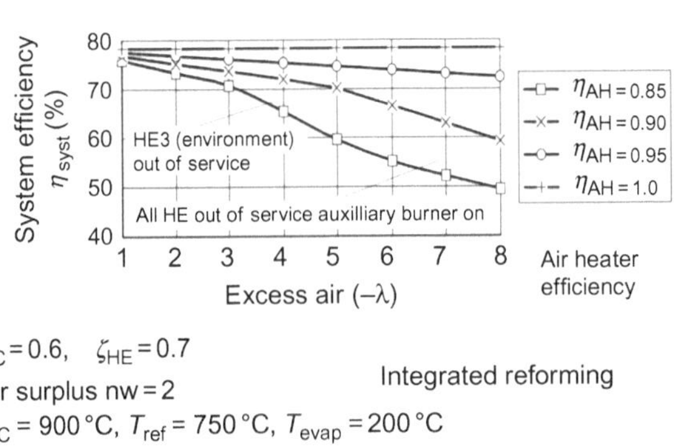

Figure 3.13 The influence of the excess air λ and the efficiency η_{AH} of the heat transfer in the air heater on the system efficiency η_{syst} of the SOFC–heat engine hybrid cycle.

compensate the increasing heat loss of the air heater. This causes the sharper decrease of the system efficiency η_{syst} with an increasing λ in the region $3<\lambda<6$ by a decreasing supply of work by HE1 and HE2. In the region $\lambda>6$, there is no heat engine in operation. All further heat losses (increasing with increasing excess air λ) must be compensated by the auxiliary burner. η_{syst} drops to values lower than 50% as shown in Figure 3.13. These results show that it is important to assure a good heat recovery in the air heater system and to avoid a very high excess air λ.

The system efficiency η_{syst} is influenced by the exergetic efficiency of the heat engines (HE1, HE2, HE3) ζ_{HE1}, ζ_{HE2}, ζ_{HE3} as well. η_{syst} is plotted against ζ_{HE} for each of the three heat engines in Figure 3.12. The maximum difference of about 9% in η_{syst} occurs if ζ_{HE3} (heat sink environment) is varied between 0 and 1. The difference is only about 2% if ζ_{HE1} (heat sink reformer) is varied. The variation of ζ_{HE2} (heat sink evaporator) leads to differences in η_{syst} of about 8%. The order of magnitude of these differences corresponds to the difference of the respective temperature differences between the SOFC and the heat sinks. This result confirms that entropy recycling by integrated reforming is more important to achieve high efficiencies than the power generation in the heat engine HE1. But the heat engine between the SOFC and the evaporator is important especially with a good exergetic efficiency ζ_{HE2}. The shape of η_{syst} of the variation of ζ_{HE2} seems to be unexpected compared with the variations of ζ_{HE1} and ζ_{HE3}, see Figure 3.14.

But it can be explained easily if we look at the conditions that lead to the needed amount of the SOFC's waste heat to supply the evaporator or the reformer. We obtain for the necessary waste heat Q_{SOFC} to operate the heat engines (HE1, HE2 = HEprocess) and to supply the processes (reformer or evaporator) with the necessary heat $Q_{process}$

$$Q_{SOFC} = \frac{Q_{process}}{1-\zeta_{HEprocess}+\zeta_{HEprocess}\cdot {}^{T_{process}}/_{T_{SOFC}}} \tag{3.94}$$

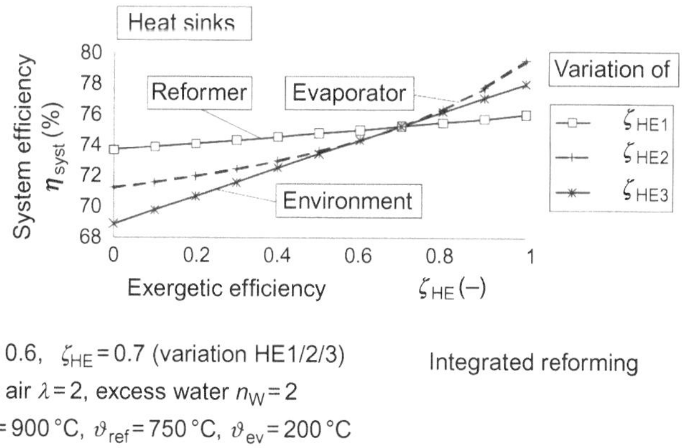

Figure 3.14 The influence of the heat engine design on the system efficiency η_{syst} of SOFC–heat engine hybrid cycles.

This result is governed by the exergetic efficiency of the heat engine $\zeta_{HEprocess}$ and the relation of the temperature of the heat sink $T_{process}$ and the temperature of the heat source T_{SOFC} (T in K). Table 3.3 shows that the relation $T_{process}/T_{SOFC}$ is about 0.4 in the case of the evaporator and about 0.9 in the case of the reformer. Recycling of the anode outlet flow is another option to supply the reformer with steam.

3.6 Design principles of SOFC Hybrid systems

The process models as given in Figures 3.10 and 3.12 can be realised by different heat engines; Figure 3.15 gives an example. The heat source can be the flue gas or the stack heat. Obviously, a gas turbine (GT) or a waste heat boiler of a steam cycle can utilise the heat from the flue gas. But the total system integration may utilise the waste heat of the cell directly in both cases, as expressed by the dotted line in Figure 3.15. The different cycles based on a Carnot cycle with a separate process flow are other options for direct stack cooling. The use of a Stirling engine might be one option as the latest developments indicate [14]. A further option might be the conversion of heat to electricity by an AMTEC process [15]. The thermoelectric conversion might be a possibility to extract heat for electricity generation in smaller units as, for example, for defence applications [16]. The direct power generation in the last two options might be of specific interest for the electric system integration. Finally, there is a further option to use the cell entropy in the sense of the second law of thermodynamics. Any endothermic process needs a transfer of heat at a certain temperature, and thus a certain supply of entropy [17]. This amount of entropy is a thermodynamic process requirement different from, for example, a heat supply for room heating that can be clearly reduced by a better heat recovery and a better insulation. However, CHP for room heating might be the better commercial solution.

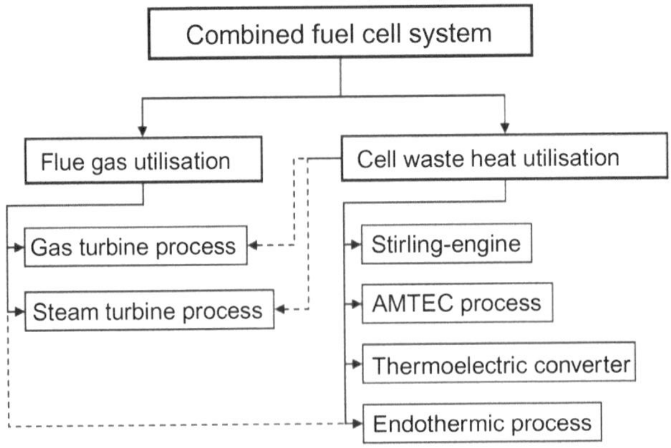

Figure 3.15 Possibilities of system integration in SOFC–heat engine hybrid cycles.

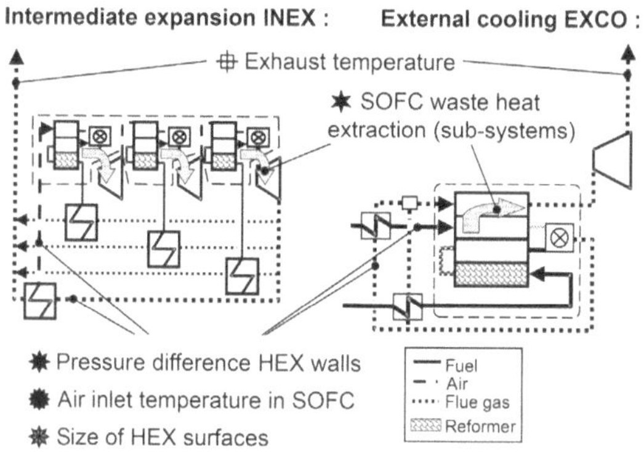

Figure 3.16 Cooling strategies for SOFC modules by GT cycles.

The SOFC-GT system is very interesting for high efficiency power generation [18–20]. Any successful cooling strategy for SOFC of an SOFC-GT system must avoid high excess air at the system's outlet as shown above (see Figure 3.13). Figure 3.16 shows the possible strategies. The SOFC module can be divided in sub-modules, and the heat of the SOFC module is extracted by cooling the waste air of the first sub-module to the inlet temperature of the cathode of the following sub-module by the power generation by a GT. This intermediate expansion (INEX) can be carried on until the last GT delivers the waste gas for the heat exchangers (HEX) to heat the air and the fuel.

The other strategy is the SOFC cooling by an external cooler (EXCO) fed with the flue gas that has been cooled by the heating of air and fuel. The SOFC module is the heat source for the GT cycle, and the air is heated by the flue gas as in the generalised model. The integrated gas heater can be heated by radiation and allows an optimisation of the temperature level of the SOFC cooling together with an integrated air heater and avoids unacceptable thermal stresses. The main differences between INEX and EXCO are marked in Figure 3.16 and compared in Table 3.4 [21,22].

Table 3.4 **Comparison of the INEX and the EXCO design**

		INEX	EXCO
1	SOFC waste heat extraction (sub-systems)	$2-n$ pressure levels (systems)	1 pressure level (system)
2	Pressure difference HEX walls	Maximal pressure difference	Only pressure loss
3	Limit for air inlet temperature in SOFC	Gas turbine outlet temperature	SOFC temperature
4	Size of HEX surfaces	Min. 1/2.5 of ambient system	Min. 1/7 of ambient system
5	Exhaust temperature	~200 °C	500–600 °C

(1) The waste heat is extracted at one pressure level in the EXCO design and at up to *n* pressure levels in the INEX design, depending on the allowable temperature difference of the cathode. The number of pressure levels is equal to the number of the pressurised sub-systems.
(2) The pressure difference on the HEX walls of the air heater is the maximum pressure in the INEX design and the only pressure loss in the module in the EXCO design.
(3) The air inlet temperature of the SOFC module is limited by the SOFC (module outlet) temperature in the EXCO design and by the lower GT outlet temperature in the INEX design.
(4) The size of the HEX of an on one side pressurised INEX design is about 2.5 times smaller than under ambient conditions, but the on both sides pressurised EXCO design has up to 7 times smaller HEX surfaces.
(5) The electric efficiency of an INEX design with two turbines is about 70% [23], similar to the EXCO design. The exhaust temperature in the INEX design is about 200 °C and that of the EXCO design is about 500–600 °C depending on the individual parameters.

The EXCO design has thus the potential for a combination with a steam turbine cycle (ST) that could be, for example, a Cheng cycle. This leads to an electric efficiency of about 75% [24]. The first studies [24] of the EXCO design included a reheat cycle with an additional heat exchanger within the SOFC module. This design seemed to be too complicated. But a comparison of both designs shows that the benefit of the EXCO design to reduce the excess air in one process step at one pressure level with small HEXs can be combined with the benefit of the INEX design to allow a simple cascading of GT cycles as needed for a reheat GT-cycle. This led to the proposal of the reheat SOFC-GT cycle combined with a steam turbine (ST) cycle which reaches slightly more than 80% as the calculated efficiency [25]. But if we compare the slight increase of efficiency compared with the basic EXCO design, it seems questionable if this technology is economically feasible.

3.7 Summary

Thermodynamic considerations are used to understand the processes of energy conversion in SOFCs. Such theoretical studies of the behaviour of the reversible processes have a high practical value in helping to understand complex systems. The reversible work of a fuel cell is defined by the free or Gibbs enthalpy of the reaction. If we use the assumption of the ideal gas, we immediately get the equation of the Nernst voltage from the Gibbs enthalpy of the reaction. The consideration of the electrical effects shows that the molar flow of the spent fuel is proportional to the electric current and the reversible work is proportional to the reversible voltage. A coupling between the thermodynamic data and the electrical data is only possible using the quantities power or heat flow and not by using work and heat. This is caused by the fact that we use a mass or substance transport as the basis for thermodynamic considerations and we use a charge transport to describe electrical phenomena.

Irreversible losses cause a difference in the efficiency of reversible and real processes. These losses can be described and quantified by their irreversible entropy production. The consideration of the ohmic losses shows that the irreversible entropy production in an SOFC is smaller than in another low-temperature fuel cell. This is caused by the lower irreversible entropy production of the heat dissipated at a higher temperature. The changes in the Nernst voltage can be understood by the analysis of the fuel utilisation.

Because all the fuel cannot be fully reacted in practice within the fuel cell, the SOFC stack can be treated like a power-generating burner so as to integrate it easily into a system model. The stack cooling depends on the amount of excess air.

The combination of an SOFC with a heat engine allows an extremely high electric efficiency. Any real combination of an SOFC and a heat engine is based on a reversible system, but a simplified version can be used to analyse the principles of the design of a combined SOFC–heat engine. It is important that the cell itself and not the flue gas is considered as the heat source. Integration of fuel processing is another important factor in achieving a high efficiency since the embedded fuel processor can be supplied with a certain amount of entropy from the cell heat. This entropy is used for the reforming reaction and need not be transported to the environment as an entropy loss.

The theoretical SOFC analysis presented here has found a number of broad applications in recent years. Mostly the improvements generated have been system integrations, but there are also innovations relating to cell behavior, materials research and energy economic principles. A few examples will be given below for introducing further reading.

The results of field experience are the best confirmation of theory. The principal shown importance of heat recovery and loss minimisation has been proven by an unexpected high electrical efficiency of 60% of micro SOFC-CHP systems in field tests [26]. A number of investigations concentrated on SOFC-GT systems known as SOFC hybrids, dealing with specific components behaviour in testing and design [27,28], modelling tools and controlling [29,30], specific investigations in aircraft and maritime applications [31,32] gave a view on possible future markets. An overview about publications can also be found in Ref. [33]. But also fuel processing of unconventional fuels are topics of thermodynamic system analysis [34]. SOFC operation and cell modelling are further aspects for applications of thermodynamic models [35–37], further impacts can be found in material research [38] and general new energy studies [39]

References

[1] W. Winkler, Brennstoffzellenanlagen, Springer Verlag, Berlin, 2002.

[2] K. Wark, Advanced Thermodynamics for Engineers, McGraw-Hill, New York, ISBN: 0-07-068292-5, 1995.

[3] U. Bossel, Facts and figures. Final report on SOFC data. IEA programme of R, D&D on advanced fuel cells, Annex II: modelling & evaluation of advanced SOFC. Swiss Federal Office of Energy; Operating Agent Task II. Berne, April 1992.

[4] W. Winkler, Reversible SOFC Model as a Base of Design and Testing, ECS Transactions, 7, The Electrochemical Society, 2007, ISBN: 978-1-56677-554-0, pp. 2021–2029.

[5] Reversible fuel cell models as a base of fuel cell modeling and design. International Colloquium on Environmentally Preferred Advanced Power Generation. January 29-31, 2008, Fairmont Newport Beach, Newport Beach, California. Conference-CD.

[6] W. Winkler, Der Einfluß der Prozeßkonfiguration auf das Arbeitsvermögen von Verbrennungskraftprozessen, Brennstoff-Wärme-Kraft 46 (1994) 334–340, Heft 7/8. S.
[7] W. Winkler, Sustainable product development based on second law of thermodynamics, Appl. Energy 88 (2011) 3248–3256.
[8] W. Winkler, Analyse des Systemverhaltens von Kraftwerksprozessen mit Brennstoffzellen, Brennstoff-Wärme-Kraft 45 (1993) 302–307, Heft 6. S.
[9] U.S. Department of Energy: Broad Agency Announcement (BAA) NO. DE-BA26-99FT40274 for research entitled: "Multi-Layer Ceramic, Fuel Cell Research", 7th June 1999.
[10] M.C. Williams, Status of solid oxide fuel cell development and commercialization in the U.S. solid oxide fuel cells (SOFC VI), in: S.C. Singhal, M. Dokiya (Eds.), Proceedings of the Sixth International Symposium, Electrochemical Society Proceedings, vols. 99–19, ISBN: 1-56677-242-7, 1999, pp. 3–9, Pennington.
[11] W. Winkler, SOFC-Integrated Power Plants for Natural Gas, in: Proceedings First European Solid Oxide Fuel Cell Forum, 3–7 October 1994, Ulf Bossel, Lucerne, 1994, pp. 821–848.
[12] W. Winkler, Lay out principles of the integration of fuel preparation in fuel cell systems, in: Proceedings 2nd IFCC (International Fuel Cell Conference), NEDO, Kobe, Japan, 1996, pp. 397–400.
[13] A. Gubner, Grundlagen der Modellbildung für die Methanbildung in Hochtemperaturbrennstoffzellen, Thesis, University of Applied Sciences Hamburg, 1992.
[14] 10th International Stirling Engine Conference 2001 (10th ISEC), 24–26 September 2001, Osnabrück, VDI Gesellschaft für Energietechnik, 2001.
[15] J.-M. Tournier, M.S. El-Genk, L. Huang, Experimental investigations, modeling, and analysis of high-temperature devices for space applications. Final report. AFRL-VS-PS-TR-1998-1108. (U.S.) Air Force Research Laboratory, Kirtland Air Force Base, January 1999.
[16] R.J. Nowak, A DARPA perspective on small fuel cells for the military, in: Solid State Energy Conversion Alliance (Seca) Workshop, Arlington, VA, 29 March, 2001.
[17] R. Kikuchi, K. Sasaki, K. Eguchi, Design of energy and chemical co-production systems using solid oxide fuel cell technology, in: H. Yokogawa, S.C. Singhal (Eds.), Proceedings 7th International Symposium on Solid Oxide Fuel Cells in Tsukuba, Japan, vol. 2001-16, The Electrochemical Society, Inc., Pennington, NJ, USA, 2001, pp. 214–223.
[18] M.C. Williams, C.M. Zeh, Workshop on very high efficiecy fuel cell/gas turbine power cycles, in: Proceedings. US Department of Energy – Office of Fossil Energy Morgantown Energy Technology Center, 1995.
[19] R.A. George, Status of tubular SOFC field demonstration, J. Power Sources 86 (2000) 134–139.
[20] M. Koyama, H. Komiyama, K. Tanaka, K. Yamada, Evaluation of a solid oxide fuel cell and gas turbine combined cycle with different cell component materials, in: H. Yokogawa, S.C. Singhal (Eds.), Proceedings 7th International Symposium on Solid Oxide Fuel Cells in Tsukuba, Japan, vol. 2001-16, The Electrochemical Society, Inc., Pennington, NJ, USA, 2001, pp. 234–243.
[21] W. Winkler, Thermodynamic influences on the cost efficient design of combined SOFC cycles, in: Philippe Stevens (Ed.), Proceedings 3rd European Solid Oxide Fuel Cell Forum in Nantes, Oral Presentations, 1998, pp. 525–553.
[22] W. Winkler, Cost effective design of SOFC-GT, in: S.C. Singhal, M. Dokiya (Eds.), Proceedings 6th International Symposium on Solid Oxide Fuel Cells in Honolulu, USA, The Electrochemical Society, Inc., Pennington, NJ, USA, 1999, pp. 1150–1159.

[23] W.G. Parker, F.P. Bevc, SureCELLTM integrated solid oxide fuel cell/oxidation turbine power plants for distributed power applications, in: Proceedings 2nd International Fuel Cell Conference February 5–8, 1996, Kobe, Japan, 1996, pp. 275–278.
[24] W. Winkler, Möglichkeiten der Auslegung von Kombikraftwerken mit Hochtemperaturbrennstoffzellen, Brennstoff-Wärme-Kraft 44 (1992) 533–538, Heft 12. S.
[25] W. Winkler, H. Lorenz, Layout of SOFC-GT cycles with electric efficiencies over 80%, in: Proceedings 4th European Solid Oxide Fuel Cell Forum in Lucerne, July, 2000, pp. 413–420.
[26] K. Föger, F. Dahlmanns, Highly Efficient μ-CHP System as Part of a Virtual Power Plant, vol. 9, VGB Power Tech, Germany, 2012, p. 97.
[27] W. Winkler, P. Nehter, M.C. Williams, D. Tucker, R. Gemmen, General fuel cell hybrid synergies and hybrid system testing status, J. Power Sources 159 (1) (2006) 656–666.
[28] A. Traverso, L. Magistri, A.F. Massardo, Turbomachinery for the air management and energy recovery in fuel cell gas turbine hybrid systems, Energy 35 (2010) 764–777.
[29] C. Pianese, M. Sorrentino, Development of Modeling Tools for Optimal Design, Control and Diagnosis of Solid Oxide Fuel Cell APUs, Transaction in Modeling, Control and Energy, (2010), ISSN: 1985-9406, Online Publication, June, pp. 10–16.
[30] P. Chinda, P. Brault, The hybrid solid oxide fuel cell (SOFC) and gas turbine (GT) systems steady state modeling, Int. J. Hydrogen Energy 37 (2012) 9237–9248.
[31] J.E. Freeh, J.W. Pratt, J. Brouwer, Gas Turbine Hybrid System Model for Aerospace Applications, NASA/TM-2004-213054, May 2004.
[32] Y.M.A. Welaya, M. Mosleh, N.R. Ammar, Thermodynamic analysis of a combined gas turbine power plant with a solid oxide fuel cell for marine applications, Int. J. Nav. Arch. Ocean Eng. 5 (2013) 529–545.
[33] F. Zabihian, A. Fung, A review on modeling of hybrid solid oxide fuel cell systems, Int. J. Engineering (IJE) 3 (2) (2009) 85–119.
[34] W. Winkler, P. Nehter, M.C. Williams, Basic thermodynamic studies and exergetic analysis of fuel cell steam reforming for natural gas in waste landfill and anaerobic digestor gas, Resour. Process. 57 (1) (2010) 3.
[35] M.C. Williams, The use of exergetic efficiency rate of change to evaluate fuel cell performance, ECS Trans. 26 (2010) 3–17.
[36] M.C. Williams, K. Reifsnider, Exergetic efficiency as a basis for durability analysis and prognosis, Int. J. Hydrogen Energy 36 (2011) 10426–10433.
[37] F. Miao, Impact on fuel transport efficiency in anode of planar solid oxide fuel cells, Int. J. Electrochem. Sci. 8 (2013) 11814–11822.
[38] S. Wang, L. Wu, J. Gao, Q. He, M. Liu, Oxygen ion transference number of doped lanthanum gallate, J. Power Sources 185 (2008) 917–921.
[39] M.J. Hermida-Castro, D. Hermida-Castro, X.M.V. Martínez, J.A. Orosa, Comparative study of fuel cell applications and future plant conservation applications, Int. J. Energy Science (IJES) 3 (5) (2013) 357–361.

Electrolytes

4

J.A. Kilner, J. Druce, T. Ishihara
International Institute for Carbon-Neutral Energy Research (I2CNER),
Nishi-Ku, Fukuoka, Japan

4.1 Introduction

The electrolyte for a solid oxide fuel cell (SOFC) must meet a very exacting combination of electrical, chemical and mechanical requirements in order to be suitable for practical application. For example, it has to be stable in both reducing and oxidising environments and must have sufficiently high ionic conductivity with low electronic conductivity at the cell operating temperature. In addition, it must be possible to form the material into a thin, strong gas-tight layer. The most usual oxides of choice for this application are those possessing the fluorite structure, such as yttria-stabilised zirconia, the most common electrolyte for SOFCs. Other fluorite oxide ion conductors, such as doped ceria, have also been proposed as electrolyte materials for SOFCs, especially to attain the goal of reduced temperature operation (600–800 °C). More recently, a number of other materials, including perovskites, brownmillerites and hexagonal-structured oxides, have also been found to possess good ionic conductivity at these lower temperatures. This chapter first describes fundamental and practical aspects of fluorite electrolytes and then discusses the structure and properties of perovskite and other ion conductors.

4.2 Fluorite-structured electrolytes

Oxide ion conductivity was first observed in ZrO_2 containing 15 wt% Y_2O_3 (yttria-stabilised zirconia or YSZ) by Nernst in the 1890s. In 1937, Baur and Preis [1] constructed the first SOFC using this electrolyte. Since that time, many oxide systems have been examined as potential electrolytes for SOFCs, and there are many reviews of oxygen ion conductivity in books, book chapters, for example [2–5] and journal articles [6–8]. Figure 4.1 shows the temperature dependence of the ionic conductivity for several candidate electrolytes, indicating that YSZ, despite its widespread use, does not have the highest oxygen ion conductivity over the temperature range of interest, reinforcing the comment above that conductivity is only one of the criteria for the selection of a solid electrolyte.

Bismuth oxide compositions [3] show the highest conductivity, and several other materials are also superior to YSZ, particularly at temperatures below 600 °C; however, these other oxides have disadvantages such as significant electronic conductivity, high cost or difficulties in processing which make them difficult to use.

High-temperature Solid Oxide Fuel Cells for the 21st Century. http://dx.doi.org/10.1016/B978-0-12-410453-2.00004-X

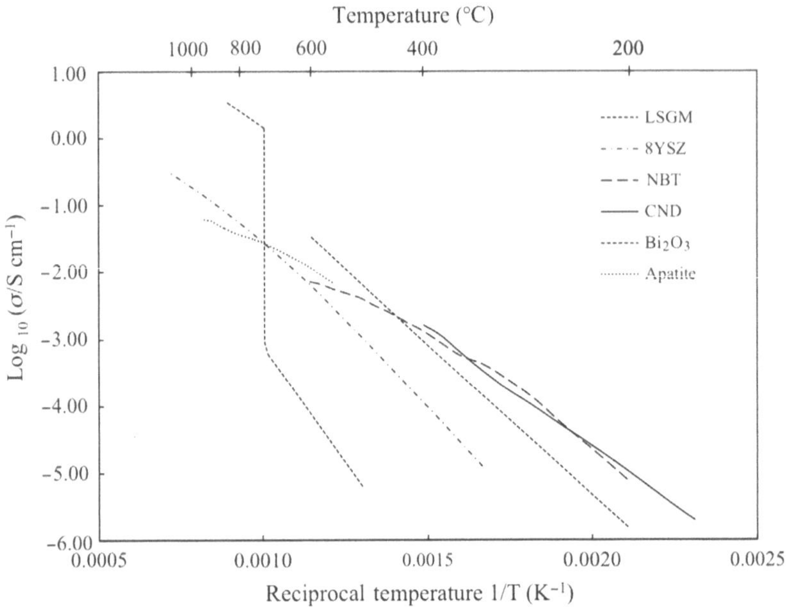

Figure 4.1 Conductivity of oxygen ion conductors including candidate SOFC electrolytes. $La_{0.8}Sr_{0.2}Ga_{0.9}Mg_{0.1}O_{3-\delta}$ (LSGM) [9], $Zr_{0.84}Y_{0.16}O_{1.92}$ (8YSZ) [10], $Na_{0.5}Bi_{0.49}Ti_{0.98}Mg_{0.02}O_3$ (NBT) [11], $Ce_{0.9}Nd_{0.1}O_{1.95}$ (CND) [12], Bi_2O_3 [13] and $La_{10}Si_{5.8}Mg_{0.3}O_{26.8}$ (Apatite) [14].

The candidate electrolytes discussed in this chapter can be divided into three major groups, those with the fluorite structure such as YSZ discussed here, the perovskite structure such as lanthanum gallate discussed later in Section 4.3 and newer materials such as those with the apatite structure discussed later in Section 4.4.

The fluorite structure (MO_2) is formed by tetravalent metal cations such as Zr^{4+}, Ce^{4+} and Th^{4+}. The fluorite structure, shown in Figure 4.2, is a face-centred cubic arrangement of M^{4+} cations with the oxygen ions occupying all the tetrahedral sites, leading to a large number of octahedral interstitial voids. Thus this structure is a rather open one and rapid ion diffusion might be expected.

Zirconia (ZrO_2) only forms the fluorite structure at high temperatures (see phase diagram in Figure 4.5), but this can be stabilised to lower temperatures by the addition of divalent or trivalent (i.e. aliovalent) cations such as Ca^{2+} or Y^{3+}. Ceria (CeO_2) also adopts the fluorite structure, but does not need to be stabilised by the addition of aliovalent cations to do so, and has the added advantage of displaying extensive solid solutions with the rare earth sesquioxides.

Oxide ion conduction is provided by point defects on the oxygen sublattice, either by oxide ion vacancies or interstitial oxide ions. To move through the crystal, the ions must be able to move into an unoccupied equivalent site with a minimum of hindrance. The mobile species is often considered to be the defect, for example, the vacancy. Intrinsic defect concentrations are fixed by thermodynamic equilibrium in pure compounds and are extremely low in most oxide materials, whilst extrinsic defects are established by the

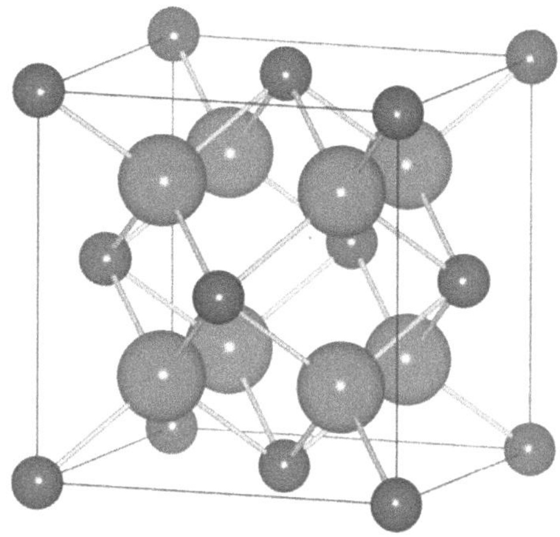

Figure 4.2 The cubic fluorite structure. Key red (light grey in the print version) spheres oxygen ions and purple (dark grey in the print version) spheres M^{4+} cations.
Reproduced from Ref. [8] with permission from The Royal Society of Chemistry.

presence of aliovalent substitutionals and can reach levels of a few mol%. To maintain electroneutrality, a soluble aliovalent ion in an ionic compound can be compensated by an increase in the concentration of an ionic defect such as a vacancy or interstitial. In cases where the oxide contains multivalent cations, charge compensation can also occur by the formation of electronic defects giving rise to mixed or even electronic conductivity. In the case of pure ZrO_2 and CeO_2, the ionic conductivity is very low because the concentration of the intrinsic defects is low. However, as aliovalent subtitutional ions (often known as dopants) such as Y^{3+} are added, the defect concentration, and hence the conductivity, increases.

As an example, the dissolution of yttria into the fluorite phase of ZrO_2 can be written by the following defect equation in Kröger–Vink notation:

$$Y_2O_3 \xrightarrow{ZrO_2} 2Y'_{Zr} + 3O_O^x + V_O^{\bullet\bullet} \tag{4.1}$$

Each additional mole of yttria creates one oxygen vacancy. The concentration of the vacancies is given simply by the electrical neutrality condition for this case, $[Y'_{Zr}] = 2[V_O^{\bullet\bullet}]$, implying that the vacancy concentration is linearly dependent on the substitution level. The ionic conductivity, σ, can be expressed by;

$$\sigma = nq\mu \tag{4.2}$$

where n is the number of charge carriers, μ their thermally activated mobility, and q their charge. In the case of oxide ion conductors such as substituted zirconia and ceria, Equation (4.2) leads to Equation (4.3) where the fraction of mobile oxide ions is the product of the fraction of vacancies, $[V_O^{\bullet\bullet}]$, and the fraction of occupied oxygen sites,

$1-[V_O^{\bullet\bullet}]$. E_a is the activation energy for conduction process, R the gas constant, T absolute temperature and A, a pre-exponential constant [15].

$$\sigma T = A[V_O^{\bullet\bullet}](1-[V_O^{\bullet\bullet}])\exp\left\{\frac{-E_a}{RT}\right\} \tag{4.3}$$

The conductivity of substituted zirconia and ceria varies as a function of substitutional concentration and shows a maximum at a specific concentration. However, this maximum occurs at an oxygen vacancy concentration much lower than the 50% (by site fraction) expected from Equation (4.3) and shows an unexpected variation with the type of trivalent cation. An example of this behaviour is shown in Figure 4.3 as reported by Arachi et al. [16] for the ZrO_2–M_2O_3 (M = Sc, Yb, Y, Dy, Gd or Eu) systems.

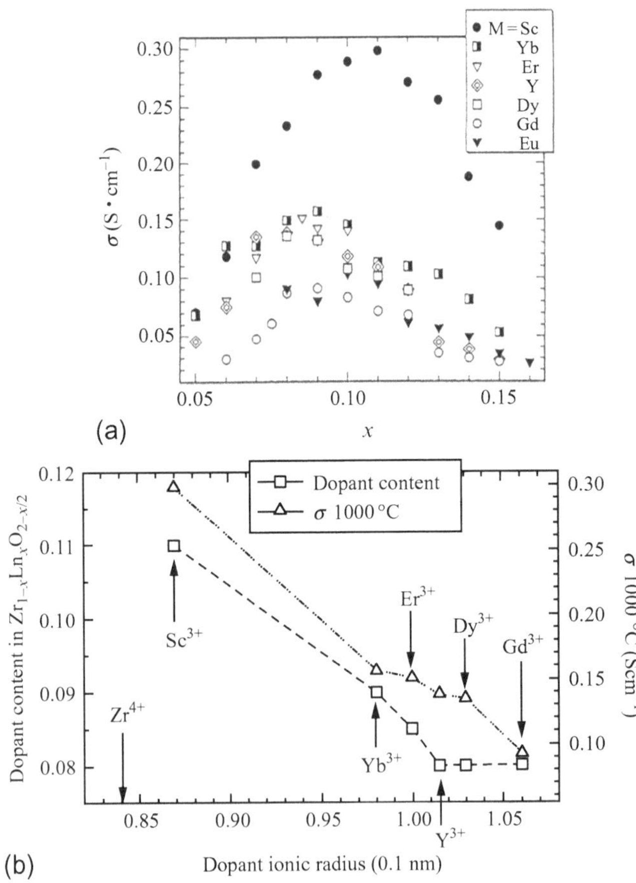

Figure 4.3 Composition dependence of the electrical conductivity at 1000 °C for ZrO_2–M_2O_3 compositions (a) and Dopant concentration exhibiting the highest conductivity dependence on dopant cation radius in ZrO_2–M_2O_3 systems (b).
Data adapted from Ref. [16].

Attempts to explain the origin of this maximum have been made by numerous authors for both the zirconia [16] and the ceria systems [15,17,18], usually in terms of the mismatch in size between the aliovalent substitutional ion and the host cation. The relation between the substitutional concentration with the highest conductivity at 1000 °C and the substitutional ionic radius in the ZrO_2–M_2O_3 system is shown in Figure 4.3 [16]. The content of substitutional giving the highest conductivity decreases with increasing radius of substitutional ion. The substitutionals, Dy^{3+} and Gd^{3+}, with higher ionic radii show a limiting value of 8 mol%. The substitutional Sc^{3+}, which has the closest ion radius to the host ion, Zr^{4+}, shows the highest conductivity and the highest substitutional content at the maximum conductivity. A similar conductivity dependence has been observed in the CeO_2 system where a much wider range of substitutional ions is possible; this is explored below.

The concentration of mobile oxide ion vacancies is affected by the elastic strain energy, which is related to the size mismatch between the host and substituted cations. The importance of defect pairs formed due to interaction between the oxide ion vacancies, $V_O^{\bullet\bullet}$, and aliovalent cations, M'_{Ce}, in CeO_2 has been stressed by a number of authors [15,17–25]. This association reaction is shown below for an associated defect pair;

$$V_O^{\bullet\bullet} + M'_{Ce} \leftrightarrow \{V_O^{\bullet\bullet}M'_{Ce}\}^{\bullet} \tag{4.4}$$

higher-order clusters are possible, for example, the trimer.

$$V_O^{\bullet\bullet} + 2M'_{Ce} \leftrightarrow \{M'_{Ce}V_O^{\bullet\bullet}M'_{Ce}\}^{x} \tag{4.5}$$

These can build to larger clusters involving many aliovalent cations and vacancies. It has been suggested that an isolated pair, for example, $\{V_O^{\bullet\bullet}M'_{Ce}\}^{\bullet}$ is more likely to occur because of the expected random distribution of M'_{Ce} [26]. However, more recent studies have shown that, depending upon the type of aliovalent cation and its concentration, quite complex clusters can be formed [27 31].

The effect of the binding enthalpy of an associated ion (to whatever size of cluster) can significantly affect the population of free vacancies at lower temperatures, where association is almost complete. For example, in the case of simple pair formation, that is, at very low aliovalent cation concentrations, the following will apply.

$$[\{V_O^{\bullet\bullet}M'_{Ce}\}^{\bullet}] \gg [V_O^{\bullet\bullet}] \tag{4.6}$$

and

$$[V_O^{\bullet\bullet}] = \frac{B}{T}\exp\left\{\frac{\Delta H_a}{RT}\right\} \tag{4.7}$$

where ΔH_a is the association enthalpy and B a constant. This means that both the concentration of the mobile vacancies and their migration are thermally activated, and that when the association enthalpy is large, this will have a detrimental effect upon the ionic conductivity, especially at low temperatures. Thus strategies to minimise this association term are important in optimising ionic conductivity in the fluorite oxides. Kilner [18] has pointed out that both calculated and experimental data for the association enthalpies could be correlated with the ionic radius of the dopant and its mismatch with the host cation radius, with a minimum occurring close to the host cation radius. Since this early work, more sophisticated atomistic [32] and density functional theory (DFT) [33] calculations have been carried out that show this minimum exists and is the result of a switch in the position of the oxygen vacancy from the nearest neighbour to the next-nearest neighbour position, as seen in Figure 4.4. The minimum association enthalpy occurs when the ionic radii of the host and the aliovalent cation are close to each other such that the lattice relaxations, and hence the lattice energy, are minimised. This minimum occurs when the cation is slightly larger than the host; this may be related to the recent observation that an oxygen vacancy is effectively smaller than an oxygen ion [34–36] with the net result of a minimum total disturbance to the lattice.

The DFT calculations of Andersson and co-workers [33] suggested that it might not be possible to find a cation that has the optimum radius and suggested the use of a combination of cations to obtain the optimum, similar to an earlier suggestion of Ralph et al. [37]. This has led to a number of studies of so-called double or co-doping in order to find the optimum cation substitution [38–44]. Despite some interesting work by Omar and colleagues [42–44], the overall improvement in conductivity is rather minimal.

Zirconia solid solutions behave in a similar manner. Although the minimum with aliovalent cation size is harder to observe experimentally, it has been shown to occur

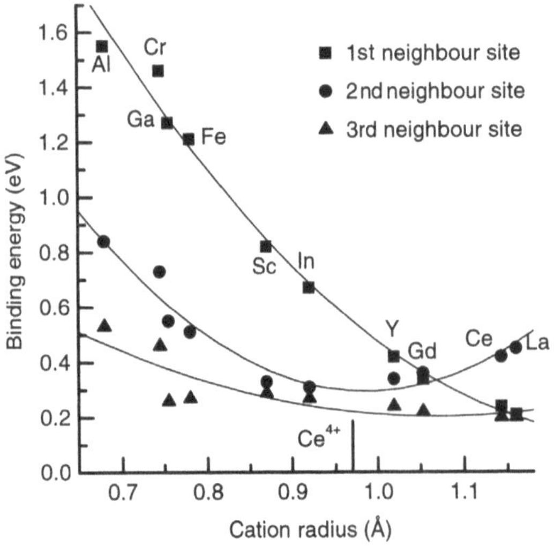

Figure 4.4 Binding energies of M^{3+} cations with an oxygen vacancy in ceria.
Reprinted from Ref. [32] with permission from Elsevier.

by atomistic simulation [45]. These observations have led to the hypothesis that this effect of the matching of the aliovalent cation size with that of the host is a universal criterion for optimising the ionic conductivity of oxide ion conductors where aliovalent substitution is needed. Although this is a tempting inference, calculations by Burbano et al. [46] have shown that defect interactions between the oxygen vacancies may also play a role in determining the conductivity maximum and that simply focusing on the cation sublattice may not describe all the phenomena determining the ionic conductivity.

4.2.1 Zirconia-based oxide ion conductors

Zirconia exhibits three polymorphs. It has a monoclinic structure at room temperature, changing to tetragonal above 1170 °C and to the cubic fluorite structure above 2370 °C. The addition of yttria can stabilise the fluorite and tetragonal phases down to room temperature. Yashima et al. [47] surveyed the phase diagrams of doped zirconia systems including those for the ZrO_2–Y_2O_3 (shown in Figure 4.5) and ZrO_2–Sc_2O_3 systems [48,49].

Yttria stabilised (YSZ), which has been used extensively in SOFCs, shows a conductivity of circa 0.14 S cm^{-1} at 1000 °C for 8 mol% of yttria addition. However,

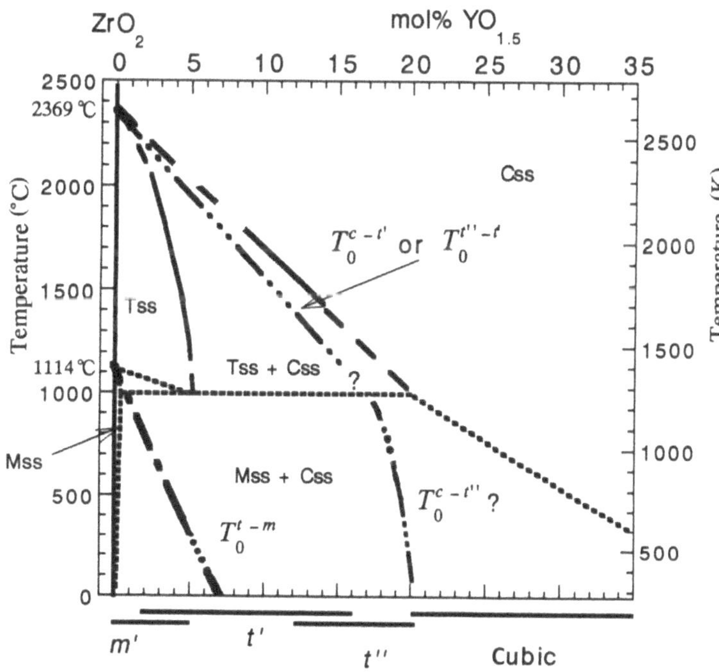

Figure 4.5 Phase diagram of the ZrO_2–Y_2O_3 system.
Reproduced from Ref. [47], copyright 1996, with permission from Elsevier.

with the move towards lower operating temperatures for SOFCs scandia-stabilised zirconia (ScSZ) has been widely investigated. Scandia–zirconia solid solutions are also attractive as electrolytes for SOFCs, especially for intermediate-temperature (600–800 °C) application. The phase diagram for this system was developed by Ruh et al. [49], who found that for significant scandia addition (>~10 mol%), the system was cubic at higher temperatures but became rhombohedral below ~600 °C. The rhombohedral phase has a lower conductivity than the cubic, which is evident in the temperature dependence of the conductivity for the zirconia with 10 mol% Scandia (10ScSZ) shown in Figure 4.6. ScSZ with lower scandia contents undergo ageing on long-term exposure to high temperatures. ZrO_2 with compositions with less than 10–11 mol% Sc_2O_3 exhibited a significant ageing effect with annealing at 1000 °C [51]. The conductivity of a solid solution containing 9 mol% Sc_2O_3 was around 0.28 S cm^{-1} at 950 °C in the as-sintered condition but decreased to 0.12 S cm^{-1} after ageing at 1000 °C for 1000 h. Note that the conductivity after ageing is comparable to that of ZrO_2 with 9 mol% Y_2O_3. On the other hand, ZrO_2 with 11 mol% Sc_2O_3 showed no ageing effect after annealing at 1000 °C for more than 1000 h.

Because of the good ionic conductivity of the zirconia–scandia system, particularly at the lower temperatures, there has been a resurgence of interest in this material,

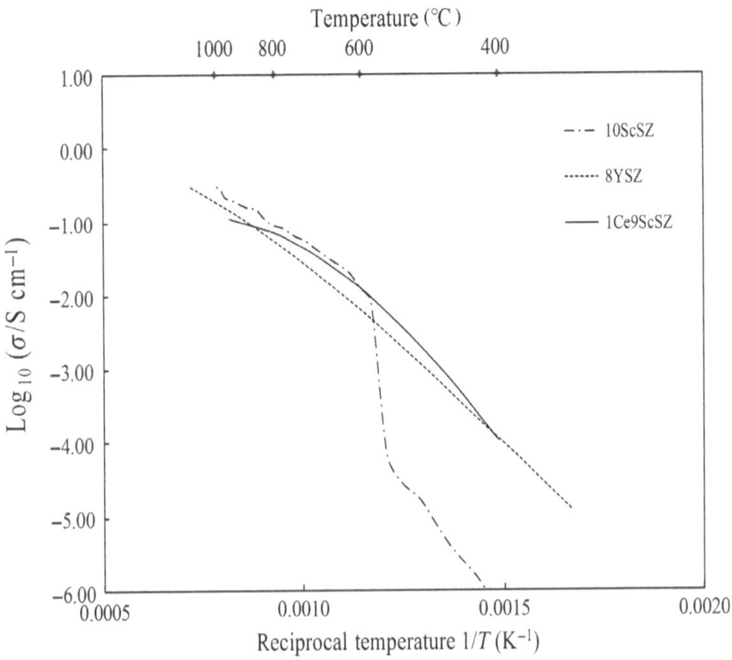

Figure 4.6 Arrhenius plot of the ionic conductivity of ZrO_2–Sc_2O_3 based solid solutions compared to 8 mol% YSZ [10]. Data for 10 mol% Sc_2O_3 (10ScSz) [16], 1 mol%CeO_2–9 mol% Sc_2O_3 ($1Ce_9ScSZ$) [50].

focusing on the effects of substituting other elements to stabilise the cubic phase to lower temperatures. The cubic phase can be stabilised at room temperature by the addition of a small amount of Al_2O_3 [51]. Lee and co-workers [52] have shown that co-substitution (co-doping) with scandia and ceria is beneficial and gives good stability and conductivity. In particular, 1 mol%CeO_2–10 mol%Sc_2O_3 gave high conductivity values with no discernable ageing. Other authors have tried different strategies including co-substitution with a range of trivalent cations [53] including yttria [54–56], ytterbia [57] and holmia [58]. The temperature dependence of the ionic conductivity of selected compositions of these zirconia-based solid solutions is shown earlier in Figure 4.6.

In addition to high ionic conductivity, high strength and toughness are also desirable properties for an electrolyte material, especially when it is employed as part of a planar cell configuration. The bending strengths of zirconia-based electrolytes along with their thermal expansion coefficients are also shown in Table 4.1. ScSZ shows a good set of mechanical properties similar to YSZ. ZrO_2 with 11 mol% Sc_2O_3 and 1 wt% Al_2O_3 appears as one of the best candidates for an intermediate-temperature SOFC because of its high oxide ion conductivity, phase stability and excellent mechanical properties.

4.2.2 *Ceria-based oxide ion conductors*

Electrolytes based on ceria (CeO_2) have been suggested as alternative electrolytes for intermediate-temperature and low-temperature SOFCs [59]. Reviews on the electrical conductivity and conduction mechanism in ceria-based electrolytes have been

Table 4.1 Electrical conductivity, bending strength, and thermal expansion coefficient of zirconia-based electrolytes

Electrolyte	Conductivity at 1000 °C S cm^{-1}		Bending strength (MPa)	Thermal expansion coefficient ($1\ K^{-1} \times 10^4$)
	As sintered	After annealing		
ZrO_2-3 mol%Y_2O_3	0.059	0.050	1200	10.8
ZrO_2-3 mol%Yb_2O_3	0.063	0.09		
ZrO_2-2.9 mol%Sc_2O_3	0.090	0.063		
ZrO_2-8 mol%Y_2O_3	0.13	0.09	230	10.5
ZrO_2-9 mol% Y_2O_3	0.13	0.12		
ZrO_2-8 mol%Yb_2O_3	0.20	0.15		
ZrO_2-10 mol%Yb_2O_3	0.15	0.15		
ZrO_2-8 mol%Sc_2O_3	0.30	0.12	270	10.7
ZrO_2-11 mol%Sc_2O_3	0.30	0.30	255	10.0
ZrO_2-11 mol%Sc_2O_3-1 wt% Al_2O_3	0.26	0.26	250	

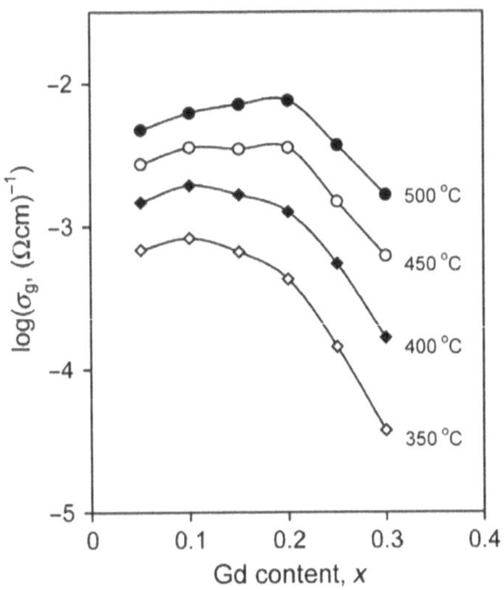

Figure 4.7 Bulk conductivity of the ceria gadolinia system ($Ce_{1-x}Gd_xO_{2-x/2}$). Reprinted from Ref. [62] with permission from Elsevier.

presented in Refs. [15,23,60,61]. Ceria possesses the same fluorite structure as stabilised zirconia. Mobile oxygen vacancies are introduced by substituting Ce^{4+} with trivalent rare earth ions such as Gd^{3+}, Sm^{3+} and La^{3+}, as shown in Equation (4.1). The conductivity of doped ceria systems depends on the kind of dopant and its concentration. A typical dopant concentration dependence of the bulk electrical conductivity in the CeO_2–Gd_2O_3 system as reported by Tianshu et al. [62] is shown in Figure 4.7.

The maximum conductivity is observed at around 10 mol% Gd_2O_3. The conductivity of the CeO_2–M_2O_3 system depends on the dopant ionic radius, mainly due to the dependence of the activation energy on both the size and concentration of the trivalent substitutional [63]. This is illustrated in Figure 4.8.

Clearly, the materials with the lowest activation energy for conduction will show some of the highest conductivity at low temperatures. As mentioned above, the calculations of Andersson et al. [33] instigated a series of investigations of double substitution in the ceria-based materials in the hope of finding a pair of substitutional ions that would average any ionic radius effects to yield an optimum effective radius, and hence improve the ionic conductivity [42–44]. This may not prove to be the case according to work by Kasse and Niño [38], who studied double substitution with both samaria and neodymia finding that the behaviour was intermediate between the two singly substituted materials. The conductivity of selected ceria-based materials at 500 and 700 °C are given in Table 4.2.

Ceria-based oxide ion conductors are reported to have purely ionic conductivity at high oxygen partial pressures and low temperatures. At the lower-oxygen partial pressures prevalent on the anode side of an SOFC, these materials become partially reduced. This leads to electronic conductivity in a large volume fraction of the electrolyte

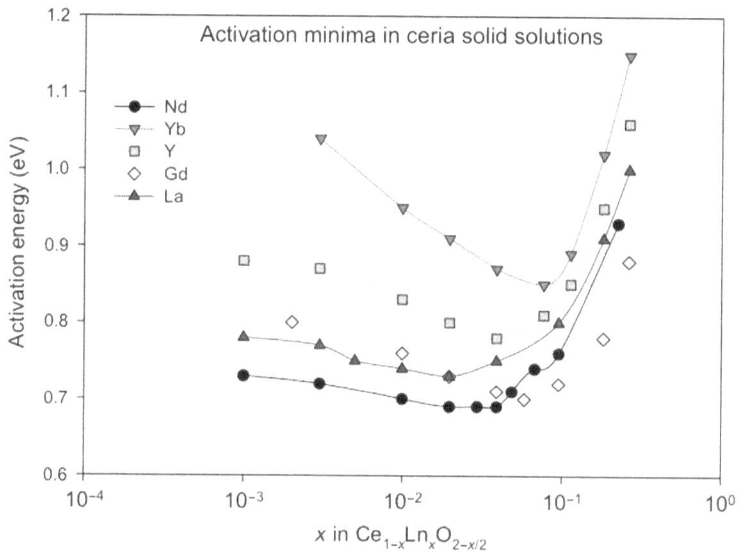

Figure 4.8 Minima in activation energy for ionic conduction in $Ce_{1-x}Ln_xO_{2-x/2}$ solid solutions (Ln = Nd, Yb, Y, Gd, La).
Data adapted from Faber et al. [64].

Table 4.2 Electrical conductivity data for CeO_2–M_2O_3

M_2O_3		Conductivity ($S\ cm^{-1}$)		Activation energy (kJ/mol)	References
	Mol %	700 °C	500 °C		
Sm_2O_3	10	3.5×10^{-2}	2.9×10^{-3}	68	[33]
	10	4.0×10^{-2}	5.0×10^{-3}	75	[31]
Gd_2O_3	10	3.6×10^{-2}	3.8×10^{-3}	70	[33]
Y_2O_3	10	1.0×10^{-2}	0.21×10^{-3}	95	[31]
CaO	5	2.0×10^{-2}	1.5×10^{-3}	80	[33]

extending from the anode side. When a cell is constructed with such an electrolyte showing electronic conduction, electronic current flows through the electrolyte even at open circuit, and the terminal voltage is somewhat lower than the theoretical value. In Figure 4.9, total electrical conductivity (ionic and electronic) of $Ce_{0.8}Sm_{0.2}O_{1.9-\delta}$ at different temperatures is shown as a function of oxygen partial pressure.

Godickemeier and Gauckler [66] analysed the efficiency of cells with $Ce_{0.8}Sm_{0.2}O_{1.9}$ by consideration of the electronic conduction. The maximum efficiency based on Gibbs free energy was 50% at 800 °C and 60% at 600 °C. Therefore, an SOFC with $Ce_{0.8}Sm_{0.2}O_2$ electrolyte should be operated at temperatures below about 600 °C to avoid such efficiency loss due to electronic leakage. For example, ceria-based electrolytes have been used in SOFCs operating at 550 °C and lower.

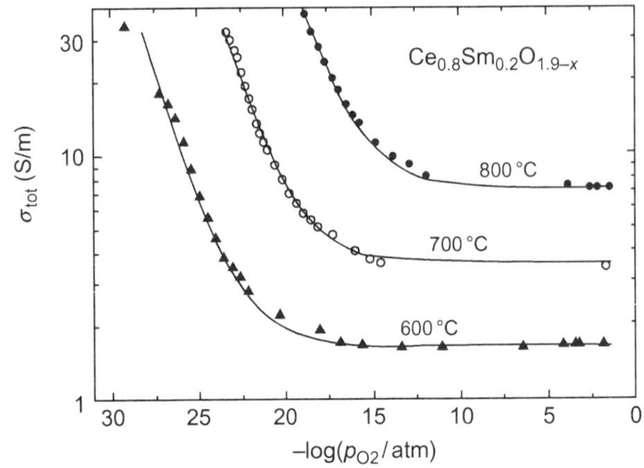

Figure 4.9 Total electrical conductivity of $Ce_{0.8}Sm_{0.2}O_{1.9-\delta}$ as a function of oxygen partial pressure at 600, 700 and 800 °C [65]. Reproduced by permission of The Electrochemical Society.

If higher temperature operation is required, then the electronic conduction can be prevented by protecting the ceria electrolyte with a thin coating of a zirconia on the anode side [65].

4.3 Perovskite and perovskite-related electrolytes

In addition to the fluorite-structured electrolytes, such as stabilised zirconia and ceria discussed above, which are the most commonly used oxide ion conductors, there are many non-fluorite-structured oxides which are potentially attractive for application as electrolytes for lower-temperature SOFCs. These include perovskites such as lanthanum gallate and, to a lesser degree, lanthanum titanate. Defect perovskite structures such as K_2NiF_4-type oxides (e.g. $LaSrGaO_4$) or Brownmillerite (e.g. $Ba_2In_2O_5$) are also reported to be promising oxide ion conductors. Other candidate materials with perovskite-related structures are the pyrochlores such as yttrium zirconate ($Y_2Zr_2O_7$) and gadolinium titanate ($Gd_2Ti_2O_7$) [67,68], but these are suitable only in very limited oxygen partial pressure ranges. Therefore, the main discussion here will focus on the perovskites and perovskite-related oxides. Section 4.4 will discuss some of the less widely used electrolyte materials based on neither the fluorite nor the perovskite structure.

The perovskite-based oxides, with the general formula ABO_3, comprise a rich family of compounds with important applications in SOFCs, ferroelectrics, superconducting materials and oxidation catalysis [69]. They find such broad applications because of the high stability of the crystal structure and variety of cations that can be accommodated within it; whilst the total charge of the A and B cations must be +6, this can be achieved by the combination of 1+5 (e.g. $K^{1+}Ta^{5+}O_3$), 2+4 (e.g. $Sr^{2+}Ti^{4+}O_3$) and 3+3 (e.g. $La^{3+}Al^{3+}O_3$), and also in more complex ways as in $Pb(B'_{1/2}B''_{1/2})O_3$, where B′ = Sc or Fe and B″ = Nb or Ta, or $A'_{1/2}A''_{1/2}TiO_3$ where A′ = Li, Na or A″ = La, Pr, etc. Many of these compounds display both ionic and electronic conductivity and so are useful as electrodes in SOFCs [70].

Only a few perovskites are predominantly ionic in their conduction behaviour. The oxide ionic transport properties of selected ionic-conducting perovskites are discussed below.

4.3.1 $LaAlO_3$

Since they first reported fast ion conductivity in Ti- and Al- based compositions, Takahashi and Iwahara have continued to measure oxide ion conductivity in a range of perovskite formulations as shown in Figure 4.10 [71]. From this figure, it is clear that Al- or Mg-doped $CaTiO_3$ exhibits the highest conductivity.

Figure 4.10b shows the transport number of oxide ions estimated from H_2–O_2 gas concentration cell experiments. Although a high transport number is obtained in both doped and undoped $CaTiO_3$ at intermediate temperatures between 500 and 800 °C in oxidising conditions (O_2 and air), Ca-doped $LaAlO_3$ is suggested as an attractive candidate SOFC electrolyte, because it displays a similarly high transport number, always greater than 0.9, in the same temperature range (500–800 °C) and gives no electronic conduction in reducing atmospheres.

Since the initial work of Takahashi and Iwahara, many other researchers have investigated the oxide ion conductivity in $LaAlO_3$-based materials. For example, Mizusaki et al. [72] reported the oxide ion conductivity and defect chemistry of $La_{1-x}Ca_xAlO_3$ single crystal samples with $x=0.0027$–0.008, prepared by a floating zone technique. They reported that the variation in oxygen content with pO_2 and temperature was

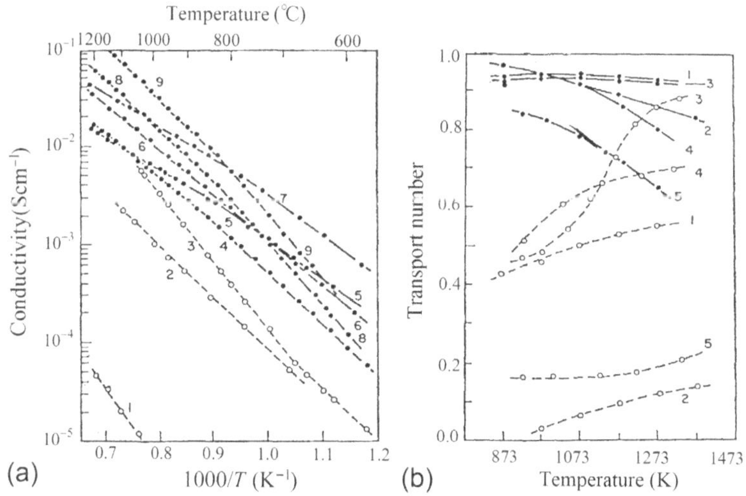

Figure 4.10 (a) Arrhenius plots of the oxide ion conductivity in perovskites in air and (b) transport number of oxide ion in perovskites (broken and solid lines indicate data obtained using O_2/air and O_2/humidified H_2 cells respectively). The labels correspond to the following compounds; (1) $LaAlO_3$, (2) $CaTiO_3$, (3) $SrTiO_3$, (4) $La_{0.7}Ca_{0.3}AlO_3$, (5) $La_{0.9}Ba_{0.1}AlO_3$, (6) $SrTi_{0.9}Al_{0.1}O_3$, (7) $CaTi_{0.95}Mg_{0.05}O_3$, (8) $CaTi_{0.5}Al_{0.5}O_3$, (9) $CaTi_{0.9}Al_{0.3}O_3$. Adapted from Ref. [71]with permission from Elsevier.

negligible, which suggests that oxygen vacancies are the major defects in this material, and electron holes the minor. The conductivity due to electron holes increases with $pO_2^{1/4}$ and the activation energy for migration of oxide ions is 0.74 ± 0.05 eV. Perovskite-structured $LaScO_3$ has also been reported to be an oxide ion conductor [73–75]; however, recently, this oxide is also reported as high-temperature proton conductor, as will be discussed in Section 4.3.4. Although the unit lattice volume of $LaScO_3$ is similar to that of $LaGaO_3$, hole conductivity is significant at high oxygen partial pressures. However, the oxide ion conductivity in all reported perovskite oxides is lower than that of YSZ, since the solid solubility of most dopants in these oxides is limited without co-substitution.

In ABO_3 perovskites, it was originally believed that the electric and dielectric properties were strongly dependent on the nature of the B-site cations. However, migrating oxide ions have to pass through a triangular space consisting of two large A and one small B-site cations in the crystal lattice, the so-called critical triangle [22]. Theoretical calculations now suggest that the enlargement in size of this triangular space is important for easing the migration of oxide ions in the crystal lattice [76]. From this, it follows that the ionic size of the A-site cation is also significant in determining the oxide ion conductivity. This effect is illustrated in Figure 4.11, which shows the effect of the atomic radius of the A-site cation on the oxide ion conductivity and its apparent activation energy in $Ln_{0.9}Ca_{0.1}AlO_3$ (Ln = Yb, Y, Gd, Sm, Na, Pr, La) [77]. It is clear that the ionic radius of the A-site cation, which dictates the size of the triangular space, has a strong effect on the oxide ion mobility. Oxide ion conductivity and activation energy increase and decrease, respectively, with increasing ionic radius, with $PrAlO_3$ exhibiting the highest conductivity among the Al-based perovskites. These results suggest that enlargement of the size of the triangular space in the lattice can be an effective means to enhance ionic conduction, and that higher oxide ion conductivity is to be expected in perovskites with larger unit lattice dimensions [78].

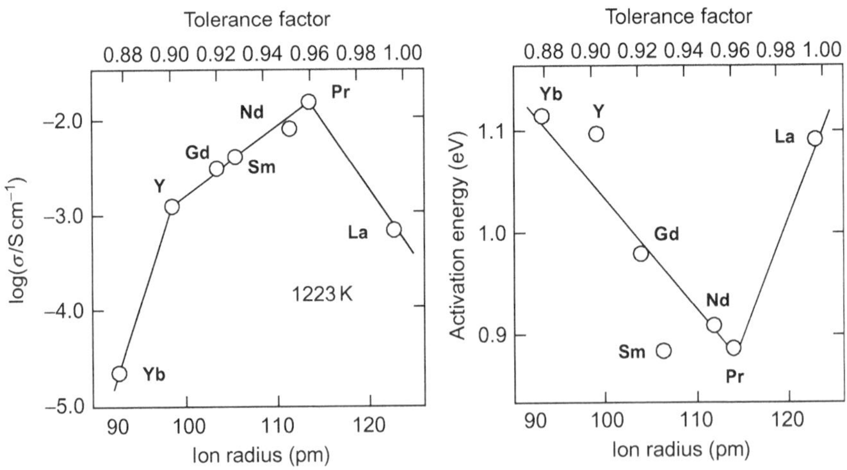

Figure 4.11 Oxide ion conductivity at 1223 K and activation energy for ionic conduction as a function of A- site cation ionic radius of Al-based perovskite oxide, $Ln_{0.9}Ca_{0.1}AlO_3$.

4.3.2 $LaGaO_3$-doped with Ca, Sr and Mg

Arrhenius plots of the electrical conductivity [79] of Ca-doped $LnGaO_3$ (Ln = La, Pr, Nd, Sm) are shown in Figure 4.12. The oxide ion conductivity strongly depends on the particular cation on the A-site, and increases in the order, Pr > La > Nd > Sm. The electrical conductivity of all Ga-based perovskites is almost independent of the oxygen partial pressure, indicating that oxide ion conduction is dominant in these materials.

Substitution of lower valence cations in place of La generally results in the formation of oxygen vacancies in order to maintain electrical neutrality, increasing the oxide ion conductivity, in a similar way to that shown for Y_2O_3 doping of fluorite ZrO_2 in Equation (4.1). This is illustrated in Figure 4.13 which shows the effect of partial substitution of La by various alkaline earth cations [80].

Electrical conductivity depends strongly on the particular alkaline earth cation doped onto the La site and increases in the order Sr > Ba > Ca. Therefore, strontium appears to be the most suitable dopant for $LaGaO_3$. Theoretically, increasing the amount of Sr increases the number of oxygen vacancies and the oxide ion conductivity should, therefore, increase. However, the solid solubility of Sr on the La site of $LaGaO_3$ is low, and $Sr_2Ga_2O_5$ or La_4SrO_7 secondary phases form when the Sr content is higher than 10 mol%. Thus there is only a limited gain in conductivity from increasing the dopant level.

Oxygen vacancies can also be formed by substitution of an aliovalent cation onto the Ga site in addition to the La site. Doping with Mg on the Ga site increases the ionic conductivity substantially, reaching a compositional maximum with 20 mol% Mg doping on the Ga site. The lattice parameter also increases by substitution of Mg for Ga, the ionic radius of Mg being larger than that of Ga. Although the solid solubility of Sr in $LaGaO_3$ is around 10 mol% in the absence of Mg, it can reach up to 20 mol% when Mg is also present. Such an increase in Sr solid solubility, brought

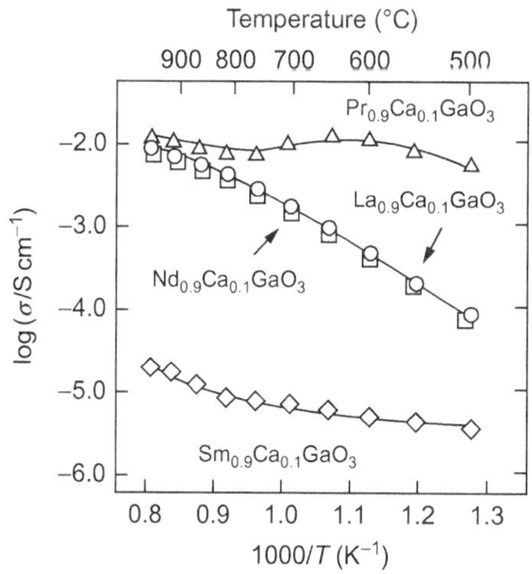

Figure 4.12 Arrhenius plots of the electrical conductivity of Ca-doped $LnGaO_3$ (Ln = La, Pr, Nd, Sm).

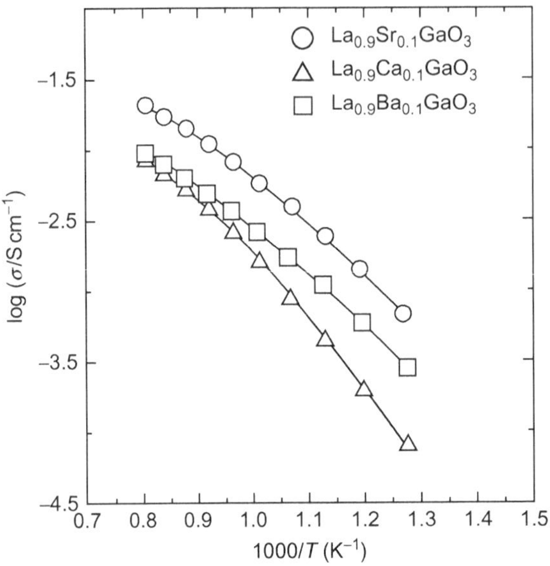

Figure 4.13 Effect of alkaline earth cation (M) in La site on electrical conductivity of $La_{0.9}M_{0.1}GaO_3$.
Reprinted with permission from Ref. [80]. Copyright 1994 American Chemical Society.

about by the enlarged crystal lattice, has also been reported by Majewski et al. [81]. It was shown in Ref. [82] that the highest oxide ion conductivity in these oxides could be obtained with the composition $La_{0.8}Sr_{0.2}Ga_{0.8}Mg_{0.2}O_3$ (LSGM).

Since the initial work [80], $LaGaO_3$-based electrolytes have been studied by various groups, examining the effects of substitution with a wide range of cation dopants [83]. Huang and Petric [84] reported the highest oxide ion conductivity at the composition $La_{0.8}Sr_{0.2}Ga_{0.85}Mg_{0.15}O_3$. On the other hand, Huang et al. [85] reported the highest conductivity of 0.166 S cm^{-1} at 800 °C for Sr = 0.2, Mg = 0.17. The composition with the highest ionic conductivity as found by the three groups [82,84,85] lies between $y = 0.15$ and 0.2 in $La_{0.8}Sr_{0.2}Ga_{1-y}Mg_yO_3$. Although the optimum composition is still controversial, the optimum composition found by the various workers is almost the same.

It was shown in Figure 4.1 above that, dependent upon temperature, the oxide ion conductivity of $LaGaO_3$ doubly doped with Sr and Mg is higher than that of typical YSZ- or ceria-based materials and somewhat lower than Bi_2O_3-based oxides. However, $LaGaO_3$-based electrolytes do not suffer from the electronic conduction and thermal instability problems from which Bi-based electrolytes do. Therefore, doubly doped $LaGaO_3$ formulations are very promising electrolytes for SOFCs in terms of ionic conductivity and stability.

However, to complicate matters, X-ray diffraction analysis of doubly doped $LaGaO_3$ always reveals the formation of a secondary phase. Although the crystal structure of such a secondary phase is not confirmed, it appears to be $LaSrGaO_4$ [85]. Some additional phases such as $LaSrGa_3O_7$ have also been reported. Majewski et al. [81] and Hrovat et al. [86] have investigated the phase diagram in the system La_2O_3–SrO–MgO–Ga_2O_3 at 1623–1673 K in air. A section of this for the plane

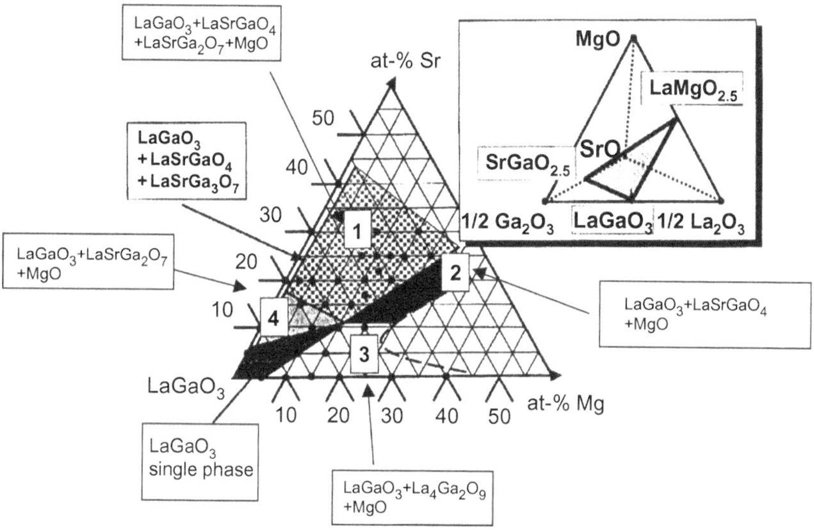

Figure 4.14 Phase diagram in the system La_2O_3–SrO–MgO–Ga_2O_3. Adapted from Ref. [81], copyright 2001, with permission from Elsevier.

$LaGaO_3$–$SrGaO_3$–$LaMgO_3$ is shown in Figure 4.14. The La_2O_3–SrO–MgO–Ga_2O_3 system exhibits complex phase relations. Compared to the ternary La_2O_3–SrO–Ga_2O_3 system, the combined solid solubility of Sr and Mg in $LaGaO_3$ is significantly enhanced in the quaternary La_2O_3–SrO–MgO–Ga_2O_3 system. In order to obtain single-phase LSGM, great attention is required in its preparation. Huang et al. have investigated the synthesis of $La_{0.9}Sr_{0.1}Ga_{0.8}Mg_{0.2}O_3$ by the alkoxide method [87] and reported that almost single-phase LSGM can be obtained at a temperature as low as 1270 °C by this approach.

Two-dimensional oxide ion conductivity has also been reported in melilite-structured $LaSrGa_3O_7$ [88]. As shown in Figure 4.14, this melilite oxide is often observed as a secondary phase during $LaGaO_3$ synthesis. However, the interstitial oxygen introduced to charge compensate excess La in this phase shows high mobility. Kuang et al. [88] reported that $La_{1.54}Sr_{0.46}Ga_3O_7$ shows high oxide ion conductivity over a wide pO_2 range. The oxide ion conductivity of this melilite oxide is 2×10^{-2} S cm^{-1} at 873 K, which is higher than that of YSZ and close to LSGM, and it is also stable in the pO_2 range from 1 to 10^{-23} atm. The estimated oxide ion transport number is higher than 0.9, and so, this oxide may also be used as the electrolyte in an SOFC [89]. Tealdi et al. also simulated the oxide ion conduction pathway of this melilite oxide and found that oxygen diffusion occurred through an interstitial or cooperative-type mechanism involving the concerted knock-on motion of interstitial and lattice oxygen [90]. Oxide ion diffusion in this oxide is highly anisotropic and restricted to within the layers of corner sharing GaO_4 tetrahedra. Facile rotation of the GaO_5 units facilitates the migration of oxygen interstitials to adjacent pentagonal rings. Although the crystal structure of this oxide is not itself perovskite, the high oxide ion conductivity of melilite is very interesting.

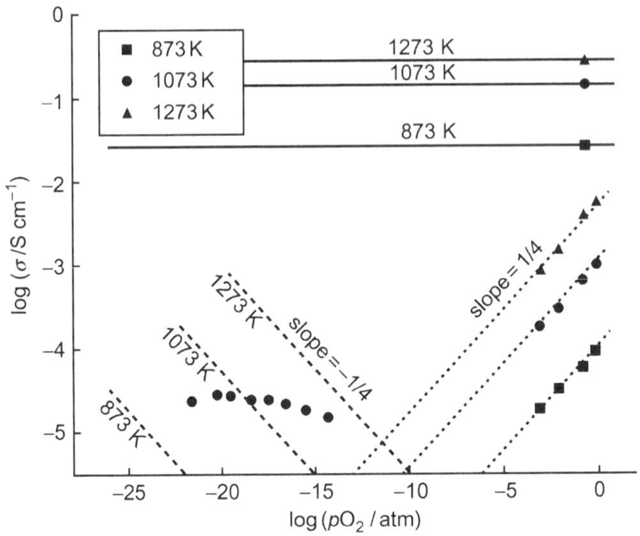

Figure 4.15 Estimated hole and electron conductivity (dotted and dashed lines, respectively) and oxide ion conduction (solid lines) in LSGM as a function of oxygen partial pressure.

Measurements of the conductivity due to hole and electron minority carriers and oxide ion transport numbers in $LaGaO_3$-based oxides have been reported by Baker et al. [91], Yamaji et al. [92] and Kim et al. [93] based on the polarisation method. Figure 4.15 shows the estimated hole and electron conductivity in LSGM as a function of oxygen partial pressure [91]. It is seen that pO_2 dependences of hole and electron conductivities are $pO_2^{1/4}$ and $pO_2^{-1/4}$, respectively, obeying the Hebb–Wagner theory. These results show that $LaGaO_3$ exhibits almost pure oxide ion conduction over a wide range of oxygen partial pressures ($10^5 > pO_2 > 10^{-25}$ atm).

Kim et al. [93] also investigated the temperature dependence of hole and electronic conductivity in $La_{0.9}Sr_{0.1}Ga_{0.8}Mg_{0.2}O_3$ using the polarisation method. Figure 4.16 compares the evaluated boundaries of the electrolytic domain of $La_{0.9}Sr_{0.1}Ga_{0.8}Mg_{0.2}O_3$ in the plane of $\log(pO_2/\text{atm})$ versus reciprocal temperature with data for other oxide ion conductors. The lower boundary of the electrolytic domain (defined as $t_{ion} > 0.99$) for LSGM is 10^{-23} atm at 1000 °C, further indicating its suitability as a SOFC electrolyte. Indeed, the stability of LSGM under the reducing conditions of the SOFC anode is much higher than that of YSZ.

Yokokawa et al. [94] estimated the electrolyte efficiency of LSGM when used as the electrolyte in an SOFC (Figure 4.17). Electrolyte efficiency was given by a function of fuel utilisation and internal resistance [92]. When the thickness of the electrolyte was too low, chemical leakage of oxygen due to electronic conduction became significant, and the extra fuel consumption resulted in decreased electrolyte efficiency. On the other hand, with increasing electrolyte thickness, the internal resistance of the cell increased, also resulting in decreased electrolyte efficiency. Consequently, an optimum thickness exists for each electrolyte material at a given temperature and

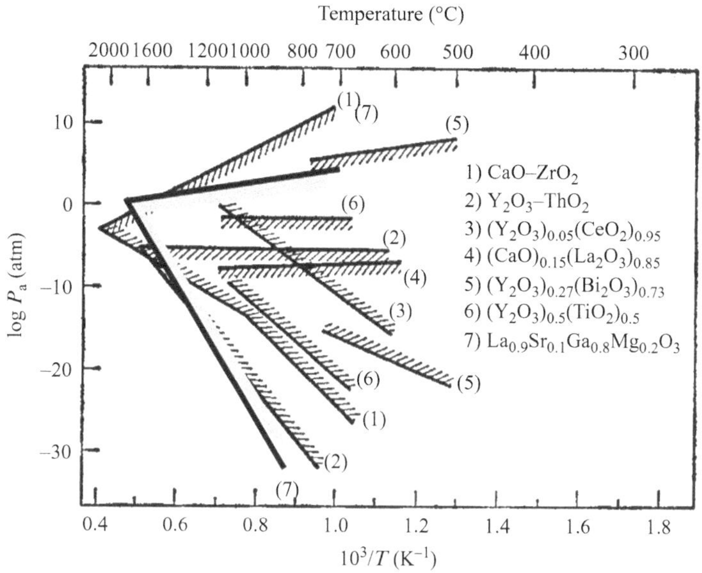

Figure 4.16 Comparison of evaluated electrolytic domain ($t_i > 0.99$) of LSGM [93] with other conductors by polarisation method.

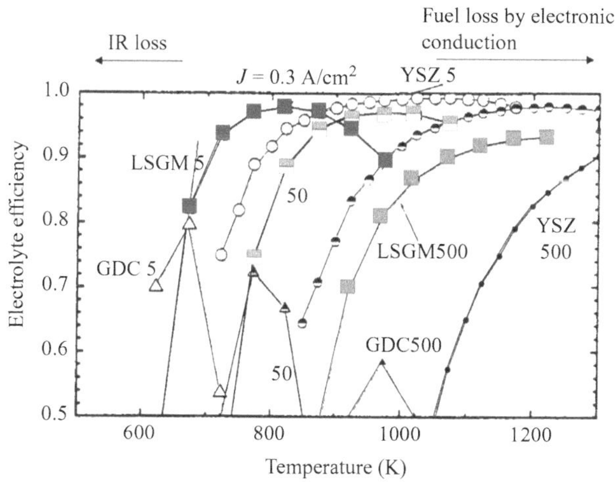

Figure 4.17 Gibbs-energy-based conversion loss occurring in electrolyte due to Joule effect and shorting effects as a function of temperature for a given current density (0.3 Acm^{-2}). Numbers after composition indicate electrolyte thickness, e.g. LSGM 5 = 5 μm thick LSGM electrolyte.

Adapted from Ref. [94] with permission.

current density. For example, YSZ operating at 700 °C has an optimum thickness of about 10 μm. The benefit of LSGM is that its highest efficiency with a thickness near 5 μm is achieved around 450 °C. Thus LSGM appears to be the best electrolyte discovered so far for low-temperature operation.

Several groups have investigated the electrochemical performance of cells with LSGM electrolytes [95–99]. The observed open circuit voltage (OCV) tends to increase with decreasing temperature and is in good agreement with theoretical values estimated from the Nernst equation. The maximum power densities (MPDs) were greater than 1.0 Wcm^{-2} at 1000 °C and about 0.1 Wcm^{-2} at 600 °C for a cell with a 0.5 mm thick LSGM electrolyte, a Ni anode and an $Sm_{0.5}Sr_{0.5}CoO_3$ cathode [100]. In other investigations using $LaGaO_3$-based electrolytes [95,96], similar large power densities have been reported at intermediate temperatures with an $La_{0.6}Sr_{0.4}CoO_3$ cathode and Ni–(Ce, La)O_2 cermet anode.

On the other hand, the fabrication of thin films of LSGM by both wet chemical and by "dry" (i.e. physical) processing methods is also being extensively studied. Since LSGM tends to react with NiO, preventing this reaction is one of the key issues for producing thin films [101]. In comparison to conventional wet processing methods, physical vapour deposition approaches can be used at lower temperatures and so LSGM films can be formed onto NiO substrates without reaction by means of pulsed laser deposition (PLD). Ju et al. reported preparation of a 5 μm LSGM film on a metallic Ni–Fe anode substrate [102,103]. The cell showed an OCV close to the theoretical value, and reasonably large power densities at low temperature (ca. 2 Wcm^{-2} at 700 °C and 0.15 Wcm^{-2} at 400 °C, Figure 4.18). Therefore, as suggested by Yokokawa et al. (Figure 4.17) [94], LSGM film electrolytes with a few micrometres in thickness are highly promising for SOFCs operating at low temperatures with high efficiencies.

The reactivity of $LaGaO_3$ with common electrode materials has also been investigated [95,104]. Platinum seems to react readily with gallium oxide to reduce Ga^{3+} to volatile Ga^{+}. It is also reported that nickel from Ni-based anodes can easily diffuse into

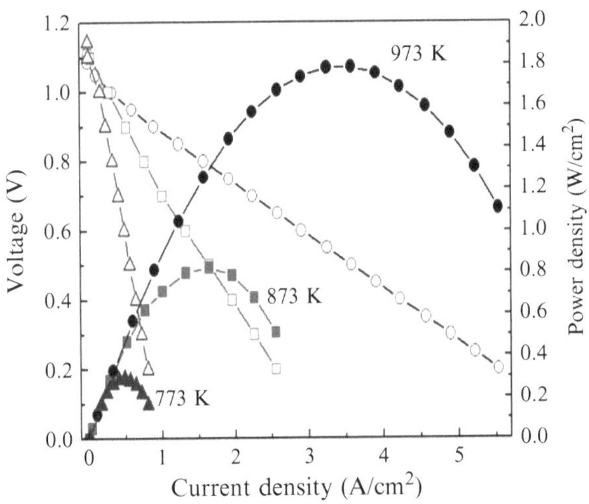

Figure 4.18 Power generation properties of LSGM/SDC bi-layer film on porous Ni–Fe bimetallic substrate reduced at 973 K. Reproduced from Ref. [102] with permission of The Royal Society of Chemistry.

LSGM, resulting in partial electronic conductivity in the LSGM electrolyte. Therefore, effective prevention of this diffusion is necessary when making LSGM film by conventional wet methods such as screen printing. Huang et al. [105] reported that $Ce_{0.6}La_{0.4}O_2$ (LDC) makes an effective barrier to prevent the diffusion of Ni into LSGM. However, because of the low electrical conductivity and poor sinterability of LDC, the internal resistance of LSGM thin film-based cells prepared by co-sintering with LDC is usually high despite the low film thickness. Recently, it has been reported that electrical conductivity as well as sinterability of LDC can be much increased by the addition of Ti or Co because of improved mass transport [106]. The MPD of a cell using a 50 μm thick LSGM electrolyte, which was prepared by screen printing followed by co-sintering at 1500 °C, was 1.2 Wcm^{-2} at 700 °C. Therefore, for the preparation of LSGM film by conventional sintering methods, the use of a buffer layer is very important.

Figure 4.19 shows a schematic of an LSGM microtubular cell developed by TOTO Co. Ltd. for application in a residential CHP system [107]. This cell uses LDC as the buffer layer and was prepared by a dip coating and co-sintering method. The figure

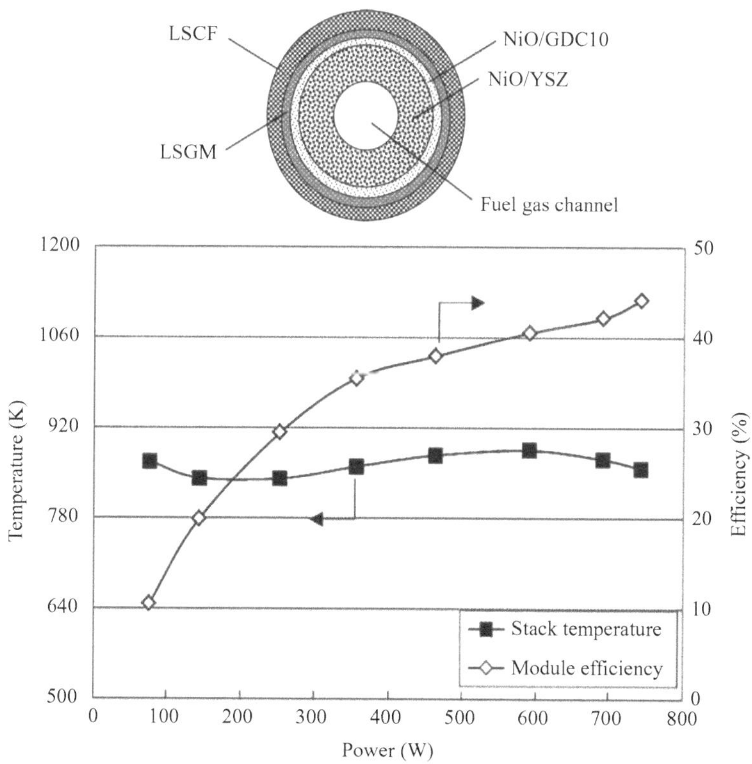

Figure 4.19 Micro-tubular type SOFC and operating temperature energy conversion efficiency of 1 kW unit consisting of tubular cells using LSGM thin film electrolyte.
Adapted from Ref. [107] copyright 2012, with permission from Elsevier.

also shows the stack temperature and efficiency of the developed module using an LSGM electrolyte. The cell was operated around 600 °C, and the efficiency increased with increasing power drawn from the module. A conversion efficiency higher than 50% was achieved when the module power became higher than the designed power of 700 W. The power density of the developed module was stable over 5000 h, demonstrating that LSGM is also a promising electrolyte for low-temperature SOFC application (<600 °C).

The diffusivity of oxide ions in LSGM has been studied through ^{18}O tracer diffusion measurements [108,109]. LSGM exhibits a high oxygen self-diffusion coefficient because of the higher mobility of oxide ions compared to that in the fluorite-structured oxides (Table 4.3). The perovskite structure has a large free lattice volume, which allows a high diffusivity of oxide ions and thus high ionic conductivity.

The transport properties of LSGM may also be modified by doping in several ways. Larger effects were observed when transition metals such as Co were introduced at levels below 10 mol% [110]. Baker et al. [91] investigated the effects of Cr and Fe on the oxide ion conductivity of $LaGaO_3$; doping Cr or Fe on the Ga site induced hole conduction in the $LaGaO_3$, resulting in decreased stability against reduction. On the other hand, Ishihara et al. [110] found that doping with small amounts of transition metals, particularly Co or Ni, increased the oxide ion conductivity in LSGM.

Figure 4.20 shows an Arrhenius plot of electrical conductivity for $LaGaO_3$ doped with various transition metal cations on the Ga site. The conductivity increases with Co, Ni and Fe doping and decreases by doping with Cu and Mn. Doping with Mn and Ni greatly enhances n-type conductivity, whilst p-type conduction is increased by doping with Cu. The effects of substitution of other transition metals for Ga have also been reported [111,112], but Co appears to be the most interesting dopant to date for SOFC electrolytes.

4.3.3 $ATiO_3$-based perovskite (A = alkaline or alkaline earth)

As mentioned above, $CaTiO_3$ doped with Mg or Al shows reasonable oxide ion conductivity [71]. However, because of the low solubility of dopants in the $CaTiO_3$ host lattice, the oxide ion conductivity in this oxide is lower than that of YSZ. In addition, n-type electronic conduction is observed at low oxygen partial pressures. Recently, it has been reported that $Na(Bi)TiO_3$ (NBT) shows reasonably high oxide ion conductivity [11]. Figure 4.1 showed a comparison of the oxide ion conductivity of this oxide with those of other more commonly used oxide ion electrolytes. The highest oxide ion conductivity was achieved in $Na_{0.5}Bi_{0.49}Ti_{0.98}Mg_{0.02}O_3$, which has a conductivity equal to that of YSZ at 600 °C ($\sigma = 0.01\ S\ cm^{-1}$ at 600 °C), and superior at temperatures below this. However, the pO_2 region in which electrolytic conduction dominates is limited for this oxide and so in terms of potential application as an SOFC electrolyte, NBT has similar limitations to doped Bi_2O_3.

Table 4.3 Comparison of mobility of oxide ions in selected fluorite and LSGM oxides at 1073 K. D_T, tracer diffusion coefficient. D_{self}, self-diffusion coefficient [109]

Composition	D_T ($cm^2\ s^{-1}$)	E_a (eV)	δ	$[V_O^{\bullet\bullet}]$ (cm^{-3})	D_{self} ($cm^2\ s^{-1}$)	μ ($cm^2\ V^{-1}\ s^{-1}$)
$Zr_{0.81}Y_{0.19}O_{2-\delta}$	6.2×10^{-8}	1.0	0.10	2.95×10^{21}	1.31×10^{-6}	1.41×10^{-5}
$Zr_{0.858}Ca_{0.142}O_{2-\delta}$	7.54×10^{-9}	1.53	0.142	4.19×10^{21}	1.06×10^{-7}	1.15×10^{-6}
$Zr_{0.85}Ca_{0.15}O_{2-\delta}$	1.87×10^{-8}	1.22	0.15	4.43×10^{21}	2.49×10^{-7}	2.69×10^{-6}
$Ce_{0.9}Gd_{0.1}O_{2-\delta}$	2.7×10^{-8}	0.9	0.05	1.26×10^{21}	1.08×10^{-6}	1.17×10^{-5}
$La_{0.9}Sr_{0.1}Ga_{0.8}Mg_{0.2}O_{3-\delta}$	3.24×10^{-7}	0.74	0.15	2.53×10^{21}	6.4×10^{-6}	6.93×10^{-5}
$La_{0.8}Sr_{0.2}Ga_{0.8}Mg_{0.2}O_{3-\delta}$	4.13×10^{-7}	0.63	0.20	3.34×10^{21}	6.12×10^{-6}	6.62×10^{-5}
$La_{0.8}Sr_{0.2}Ga_{0.8}Mg_{0.125}Co_{0.085}O_{3-\delta}$	4.50×10^{-7}	0.42	0.165	2.78×10^{21}	8.21×10^{-6}	8.89×10^{-5}

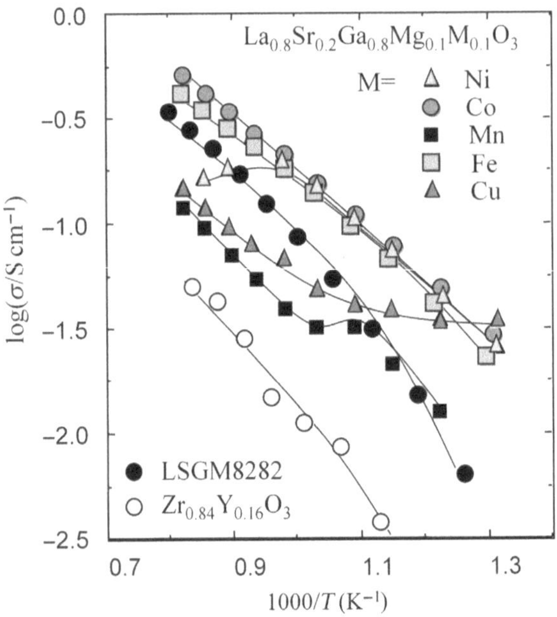

Figure 4.20 Arrhenius plots of electrical conductivity of $LaGaO_3$-based oxides doped with various transition metal cations on the Ga site.

4.3.4 High-temperature proton conducting perovskites

Some groups of perovskite oxide show high-temperature proton conductivity; typical examples are $BaZrO_3$ and $BaCeO_3$ doped with lower valence cations on the B-site. In this type of oxide proton conductor, interstitial protons are introduced in accordance with reaction (4.8).

$$V_O^{\bullet\bullet} + H_2O \leftrightarrow O_O^x + 2H_i^{\bullet} \tag{4.8}$$

Therefore, the degree of proton incorporation is related to the oxygen vacancy concentration. Figure 4.21 shows a comparison of proton conductivity for the perovskite oxides [113]. Compared to the low-temperature proton conductors (e.g. polymeric materials), perovskite oxide proton conductors only show high proton conductivities at higher temperatures (>400 °C). Since the activation energy for migration is lower for protons than for oxide ions, higher conductivities are achieved in the proton conducting perovskites than in the oxide ion conducting perovskites, and so it is expected that the application of proton conducting perovskites and electrolytes for low temperature (proton conducting) SOFCs is promising, although the fuel is limited to hydrogen in these cases. On the other hand, $ACeO_3$ and $AZrO_3$ (A = Ba or Sr) are not very stable, particularly against CO_2 and H_2O, and so phase separation easily occurs [114]. In contrast, $BaCe(Zr)O_3$ shows much higher stability, and $BaZr_{0.4}Ce_{0.4}Sc_{0.2}O_3$ is reported to be the optimal composition from the point of view of a balance between chemical stability and proton conductivity (see Figure 4.22) [115].

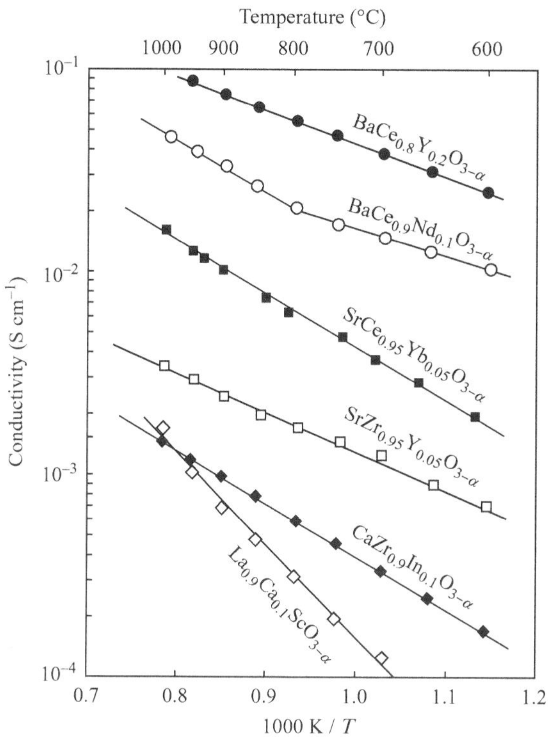

Figure 4.21 Comparison of proton conduction in perovskite oxides.

4.3.5 Oxides with perovskite-related structures: Brownmillerites (e.g. $Ba_2In_2O_5$)

Another perovskite-related structure which is interesting from the viewpoint of oxide ion conduction is brownmillerite, with a general formula of $A_2B'B''O_5$ or $A_2B_2O_5$. This structure can be viewed as a perovskite with oxygen vacancies ordered along the [101] direction in alternate layers. Such vacancy ordering results in an enlarged unit cell relative to the perovskite. In other words, the lattice parameters of the *a* and *c* axes of the ideal brownmillerite oxide are larger than those of the ideal perovskite oxide by a factor of $\sqrt{2}$, whilst the *b*-axis of brownmillerite is the same as that of perovskite. In some cases, the oxygen vacancies do not order, which results in a perovskite structure with a statistical distribution of oxygen vacancies on the oxygen sites. Therefore, high oxide ion conductivity is also expected in the brownmillerite-type oxides.

Goodenough et al. [116] have reported high oxide ion conductivity in several brownmillerite oxides. A listing is given in Table 4.4 [117]. All these brownmillerites exhibit oxide ion conductivities which are rather high compared with those of fluorite-structured oxides. Among these, $Ba_2In_2O_5$ is attracting considerable interest as a potential SOFC electrolyte material.

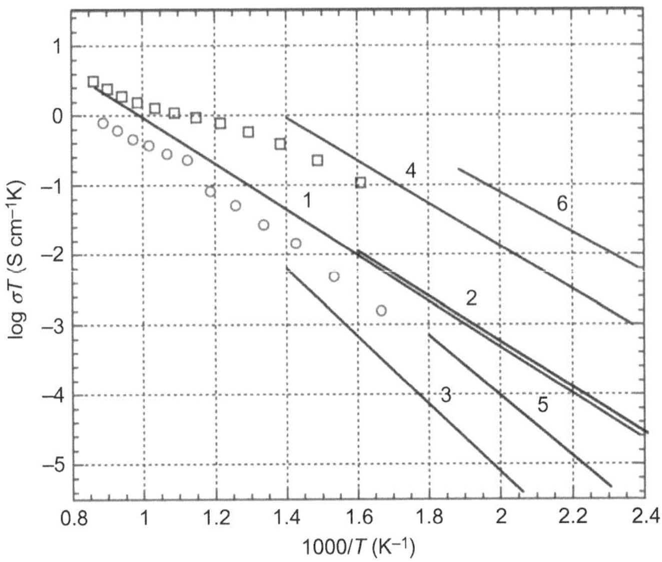

Figure 4.22 Comparison of conductivities of $Ba(Ce,Zr)_{1-x}Sc_xO_{3-\delta}$ ($x=0.1, 0.2$) with reported barium cerate and zirconate data: (1) $BaCe_{0.9}Y_{0.1}O_{2.95}$ σ_{total} in wet 5% H_2; (2) $BaCe_{0.7}Zr_{0.2}Nd_{0.1}O_{2.95}$ σ_{bulk} in wet Ar; (3) $BaCe_{0.7}Zr_{0.2}Nd_{0.1}O_{2.95}$ σ_{gb} in wet Ar; (4) $BaCe_{0.9}Gd_{0.1}O_{2.95}$ σ_{bulk}; (5) $BaCe_{0.9}Gd_{0.1}O_{2.95}$ σ_{gb}; (6) $BaCe_{0.9}Y_{0.1}O_{2.95}$ σ_{total}; (○) $BaCe_{0.45}Zr_{0.45}Sc_{0.1}O_{3-\delta}$ σ_{total} in wet 5% H_2 and (□) $BaCe_{0.4}Zr_{0.4}Sc_{0.2}O_{3-\delta}$ σ_{total} in wet 5% H_2. Reproduced from Ref. [115], copyright 2007, with permission from Elsevier.

Figure 4.23 shows an Arrhenius plot comparing oxide ion conductivity in $Ba_2In_2O_5$ and Ce-doped $Ba_2In_2O_5$ in reducing atmospheres. The pO_2 dependence of $Ba_2In_2O_5$ suggests that hole conduction is dominant in this oxide at pO_2 higher than 10^{-3} atm, whereas oxide ion conductivity becomes dominant with decreasing pO_2. $Ba_2In_2O_5$ shows a large jump in conductivity around 900 °C, which can be explained by an order/disorder transition of the arrangement of oxygen vacancies, namely, from oxygen vacancy-ordered brownmillerite to oxygen vacancy-disordered pseudo-perovskite. A similar jump has been observed in Bi_2O_3 electrolytes [13,118–120]. The oxide ion conductivity of $Ba_2In_2O_5$ is comparable to that of YSZ at temperatures above 900 °C.

In an attempt to eliminate this order/disorder transition and to stabilise the structure down to lower temperatures, the effects of dopants on $Ba_2In_2O_5$ have been studied by several groups. For example, Figure 4.23 shows the temperature dependence of the conductivity of Ce-doped $Ba_2In_2O_5$. The large jump around 900 °C was suppressed by doping with Ce, but the conductivity at high temperature was reduced. Such efforts to stabilise the disordered phase at lower temperatures by selective doping of A- and/or B-sites have met with some success and rather high oxide ion conductivity has been observed in some phases as shown by Kendall et al. [117]. For $Ba_2In_2O_5$, La doping on the Ba site is the most promising for increasing oxide ion conductivity. The temperature dependence of the ionic conductivity in $(Ba_{1-x}La_x)_2In_2O_5$ indicates an effect similar to Ce doping, the jump in the Arrhenius plots shifting to lower temperature with increasing amount of La and disappearing around $x=0.2$ [121]. Since the amount

Table 4.4 Oxide ion conductivity for selected brownmillerite compounds [117]

Compound	*T* (K)	σ (S cm^{-1})
$Ba_2In_2O_5$	973	5×10^{-3}
	1223	1×10^{-1}
$BaZrO_3$	973	1×10^{-6}
$BaZr_{0.5}In_{0.5}O_{2.75}$	973	1×10^{-2}
$Ba_3In_2ZrO_8$	973	5×10^{-3}
$Ba_3In_{1.7}Zr_{1.3}O_{8.15}$	973	5×10^{-2}
$Ba_2In_{1.33}Zr_{0.67}O_{5.33}$	973	1×10^{-3}
$Ba_2In_{1.75}Ce_{0.25}O_{5.125}$	973	9×10^{-3}
	1223	6×10^{-2}
$Ba_3In_2TiO_8$	973	7×10^{-4}
$Ba_3In_2ZrO_8$	673	6.8×10^{-3}
$Ba_3In_2CeO_8$	673	1.5×10^{-3}
$Ba_3In_2HfO_8$	673	1.0×10^{-3}
$Sr_3In_2HfO_8$	973	1×10^{-4}
$Ba_3Sc_2ZrO_8$	973	7×10^{-3}
$Ba_2GdIn_{0.8}Ga_{0.2}O_5$	873	5×10^{-3}
$Ba_2GdIn_{0.6}Ga_{0.4}O_5$	873	5×10^{-3}
$Ca_2Cr_2O_5$	973	5×10^{-3}
Sr_2ScAlO_5	973	1×10^{-5}
$Sr_2Sc_{1.3}Al_{0.7}O_5$	973	1×10^{-3}
$Sr_2ScAl_{0.8}Mg_{0.2}O_{4.9}$	973	5×10^{-4}
$Ar_2ScAl_{0.8}Zn_{0.2}O_{4.9}$	973	2×10^{-4}
$Sr_2Sc_{0.8}Y_{0.2}AlO_5$	973	1×10^{-4}
$Sr_{1.8}Ba_{0.2}ScAlO_5$	973	1×10^{-4}

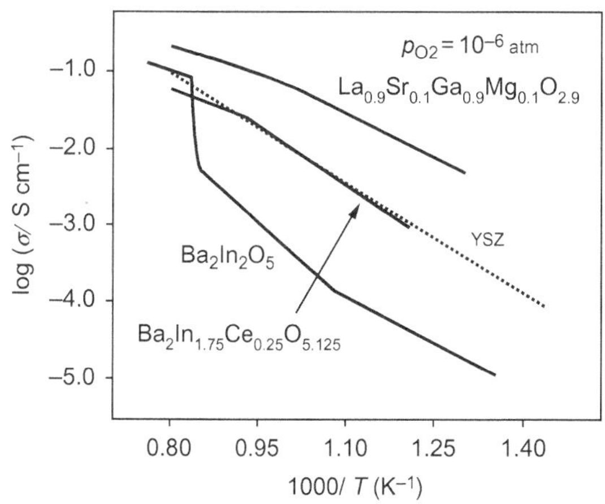

Figure 4.23 Arrhenius plots of the oxide ion conductivity in $Ba_2In_2O_5$ and Ce-doped $Ba_2In_2O_5$. Adapted in part from Ref. [116].

of lattice oxygen decreases with increasing La content, the oxygen-disordered structure may be stabilised by doping with La. The electrical conductivity at high temperatures further increases by doping with La on the Ba site. The highest oxygen ion conductivity is achieved at $x=0.6$ in this system, and this conductivity is higher than that of 8 mol% Y_2O_3-stabilised ZrO_2.

The simulations of Islam et al. [122] on the defect energies of $Ba_2In_2O_5$ by atomistic modelling techniques show that oxygen Frenkel pairs (vacancies and interstitials) are the dominant intrinsic defects in the low-temperature phase, with a preference for oxide ion diffusion via these defects in the [001] direction. $Ba_2In_2O_5$ is also predicted to be more stable against reduction than against oxidation, with reduction involving removal of oxygen from octahedral layers and oxidation involving insertion of oxide ions into the interstitial sites. On the other hand, simulations also suggest that this oxide may exhibit proton conductivity. In fact, Zhang et al. [123] and Schober et al. [124] reported that $Ba_2In_2O_5$ exhibits proton conduction at temperatures lower than 400 °C. Protonic conduction in the crystal phase of $Ba_2In_2O_5 \cdot H_2O$ has been confirmed by several groups.

Another interesting oxygen-deficient perovskite-related material is $LaSrGaO_4$, which may be viewed as a series connection of $LaGaO_3$ perovskite blocks connected with SrO rock salt blocks [125]. Reasonable oxide ion conductivity is reported for Mg-doped $LaSrGaO_4$, with a conductivity of $\sigma = 10^{-3}$ S cm^{-1} at 750 °C. Given the promising results to date, and the possibilities of substitution to tailor their properties, it is expected that further compositions in the rich family of perovskite and perovskite-related oxides will be found to show superior oxide ion conductivity in the next several years.

4.4 Alternative-structured electrolyte materials

The most commonly applied electrolytes for SOFCs typically take either the fluorite (e.g. YSZ and substituted-CeO_2) or the perovskite (e.g. LSGM) crystal structures. However, materials based on other (often non-cubic) structures can offer electrical transport properties (high ionic and low electronic conductivities) competitive with more “conventional” materials, making them candidates for application as SOFC electrolytes. For comparison, some of these were shown in Figure 4.1.

These types of materials are attractive for two reasons; firstly, as the more conventional electrolytes have been studied for a long time, their performance may well be close to optimal. The discovery of new structures showing oxide ion conductivity could yield a leap forward in terms of achievable ionic conductivity. The second aspect is that from a scientific perspective, these materials often show unique transport properties (e.g. anisotropic interstitial conduction in apatite silicates), which are interesting from a fundamental point of view.

The layered Aurivillius-type $Bi_2Me_xV_{1-x}O_{5.5-3x/2-\delta}$ (“BIMEVOX”, where Me is a metal cation such as Cu, Ni, Co or Mg) materials had previously attracted much interest. This, however, has been declining in recent years because of concerns about

Bi volatilisation at intermediate temperatures, as well as their thermodynamic instability in the reducing atmospheres to which they are exposed during SOFC operation.

In this section, we give an overview of two of the more commonly studied materials families, namely, the lanthanum apatite silicates and LAMOX compounds, with particular regard to their application in SOFCs. The interested reader is referred to recent reviews for a more detailed discussion of the materials chemistry behind these potential electrolytes [8,126,127].

4.4.1 Lanthanum silicate apatite-based electrolytes

Lanthanum silicates with the apatite structure were first reported by Nakayama in the mid 1990s [128–130]. They show oxide ionic conductivity comparable to YSZ at high temperatures (see Figure 4.1), but the lower activation energy for ionic conduction in the apatite phases makes them superior by around half an order of magnitude as the temperature is decreased. In the interests of space, this section will provide only a brief summary of the present state-of-the art; the interested reader is referred to recent reviews by Kendrick [126], Orera and Slater [127] and Malavasi et al. [8].

These materials have the general formula $A_{10}M_6O_{26\pm y}$, where A is a rare or alkaline earth cation and M is Si, Ge or P. The phosphate analogues do not show appreciable ionic conductivity [126,131,132], and whilst the germanate analogues show excellent ionic conductivities (e.g. $\sigma_i = 0.14$ S cm^{-1} at 950 °C for $La_9SrGe_6O_{26.5}$ [133]), the tendency for Ge to volatilise or form glasses, and the high cost of GeO_2, mean they are not as attractive [126,134]. In contrast, the silicate apatites are attractive because the ready availability of silicon renders them inexpensive; these are in fact the most studied of this family of materials.

The chemical formula for the apatites may also be written $A_{10}(MO_4)_6O_{2+y}$, which reflects the structure (seen Figure 4.24), which consists of a framework of isolated

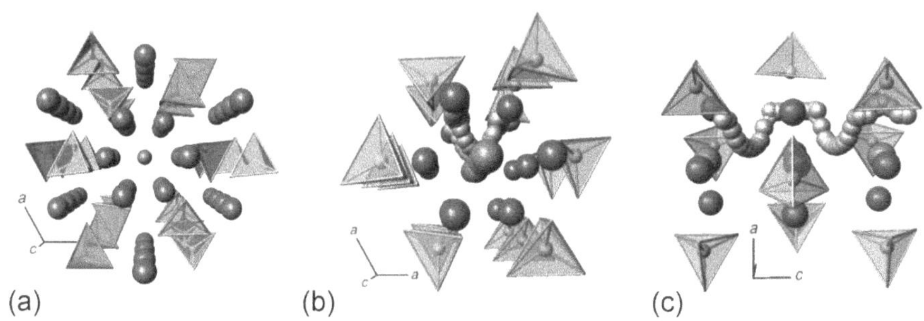

Figure 4.24 (a) Apatite crystal structure viewed along c-axis, showing conductivity channels and (b and c) oxygen migration pathway viewed along c-axis and b-axis, respectively. MO_4 tetrahedra shown in light blue (light grey in the print version), A cations in purple (dark grey in the print version) and (stationary) oxygen in red (grey in the print version). Migrating oxygen shown in orange (light grey in the print version).
Adapted from Ref. [126] with permission from The Royal Society of Chemistry.

MO_4 tetrahedra (light blue, light grey in the print version) in a hexagonal symmetry, forming two distinct channels; the smaller channel is filled by A cations (and A cation vacancies when present), and the larger (in the centre of the hexagonal cell) contains rings of lanthanum cations with a column of oxygen ions running down the centre.

The electrical properties of the apatites are dominated by their ionic conductivity, with the electronic component of the total electrical conductivity generally quite low even to low pO_2s at moderate temperatures [7,126,135,136]. Oxygen transport in the apatites occurs by an interstitial conduction mechanism. Oxygen ions follow a complex sinusoidal pathway involving energetically favourable interstitial oxygen positions at the edge of the open conducting channels [137,138], illustrated in Figure 4.24b [7,139–141]. This mechanism is inherently anisotropic, favouring transport along the c-axis of the materials, which is evident in measurements on single crystals [142,143] and textured thin films [144], but not as significant as may first be assumed from inspection of the structure; in fact, the c-axis conductivity is only around one order of magnitude higher than the conductivity in the ab plane (e.g. $\sigma_c = 1.5 \times 10^{-3}$ S cm^{-1} c.f. $\sigma_{ab} = 2.4 \times 10^{-4}$ S cm^{-1} for $Sm_{9.33}(SiO_4)_6O_2$ at 300 °C) [142].

As shown in Figure 4.25, the conductivity increases with the radius of the A-site cation, with a maximum for La, and high conductivities also seen for Pr and Nd [127,128,130]. The apatite structure is also flexible in terms of accommodating cation substitution on both the A-site (e.g. Ca, Sr, Ba) and M-site (e.g. Al, Fe, B, Ga, Zn, Ti, V) [126,145–147]. Interestingly, magnesium may occupy both A- and M-sites [145]. The effect of substitution of some of these cations on the oxygen transport properties is summarised in Figure 4.26. It is seen that the more promising strategy seems to be substitution of cations such as Mg and Al on the M site, although A-site substitution does also improve conductivity compared to the undoped material, especially at higher temperatures.

Another interesting means by which to manipulate the conductivities of the apatite electrolytes is by the deliberate introduction of cation deficiency on the A-site (i.e. V_{La}'''). Figure 4.26 shows this effect for $La_{9.33-2x/3}Sr_xV_{La0.67-x/3}'''Si_6O_{26}$, where Sr is added to maintain constant oxygen constant. Increasing the concentration of V_{La}''' increases the conductivity, which passes through a compositional maximum at 0.5 cation vacancies per unit cell before beginning to decrease again. This A-cation non-stoichiometry affects the unit cell volume, displacing the channel oxygens into the interstitial positions, increasing the ionic conductivity. As such, this may be regarded as enhancing anti-Frenkel disorder, and promoting the "intrinsic" defect processes by manipulating the composition and structure, rather than by the more common approach of aliovalent doping [129,131,132,150,151].

A significant drawback of apatite-based electrolytes is the high temperatures (in excess of 1300 °C) required for the synthesis of the material by conventional solid-state routes, and sintering the material to form a sufficiently gas-tight electrolyte membrane (typically greater than 1600 °C) [7,152]. Recent research efforts are targeting alternative approaches to decrease the temperatures required for synthesis and processing, mainly by wet chemistry methods such as Pechini, sol–gel and homogeneous precipitation routes [147,153–156]). Some researchers have even reported synthesis of apatite phases by room temperature mechanical milling of appropriate precursors [157,158].

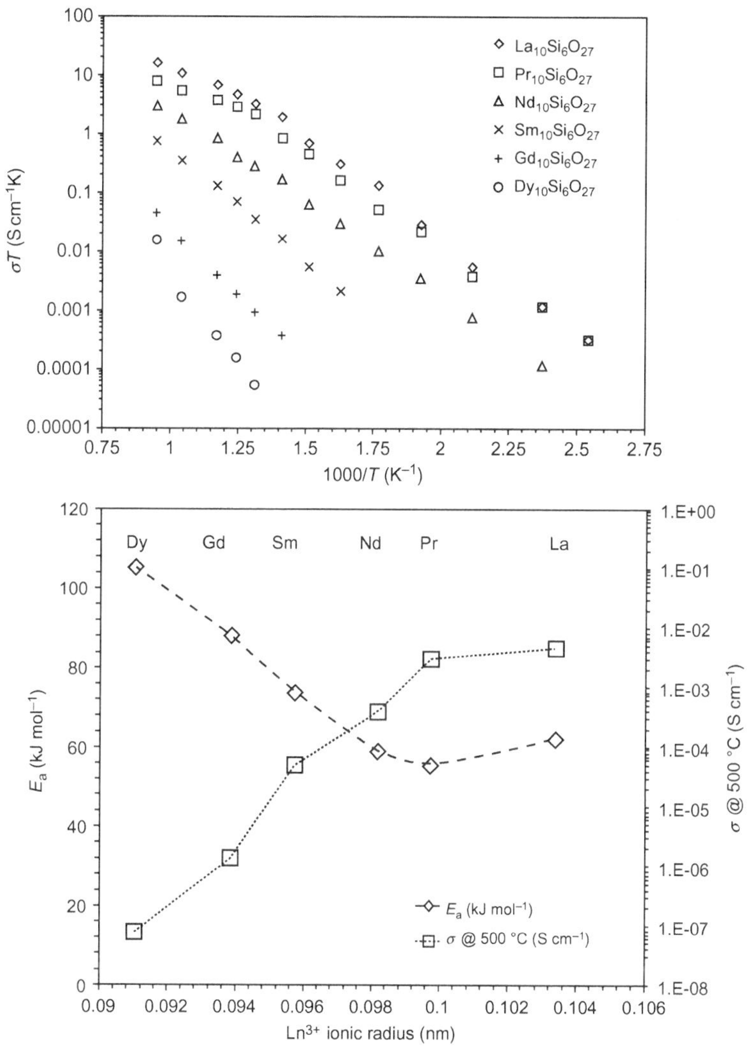

Figure 4.25 Effect of Ln ionic radius on conductivity in $Ln_{10}Si_6O_{27}$ (Ln = La, Pr, Nd, Sm, Gd, Dy) compounds [128].

Approaches to reduce sintering temperatures include increasing the sinterability of the starting powder by refinement of the particle size by altering the synthesis method (e.g. Ref. [156]) and the use of transition metal sintering aids such as iron, cobalt or manganese [159,160]; significant substitution is possible whilst still retaining high ionic transference numbers as the apatite structure stabilises the oxidation state of variable valence cations such as Fe and Pr [149,161]. For example, up to 30% Fe substitution on the M-site is possible in $La_{10}Si_4Fe_2O_{26}$ retaining ionic transference numbers greater than 0.96 [159]. Aside from more "conventional" sintering approaches, other

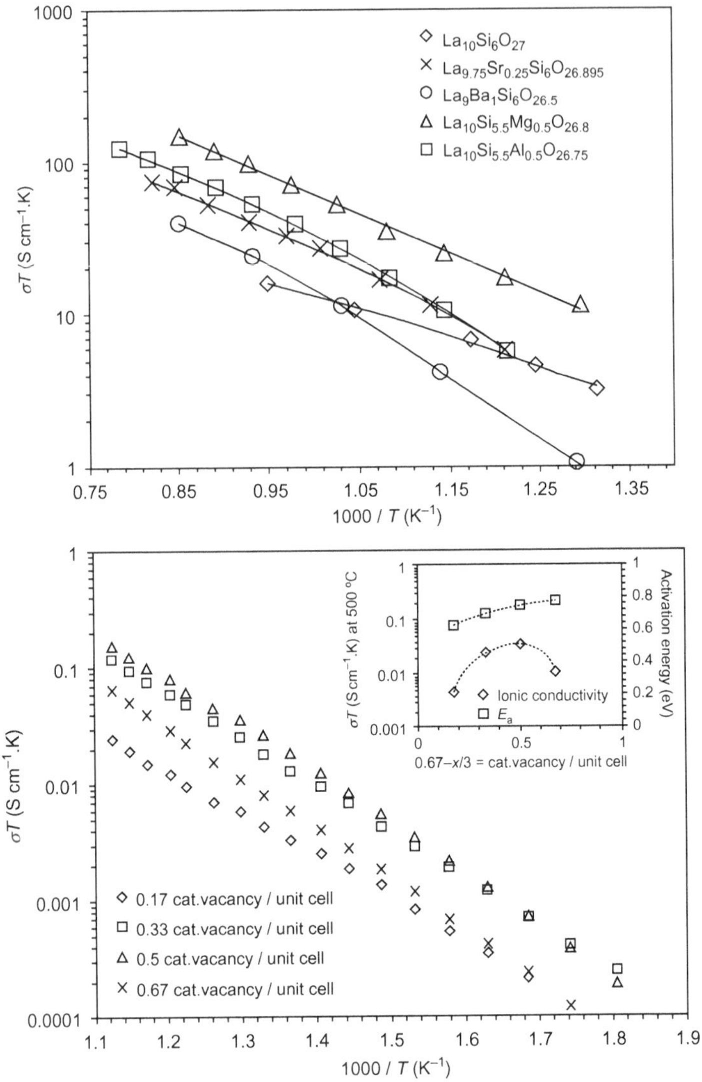

Figure 4.26 Effect of (a) cation substitution in lanthanum apatite silicates [14,128,133,148,149] and (b) cation vacancies in $La_{9.33-2x/3}Sr_xV'''_{La0.67-x/3}Si_6O_{26}$ (V'''_{La} denotes a cation vacancy on the La site) on ionic conductivity [150].

processing methods are being investigated, including deposition of thin films by reactive magnetron sputtering [144,162] and plasma spraying [163]. These latter approaches have the additional benefit of reducing the contribution of the Ohmic resistance of the electrolyte to the total electrical resistance cell, where this may be rate limiting.

Although chemical compatibility studies of apatite electrolytes with common manganite, ferrite, nickelate and cobaltite perovskite cathode materials reveal no secondary-phase formation, they do show evidence for inter-diffusion of cations between the electrode and electrolyte [160,164–166]. Marrero-Lopez et al. [164] reported increasing ASR with electrode processing temperature, suggesting cation inter-diffusion affected charge transfer at the electrode–electrolyte interface. The extent of inter-diffusion is found to be lower with nickel-based Ruddlesden–Popper type cathode materials [165]. For long-term operation and durability of the cells, diffusion of silicon from the apatite to the surface of the cathode may also be a concern as it could lead to poisoning of the electrode and increase in the cathodic overpotential.

SOFCs based on La silicates have been demonstrated by various groups [14,163,164,167], but the absolute performances are not yet competitive with the more traditional materials discussed above. In most cases, this is attributable to non-optimal selection of electrode materials and cell geometry, rather than the electrolyte itself. For example, low MPDs of 20 $mWcm^{-2}$ at 800 °C [167], attributed mainly to electrode overpotentials associated with the selection of Pt electrodes, were improved to 120 $mWcm^{-2}$ by choosing LSC and Ni–SDC as cathode and anode [14]. More recent work applied LSCF electrodes, increasing the MPD to 195 $mWcm^{-2}$ [164]. In this case, the Ohmic contribution to the cell was dominated by the 1 mm thick electrolyte, and decreasing the electrolyte thickness to 100 μm was projected to give an improved MPD of 840 $mWcm^{-2}$ at 850 °C [164]. It was also necessary to use a CGO buffer layer to prevent cation inter-diffusion. A cell taking advantage of an anode-supported geometry with an LSCF electrode on a thin plasma-sprayed Mg-doped $La_{9.8}Si_6O_{26+y}$ electrolyte layer achieved a MPD of 79.9 $mWcm^{-2}$ at 800 °C [163].

4.4.2 $La_2Mo_2O_9$: LAMOX

The applicability of materials based on $La_2Mo_2O_9$ (LAMOX) to SOFCs was first reported by Lacorre et al. in 2000 [168,169]. The high temperature (β) form, which shows oxygen ionic conductivity better than that of YSZ, has a cubic unit cell [169] but undergoes a phase transition around 580 °C to the monoclinic α-$La_2Mo_2O_9$ form [170], accompanied by a drop in conductivity of one to two orders of magnitude, as seen in the conductivity plots in Figure 4.27. This drop in (bulk) conductivity is correlated both with a change in unit cell volume, as well as a change from static to dynamic arrangement of oxygen defects associated with the phase change [174,175].

The parent $La_2Mo_2O_9$ material may be considered as a form of β-$SnWO_4$ substituted with La and Mo, illustrating the concept of "lone-pair substitution" proposed by Lacorre et al. [168]. In β-$SnWO_4$, the Sn^{2+} lone electron pairs project into neighbouring vacant oxygen sites, occupying roughly the same volume and effectively stabilising the vacancy. By substitution of the Sn^{2+} cations with a cation of approximately the same size, but without a lone electron pair, such as La^{3+}, the two oxygen sites formerly occupied by the Sn^{2+} lone pair are now vacant, effectively introducing oxygen vacancies into the structure. However, in order to maintain charge neutrality and compensate the introduction of La^{3+} for Sn^{2+}, one extra oxide ion per

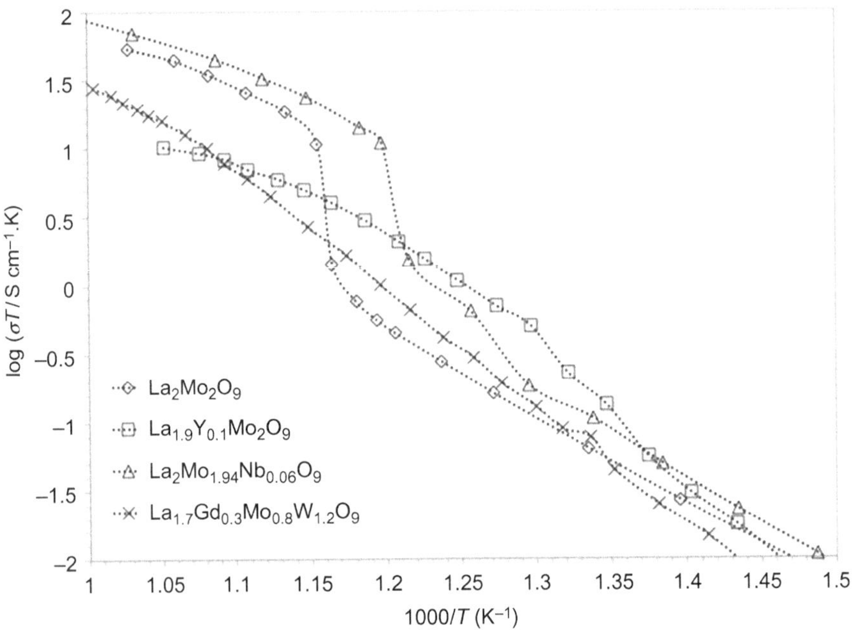

Figure 4.27 Temperature variation of electrical conductivity of various LAMOX-based materials.
Data adapted from Ref. [171–173].

formula unit must be incorporated, meaning the overall effect of the substitution is that two oxygen vacancies stabilised by lone electron pairs become one lattice oxygen and one oxygen vacancy.

The unit cell of LAMOX is complex, with Mo in several variable coordination environments, apparently vital for high ionic conductivity. There is some evidence that oxide ion diffusion occurs via a cooperative mechanism [176]. Oxygen tracer diffusion experiments show a high D^* value of 1.4×10^{-6} cm^2 s^{-1} at 800 °C [177]. Although experiments comparing exchange in dry $^{18}O_2$-enriched gas and D_2O vapour suggested that there may be some degree of hydroxyl incorporation and diffusion, no protonic conductivity could be identified from AC impedance in wet atmosphere [178].

In air, LAMOX shows *n*-type electronic conductivity of an extent comparable to the *p*-type electronic conductivity of LSGM [134]. However, the electronic transference number, t_e, increases significantly with temperature and reducing atmospheres due to the reduction of Mo.

The properties of the LAMOX family of materials may be tailored by further substitution starting from the $La_2Mo_2O_9$ material. There are two main areas targeted by these extra substitutions. Because of the two order of magnitude drop in ionic conductivity associated with the phase change mentioned above, undoped $La_2Mo_2O_9$ is only practical (i.e. with σ_i better than YSZ [168]) at high temperatures. Therefore, one focus of research is doping to stabilise the high-temperature beta phase to lower temperatures. This is typically done by substituting lanthanum with elements drawn from the rare earths, alkali earths, alkali metals and bismuth. For example, whilst the alpha form is most stable at room temperature for a wide range of Nd substitutions, even small substitution of Gd and Y can stabilise the beta form at room temperature [7,171]. The effect of this stabilisation on the temperature dependence of the electrical conductivity is shown in Figure 4.27. Although the sharp drop associated with the phase transition is not evident for the $La_{1.9}Y_{0.1}Mo_2O_9$ composition, the conductivity data do show some curvature; this has been interpreted as a change in conduction mechanism, related to a transition from a static (at low temperature) to dynamic ordering of the oxide ion vacancies [171].

As well as increasing the *p*-type electronic conductivity, the reducibility of Mo limits the phase stability of the materials in reducing conditions [179–181]. This has led to investigations on the effect of replacing Mo with less reducible cations including W, V, S, Cr, Nb and Ta [171,182,183]. Substitution of niobium for Mo can even increase the ionic conductivity higher than the undoped β-form, giving $\sigma_i = 0.11$ S cm^{-1} at 800 °C [173]. Most focus so far has been on doping with W, which seems to increase material stability under reducing conditions. However, work by Vega-Castillo et al. [184] suggests that W substitution does not enhance thermodynamic stability but slows down the kinetics of reduction, suggesting that long-term degradation due to reduction of W-substituted LAMOX may still be a concern.

Higher levels of W substitution also decrease the ionic conductivity, although $La_2Mo_{2-x}W_xO_9$, $x \leq 1.6$ has ionic conductivity sufficient for operation at 600 °C [181], whilst additional gadolinium doping in $La_{1.7}Gd_{0.3}Mo_{0.8}W_{1.2}O_9$ also gives a promising compromise between stability against reduction and ionic performance, whilst also suppressing the phase change to some extent [172], as seen in Figure 4.27.

The volatility of molybdenum at the high temperatures used for synthesis and sintering of LAMOX is also a concern, so efforts are being made to find lower-temperature processing routes, such as combustion spray pyrolysis [185,186]. Ceramics with greater than 95% relative density have been obtained using sintering temperatures 300 °C lower than needed for conventional solid-state synthetic approaches [187,188].

The selection of compatible electrode materials is also a challenge, as LAMOX electrolytes have higher coefficients of thermal expansion than most electrode materials ($15–20 \times 10^{-6}$ K^{-1} [181]), which causes issues with thermomechanical incompatibility. High reactivity with cathode materials is also reported, in the form of Mo diffusion into the cathode [181]. Although LAMOX electrolytes appear stable with $(La,Ca)FeO_3$ cathode material at representative operating temperatures of 600 °C, they do react at temperatures around those required for sintering (>1000 °C), with a tendency to form scheelite-type oxides ($AMoO_4$ or AWO_4) [189]. Reaction between Ruddlesden–Popper La_2NiO_{4+d} phases and $La_2Mo_2O_9$, forming La_2MoO_6 and NiO, starts at 600 °C and is completed by 1000 °C [190].

On the other hand, LAMOX compositions are stable with CeO_2-based electrolytes, which present the possibility of using a thin layer of a doped ceria as a barrier to prevent cation diffusion and reaction between a LAMOX electrolyte and cobaltite or ferrite electrode [191].

The selection of suitable anode materials appears to be less of a challenge, as chemical compatibility studies suggest LAMOX materials are stable with NiO/Ni as anodes [189,190]. A conflicting report suggested some unexplained formation of the poorer-conducting α-$La_2Mo_2O_9$ when calcined at 800 °C with NiO, and that CuO was more compatible as an anode material [181].

Perhaps due in part to many of these challenges, reports of electrochemical performances of SOFCs based on LAMOX electrolytes are relatively few. A MPD of 4 $mWcm^{-2}$ at 650 °C has been reported for an 1.5 mm thick electrolyte-supported cell with Pt as both cathode and anode [189]. Undoubtedly, there is scope for improvement in this performance by reducing the Ohmic contribution by decreasing the electrolyte thickness, or choosing an alternative geometry to support the cell, as well as by the application of an improved electrode.

The application of LAMOX in a single-chamber fuel cell may avoid some of these challenges; the mixing of the fuel and oxidant streams means the environment is not as reducing, and the lack of requirement for gas sealing avoids any potential issues with TEC mismatch between the cell and glass seals. An anode-supported cell using a 60 μm thick dysprosium-doped LAMOX electrolyte with Ni–GDC cermet anode and an LSCF-6482 cathode with Fe–GDC interlayer achieved a MPD of 200 $mWcm^{-2}$ at 650 °C in single-chamber fuel cell tests [192].

4.5 Summary

Many oxide formulations have been explored in the search for candidate SOFC electrolyte materials, especially over the last few decades, where higher conductivities have been sought at temperatures $\leq$ 750 °C. Substantial oxide ion conductivity is a rare phenomenon in complex oxides, especially at low temperatures, and so the number of materials that display conductivity adequate for application combined with the restrictions of mechanical strength, high stability, ease of processing and low cost is extremely limited. The search for new materials will continue; however, the list of candidate materials will remain low, and thus the development of existing materials is essential.

Zirconia-based compositions remain some of the best electrolytes owing to their good stability under reducing atmospheres, low electronic conductivity and acceptable oxide ion conductivity above 800 °C. The recent trend of SOFC development is towards operation at lower temperatures. The lowest operation temperature limit of the cell, for thin YSZ electrolytes, is estimated to be about 700 °C from YSZ conductivity and mechanical property data. Scandia-doped zirconia, which shows a higher conductivity than that of YSZ, could be preferred at temperatures below 700 °C, if the costs of scandia are acceptable.

Ceria-based electrolytes could be used at 550 °C or below where the electronic conductivity is acceptable for application in an SOFC. To operate at higher temperatures, a dual-layer electrolyte, with a thin YSZ layer on CGO, has been proposed to avoid the electronic current leakage. Inter-diffusion at the interface becomes important in this case for durable electrolytes with long lifetimes. Despite these problems, ceria-based electrolytes are being developed commercially for metal supported thin film SOFCs. [193]

The perovskite and perovskite-related structures offer a rich compositional space for the development of new materials, as the two cation sites can accommodate a broad range of cations with different valence states on the A- and B-sites. The most promising candidate oxide with the perovskite structure is $LaGaO_3$ substituted on the A- and B-sites with Sr and Mg. This material shows excellent oxygen ionic conductivity and has been successfully employed in cells operating at low temperatures. Other possible materials are the wealth of perovskite-related structures, such as the brownmillerites of which $Ba_2In_2O_5$ doped with Ce or La is an example.

Other ionic-conducting oxides have also been found including $La_{10}Si_6O_{26}$ composition that could be interesting from the point of view of low-cost, high abundance materials. The high-temperature phase of $La_2Mo_2O_9$ (LAMOX) shows high conductivity, but the phase transition around 580 °C is associated with an order of magnitude drop in conductivity. Unfortunately, a drawback of both of these materials is the high temperatures required for their synthesis and densification to the gas-tightness required as an SOFC electrolyte. Therefore, alternative synthesis and densification methods (including deposition as thin films) are key areas of development for their application.

Finally, proton conductors such as $SrCeO_3$, $SrZrO_3$ or $BaCeO_3$ doped with Y or Yb may also be effective electrolytes in proton conducting fuel cells, but the reactivity of many of these materials with CO_2 has to be resolved first.

References

[1] E. Baur, Z. Preis, Über Brennstoff-Ketten mit Festleitern, Z. Elektrochem. 43 (9) (1937) 727–732.

[2] K. Huang, J.B. Goodenough, Solid oxide fuel cell technology; principles, performance and operations, Woodhead Energy Series, Woodhead Publishing Ltd., Cambridge, 2009, 344 p.

[3] N. Preux, A. Rolle, R.N. Vannier, Electrolytes and ion conductors for solid oxide fuel cells (SOFCs), in: J.A. Kilner et al., (Ed.), Functional Materials for Sustainable Energy Applications, Woodhead, Oxford, 2012.

[4] V.V. Kharton, M.B. Marques, J.A. Kilner, A. Atkinson, Oxygen ion conducting materials, in: V.V. Kharton (Ed.), Solid State Electrochemistry, Wiley-VCH, Weinheim, Germany, 2009, pp. 310–334.

[5] T. Ishihara, Oxide ion conductivity in perovskite oxide for SOFC electrolyte, in: T. Ishihara (Ed.), Perovskite Oxide for Solid Oxide Fuel Cells, Springer, New York, 2009.

[6] A. Aguadero, L. Fawcett, S. Taub, R. Woolley, K.T. Wu, N. Xu, J.A. Kilner, S.J. Skinner, Materials development for intermediate-temperature solid oxide electrochemical devices, J. Mater. Sci. 47 (9) (2012) 3925–3948.

[7] A.J. Jacobson, Materials for solid oxide fuel cells, Chem. Mater. 22 (3) (2010) 660–674.

[8] L. Malavasi, C.A.J. Fisher, M.S. Islam, Oxide-ion and proton conducting electrolyte materials for clean energy applications: structural and mechanistic features, Chem. Soc. Rev. 39 (11) (2010) 4370–4387.

[9] C. Haavik, E.M. Ottesen, K. Nomura, J.A. Kilner, T. Norby, Temperature dependence of oxygen ion transport in Sr plus Mg-substituted $LaGaO_3$ (LSGM) with varying grain sizes, Solid State Ionics 174 (1-4) (2004) 233–243.

[10] M. Weller, R. Herzog, M. Kilo, G. Borchardt, S. Weber, S. Scherrer, Oxygen mobility in yttria-doped zirconia studied by internal friction, electrical conductivity and tracer diffusion experiments, Solid State Ionics 175 (1-4) (2004) 409–413.

[11] M. Li, M.J. Pietrowski, R.A. De Souza, H. Zhang, I.M. Reaney, S.N. Cook, J.A. Kilner, D.C. Sinclair, A family of oxide ion conductors based on the ferroelectric perovskite $Na_{0.5}Bi_{0.5}TiO_3$, Nat. Mater. 13 (1) (2014) 31–35.

[12] I.E.L. Stephens, J.A. Kilner, Ionic conductivity of $Ce_{1-x}Nd_xO_{2-x/2}$, Solid State Ionics 177 (7-8) (2006) 669–676.

[13] R.D. Bayliss, S.N. Cook, S. Kotsantonis, R.J. Chater, J.A. Kilner, Oxygen ion diffusion and surface exchange properties of the α- and δ-phases of Bi_2O_3, Adv. Energy Mater. 4 (10) (2014) 1301575.

[14] H. Yoshioka, Y. Nojiri, S. Tanase, Ionic conductivity and fuel cell properties of apatite-type lanthanum silicates doped with Mg and containing excess oxide ions, Solid State Ionics 179 (38) (2008) 2165–2169.

[15] J.A. Kilner, Fast oxygen transport in acceptor doped oxides, Solid State Ionics 129 (1) (2000) 13–23.

[16] Y. Arachi, H. Sakai, O. Yamamoto, Y. Takeda, N. Imanishai, Electrical conductivity of the ZrO_2-Ln_2O_3 (Ln = lanthanides) system, Solid State Ionics 121 (1-4) (1999) 133–139.

[17] D.J. Kim, Lattice parameters, ionic conductivities, and solubility limits in fluorite-structure $Hf^{4+}O_2$, $Zr^{4+}O_2$, $Ce^{4+}O_2$, $Th^{4+}O_2$, $V^{4+}O_2$ oxide solid-solutions, J. Am. Ceram. Soc. 72 (8) (1989) 1415–1421.

[18] J.A. Kilner, Fast anion transport in solids, Solid State Ionics 8 (3) (1983) 201–207.

[19] J.A. Kilner, Defects and conductivity in ceria-based oxides, Chem. Lett. 37 (10) (2008) 1012–1015.

[20] J.A. Kilner, The role of dopant size in determining oxygen ion conductivity in the fluorite structure oxides, in: R. Metselaar, H.J.M. Heijligers, J. Schoonman (Eds.), Studies in Inorganic Chemistry, Elsevier Scientific Publishing Company, Amsterdam, 1983, pp. 189–192.

[21] J.A. Kilner, B.C.H. Steele, The effect of ion size on the energy of association between oxygen vacancies and dopant cations in oxide solid electrolytes, J. Electrochem. Soc. 129 (3) (1982) C143.

[22] J.A. Kilner, R.J. Brook, A study of oxygen ion conductivity in doped nonstoichiometric oxides, Solid State Ionics 6 (3) (1982) 237–252.

[23] H. Inaba, H. Tagawa, Ceria-based solid electrolytes – Review, Solid State Ionics 83 (1-2) (1996) 1–16.

[24] D.Y. Wang, D.S. Park, J. Griffith, A.S. Nowick, Oxygen-ion conductivity and defect interactions in yttria-doped ceria, Solid State Ionics 2 (2) (1981) 95–105.

[25] R. Gerhardtanderson, A.S. Nowick, Ionic-conductivity of CeO_2 with trivalent dopants of different ionic-radii, Solid State Ionics 5 (1981) 547–550.

[26] P.S. Manning, J.D. Sirman, R.A. DeSouza, J.A. Kilner, The kinetics of oxygen transport in 9.5 mol% single crystal yttria stabilised zirconia, Solid State Ionics 100 (1-2) (1997) 1–10.

[27] F. Ye, T. Mori, D.R. Ou, A.N. Cormack, R.J. Lewis, J. Drennan, Simulation of ordering in large defect clusters in gadolinium-doped ceria, Solid State Ionics 179 (35-36) (2008) 1962–1967.

[28] D.R. Ou, T. Mori, F. Ye, J. Zou, G. Auchterlonie, J. Drennan, Oxygen-vacancy ordering in lanthanide-doped ceria: dopant-type dependence and structure model, Phys. Rev. B: Condens. Matter Mater. Phys. 77 (2) (2008) 024108.
[29] D.R. Ou, T. Mori, F. Ye, T. Kobayashi, J. Zou, G. Auchterlonie, J. Drennan, Oxygen vacancy ordering in heavily rare-earth-doped ceria, Appl. Phys. Lett. 89 (17) (2006) 171911–171913.
[30] S. Yamazaki, T. Matsui, T. Ohashi, Y. Arita, Defect structures in doped CeO_2 studied by using XAFS spectrometry, Solid State Ionics 136–137 (2000) 913–920.
[31] H. Deguchi, H. Yoshida, T. Inagaki, M. Horiuchi, EXAFS study of doped ceria using multiple data set fit, Solid State Ionics 176 (23-24) (2005) 1817–1825.
[32] L. Minervini, M.O. Zacate, R.W. Grimes, Defect cluster formation in M_2O_3-doped CeO_2, Solid State Ionics 116 (3-4) (1999) 339–349.
[33] D.A. Andersson, S.I. Simak, N.V. Skorodumova, I. A. Abrikosov, B. Johansson, Optimization of ionic conductivity in doped ceria, Proc. Natl. Acad. Sci. U. S. A. 103 (10) (2006) 3518–3521.
[34] D. Marrocchelli, S.R. Bishop, J. Kilner, Chemical expansion and its dependence on the host cation radius, J. Mater. Chem. A 1 (26) (2013) 7673–7680.
[35] S.R. Bishop, D. Marrocchelli, N.H. Perry, H.L. Tuller, G. Watson, B. Yildiz, K. Amezawa, J.A. Kilner, Chemical expansion in SOFC materials: ramifications, origins, and mitigation, ECS Trans. 57 (1) (2013) 643–648.
[36] D. Marrocchelli, S.R. Bishop, H.L. Tuller, B. Yildiz, Understanding chemical expansion in non-stoichiometric oxides: ceria and zirconia case studies, Adv. Funct. Mater. 22 (9) (2012) 1958–1965.
[37] J.M. Ralph, J. Przydatek, J.A. Kilner, T. Seguelong, Novel doping systems in ceria, Ber. Bunsen-Ges. Phys. Chem. Chem. Phys. 101 (9) (1997) 1403–1407.
[38] R.M. Kasse, J.C. Nino, Ionic conductivity of $Sm_xNd_yCe_{0.9}O_{2-\delta}$ co-doped ceria electrolytes, J. Alloys Compd. 575 (2013) 399–402.
[39] K. Park, H.K. Hwang, Electrical conductivity of $Ce_{0.8}Gd_{0.2-x}Dy_xO_{2-\delta}$ ($0<=x<=0.2$) co-doped with Gd^{3+} and Dy^{3+} for intermediate-temperature solid oxide fuel cells, J. Power Sources 196 (11) (2011) 4996–4999.
[40] P.P. Dholabhai, J.B. Adams, P.A. Crozier, R. Sharma, In search of enhanced electrolyte materials: a case study of doubly doped ceria, J. Mater. Chem. 21 (47) (2011) 18991–18997.
[41] W. Zajac, J. Molenda, Electrical conductivity of doubly doped ceria, Solid State Ionics 179 (1-6) (2008) 154–158.
[42] S. Omar, E.D. Wachsman, J.C. Nino, Higher conductivity Sm^{3+} and Nd^{3+} co-doped ceria-based electrolyte materials, Solid State Ionics 178 (37-38) (2008) 1890–1897.
[43] S. Omar, E.D. Wachsman, J.C. Nino, Higher ionic conductive ceria-based electrolytes for solid oxide fuel cells, Appl. Phys. Lett. 91 (14) (2007) 144106.
[44] S. Omar, E.D. Wachsman, J.C. Nino, A co-doping approach towards enhanced ionic conductivity in fluorite-based electrolytes, Solid State Ionics 177 (35-36) (2006) 3199–3203.
[45] M.O. Zacate, L. Minervini, D.J. Bradfield, R.W. Grimes, K.E. Sickafus, Defect cluster formation in M_2O_3-doped cubic ZrO_2, Solid State Ionics 128 (1-4) (2000) 243–254.
[46] M. Burbano, S.T. Norberg, S. Hull, S.G. Eriksson, D. Marrocchelli, P.A. Madden, G. W. Watson, Oxygen vacancy ordering and the conductivity maximum in Y_2O_3-doped CeO_2, Chem. Mater. 24 (1) (2012) 222–229.

[47] M. Yashima, M. Kakihana, M. Yoshimura, Metastable-stable phase diagrams in the zirconia-containing systems utilized in solid-oxide fuel cell application, Solid State Ionics 86–8 (1996) 1131–1149.
[48] R.A. Miller, J.L. Smialek, R.G. Garlick, Phase stability in yttria stabilized ZrO2, in: A.H. Heuer, L.W. Hobbs (Eds.), Science and Technology of Zirconia, American Ceramic Society, Columbus, Ohio, 1981, pp. 1403–1405.
[49] R. Ruh, H.J. Garrett, R.F. Domagala, V.A. Patel, System zirconia-scandia, J. Am. Ceram. Soc. 60 (9-10) (1977) 399–403.
[50] H.A. Abbas, C. Argirusis, M. Kilo, H.D. Wiemhofer, F.F. Hammad, Z. M. Hanafi, Preparation and conductivity of ternary scandia-stabilised zirconia, Solid State Ionics 184 (1) (2011) 6–9.
[51] C. Haering, A. Roosen, H. Schichl, M. Schnoller, Degradation of the electrical conductivity in stabilised zirconia system Part II: scandia-stabilised zirconia, Solid State Ionics 176 (3-4) (2005) 261–268.
[52] D.S. Lee, W.S. Kim, S.H. Choi, J. Kim, H.W. Lee, J.H. Lee, Characterization of ZrO_2 co-doped with Sc_2O_3 and CeO_2 electrolyte for the application of intermediate temperature SOFCs, Solid State Ionics 176 (1-2) (2005) 33–39.
[53] S. Omar, W. Bin Najib, W.W. Chen, N. Bonanos, Electrical conductivity of 10 mol% Sc_2O_3-1 mol% M_2O_3-ZrO_2 ceramics, J. Am. Ceram. Soc. 95 (6) (2012) 1965–1972.
[54] H. Yasuda, M. Uehara, N. Enomoto, J. Hojo, Electrical and mechanical properties of cubic ZrO_2 stabilized with Sc_2O_3 and Y_2O_3, J. Ceram. Process. Res. 5 (1) (2004) 80–83.
[55] T.I. Politova, J.T.S. Irvine, Investigation of scandia-yttria-zirconia system as an electrolyte material for intermediate temperature fuel cells-influence of yttria content in system$(Y_2O_3)_x(Sc_2O_3)_{11-x}(ZrO_2)_{89}$, Solid State Ionics 168 (1-2) (2004) 153–165.
[56] A. Spirin, V. Ivanov, A. Nikonov, A. Lipilin, S. Paranin, V. Khrustov, A. Spirina, Scandia-stabilized zirconia doped with yttria: synthesis, properties, and ageing behavior, Solid State Ionics 225 (2012) 448–452.
[57] F. Yuan, J.X. Wang, H. Miao, C.X. Guo, W.G. Wang, Investigation of the crystal structure and ionic conductivity in the ternary system $(Yb_2O_3)_x$-$(Sc_2O_3)_{0.11-x}$-$(ZrO_2)_{0.89}$ $(x=0\text{-}0.11)$, J. Alloys Compd. 549 (2013) 200–205.
[58] J.C. Tao, Y. Hao, J. Wang, The research of crystal structure and electrical properties of a new electrolyte material: scandia and holmia stabilized zirconia, J. Ceram. Soc. Jpn. 121 (1411) (2013) 317–325.
[59] B.C.H. Steele, Appraisal of $Ce_{1-y}Gd_yO_{2-y/2}$ electrolytes for IT-SOFC operation at 500 °C, Solid State Ionics 129 (1-4) (2000) 95–110.
[60] M. Mogensen, N.M. Sammes, G.A. Tompsett, Physical, chemical and electrochemical properties of pure and doped ceria, Solid State Ionics 129 (1-4) (2000) 63–94.
[61] H. Tuller, A.S. Nowick, Doped ceria as a solid oxide electrolyte, J. Electrochem. Soc. 122 (2) (1975) 255–259.
[62] Z. Tianshu, P. Hing, H. Huang, J. Kilner, Ionic conductivity in the CeO_2-Gd_2O_3 system $(0.05 \leq Gd/Ce \leq 0.4)$ prepared by oxalate coprecipitation, Solid State Ionics 148 (3-4) (2002) 567–573.
[63] J.A. Kilner, M. Burriel, Materials for intermediate-temperature solid-oxide fuel cells, Annu. Rev. Mater. Res. 44 (1) (2014) 365–393.
[64] J. Faber, C. Geoffroy, A. Roux, A. Sylvestre, P. Abelard, A systematic investigation of the Dc-electrical conductivity of rare-earth doped ceria, Appl. Phys. Mater. Sci. Process. 49 (3) (1989) 225–232.
[65] M. Godickemeier, L.J. Gauckler, Engineering of solid oxide fuel cells with ceria-based electrolytes, J. Electrochem. Soc. 145 (2) (1998) 414–421.

[66] M. Godickemeier, K. Sasaki, L.J. Gauckler, I. Riess, Electrochemical characteristics of cathodes in solid oxide fuel cells based on ceria electrolytes, J. Electrochem. Soc. 144 (5) (1997) 1635–1646.

[67] M. Pirzada, R.W. Grimes, L. Minervini, J.F. Maguire, K.E. Sickafus, Oxygen migration in $A_2B_2O_7$ pyrochlores, Solid State Ionics 140 (3-4) (2001) 201–208.

[68] S.A. Kramer, H.L. Tuller, A novel titanate-based oxygen ion conductor: $Gd_2Ti_2O_7$, Solid State Ionics 82 (1–2) (1995) 15–23.

[69] M. O'Connell, A.K. Norman, C.F. Huttermann, M.A. Morris, Catalytic oxidation over lanthanum-transition metal perovskite materials, Catal. Today 47 (1-4) (1999) 123–132.

[70] R.A. De Souza, J.A. Kilner, Oxygen transport in $La_{1-x}Sr_xMn_{1-y}Co_yO_{3\pm\delta}$ perovskites: Part I. Oxygen tracer diffusion, Solid State Ionics 106 (3–4) (1998) 175–187.

[71] T. Takahashi, H. Iwahara, Ionic conduction in perovskite-type oxide solid solution and its application to solid electrolyte fuel cell, Energy Convers. 11 (3) (1971) 105.

[72] J. Mizusaki, I. Yasuda, J. Shimoyama, S. Yamauchi, K. Fueki, Electrical-conductivity, defect equilibrium and oxygen vacancy diffusion-coefficient of $La_{1-X}Ca_xAlO_{3-\delta}$ single-crystals, J. Electrochem. Soc. 140 (2) (1993) 467–471.

[73] D. Lybye, F.W. Poulsen, M. Mogensen, Conductivity of A- and B-site doped $LaAlO_3$, $LaGaO_3$, $LaScO_3$ and $LaInO_3$ perovskites, Solid State Ionics 128 (1-4) (2000) 91–103.

[74] D. Lybye, N. Bonanos, Proton and oxide ion conductivity of doped $LaScO_3$, Solid State Ionics 125 (1-4) (1999) 339–344.

[75] K. Nomura, S. Tanase, Electrical conduction behavior in $La_{0.9}Sr_{0.1}M^{III}O_{3-\delta}$ ($M^{III}=Al$, Ga, Sc, In, and Lu) perovskites, Solid State Ionics 98 (3–4) (1997) 229–236.

[76] M. Cherry, M.S. Islam, C.R.A. Catlow, Oxygen-ion migration in perovskite-type oxides, J. Solid State Chem. 118 (1) (1995) 125–132.

[77] T. Ishihara, H. Matsuda, Y. Takita, Oxide ion conductivity in doped $NdAlO_3$ perovskite-type oxides, J. Electrochem. Soc. 141 (12) (1994) 3444–3449.

[78] T. Ishihara, H. Matsuda, Y. Mizuhara, Y. Takita, Improved oxygen-ion conductivity of $NdAlO_3$ perovskite-type oxide by doping with Ga, Solid State Ionics 70 (1994) 234–238.

[79] T. Ishihara, H. Matsuda, Y. Takita, Doped LaGaO3 perovskite type oxide as a new oxide ionic conductors, Second International Symposium on Ionic and Mixed Conducting Ceramics, The Electrochemical Society, Inc., Pennington, 1994.

[80] T. Ishihara, H. Matsuda, Y. Takita, Doped $LaGaO_3$ perovskite-type oxide as a new oxide ionic conductor, J. Am. Chem. Soc. 116 (9) (1994) 3801–3803.

[81] P. Majewski, M. Rozumek, F. Aldinger, Phase diagram studies in the systems La_2O_3-SrO-MgO-Ga_2O_3 at 1350-1400 °C in air with emphasis on Sr and Mg substituted $LaGaO_3$, J. Alloys Compd. 329 (1-2) (2001) 253–258.

[82] T. Ishihara, H. Matsuda, Y. Takita, Effects of rare-earth cations doped for La site on the oxide ionic-conductivity of $lagao_3$-based perovskite-type oxide, Solid State Ionics 79 (1995) 147–151.

[83] M. Feng, J.B. Goodenough, A superior oxide-ion electrolyte, Eur. J. Solid State Inorg. Chem. 31 (8-9) (1994) 663–672.

[84] P.N. Huang, A. Petric, Superior oxygen ion conductivity of lanthanum gallate doped with strontium and magnesium, J. Electrochem. Soc. 143 (5) (1996) 1644–1648.

[85] K.Q. Huang, R.S. Tichy, J.B. Goodenough, Superior perovskite oxide-ion conductor; strontium- and magnesium-doped $LaGaO_3$: I, phase relationships and electrical properties, J. Am. Ceram. Soc. 81 (10) (1998) 2565–2575.

[86] M. Hrovat, Z. Samardzija, J. Holc, S. Bernik, Subsolidus phase equilibria in the La_2O_3-Ga_2O_3-CeO_2 system, J. Mater. Res. 14 (12) (1999) 4460–4462.
[87] K.Q. Huang, M. Feng, J.B. Goodenough, Sol-gel synthesis of a new oxide-ion conductor Sr- and Mg-doped $LaGaO_3$ perovskite, J. Am. Ceram. Soc. 79 (4) (1996) 1100–1104.
[88] X. Kuang, M.A. Green, H. Niu, P. Zajdel, C. Dickinson, J.B. Claridge, L. Jantsky, M. J. Rosseinsky, Interstitial oxide ion conductivity in the layered tetrahedral network melilite structure, Nat. Mater. 7 (6) (2008) 498–504.
[89] C.I. Thomas, X.J. Kuang, Z.Q. Deng, H.J. Niu, J.B. Claridge, M.J. Rosseinsky, Phase stability control of interstitial oxide ion conductivity in the $La_{1+x}Sr_{1-x}Ga_3O_{7+x/2}$ melilite family, Chem. Mater. 22 (8) (2010) 2510–2516.
[90] C. Tealdi, P. Mustarelli, M.S. Islam, Layered $LaSrGa_3O_7$-based oxide-ion conductors: cooperative transport mechanisms and flexible structures, Adv. Funct. Mater. 20 (22) (2010) 3874–3880.
[91] R.T. Baker, B. Gharbage, F.M.B. Marques, Ionic and electronic conduction in Fe and Cr doped (La, Sr)$GaO_{3-\delta}$, J. Electrochem. Soc. 144 (9) (1997) 3130–3135.
[92] K. Yamaji, T. Horita, M. Ishikawa, N. Sakai, H. Yokokawa, M. Dokiya, Some characteristics for fabrication of $LaGaO_3$-based electrolyte, in: Proc. 5th International Symposium on Solid Oxide Fuel Cells, PV97-40, The Electrochemical Society, 1997, pp. 1041–1050.
[93] J.H. Kim, H.I. Yoo, Partial electronic conductivity and electrolytic domain of $La_{0.9}Sr_{0.1}Ga_{0.8}Mg_{0.2}O_{3-\delta}$, Solid State Ionics 140 (1-2) (2001) 105–113.
[94] H. Yokokawa, N. Sakai, T. Horita, K. Yamaji, Recent developments in solid oxide fuel cell materials, Fuel Cells 1 (2) (2001) 117–131.
[95] K.Q. Huang, J.H. Wan, J.B. Goodenough, Increasing power density of LSGM-based solid oxide fuel cells using new anode materials, J. Electrochem. Soc. 148 (7) (2001) A788–A794.
[96] K.Q. Huang, M. Feng, J.B. Goodenough, M. Schmerling, Characterization of Sr-doped $LaMnO_3$ and $LaCoO_3$ as cathode materials for a doped $LaGaO_3$ ceramic fuel cell, J. Electrochem. Soc. 143 (11) (1996) 3630–3636.
[97] Q.J. Zhou, T.M. He, Y. Ji, $SmBaCo_2O_{5+x}$ double-perovskite structure cathode material for intermediate-temperature solid-oxide fuel cells, J. Power Sources 185 (2) (2008) 754–758.
[98] N. Watanabe, A. Inoishi, T. Ooe, A. Kawakami, H. Shigefuji, T. Ishihara, Effects of fuel velocity in anode channel on power generation property of tubular solid oxide fuel cells using $LaGaO_3$ electrolyte film, Electrochim. Acta 97 (2013) 86–91.
[99] W.B. Wang, Z.J. Yang, H.T. Wang, G.L. Ma, W.J. Gao, Z.F. Zhou, Desirable performance of intermediate-temperature solid oxide fuel cell with an anode-supported $La_{0.9}Sr_{0.1}Ga_{0.8}Mg_{0.2}O_{3-\delta}$ electrolyte membrane, J. Power Sources 196 (7) (2011) 3539–3543.
[100] T. Ishihara, M. Honda, T. Shibayama, H. Minami, H. Nishiguchi, Y. Takita, Intermediate temperature solid oxide fuel cells using a new $LaGaO_3$ based oxide ion conductor – I. Doped $SmCoO_3$ as a new cathode material, J. Electrochem. Soc. 145 (9) (1998) 3177–3183.
[101] C.N. Munnings, S.J. Skinner, G. Amow, P.S. Whitfield, I.J. Davidson, Stability and reactivity of LSGM electrolytes with nickel-based ceramic cathodes, J. Fuel Cell Sci. Technol. 2 (1) (2005) 34–37.
[102] Y.W. Ju, S. Ida, T. Ishihara, A Ce(Mn, Fe)O_2 dense nanofilm as an improved active anode for metal-supported solid oxide fuel cells, RSC Adv. 3 (26) (2013) 10508–10515.

[103] Y.-W. Ju, H. Eto, T. Inagaki, S. Ida, T. Ishihara, Preparation of Ni-Fe bimetallic porous anode support for solid oxide fuel cells using $LaGaO_3$ based electrolyte film with high power density, J. Power Sources 195 (19) (2010) 6294–6300.
[104] K. Yamaji, T. Horita, M. Ishikawa, N. Sakai, H. Yokokawa, Compatibility of $La_{0.9}Sr_{0.1}Ga_{0.8}Mg_{0.2}O_{2.85}$ as the electrolyte for SOFCs, Solid State Ionics 108 (1-4) (1998) 415–421.
[105] K.Q. Huang, J.B. Goodenough, A solid oxide fuel cell based on Sr- and Mg-doped $LaGaO_3$ electrolyte: the role of a rare-earth oxide buffer, J. Alloys Compd. 303 (2000) 454–464.
[106] J.E. Hong, T. Inagaki, S. Ida, T. Ishihara, Titania-added $Ce_{0.6}La_{0.4}O_{2-\delta}$ for the buffer layer of high-performance solid oxide fuel cells using doped lanthanum gallate electrolyte film, J. Am. Ceram. Soc. 95 (11) (2012) 3588–3596.
[107] N. Watanabe, T. Ooe, T. Ishihara, Design of thermal self supported 700 W class, solid oxide fuel cell module using. LSGM thin film micro tubular cells, J. Power Sources 199 (2012) 117–123.
[108] T. Ishihara, J.A. Kilner, M. Honda, N. Sakai, H. Yokokawa, Y. Takita, Oxygen surface exchange and diffusion in $LaGaO_3$ based perovskite type oxides, Solid State Ionics 115 (1998) 593–600.
[109] T. Ishihara, J.A. Kilner, M. Honda, Y. Takita, Oxygen surface exchange and diffusion in the new perovskite oxide ion conductor $LaGaO_3$, J. Am. Chem. Soc. 119 (11) (1997) 2747–2748.
[110] T. Ishihara, H. Furutani, M. Honda, T. Yamada, T. Shibayama, T. Akbay, N. Sakai, H. Yokokawa, Y. Takita, Improved oxide ion conductivity in $La_{0.8}Sr_{0.2}Ga_{0.8}Mg_{0.2}O_3$ by doping Co, Chem. Mater. 11 (8) (1999) 2081–2088.
[111] V.V. Kharton, A.A. Yaremchenko, A.V. Kovalevsky, A.P. Viskup, E.N. Naumovich, P.F. Kerko, Perovskite-type oxides for high-temperature oxygen separation membranes, J. Membr. Sci. 163 (2) (1999) 307–317.
[112] V. Thangadurai, A.K. Shukla, J. Gopalakrishnan, $La_{0.9}Sr_{0.1}Ga_{0.8}Mn_{0.2}O_{2.85}$: a new oxide ion conductor, Chem. Commun. 23 (1998) 2647–2648.
[113] H. Iwahara, High temperature type proton conducting ceramics and their applications, Bull. Ceram. Soc. Jpn. 27 (2) (1992) 112.
[114] H. Matsumoto, T. Sakai, Y. Okuyama, Proton-conducting oxide and applications to hydrogen energy devices, Pure Appl. Chem. 85 (2) (2013) 427–435.
[115] A.K. Azad, J.T.S. Irvine, Synthesis, chemical stability and proton conductivity of the perovksites $Ba(Ce, Zr)_{1-x}Sc_xO_{3-\delta}$, Solid State Ionics 178 (7-10) (2007) 635–640.
[116] J.B. Goodenough, J.E. Ruizdiaz, Y.S. Zhen, Oxide-ion conduction in $Ba_2In_2O_5$ and $Ba_3In_2CeO_8$, $Ba_3In_2HfO_8$, or $Ba_3In_2ZrO_8$, Solid State Ionics 44 (1-2) (1990) 21–31.
[117] K.R. Kendall, C. Navas, J.K. Thomas, H.C. ZurLoye, Recent developments in perovskite-based oxide ion conductors, Solid State Ionics 82 (3-4) (1995) 215–223.
[118] M.J. Verkerk, K. Keizer, A.J. Burggraaf, High oxygen ion conduction in sintered oxides of the Bi_2O_3-Er_2O_3 system, J. Appl. Electrochem. 10 (1) (1980) 81–90.
[119] A. Ahi, A. Mellergard, S.G. Eriksson, Local and average atomic order of ion-conducting $Bi_{0.775}La_{0.225}O_{1.5}$ studied by neutron scattering and reverse Monte Carlo simulations, Solid State Ionics 177 (3-4) (2006) 289–297.
[120] M. Drache, S. Obbade, J.P. Wignacourt, P. Conflant, Structural and conductivity properties of $Bi_{0.775}Ln_{0.225}O_{1.5}$ oxide conductors (Ln = La, Pr, Nd, Sm, Eu, Gd, Tb, Dy) with rhombohedral Bi-Sr-O type, J. Solid State Chem. 142 (2) (1999) 349–359.

[121] K. Kakinuma, H. Yamamura, H. Haneda, T. Atake, Oxide-ion conductivity of $(Ba_{1-x}La_x)_2In_2O_{5+x}$ system based on brownmillerite structure, Solid State Ionics 140 (3-4) (2001) 301–306.
[122] C.A.J. Fisher, M.S. Islam, Defect, protons and conductivity in brownmillerite-structured $Ba_2In_2O_5$, Solid State Ionics 118 (3-4) (1999) 355–363.
[123] G.B. Zhang, D.M. Smyth, Protonic conduction in $Ba_2In_2O_5$, Solid State Ionics 82 (3-4) (1995) 153–160.
[124] T. Schober, J. Friedrich, F. Krug, Phase transition in the oxygen and proton conductor $Ba_2In_2O_5$ in humid atmospheres below 300 °C, Solid State Ionics 99 (1-2) (1997) 9–13.
[125] N. Kitamura, N. Hamao, S.C. Vogel, Y. Idemota, Oxide-ion conduction, average and local structures of $LaSrGa_{1-x}Mg_xO_{4-\delta}$ with layered perovskite structure, Electrochemistry 81 (6) (2013) 448–453.
[126] E. Kendrick, M.S. Islam, P.R. Slater, Developing apatites for solid oxide fuel cells: insight into structural, transport and doping properties, J. Mater. Chem. 17 (30) (2007) 3104–3111.
[127] A. Orera, P.R. Slater, New chemical systems for solid oxide fuel cells, Chem. Mater. 22 (2010) 675–690.
[128] S. Nakayama, M. Sakamoto, M. Higuchi, K. Kodaira, M. Sato, S. Kakita, T. Suzuki, K. Itoh, Electrical properties of new type high oxide ionic conductor $RE_{10}Si_6O_{27}$ (RE = La, Pr, Nd, Sm, Gd, Dy), J. Eur. Ceram. Soc. 18 (10) (1998) 1413–1418.
[129] S. Nakayama, M. Sakamoto, Oxide ionic conductivity of apatite type Nd9.33(SiO4)6O2 single crystal, J. Eur. Ceram. Soc. 19 (4) (1999) 507–510.
[130] S. Nakayama, H. Aono, Y. Sadaoka, Ionic conductivity of Ln10(SiO4)6O3 (Ln = La, Nd, Sm, Gd and Dy), Chem. Lett. 4 (6) (1995) 431–442.
[131] J.E.H. Sansom, D. Richings, P.R. Slater, A powder neutron diffraction study of the oxide-ion-conducting apatite-type phases, $La_{9.33}Si_6O_{26}$ and $La_8Sr_2Si_6O_{26}$, Solid State Ionics 139 (3-4) (2001) 205–210.
[132] J.R. Tolchard, M.S. Islam, P.R. Slater, Defect chemistry and oxygen ion migration in the apatite-type materials $La_{9.33}Si_6O_{26}$ and $La_8Sr_2Si_6O_{26}$, J. Mater. Chem. 13 (8) (2003) 1956–1961.
[133] A. Arikawa, H. Nishiguchi, T. Ishihara, Y. Takita, Oxide ion conductivity in Sr-doped La Ge O apatite oxide, Solid State Ionics 136–137 (2000) 31–37.
[134] V.V. Kharton, F.M.B. Marques, A. Atkinson, Transport properties of solid oxide electrolyte ceramics: a brief review, Solid State Ionics 174 (2004) 135–149.
[135] P.R. Slater, J.E.H. Sansom, J.R. Tolchard, Development of apatite-type oxide ion conductors, Chem. Rec. 4 (6) (2004) 373–384.
[136] T. Iwata, E. Bechade, K. Fukuda, O. Masson, I. Julien, E. Champion, P. Thomas, Lanthanum- and oxygen-deficient crystal structures of oxide-ion conducting apatite-type silicates, J. Am. Ceram. Soc. 91 (11) (2008) 3714–3720.
[137] M.S. Islam, J.R. Tolchard, P.R. Slater, An apatite for fast oxide ion conduction, Chem. Commun. 13 (2003) 1486–1487.
[138] R. Ali, M. Yashima, Y. Matsushita, H. Yoshioka, K. Ohoyama, F. Izumi, Diffusion path of oxide ions in an apatite-type ionic conductor $La_{9.69}(Si_{5.70}Mg_{0.30})O_{26.24}$, Chem. Mater. 20 (16) (2008) 5203–5208.
[139] S. Guillot, S. Beaudet-Savignat, S. Lambert, R.N. Vannier, P. Roussel, F. Porcher, Evidence of local defects in the oxygen excess apatite $La_{9.67}(SiO_4)O_{2.5}$ from high resolution neutron powder diffraction, J. Solid State Chem. 182 (2009) 3358–3364.

[140] E. Bechade, O. Masson, T. Iwata, I. Julien, K. Fukuda, P. Thomas, E. Champion, Diffusion path and conduction mechanism of oxide ions in apatite-type lanthanum silicates, Chem. Mater. 21 (2009) 2508–2517.

[141] A. Jones, P.R. Slater, M.S. Islam, Local defect structures and ion transport mechanisms in the oxygen-excess apatite $La_{9.67}(SiO_4)_6O_{2.5}$, Chem. Mater. 20 (2008) 5055–5060.

[142] S. Nakayama, M. Higuchi, Electrical properties of apatite-type oxide ionic conductors $RE_{9.33}(SiO_4)_6O_2$ (RE = Pr, Nd and Sm) single crystals, J. Mater. Sci. Lett. 20 (2001) 913–915.

[143] S. Nakayama, Y. Higuchi, M. Sugawara, A. Makiya, K. Uematsu, M. Sakamoto, Fabrication of c-axis oriented apatite-type polycrystalline $La_{10}Si_6O_{27}$ ceramic and its anisotropic oxide ionic conductivity, Ceram. Int. 40 (2014) 1221–1234.

[144] M.M. Vieira, J.C. Oliveira, A.K. Shaula, B. Trindade, A. Cavaleiro, Structure and ionic conductivity of reactively sputtered apatite-type lanthanum silicate thin films, Surf. Coat. Technol. 247 (2014) 14–19.

[145] J.E.H. Sansom, E. Kendrick, J.R. Tolchard, M.S. Islam, P.R. Slater, A comparison of the effect of rare earth vs Si site doping on the conductivities of apatite-type rare earth silicates, J. Solid State Electrochem. 10 (2006) 562–568.

[146] S.-F. Wang, Y.-F. Hsu, W.-J. Lin, K. Kobayashi, Transition metal-doped lanthanum germanate apatites as electrolyte materials of solid oxide fuel cells, Solid State Ionics 247–248 (48-55) (2013) 48.

[147] A. Najib, J.E.H. Sansom, J.R. Tolchard, P.R. Slater, M.S. Islam, Doping strategies to optimise the oxide ion conductivity in apatite-type ionic conductors, Dalton Trans. 33 (19) (2004) 3106–3109.

[148] A. Vincent, S.B. Savignat, F. Gervais, Elaboration and ionic conduction of apatite-type lanthanum silicates doped with Ba, $La_{10-x}Bax(SiO_4)_6O_{3-x/2}$ with x = 0.25-2, J. Eur. Ceram. Soc. 27 (2–3) (2007) 1187–1192.

[149] A.L. Shaula, V.V. Kharton, F.M.B. Marques, Oxygen ionic and electronic transport in apatite-type La(Si, Al)6O26+/-d, J. Solid State Chem. 178 (2005) 2050–2061.

[150] P.J. Panteix, I. Julien, P. Abélard, D. Bernache-Assollant, Influence of cationic vacancies on the ionic conductivity of oxyapatites, J. Eur. Ceram. Soc. 28 (4) (2008) 821–828.

[151] E.J. Abram, D.C. Sinclair, A.R. West, A novel enhancement of ionic conductivity in the cation-deficient apatite $La_{9.33}(SiO_4)_6O_2$, J. Mater. Chem. 11 (8) (2001) 1978–1979.

[152] A.L. Shaula, V.V. Kharton, F.M.B. Marques, Ionic and electronic conductivities, stability and thermal expansion of $La_{10-x}(Si, Al)_6O_{26\pm\delta}$ solid electrolytes, Solid State Ionics 177 (19-25) (2006) 1725–1728.

[153] S.W. Tao, J.T.S. Irvine, Preparation and characterisation of apatite-type lanthanum silicates by a sol-gel process, Mater. Res. Bull. 36 (7-8) (2001) 1245–1258.

[154] K. Kobayashi, Y. Matsushita, N. Igawa, F. Izumi, C. Nishimura, S. Miyoshi, Y. Oyama, S. Yamaguchi, Water-based sol-gel synthesis and crystal structure refinement of lanthanum silicate apatite, Solid State Ionics 179 (38) (2008) 2209–2215.

[155] J.E.H. Sansom, J.R. Tolchard, P.R. Slater, M.S. Islam, Synthesis and structural characterisation of the apatite-type phases $La_{10-x}Si_6O_{26+z}$ doped with Ga, Solid State Ionics 167 (1-2) (2004) 17–22.

[156] S.H. Jo, P. Muralidharan, D.K. Kim, Low-temperature sintering of dense lanthanum silicate electrolytes with apatite-type structure using an organic precipitant synthesized nanopowder, J. Mater. Res. 24 (1) (2009) 237–244.

[157] R. Serra, C. Alves, F.A.C. Oliveira, T. Marcelo, J. Mascarenhas, B. Trindade, Enhanced sinterability of mechanical alloyed $La_{9.33}Si_2Ge_4O_{26}$ oxyapatite powders for IT-SOFC electrolytes, Ceram. Int. 38 (2012) 5355–5361.

[158] A.F. Fuentes, E. Rodriguez-Reyna, L.G. Martinez-Gonzalez, M. Maczka, J. Hanuza, U. Amador, Room-temperature synthesis of apatite-type lanthanum siluicates by mechanically milling constitutent oxides, Solid State Ionics 177 (2006) 1869–1873.

[159] A.L. Shaula, V.V. Kharton, J.C. Waerenborgh, D.P. Rojas, E.V. Tsipis, N. P. Vyshatko, M.V. Patrakkev, F.M.B. Marques, Transport properties and Mossbauer spectra of Fe-substituted La10-x(Si, Al)6O26 apatites, Mater. Res. Bull. 39 (2004) 763–773.

[160] J. McFarlane, S. Barth, M. Swaffer, J.E.H. Sansom, P.R. Slater, Synthesis and conductivities of the apatite-type systems, $La_{9.33+x}Si_{6-y}MyO_{26+z}$ (M = Co, Fe, Mn) and La_8Mn_2 Si_6O_{26}, Ionics 8 (2002) 149–154.

[161] V.V. Kharton, A.L. Shaula, M.V. Patrakeev, J.C. Waerenborgh, D.P. Rojas, N. P. Vyshatko, E.V. Tsipis, A.A. Yaremchenko, F.M.B. Marques, Oxygen ionic and electronic transport in apatite-type solid electrolytes, J. Electrochem. Soc. 151 (8) (2004) A1236–A1246.

[162] C.Y. Ma, P. Briois, J. Bohlmark, F. Lapostolle, A. Bilard, $La_{9.33}Si_6O_{26}$ electrolyte thin films for IT-SOFC application deposited by a HIPIMS/DC hybrid magnetron sputtering process, Ionics 14 (2008) 471–476.

[163] H. Yoshioka, T. Mitsui, A. Mineshige, T. Yazawa, Fabrication of anode supported SOFC using plasma-sprayed films of the apatite-type lanthanum silicate as an electrolyte, Solid State Ionics 181 (2010) 1707–1712.

[164] D. Marrero-Lopez, M.C. Martin-Sedeno, J. Pena-Martinez, J.C. Ruiz-Morales, P. Nunez, M.A.G. Aranda, J.R. Ramos-Barrado, Evaluation of apatite silicates as solid oxide fuel cell electrolytes, J. Power Sources 195 (2010) 2496–2506.

[165] A.A. Yaremchenko, V.V. Kharton, D.O. Bannikov, D.V. Znosak, J.R. Frade, V. A. Cherepanov, Performance of perovskite-related oxide cathodes in contact with lanthanum silicate electrolyte, Solid State Ionics 180 (11-13) (2009) 878–885.

[166] E.V. Tsipis, V.V. Kharton, J.R. Frade, Electrochemical behavior of mixed-conducting oxide cathodes in contact with apatite-type $La_{10}Si_5AlO_{26.5}$ electrolyte, Electrochim. Acta 52 (13) (2007) 4428–4435.

[167] H. Yoshioka, S. Tanase, Magnesium doped lanthanum silicate with apatite-type structure as an electrolyte for intermediate temperature solid oxide fuel cells, Solid State Ionics 176 (2005) 2395–2398.

[168] P. Lacorre, F. Goutenoire, O. Bohnke, R. Retouz, Y. Laligant, Designing fast oxide-ion conductors based on $La_2Mo_2O_9$, Nature 404 (2000) 856–858.

[169] F. Goutenoire, O. Isnard, R. Retoux, P. Lacorre, Crystal structure of $La_2Mo_2O_9$, a new fast oxide-ion conductor, Chem. Mater. 12 (2000) 2575–2580.

[170] I.R. Evans, J.A.K. Howard, J.S.O. Evans, The crystal structure of α-$La_2Mo_2O_9$ and the structural origin of the oxide ion migration pathway, Chem. Mater. 17 (6) (2005) 4074–4077.

[171] S. Georges, F. Goutenoire, O. Bohnke, M.C. Steil, S.J. Skinner, H.D. Wiemhofer, P. Lacorre, The LAMOX family of fast oxide-ion conductors: overview and recent results, J. New Mater. Electrochem. Syst. 7 (1) (2004) 51–57.

[172] S. Georges, O. Bohnke, F. Goutenoire, Y. Laligant, J. Fouletier, P. Lacorre, Effects of tungsten substitution on the transport properties and mechanism of fast oxide-ion conduction in $La_2Mo_2O_9$, Solid State Ionics 177 (19-25) (2006) 1715–1720.

[173] S. Basu, P.S. Devi, H.S. Maiti, Nb-doped $La_2Mo_2O_9$: a new material with high ionic conductivity, J. Electrochem. Soc. 152 (11) (2005) A2143–A2147.

[174] L. Malavasi, H. Kim, S.J.L. Billinge, T. Proffen, C. Tealdi, G. Flor, Nature of the monoclinic to cubic phase transition in the fast oxygen ion conductor $La_2Mo_2O_9$ (LAMOX), J. Am. Chem. Soc. 129 (21) (2007) 6903–6907.
[175] J. Liu, S. Hull, A. Istaq, S. Skinner, Application of combined neutron diffraction and impedance spectroscopy for in-situ structure and conductivity studies of $La_2Mo_2O_9$, Nucl. Instrum. Methods Phys. Res. Sect. B 269 (2011) 539–543.
[176] C.J. Hou, Y.D. Li, P.J. Wang, C.S. Liu, X.P. Wang, Q.F. Fang, D.Y. Sun, Oxygen-ion arrangements and concerted motion in β-$La_2Mo_2O_9$, Phys. Rev. B 76 (1) (2007) 014104.
[177] S. Georges, S.J. Skinner, P. Lacorre, M.C. Steil, Oxide ion diffusion in optimised LAMOX materials, Dalton Trans. 19 (2004) 3101–3105.
[178] J. Liu, R.J. Chater, R.J.H. Morris, S.J. Skinner, Oxygen surface exchange and diffusion studies of $La_2Mo_2O_9$ in different exchange atmospheres, Solid State Ionics 189 (1) (2011) 39–44.
[179] A. Morata, F. Peiró, J.R. Morante, Conductivity dependence on oxygen partial pressure and oxide-ion transport numbers determination for $La_2Mo_2O_9$, Electrochem. Solid-State Lett. 7 (2004) A373–A375.
[180] D. Marrero-López, J. Canales-Vázquez, J.C. Ruiz-Morales, J.T.S. Irvine, P. Núñez, Electrical conductivity and redox stability of $La_2Mo_{2-x}W_xO_9$ materials, Electrochim. Acta 50 (2005) 4385–4395.
[181] D. Marrero-Lopez, J. Pena-Martinez, J.C. Ruiz-Morales, D. Perez-Coll, K.M. Martin-Smith, P. Nunez, Applicability of $La_2Mo_{2-y}W_yO_9$ materials as solid electrolyte for SOFCs, Solid State Ionics 178 (23-24) (2007) 1366–1378.
[182] S. Georges, F. Goutenoire, Y. Laligant, P. Lacorre, Reducibility of fast oxide-ion conductors $La_{2-x}R_xMo_{2-y}W_yO_9$ (R = Nd, Gd), J. Mater. Chem. 13 (9) (2003) 2317–2321.
[183] G. Corbel, Y. Laligant, F. Goutenoire, E. Suard, P. Lacorre, Effects of partial substitution of Mo^{6+} by Cr^{6+} and W^{6+} on the crystal structure of the fast oxide-ion conductor structural effects of W^{6+}, Chem. Mater. 17 (18) (2005) 4678–4684.
[184] J.E. Vega-Castillo, U.K. Ravella, G. Corbel, P. Lacorre, A. Caneiro, Thermodynamic stability of $La_2Mo_{2-y}W_yO_9$, $La_2Mo_{2-y}W_yO_{8.96+0.02y}$ and $La_7Mo_{7(2-y)/2}W_{7y/2}O_{30}$ (y = 0, 0.5 and 1.0), Dalton Trans. 43 (2014) 2661–2669.
[185] L. Baque, J. Vega-Castillo, S. Georges, A. Caneiro, E. Djurado, Microstructural and electrical characterisations of tungsten-doped $La_2Mo_2O_9$ prepared by spray pyrolysis, Ionics 19 (2013) 1761–1774.
[186] S. Basu, P.S. Devi, N.R. Bandyopadhyaya, Sintering and densification behaviour of pure and alkaline earth (Ba^{2+}, Sr^{2+} and Ca^{2+}) substituted $La_2Mo_2O_9$, J. Eur. Ceram. Soc. 33 (2013) 79–85.
[187] S. Georges, A.R. Renata, E. Djurado, Microstructure related conductivity in $La_2Mo_2O_9$ ceramics, J. Phys. Chem. C 112 (9) (2008) 3194–3202.
[188] D. Marrero-Lopez, J.C. Ruiz-Morales, P. Nunez, J.C.C. Abrantes, J.R. Frade, Synthesis and characterization of $La_2Mo_2O_9$ obtained from freeze-dried precursors, J. Solid State Chem. 177 (7) (2004) 2378–2386.
[189] C. Tealdi, G. Chiodelli, G. Flor, S. Leonardi, Electrode stability and electrochemical performance of LAMOX electrolytes under fuel cell conditions, Solid State Ionics 181 (2010) 1456–1461.
[190] G. Corbel, P. Lacorre, Compatibility evaluation between $La_2Mo_2O_9$ fast oxide-ion conductor and Ni-based materials, J. Solid State Chem. 179 (2006) 1339–1344.

[191] H.-C. Chang, D.-S. Tsai, W.-H. Chung, Y.-S. Huang, M.-V. Le, A ceria layer as diffusion barrier between LAMOX and lanthanum strontium cobalt ferrite along with the impedance analysis, Solid State Ionics 180 (2009) 412–417.

[192] J.-C. Lo, D.-S. Tsai, Y.-C. Chen, M.-V. Le, W.-H. Chung, $La_2Mo_2O_9$-based electrolyte: ion conductivity and anode-supported cell under single chamber conditions, J. Am. Ceram. Soc. 94 (3) (2011) 806–811.

[193] R. Leah, A. Bone, A. Selcuk, D. Corcoran, M. Lankin, Z. Dehaney-Steven, M. Selby, P. Whalen, Development of highly robust, volume-manufacturable metal-supported SOFCs for operation below 600 °C, Solid Oxide Fuel Cells 12 (SOFC XII) 35 (1) (2011) 351–367.

Anodes

5

M. Cassidy, P.A. Connor, J.T.S. Irvine, C.D. Savaniu
University of St. Andrews, St. Andrews, UK

5.1 Introduction

The basic requirement of an anode (i.e. fuel oxidation electrode) in any fuel cell is to provide sufficient active sites for the oxidation of the fuel, under operating conditions. The simplest oxidation reaction at an anode is:

$$H_{2(g)} \rightarrow 2H^{+}_{(i)} + 2e^{-} \quad (5.1)$$

This equation illustrates that three phases are needed to meet in one place to produce a working anode: the gas-phase H_2, a phase to carry the H^+ ions from the electrolyte and a phase to carry electrons to the circuit. For a solid oxide fuel cell (SOFC), this means a porous anode is made of ionic and electronic conductors. The ionic and electronic conductivity can be from a single phase or from a composite of an ionic conducting phase and an electronic conducting phase. The intersection of these three phases is called the triple phase boundary (TPB). It is the critical part of the anode and needs careful microstructural engineering to give good performance. This will be discussed later in the chapter.

Forming a good TPB is basic requirements for any fuel cell anode. However, the high operating temperature of an SOFC means that there are a lot more stability requirements for an electrode to ensure the expected long lifetime of the cells. The cathode only ever experiences air, with the partial pressure of oxygen always high, only locally changing by a small amount. However, the anode will experience a large range of oxygen partial pressures (pO_2). Anodes are typically prepared and heated in air (high pO_2), but then reduced in fuel at high temperature (very low pO_2). Then during normal operation, the pO_2 can still vary over several orders of magnitude. Therefore, the anode materials need to withstand high temperatures and varying O_2 partial pressures without performance loss, microstructure change or reactions with other cell parts.

A general set of requirements for an SOFC anode is listed in Table 5.1. They broadly separate into cell performance and cell stability/lifetime. These criteria are largely the same as those in the anode chapter of the first edition of this book published in 2003. The difference now is the large body of research which has extended the possibilities of SOFC anodes during the past 12 years.

High-temperature Solid Oxide Fuel Cells for the 21st Century. http://dx.doi.org/10.1016/B978-0-12-410453-2.00005-1

Table 5.1 Requirements for SOFC electrodes

Requirements	Guide parameter [1]
Electronic conductivity	>100 S cm^{-1} [2]
Ionic conductivity	>10 mS cm^{-1} [3]
Catalyst for fuel reduction	<0.1 Ω cm^{-2} for electrode
Electrode resistance	Total < 0.15 Ω cm^{-2} [4]
Thermal stability	At operating temperature, for long time
Chemical stability	At operating temperature in contact with other materials (electrolyte, interconnect)
Thermal expansion match	Cell experiences large temperature changes
Structural stability	Must hold porosity at operating temperature, for long time
Reduction stability	Highly reducing conditions, reducing gas at high temperature

5.2 Cell performance requirements

The most important property of an anode material is to have adequate electronic conductivity to produce a low enough resistance to get the electrons out of the cell. For good performance, it is usually taken that the total sheet resistance of a component in an SOFC is less than 0.15 Ω cm^{-2} [4] which includes electronic, ionic and chemical resistances. This means each of these factors must be less than this. With the porous nature of the electrodes increasing the resistance and to cover any extra contact resistances, the conductive material would have to have a conductivity higher than 100 S cm^{-1}. With careful engineering of the anode current collection, this could be relaxed to about 1 S cm^{-1} [1].

Graded anodes which separate the chemical function of the anode from the current collection function are advantageous. This allows the use of a lower conductivity chemically active anode, but a more highly conductive material is then needed as a current collector, with the same stability criteria as below.

The ionic transport within an SOFC anode should be high in order to extend the electrochemical reaction from the TPB to the entire surface. The required ionic conductivity can be introduced into the anode in two ways. The ideal way is to use a mixed ionic and electronic conductor (MIEC) which has sufficient electronic conductivity and ionic conductivity over 1 S cm^{-1}. However, this high ionic and electronic conductivity is uncommon in materials and so leads to the other way of introducing ionic conductivity, by mixing an electronically conducting "anode" phase with an ionic conductor, usually the same material as the electrolyte. This forms a composite anode, but where the electronic and ionic parts are separately variable, and introduces more ability to tune the physical properties of the anode.

The final property needed for a good anode material is that it must be a good catalyst for fuel oxidation. The high operating temperature of an SOFC gives a kinetic

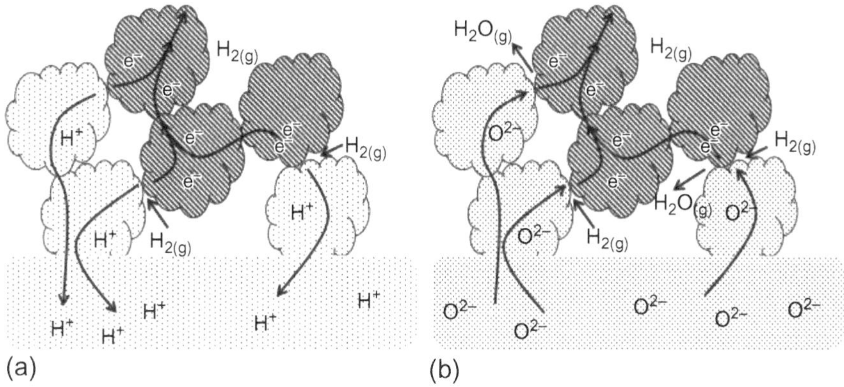

Figure 5.1 Schematic representation of hydrogen oxidation processes at composite electrodes with (a) proton and (b) oxide conducting electrolytes.

advantage to any reaction, meaning that precious metal catalysts, such as Pt, Pd, as used in low-temperature cells, are not needed and can be replaced with cheaper Ni- or oxide-based catalysts. Materials are usually chosen to have sufficient conductivity, and the temperature will make them a good enough catalyst. That being said, a better catalyst still lowers the resistance of the electrode and so the cell, which can be achieved by doping or adding active materials later.

The total reaction in a fuel cell is hydrogen + oxygen combining to give water, but the exact reaction at each electrode and direction of ion movement depend on whether the electrolyte is a protonic or oxide ion conductor (see SOFC electrolyte Chapter 4).

With a proton-conducting electrolyte, H^+ are transported across electrolyte and react with O_2 on the cathode side, leaving the half reaction on the anode side (Equation 5.1 and Figure 5.1a).

With an oxide ion conducting electrolyte, O^{2-} ions arrive from the electrolyte and react with the H^+ on the surface. Note that for this oxide ion case, as well as depleting the H_2 from the fuel stream, the product (H_2O) is added to the fuel stream, diluting it further and lowering the pO_2 of the system. (Equation 5.2 and Figure 5.1b).

$$H_{2(g)} + O^{2-}_{(i)} \rightarrow H_2O_{(g)} + 2e^- \tag{5.2}$$

For a hydrogen-fuelled cell, the only reactions possible are the hydrogen oxidation reactions (Equation 5.1 or 5.2).

5.3 Cell lifetime requirements

While there are many different designs of SOFC that operate at a variety of temperatures, they run above several hundred °C and, being made of ceramics, have normally been produced at even higher temperatures. Therefore, all of the components in an anode must be both physically and chemically stable for short times at manufacturing temperatures, and for long times (1000s of hours) at operating temperatures. It is often difficult to observe small, slow degradation mechanisms, requiring very long test times.

The anode is not a standalone component and must also be both physically and chemically compatible with the other components of the cell, especially the electrolyte and interconnect. The thermal expansion of the anode must match that of the other cell components, with no phase changes of significant volume difference, which would cause cracking of the cell on thermal cycling. Typically, the thermal expansion coefficient (TEC) is about 10–12 ppm/°C to match that of the electrolyte but can be varied a bit depending on the design of the cell. The use of composite anodes can allow for the adjustment of the TEC for a better match.

Anode materials must also not react with the other cell components at operating temperatures, at least to form deleterious phases. Again it is the very long operating times at high temperatures that can slowly form blocking phases which slowly degrade cell/stack performance. Any materials used in an anode must be stable to a wide range of reducing conditions, depending on load and fuel, at the operating temperature. It also experiences a significant range of pO_2 during operation, with very low pO_2 under low load or at the fuel input, with the pO_2 rising as the fuel is used up under high load or near the outlet of the anode. This increasing of the pO_2 is compounded by the extra water added to the fuel stream for oxide ion conductors. It is also quite likely over the lifetime of any SOFC system that it will be exposed to air several times while still hot, and so will need to withstand this oxidation. Also for practical and cost reasons, the anode will usually be sintered in air and so must be stable enough to be fabricated into a device. The best solution to all these problems is to make a material that is stable under both oxidising and reducing conditions.

5.4 Catalytic and reforming properties

The catalytic properties of the anode in a SOFC are very important in determining the overall cell performance. Usually, the catalyst is part of the anode, and the high temperature gives sufficient kinetics. However, as with using a composite electrode to gain both electronic and ionic conductivity, a small amount of catalyst may be added to the surface of the anode after it has been sintered, via impregnation for instance, to improve the electrode behaviour, see Figure 5.2. This places the catalyst more at the TPB where it is needed and less in the bulk and so it can be used at much lower levels to get the same effect as a much larger amount if used as a component of the anode. This allows the use of more expensive catalyst materials. See Section 5.7.2.2 for more details on this.

If one of the more commonly available hydrocarbon fuels is to be used in a fuel cell, then the catalyst requirements are more complicated. The catalyst must be able to cope with the oxidation of the hydrocarbon and the higher levels of impurities found in these fuels. To use a hydrocarbon directly in a fuel cell, the anode catalyst must be

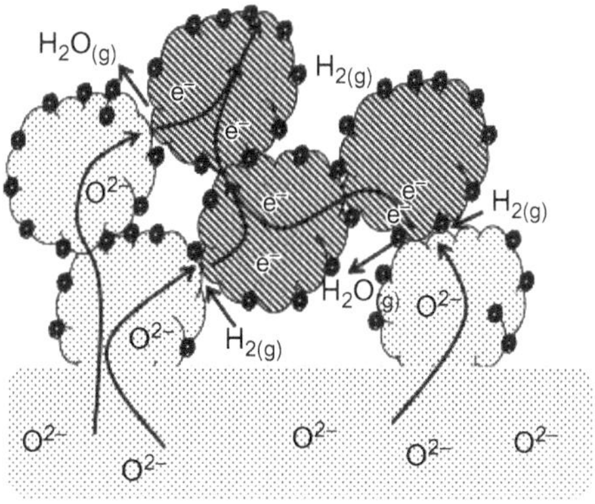

Figure 5.2 Surface decorated with catalytic particles, which may be Pt, Ni or various transition metal oxides to catalyse oxidation of hydrogen.

able to oxidise the hydrocarbons, sometimes without the need of water or reforming [5], as shown in Equation (5.3).

$$CH_{4(g)} + 4O^{2-}_{(i)} \rightarrow CO_{2(g)} + 2H_2O_{(g)} + 8e^- \tag{5.3}$$

However, this reaction is kinetically slow and so the hydrocarbons are typically reformed to form smaller molecules, ideally CO and H_2 and the anode only needs to oxidise these species. The reforming can be done externally or via internal reforming in the cells themselves reducing the external part count. The hydrocarbons are typically not only reformed with steam (Equation 5.4) but can also be reformed with small amounts of O_2/air (see Chapter 14 for more details).

$$CH_{4(g)} + H_2O_{(g)} \leftrightarrow CO_{(g)} + 3H_{2(g)} \tag{5.4}$$

This reaction is endothermic and so forms a good synergy to use the waste heat from the fuel cell to drive the reformation. Both the CO and H_2 produced from reforming can be used in an SOFC (unlike low-temperature fuel cells), but the anode needs to catalyse the oxidation of both H_2 (Equation 5.2) and CO to CO_2.

$$CO_{(g)} + O^{2-}_{(i)} \rightarrow CO_{2(g)} + 2e^- \tag{5.5}$$

The kinetics of reaction (5.5) are typically more sluggish than the hydrogen oxidation reaction [6] and in a lot of cases this reaction is bypassed by the faster gas-phase water–gas shift reaction (WGSR, Equation 5.6), reacting the CO with water, from feed or hydrogen oxidation, to form more H_2, which is then utilised by the faster hydrogen oxidation reaction.

$$CO_{(g)} + H_2O_{(g)} \leftrightarrow CO_{2(g)} + H_{2(g)} \tag{5.6}$$

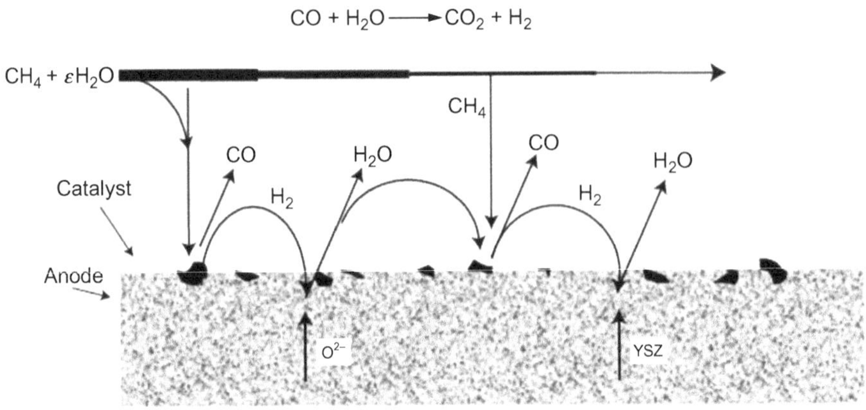

Figure 5.3 Gradual internal reforming of methane.
Reprinted from Ref. [9]. Reproduced by permission of The Electrochemical Society.

In practice, hydrocarbon fuels are normally introduced with extra water (usually with a minimum steam to carbon ratio of 2) over a steam reforming catalyst where reactions (5.4) and (5.6) take place [7]. This situation where the reforming reactions take place in the outer anode support and with only reformate products reaching the triple-phase boundaries is better described as direct utilisation rather than direct oxidation [8]. In the case where the process is enhanced when the products of the fuel cell reaction are also utilised, the reforming process is known as gradual internal reforming [9], see Figure 5.3.

Care needs to be taken in the design and control of the fuel mixture to ensure the speed of the endothermic reformation reactions does not lead to localised cooling of the inlet of the cell which in turn can create thermal and mechanical stresses which can lead to cracking and failure [10].

The reforming reactions compete with the cracking and Boudouard reactions, Equations (5.7) and (5.8), respectively.

$$CH_{4(g)} \rightleftharpoons C_{(s)} + 2H_{2(g)} \tag{5.7}$$

$$2CO_{(g)} \rightarrow CO_{2(g)} + C_{(s)} \tag{5.8}$$

This formation of solid carbon can quickly block pipes and destroy an anode, by filling up the pores and breaking up the structure of the anode. This is a particular problem for Ni-based anodes, see Section 5.6 for details.

External reforming, in a dedicated external unit, can allow better control of the reforming process [11]. The heat from the cell stack can be used to maintain the reformer temperature while providing advantageous heat management in stacks [12]. These external reformers also allow the careful management of the carbon to steam ratio and to ensure no carbon formation anywhere in the system.

The actual details of the processes happening in SOFC anodes are not well understood. Many *in situ* spectroscopic probes such as Raman and infrared spectroscopy are routinely used in catalysis research to characterise surface intermediates and reaction mechanisms. It is very difficult, however, to apply these techniques to an operating SOFC anode. Nonetheless some inroads are being made. For example, Liu and coworkers [13] pioneered the application of infrared emission spectroscopy to characterise working SOFC anodes. Although these studies are in the early stages, preliminary results appear very promising. Developments of models of anodes that allow other spectroscopic probes such as photoelectron spectroscopy and detailed kinetic measurements to be made are also needed in order to advance research in this area.

Sulphur compounds are present in many real world fuels either as a natural impurity in biogas, diesel and coal gas, sometimes at significant levels, or introduced as a deliberate additive, such as odorants in natural gas supplies for leak detection. Due to the operational temperatures employed in SOFCs, most sulphur compounds will be converted to H_2S by the time they reach either the support or the anode itself. This will react rapidly with the electrode surface, even at ppm levels to block reactions sited for both steam reforming and fuel oxidation severely degrading Ni catalyst performance [14]. Degradation is often a two-stage process with an instantaneous drop followed by a slower, longer-term continuous degradation [15] as shown in Figure 5.4. While exact mechanisms are still under discussion, it is thought that the main source of the blocking is the formation of sulphides on the surface of the metal [16]. It has also been observed that while some recovery will take place after removal of the sulphur there is often a permanent drop in performance; the anode does not fully recover, and in many circumstances, this can be fatal to the operation of a larger stack or system. A more detailed discussion of these issues and the application of fuels beyond hydrogen have been described elsewhere [17].

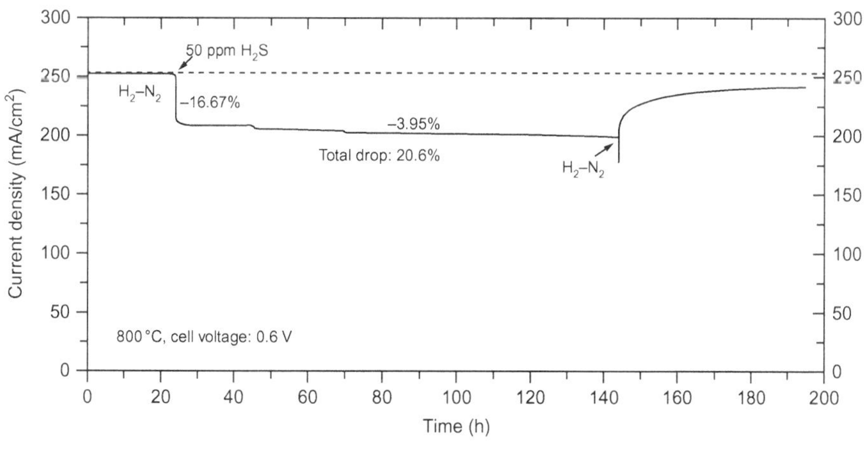

Figure 5.4 Effect of 50 ppm H_2S in fuel on cell performance showing two-stage degradation and partial recovery.
Reprinted from Ref. [15]. Reproduced by permission of The Electrochemical Society.

5.5 Anode design and engineering

In early designs of SOFC, the anode was present primarily for its electrode function with the support elements being either the electrolyte in planar systems [18] or the cathode in the case of the Siemens Westinghouse tubular system [19]. In these cases, the anode could be a relatively thin layer, say 20–30 μm and would be focussed around the electrochemistry with a finer microstructure to maximise surface area and TPB. As SOFC technology matured, there was a drive to reduce the operating temperature to ease the demands on other materials in the stack.

Target temperatures were in the range of 750 °C or below, and this would allow the use of far wider range of materials, such as ferritic stainless steels in the interconnect role. However, to maintain an acceptable ohmic loss across the electrolyte, the thickness could be no more than 15 μm which was too thin to be the structural element of the cell so required another supporting element [6]. The relative mechanical properties of the other cell materials at the time (Ni-YSZ cermet for the anode and $LaSrMnO_3$ perovskite for the cathode) dictated that the anode would make the better supporting structure in a supported planar system, and so the anode-supported cell became a prevalent design.

A number of production routes have been utilised to fabricate anode-supported cells [20]. However, in all cases, the new supporting anode was thicker (250 μm–2 mm) than supported anodes. This demanded some compromises in the design of the anode to allow for the additional support function, such as larger particle sizes and coarser porosity to give better strength, current collection and gas diffusion. This had a detrimental effect on the electrochemical function. To counter this, an additional layer of cermet was placed between the supporting anode and the electrolyte [21]. This layer is usually between 10 and 20 μm in thickness, with a finer microstructure more closely tailored around the needs of the electrochemistry, often deposited by screen printing or occasionally slip casting, and gives rise to the two-layer electrode structure we are familiar with in today's cells. Frequently termed the "anode functional layer" this should in reality be referred to as the anode, as it is in fact the true anode, with the outer layer being the anode support.

An important aspect of the anode support is current collection and distribution between the anode and the contact points of the interconnect. Although there are strong similarities between anode and anode support layers, their functions can have quite different requirements [22]. This has significant implications for the integration of advanced ceramic and non-cermet anodes, where conductivity values can be a lot lower than that of the percolating Ni network in a cermet. The actual anode itself only needs to be 10–15 μm thick as the electrochemistry generally takes place very close to the electrolyte interface, as was identified for the two-layer Ni systems described above [23]. It is in this layer that the requirements of good ionic conductivity, high electrocatalytic activity and high affinity with the electrolyte material are still required. However, if this layer is now thin, the demands of electrical conductivity are greatly reduced, as the important aspect of this property is that the ohmic drop over this layer is sufficiently low (0.1 Ω cm^2) [24]. This opens up a significant number of alternate candidate materials for the anode that would not meet conductivity requirements if based on bulk nickel cermet conductivity.

Similarly, the requirements of the support/current collector are different to that of the anode with structural stability, high electrical conductivity and open porosity being of high importance, while ionic conductivity and electrocatalytic activity are not required. Chemical compatibility with the electrolyte while still advantageous is not essential as the presence of a carefully designed and formulated anode layer can maintain physical separation. Again recognising these differing requirements between the anode and the support/current collector can open up a wider range of candidate materials for these roles. This separation of requirements can be taken one stage further by also separating the support and current collector function. This is demonstrated by systems such as the segmented planar design, where thick film active components are deposited onto a porous inert support. In this case, a highly conductive but thin (10–50 μm) current collecting layer can be deposited between the anode and the support so removing the requirement for high conductivity from the anode [25]. Indeed, this system has been used to demonstrate ceramic anodes based on $La_{0.75}Sr_{0.25}Cr_{0.5}Mn_{0.5}O_3$ in industrially relevant designs [26].

A further anode support strategy that has been well studied over a number of years has been the metal-supported SOFC. In this concept, the mechanical support element is a porous metal on the fuel side of the cell. Both planar and tubular designs have utilised this approach, with the former being more prevalent. The cells are based on either a gadolinia-doped ceria [27] or thin zirconia electrolyte with an aim of operating in the intermediate temperature regime of 500–750 °C, with the ceria-based systems targeting the lower band below 600 °C and the zirconia-based systems the higher end up to 700–750 °C. The main advantages of this approach are seen to be reduced materials costs (with standard industrial metals such as ferritic stainless steel or Hastelloy proving popular choices), improved mechanical robustness and, of course, the high conductivity support afforded by the metal [28].

Initial metal-supported cells utilised similar Ni-based cermet anodes of standard SOFC; however, processing of the metal support in conjunction with the ceramic materials could be problematic as it is vital to avoid oxidation of the metal. DLR in Germany championed the use of vacuum plasma spraying of the active ceramic layers onto Ni felts to avoid high-temperature sintering in air [29]. Other developers have attempted to utilise routes such as screen printing and tape casting to produce both supporting structures and active layers; however, thermal processing can be an issue and firing in reducing atmospheres is often required [30]. In recent iterations, developers have adopted the infiltration route, described in detail later in this chapter. Here active catalyst materials can be deposited onto the porous metal supports or thinner supported porous ceramic or cermet layers. [31] These are often ceria- and Ni-based materials with electrochemistry similar to that discussed elsewhere in this chapter. One of the significant challenges in the further development of metal-supported cells is the interdiffusion of the catalyst materials with the supporting metal substrates leading to performance degradation [32], and strategies to mitigate this are an active area of research.

5.6 Conventional nickel-based anodes

Single-phase materials were investigated in the earliest SOFC developments such as graphite, iron oxide, platinum group and transition metals [33,34]. Graphite corrodes electrochemically and platinum spalls off in service, presumably due to water vapour evolution at the metal oxide interface. Iron oxidises when the partial pressures of oxidation products in the anode compartment of an operating cell exceed a critical value, with formation of a red iron oxide. Cobalt is somewhat more stable but also more costly. Nickel has a significant thermal expansion mismatch to stabilised zirconia, and at high temperatures the metal aggregates by grain growth, finally obstructing the porosity of the anode and eliminating the three-phase boundaries required for cell operation.

This led to the introduction of the nickel–zirconia cermet anode by Spacil [35] which has been the most important SOFC anode for some 50 years. He associated nickel, the best transition metal option, with the stabilised zirconia ceramic material of the electrolyte to confront the difference of thermal expansion which would mechanically stress any nickel anode – electrolyte bond, and also the Ni aggregation in service. He proposed that the function of the ceramic in the composite was essentially structural, to retain the dispersion of the metal particles and the porosity of the anode during long-term operation. Structure and elemental distribution in a typical anode cermet are illustrated in Figure 5.5. The provision of oxide ion mobility complementary to the electronic conductivity and electrocatalytic action of the metal is a very important secondary role of the ceramic, enhancing the electrochemical performance by the delocalisation of the electrochemically active zone.

In conventional Ni-YSZ-based cermets, the Ni is commonly processed in the form of nickel oxide, which allows simple compatibility with the ceramic process routes and temperatures involved in the formation of the basic cell layers [29]. Once in the operational cell, the NiO is easily reduced to the metallic form on introduction of the fuel. Indeed, this reduction can be very fast, and the method of initial introduction of fuel gas and also the morphology and size ratios of the Ni/NiO and YSZ powders can play a significant effect on anode performance [36]. As well as providing the active catalyst surface, the reduction of the nickel oxide also provides extra porosity to the anode as there is significant volume loss on reduction. However, this can also create issues with the structural integrity of the anode and an important requirement of the YSZ portion of the cermet is to stabilise and support this volume change.

While at operational temperatures, any metallic nickel particles will have a tendency to coarsen (Figure 5.6), even with the presence of the YSZ to minimise this process, it will still occur fairly rapidly [38]. This can have several effects such as increase in polarisation resistance, reduction in electrical conductivity and most significant in this instance, it can change the distribution of the Ni relative to the supporting YSZ [37]. This can have the effect that although initial reduction can be carried out without any detrimental effects, any subsequent reoxidation is not supported by the ceramic phase and leads to physical damage of the cell. This is most significant in anode-supported systems where this will result in cracking of supported

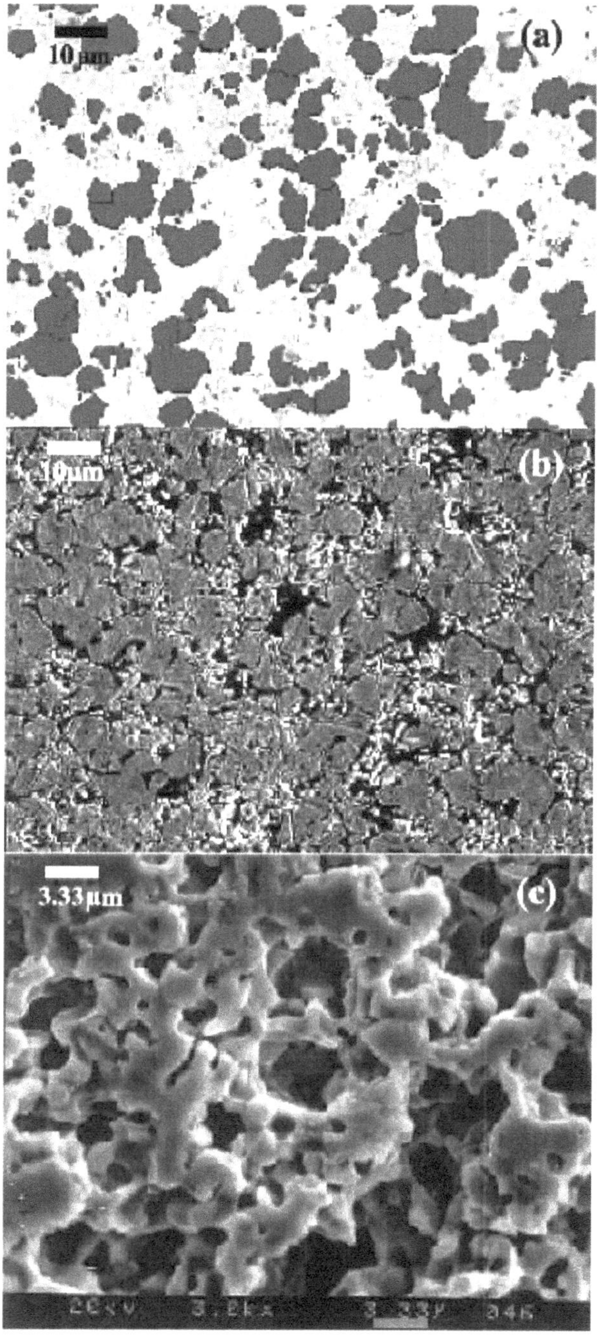

Figure 5.5 Image of a Ni-YSZ cermet by scanning electron microscopy with elemental mapping, showing (a) nickel distribution, (b) overall cermet morphology and (c) zirconia structural skeleton.
Reprinted from Ref. [2]. Copyright (2002), with permission from Elsevier.

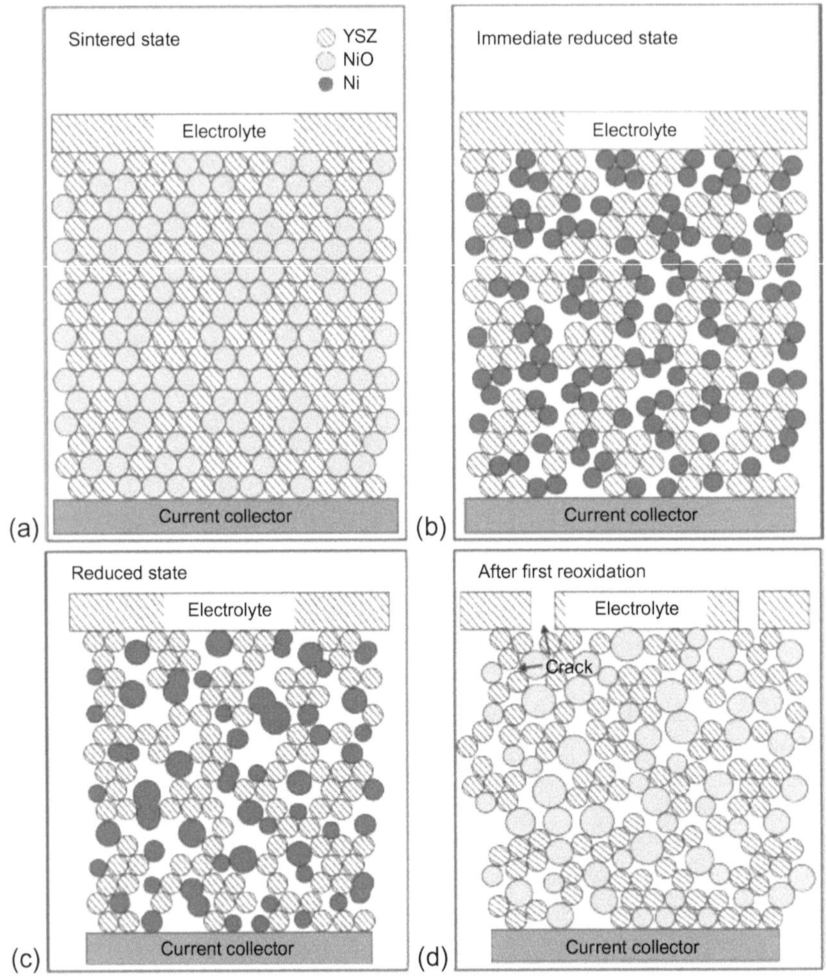

Figure 5.6 Diagramatic representation of effects of Ni coarsening during anode redox cycling. (a) Shows the as sintered anode before reduction with (b) showing the recently reduced state. (c) Shows coarsening of the Ni particles that leads to net volumetric growth on reoxidation (d), leading to cracking in the overlying electrolyte.
Reprinted from Ref. [37]. Reproduced by permission of The Electrochemical Society.

electrolyte membranes [39]. A number of attempts have been made to stabilise this effect; however, it remains a major weakness in Ni-based anode-supported systems requiring significant systems control, such as purge gases, adding complexity and cost [40,41].

Although Ni is an excellent hydrogen oxidation and methane steam reforming catalyst, it also catalyses the formation of carbon filaments from hydrocarbons under reducing conditions (see Figure 5.7a and b). The mechanism involves carbon dissolution into the bulk Ni, and precipitation of graphitic carbon from some facet of the Ni

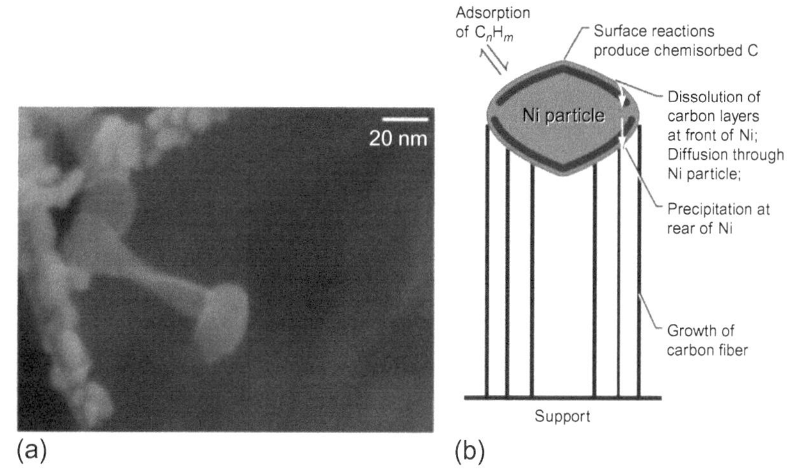

Figure 5.7 (a) SEM image showing carbon filament growth with catalyst particle at the head of the filament.
Reprinted from Ref. [42]. Reprinted from Copyright 2013, with permission from Elsevier.
(b) Schematic representation of the suggested mechanisms involved. Reprinted from Ref. [43]. Reproduced by permission of The Electrochemical Society.

particle lifting off the surface [44]. Unless sufficient amounts of steam are present along with the hydrocarbon in order to remove carbon from the Ni surface at a rate faster than that of carbon dissolution and precipitation, the anode will be destroyed. While thermodynamic calculations are often used to predict the conditions under which carbon formation is favoured [45], carbon fibres can be formed even when carbon is not thermodynamically predicted. This is particularly true in the presence of higher hydrocarbons, since these molecules form carbon on the Ni surface faster than steam can remove it. As a result, even when using methane as the fuel relatively high steam-to-carbon ratios are needed in order to suppress this deleterious reaction [46,47].

One approach to overcoming the limitations of Ni anodes with hydrocarbon fuels, which has met with some success, is to augment the oxidation activity of Ni/YSZ cermets via the addition of an oxide-based oxidation catalyst. For example, stable operation on dry methane has been reported at 650 °C in an SOFC using an yttria-doped ceria interlayer between the YSZ electrolyte and the Ni/YSZ cermet anode [48]. Ceria is a well-known oxidation catalyst and increases the activity of the anode for the electrochemical oxidation of methane.

5.7 Alternative cermet materials

5.7.1 Other cermets

In order to find other anodes with lower susceptibility to hydrocarbon cracking, alternative cermets to nickel yttria zirconia have been sought. Interest has largely focused on ceria replacing zirconia and copper replacing nickel. One approach is to alloy copper with nickel, as, for instance, Bernardo et al. [49] added copper to nickel to reduce the catalytic activity of nickel for hydrocarbon cracking. The influence of adding nickel to $(Ce,Zr)O_2$ or $(Ce,Y,Zr)O_2$ to form a cermet was also studied [50], and cerium oxide is reported as a promising support. In a series of publications, Gorte et al. [51,52] reported that noble metals (Pd, Pt, Rh) have much higher specific rates for water–gas shift, steam reforming and carbon dioxide reforming of methane when supported on cerium oxide than when supported on silica or alumina. Ramirez-Cabrera et al. [53] studied gadolinium-doped ceria, $Ce_{0.9}Gd_{0.1}O_{1.95}$ (CGO) as an anode material at 900 °C in 5% CH_4 with steam/methane ratios between 0 and 5.5. This material revealed itself to be resistant to carbon deposition, although the reaction rate was controlled by slow methane adsorption. Marina et al. [54] studied the same catalyst for methane oxidation, and it was demonstrated that ceria has a low activity for methane oxidation but a high resistance to carbon deposition. It was proposed to add a catalyst (Ni, Rh, Ru) to break the C—H bond more easily.

In contrast to the reported inactivity of acceptor-doped ceria [50], undoped ceria has been reported to be quite an effective oxidation catalyst, which can be attributed to the presence of oxygen vacancies making the doped material more difficult to reduce and hence a less effective catalyst. Ceria is a well-known oxidation catalyst and increases the activity of the anode for the electrochemical oxidation of methane. Another approach being used in the development of cermet anodes that allow the use of fairly dry hydrocarbon is to use a relatively inert metal such as Cu for electrical conductivity and a metal oxide to provide catalytic activity and ionic conductivity. Examples of this type of composite anode include $Cu/CeO_2/YSZ$ [55] and Cu/YZT (titania-doped yttria zirconia) [56]. The $Cu/CeO_2/YSZ$ anode system is particularly interesting and has been shown to be effective for the direct utilisation of a variety of hydrocarbon fuels including butane and decane and highly resistant to deactivation via carbon deposition [57]. The role of Cu in these anodes appears to be exclusively current conduction while ceria primarily acts as an oxidation catalyst. It has been shown that SOFCs that have anodes that contain only Cu and YSZ exhibit very poor performance while operating on hydrocarbons. The addition of ceria to the anode, however, produces a marked increase in performance while operating on both hydrogen and hydrocarbons. It is important to note that SOFCs with $Cu/CeO_2/YSZ$ anodes always exhibit higher current densities while operating on hydrogen compared to hydrocarbons. This indicates that catalysis plays a critical role and suggests that enhancements in the catalytic activity of the anode are still needed in order to optimise performance using hydrocarbon fuels.

5.7.2 Ceramic anodes

5.7.2.1 Perovskites

In order to overcome the problems and limitation of the Ni-cermet anodes, mixed oxides with perovskite structure have been extensively studied and considered as alternative anode materials due to their broad spectrum of favourable properties: good chemical and thermal compatibility with the SOFC adjacent components, structural and dimensional stability upon redox cycles, low activity to hydrocarbon cracking due to their intrinsic higher availability for oxygen and enhanced resistance to fuel impurities such as sulphur. Although the electronic conductivity of these oxides is much lower than that of Ni in Ni-YSZ cermet, their use as a thin current collector layer is plausible, as it requires a minimum material conductivity value of 1 S cm^{-1} in order to maintain the associated electrical losses below 0.1 Ω cm^2 as earlier discussed in this chapter. Ceramic perovskites are usually employed either as single-phase anode, anode current collector layer or as component of a composite anode.

Perovskite oxides have the general formula ABO_3 (where A is an alkali or rare earth ion and B is a transition metal ion such as Ti, V, Cr or Mn), and the availability of multiple oxidation states for the transition metal facilitates the electrocatalytic processes and provides mechanisms for electronic conductivity upon reduction. Their capacity to accommodate a large amount of oxygen vacancies and stoichiometric deviations allows doping of various ions with different oxidation states to enhance their conductivity and catalytic activity. The former aspect is indicating that several perovskites exhibit substantial ionic conduction and the prospect of a mixed electronic–ionic conductor (MIEC) material opens the pathway for oxygen ion migration also through the MIEC itself, spreading the possible reaction zone.

Among these mixed oxides with perovkite structure, chromites, manganites and titanates are the most investigated materials as fuel electrodes. Initially explored as an interconnect material, $LaCrO_3$ has drawn interest as anode material due to its high redox stability and electronic conductivity at high temperatures. Numerous studies have demonstrated that doping $LaCrO_3$ on the A- and/or B-sites could significantly influence their electrochemical properties. Although Ca and Sr substitution on the A-site were shown to improve the electrical conductivity and catalytic activity of $LaCrO_3$, a much better strategy to enhance the catalytic activity and oxygen ion conductivity is the additional B-site co-doping. Oxygen vacancies may be generated at high temperatures, in reducing atmosphere, when multi-valance transition metals (e.g. V, Ni, Fe, Co, Mn or Cu) are also introduced on the B-site [58]. Out of the various dopants, Ni (up to 10%) seems to be the most successful although nickel exsolution in fuel conditions was reported [59], explaining to a certain extent the enhanced activity observed and questioning the long-term stability of this material in fuel atmospheres.

An enhanced activity is observed in manganite-based perovskites, for example, where Mn in lanthanum manganite is replaced by Cr [60]. At low partial pressure of oxygen, reduction of Mn yields oxygen vacancies, introducing ionic conductivity. The composition $La_{0.75}Sr_{0.25}Cr_{0.5}Mn_{0.5}O_3$ (LSCM) exhibits comparable electrochemical performance to Ni/YSZ cermets [61] (see Figure 5.8). The material is stable in both fuel and air atmospheres although there is a volume change of around 1% and the electronic conductivity is quite low ($\sim$1 S cm^{-1}) in reducing atmosphere. It also shows very good electrochemical performance in humidified CH_4, without using excess steam, as catalytic studies demonstrate that it is primarily a direct oxidation catalyst for CH_4 oxidation as opposed to a reforming catalyst [62]. However, it is unstable in the presence of high concentrations of H_2S [63].

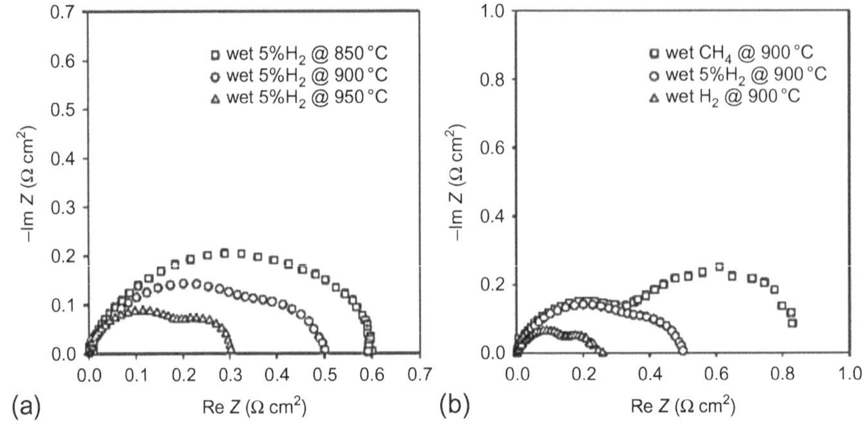

Figure 5.8 Electrode impedance of a $La_{0.75}Sr_{0.25}Cr_{0.5}Mn_{0.5}O_3$ anode in humidified 5% H/Ar at 850, 900 and 950 °C (a) and at 900 °C in different humidified (3%H_2O) fuel gas compositions (b). The electrolyte contribution has been subtracted from the overall impedance. Reprinted from Ref. [60]. Copyright 2003 Nature Publishing.

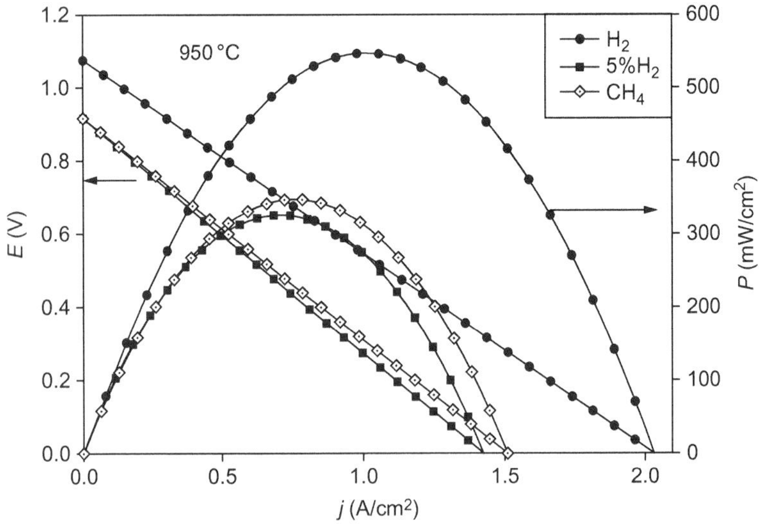

Figure 5.9 Voltage (E) and power density (P) vs. current density (j) plots for LSCM SOFC obtained at 950 °C using three different fuels.
Reprinted from Ref. [64]. Copyright 2006, with permission from Elsevier.

Due to its stability in air and fuel conditions, LSCM has been employed as air and fuel electrode in a symmetrical fuel cell design, opening the possibility of developing reversible fuel cells (see Figure 5.9) [64].

To compensate for the lack of catalytic activity of these perovskites, the addition of an ionic conductor such as YSZ, GDC or YSZ/CGO to form a composite anode is often employed. This approach greatly improves the adhesion of the LSCM material to the electrolyte, reducing, at the same time, the electrode polarisation resistance and improving the electrocatalytic activity for fuel oxidation [65]. A layered oxygen-deficient double perovskite $PrBaMn_2O_5$ has also been shown to be an efficient and stable anode for direct hydrocarbon SOFCs [66].

$SrTiO_3$-based perovskites have drawn a lot of attention lately as promising SOFC anode material candidates as they possess a very versatile, well-studied defect chemistry that allows for intricate ways to tailor the electrical and electrocatalytical properties [67]. Although not completely immune to impurities present in the fuel, it is believed that the titanates are less affected than the chromites, suggesting that the B-site cations play an important role in this tolerance.

$SrTiO_3$ exhibits *n*-type semiconducting behaviour when donor doped and/or exposed to a reducing atmosphere, where some Ti^{4+} ions will be reduced to Ti^{3+}, which will strip oxygen ions from the lattice, generating oxygen vacancies that increase the ionic conductivity and electrons which significantly increase the electronic conductivity, according to Equation (5.9), in Kroger–Vink notation:

$$2Ti_{Ti}^{x} + O_{O}^{x} \leftrightarrow 2Ti_{Ti}' + V_{O}^{\bullet\bullet} + \frac{1}{2}O_2 \qquad (5.9)$$

When doping with trivalent cations such as La^{3+} or Y^{3+} on the A-site, the extra positive charge of the donor needs to be compensated by extra oxygen beyond the normal stoichiometry or strontium vacancies (ionically) in oxidising atmospheres or conduction electrons (electronically) in reducing atmospheres. Oxygen ion vacancies or interstitials can be generated by substitution of B cations with cations of lower or higher charge. La-doped $SrTiO_3$ (LST) is generally treated in the literature as cubic, although the presence of extra oxygen beyond stoichiometry highly influences the structure and the electrocatalytic behaviour. The A-site-deficient compositions, such as air-sintered $La_xSr_{1-3x/2}TiO_{3-\delta}$, have proved to present enhanced electronic conductivity under fuel conditions compared to their stoichiometric counterparts [68], and the ionic conductivity was found to increase with the increase in A-site deficiency [69].

Attempts to improve the conductive and electrocatalytic properties of LST have been made by partially replacing Ti^{4+} with cations like Sc^{3+}, Ga^{3+} or Mn^{4+}. Sc^{3+} addition is enhancing the oxygen ion conduction and the oxidation–reduction kinetics although the electronic conductivity decreases with Sc addition to the system [70]. Ga^{3+} possesses a coordination flexibility that should develop a certain extent of ionic conductivity by generating extra oxygen vacancies for charge compensation, in addition to those created via reduction. Gallium presence was found to enhance the Ti^{4+} reduction process and redox stability [71]. Combining Mn and Ga, both accepting lower oxygen coordination numbers than Ti, therefore facilitating oxide ion migration, was an additional strategy to increase the performance of an oxygen excess anode material, $La_4Sr_8Ti_{11}Mn_{0.5}Ga_{0.5}O_{37.5}$ (LSTMG) system under methane operation [72].

Another doping strategy for improving the electronic conductivity of $SrTiO_3$ in reducing atmospheres is the partial replacement of titanium with M^{5+} cations such as Nb^{5+} or Ta^{5+} usually complemented by a decrease in Sr content, for charge compensation [73].

In many cases, in order to achieve high electronic conductivity (80–320 S cm^{-1}) for practical use as an SOFC anode, the titanates are sintered or pre-reduced at high temperatures and in reducing atmospheres prior to testing [74,75]. This will make it difficult to co-fire with the SOFC cathode. Although the chemical compatibility of strontium titanates with the usual choices of SOFC electrolytes such as of YSZ or LSGM is quite good, potential stability issues such as interdiffusion of the B-site cations into the YSZ electrolyte have been reported [76]. Therefore, for practical reasons, it is preferable to sinter the material in air and reduce it *in situ*.

Interesting anode performance was reported for LST-based anodes, especially with cerium-doped LST [77] which is thought to consist of a ceria-perovskite mixture. Stoichiometric LST compositions have also been successfully used as a conductive support for an anode functional layer consisting of a porous YSZ layer impregnated with Pd:CeO_2 [24] or a thin layer of Ni/YSZ [78], although C deposition or sulphur poisoning of the anode could not be completely avoided in these studies. An overview of the defect chemistry and electrochemical activity of various modified strontium titanate SOFC anodes is given elsewhere [79].

Other examples of perovskite mixed oxides that have been investigated as SOFC fuel electrodes include compounds with V, Mo and Fe. Although $SrVO_3$ shows excellent electronic conductivity of 1000 S cm^{-1} at 800 °C in reducing atmosphere, it has poor catalytic activity and high TEC and is unstable under a more oxidising atmosphere [80]. A material that has both adequate high temperature conductivity in reducing atmosphere and redox stability has yet to be found in the vanadates.

Molybdates were also investigated as potential SOFC anodes, especially as double perovskites – with generic formula $A_2BB'O_6$ – such as $Sr_2Mg_{1-x}Mn_xMoO_{6-\delta}$, demonstrating good electronic conductivity, redox cycling stability and fuel impurity tolerance [81]. Despite the good performance values obtained with this material, there are issues related to chemical reactivity and redox stability, especially at high operating temperatures [82]. A-site doping with Ca, Ba or La was studied for this system, and an increase in electrochemical performance concomitant with a decrease in stability in oxidising atmosphere was found [83]. B/B′-site substitution was also explored with Al or transitional elements such as Mn, Fe, Co or Ni for Mg in an attempt to introduce oxygen vacancies and decrease the cation ordering [84].

The perovskite cobaltites/ferrites such as $SrFeCo_{0.5}O_x$ or $SrCo_{0.8}Fe_{0.2}O_{3-\delta}$ type exhibit both high electronic and ionic conductivities in air and might be applicable for potential SOFC anode components although stability issues in reducing atmosphere and conflicting structural data were reported [85]. $La_{0.6}Sr_{0.4}Co_{0.2}Fe_{0.8}O_3$ was proposed as an SOFC anode at intermediate temperatures (550–700 °C) due to its catalytic activity towards CH_4 oxidation [86]. The most promising anode structures still consist of a composite with YSZ or CGO [87]. Recently, very good electrochemical performance was obtained with $La_{0.6}Sr_{0.4}Fe_{0.9}Mn_{0.1}O_3$–$Ce_{0.6}Mn_{0.}Fe_{0.1}O_2$ composite anode, in both H_2 and various hydrocarbons, with no carbon deposition, using an LSGM electrolyte [88] (Figure 5.10).

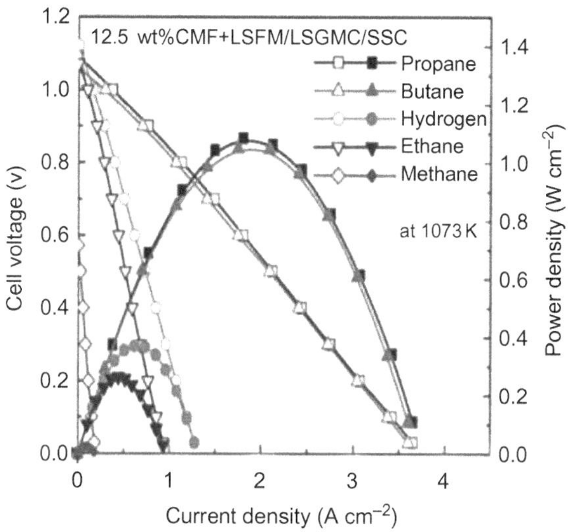

Figure 5.10 $I-V$ and power density curves at 800 °C in various fuels for SOFC containing CMF–LSFM composite anode.
Reprinted with permission from Ref. [88]. Copyright 2011 American Chemical Society.

Other oxide structures that can be explored as potential SOFC anodes are the oxides with pyrochlore, such as $Gd_2Ti_2O_7$ [89], spinels, such as Mg_2TiO_4 [90] or tungsten bronze structures, such as $Sr_{0.2}Ba_{0.4}Ti_{0.2}Nb_{0.8}O_3$ that has good conductivity but lower TEC than YSZ [91].

In conclusion, although some mixed oxides can satisfy most of the SOFC anode requirements, it is often difficult for a single compound to be successfully used as an anode, thus it is most likely that a successful anode system would be a composite. There is a strong motivation to find a catalytically active MIEC that is stable in the reducing conditions from the SOFC anode. Doping the parent perovskite with suitable cations is a suitable strategy to improve and tailor the electrochemical properties and stability by controlling their defect chemistry. One of the most important limitations of these ceramic oxides is the insufficient electronic conductivity that limits their potential use as conductive support. The lack of intrinsic ionic conductivity (probably responsible for their modest catalytic activity for oxidation of the fuels), for instance, is usually compensated by adding YSZ or ceria to form a composite anode, often at the expense of the electronic conductivity. To overcome the insufficient catalytic activity of ceramic anodes towards fuel oxidation, on another hand, an additional catalyst component needs to be introduced into the porous anode structure, preferably, in a small amount via wet impregnation, in order to avoid the traditional cermet fabrication route.

5.7.2.2 Composite anodes produced by impregnation

Although widespread in heterogeneous catalysis, the impregnation or infiltration technique did not attract significant interest in SOFC electrode processing until recent years. Wet impregnation has proved to be a very suitable technique for extending the TPB and introducing active materials into porous supporting backbones, offering superior microstructural control within the SOFC electrode. This technique avoids the requirement of high-temperature sintering steps (required in conventional preparation methods) for introducing novel electrode materials that either cannot withstand high temperatures or react with the adjacent SOFC components. Moreover, it allows for a separation of catalyst and current collection functions in the anode, along with a cost reduction by reducing the amount of expensive catalysts employed in the electrode.

An almost ideal fuel electrode would comprise an electronically conductive phase (that can be a mixed oxide), an ionic conducting component such as YSZ (in the form of a porous backbone) and small amounts of catalytically active materials such as CeO_2 and noble metals (in the range of 1–10 wt%). Cerium oxide has a self-limited grain growth [92], and noble metals have good catalytic activity for H_2 and hydrocarbon oxidation; therefore, there is a favourable synergy offered by using both electrocatalysts in enhancing the cell performance. Among other advantages, very good adherence is obtained by using the electrolyte material to produce the porous structure for impregnation. In some cases, the addition of a surfactant or binder to the impregnating solution, along with solvent polarity choice, can improve the wettability offering good coverage inside the pores. The calcination temperature for the impregnated material plays a very important role in achieving a conductive pathway within the electrode. Although the presence of nanomaterials in fuel cell operating conditions would question the long-term stability of these structures, the observation that the processing conditions (type of material, solution properties, type of substrate or calcination temperature) have a great impact on their nature and behaviour is encouraging in preventing this. The impregnation technique was successfully employed for the modification of both air and fuel electrodes, and this is covered in some comprehensive reviews in the literature [93,94].

State of the art cermet anodes such as Ni-YSZ and Ni-ceria have been modified with YSZ or ceria nanoparticles in order to improve their performance, tolerance and stability; it was found, for example, that the impregnated materials can hinder the Ni sintering and agglomeration during fuel cell operation. Ceria impregnation led to improved performance of the Ni anode, by lowering the polarisation resistance [95], enhancing the carbon deposition tolerance when tested in hydrocarbon fuels and resistance to H_2S presence in the fuel. Apart from YSZ or cerias, the impregnation of proton-conducting oxides such as strontium cerates or zirconates, as well as BaO was studied in an attempt to improve the stability and tolerance of conventional Ni-YSZ anodes [96]. The impregnation of fine particles of Pd or Ru catalysts into Ni-GDC anodes significantly contributed to a decrease in anode polarisation resistance, especially in CH_4. Although not suppressed completely, this was also increasing the resistance of the anode to carbon deposition [97] and sulphur contamination [98].

Highly dispersed Ni impregnated into YSZ or doped ceria backbones was found to be more effective in promoting the stability and electrocatalytic activity for H_2 oxidation reaction than that of mixed Ni-cermet anodes. A lower level of Ni is required to achieve sufficient conductivity, translated into faster reducibility and higher stability towards degradation [99].

As already mentioned in Section 5.7.1, since Ni is very active towards carbon deposition and prone to fuel contaminants poisoning, it was sensible to replace it with other conductive metals or oxide materials. Copper is a suitable option for this, and the impregnation method is very suitable for introducing Cu or Cu-alloys into zirconias, doped ceria or LSGM porous structures, difficult to process otherwise due to the low melting point of Cu. Due to low catalytic oxidation activity of Cu, ceria was also impregnated into the porous structures and performant anodes for direct hydrocarbon oxidation, including some aromatics, were produced in this way [100]. It is believed, though, that the high electrochemical performance of these anodes in hydrocarbons is related to the carbonaceous deposits onto Cu, creating a conductive pathway for the electrons.

High-performance SOFC anodes were also produced by infiltrating active ceramics such as LSCM into a porous YSZ scaffold. To provide sufficient electrocatalytic activity, 5 wt% CeO_2 and 0.5 wt% Pd were added after the formation of LSCM phase. An interesting evolution of the impregnate with the temperature and atmosphere was observed, from a continuous coating to nanoparticles formed around 800 °C in fuel atmosphere (Figure 5.11). This enlargement of the TPB explained the excellent electrochemical performance with this composite. Such wetting behaviour is observed only for Mn-containing compounds, pointing towards the solubility of Mn^{2+} into YSZ fluorite structure [101]. The impregnation of Pd, Cu or Ru nanoparticles into the LSCM-based anodes was beneficial for facilitating the electrochemical oxidation of hydrocarbon fuels, also preventing carbon deposition [102,103].

Strontium titanates can also be produced by impregnation inside a porous backbone to provide electronic conductivity to the composite or used as conductive backbones for catalyst impregnation. $La_{0.3}Sr_{0.7}TiO_3$ was impregnated into porous YSZ (45 wt%), followed by the addition of Pd and ceria, with good performance obtained in 3 wt% H_2O/H_2 at 800 °C [104]. An A-site-deficient anode material, $La_{0.2}Sr_{0.25}Ca_{0.45}TiO_3$, impregnated with CeO_2 and Ni led to the achievement of very promising electrochemical results: excellent redox stability over 20 redox cycles, long-term stability and polarisation resistance values comparable to the state-of-the-art Ni-YSZ in humidified H_2 [105]. Moreover, long-term fuel cell short stack tests were reported for the first time with this ceramic anode material by Hexis AG, which delivered 750 W utilising CPOx reformed natural gas [106].

$Sr_{0.88}Y_{0.08}TiO_3$-YSZ infiltrated with ceria and Ru [107] or $SrTi_{1-x}Nb_xO_3$–YSZ infiltrated with Pd and ceria [108] composite anodes were also employed to achieve remarkable power densities and tolerance towards fuel impurities. More recently, large cells were developed using $Sr_{0.895}Y_{0.07}TiO_3$-YSZ anode sintered in reducing atmosphere and impregnated with Ni and the results indicated excellent initial performance and relative stability over 200 redox cycles, with a loss of around a third of the initial performance [109]. Remarkable fuel cell performance was also obtained with a Pd-impregnated $Sr_{0.88}Y_{0.08}TiO_3$–$La_{0.4}Ce_{0.6}O_{1.8}$ composite anode and LSGM electrolyte, as the anode was also proved to be sulphur tolerant when 50 ppm H_2S was introduced in the fuel [110].

Wet impregnation was also used to fabricate and/or optimise other alternative ceramic anodes. The addition of Pd or Pd/CeO_2 was also shown to enhance the

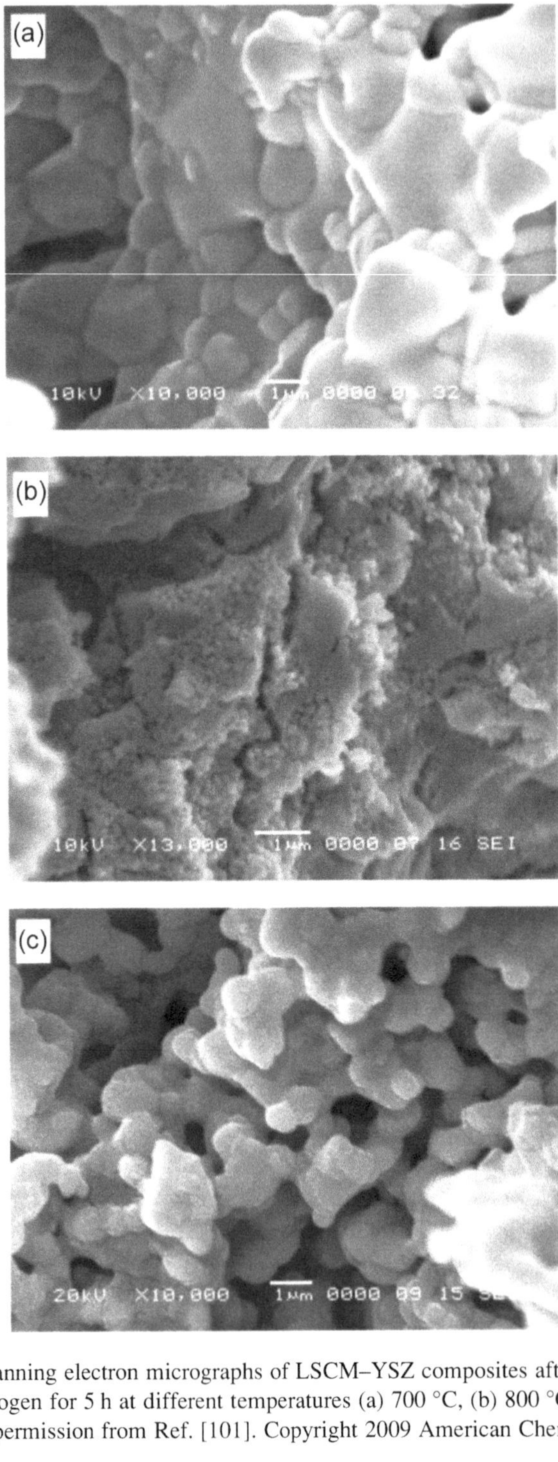

Figure 5.11 Scanning electron micrographs of LSCM–YSZ composites after reduction in humidified hydrogen for 5 h at different temperatures (a) 700 °C, (b) 800 °C and (c) 900 °C. Reprinted with permission from Ref. [101]. Copyright 2009 American Chemical Society.

performance of infiltrated V-, Mo- or W -based perovskites or spinel-based anodes such as $La_{0.7}Sr_{0.3}VO_{3.85}$ [111] or $MgCrFeO_4$ [112].

5.7.2.3 Nano-catalyst exsolution

Compared to the impregnation technique, a more convenient way to produce the catalyst on the surface of the ceramic fuel electrode is to incorporate the element into the mixed oxide structure in the oxidised state and exsolve it *in situ*, as nanoparticles, in fuel operating conditions. This process is driven by the free energy decrease associated with the reduction of the metallic oxide present in the ceramic phase to metal, in reducing atmosphere. In this way, a cermet structure containing a nucleated, non-percolating metal catalyst could be produced, offering simplicity, cost reduction as well as potential fuel selectivity by tailoring the catalyst. Redox cycling of the SOFC could help regenerating the metal surface area loss due to sintering, maintaining the long-term performance of the structure.

This idea was firstly investigated intensively in the field of automotive catalysts, where repeated precipitation and dissolution of noble metal nanoparticles has been reported for perovskite materials containing precious metals, during redox cycles. The "intelligent" catalysts used perovskite oxides (e.g. $LaFeO_3$, $LaCrO_3$) in which a catalytic noble metal substitutes a small portion of the B-site cations. Under reducing conditions, the easily reduced noble metal cations are exsolved from the lattice and precipitate as nanoparticles, decorating the surface of the perovskite. Changing the atmosphere in the catalytic converter to oxidising conditions, the noble metals undergo re-dissolution, as the cycling between environments helps maintaining activity of the metal surface [113].

It was observed in practice that metal exsolution is usually accompanied by a dramatic reduction in the anode polarisation resistance; for example, Ru nanoparticles formation from $La_{0.8}Sr_{0.2}Cr_{0.82}Ru_{0.18}O_{3-\delta}$ anodes led to a reduction in anode polarisation by 65%, although the coarsening of the nanoparticles was the cause for an increase in polarisation resistance after 300 h operation at 800 °C [114]. Other examples include $La_{0.8}Sr_{0.2}Cr_{1-x}Pd_xO_{3-\delta}$–$Ce_{0.9}Gd_{0.1}O_{2-\delta}$ (GDC) cermet (50 wt%) [115] where the initial excellent performance after activation in H_2 was followed by degradation that could be recovered only temporarily after redox cycles. As mentioned before, Ni exsolution occurs in reducing conditions when co-doping into Sr- or Ca-substituted $LaCrO_3$ and Ni nucleation was also observed after long-term operation in either H_2 or CH_4, leading to cell performance degradation. Nickel was introduced on the B-site of LSCM, along with Cr and Mn, and the exsolution was accompanied by an increase in conductivity in both air and reducing atmosphere as the compound was found to enhance the complete oxidation of methane [116]. Strontium titanates were also explored for catalyst exsolution. Cobalt was introduced into $La_{0.3}Sr_{0.7}TiO_3$ and dispersed co-nanoparticles were observed on the surface of the perovskite grains after reduction at 1400 °C; the material performed well in H_2S containing fuel and was proved to withstand redox cycles [117].

More recently, it was shown that the process can be somewhat more subtle than a simple chemical decomposition, whereby through careful control of stoichiometry of

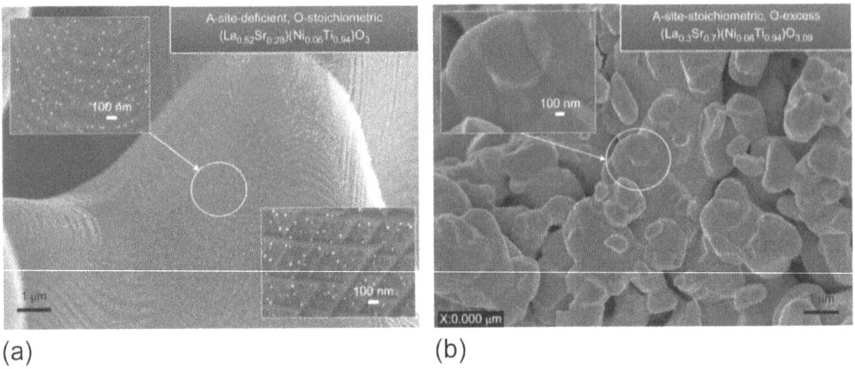

Figure 5.12 Nickel exsolution from the A-site-deficient $La_{0.52}Sr_{0.28}Ni_{0.06}Ti_{0.94}O_3$ compound (a) compared to A-site stoichiometric $La_{0.3}Sr_{0.7}Ni_{0.06}Ti_{0.94}O_{3.09}$ sample (b) after reduction at 930 °C for 20 h in 5%H_2/Ar.
Reprinted from Ref. [118]. Copyright 2013 Nature Publishing.

the parent compound intricate and stable B-site catalyst exsolution structures can be created. A-site-deficient compositions were shown to produce a wider range of nanoparticles in terms of size and surface distribution as compared to stoichiometric compounds. In these systems, the starting A-site deficiency accompanied by the generation of oxygen vacancies upon reduction makes the B-site cation exsolution (e.g. Ni, Fe, Co, Cu, etc.) an inherent, local reversion towards an ABO_3 perovskite stoichiometry, as compared to the stoichiometric compounds where exsolution is often accompanied by undesired A-site-rich phases [118] (Figure 5.12). This approach makes exsolution more tunable, instead of being dependent only on the reducibility of the cation to be exsolved.

It has also been shown that the transition metal nanoparticles (Ni, Co and Cu) could be incorporated into and removed out of the transition metal-doped vanadate ($Ce_{1-x}M_xVO_{4-0.5x}$, where M is Ni, Co, Cu) lattices by redox cycling. The initial zircon-type structure of the vanadate sintered in air undergoes a phase transition to a perovskite structure, accompanied by partial exsolution of the transition metal in reducing atmosphere. The process seemed reversible and high stability of the material upon redox cycles was observed [119].

5.8 General conclusions

Significant progress has been made in SOFC anode development since the first edition of this book. Since that time, the anode-supported SOFC has become the dominant design in both planar and tubular cell configurations. Progress in design and engineering for second-generation SOFC introduced the functionality separation into anode functional layer and current collector support, allowing the use of conductive ceramic materials. Although no single ceramic material fulfils the criteria required in an electrode-supported cell design to the extent of Ni-cermets, several materials and composites that meet individual criteria have been developed. The conventional anode materials such as Ni and YSZ have still not been replaced while the problems related to carbon coking and sulphur poisoning remain substantial. There is continuous motivation for improving and introducing new materials and concepts. These will be able to offer improved electrochemical performance and stability on redox cycling and in fuel impurity presence. Further studies are necessary to optimise these materials in terms of composition and microstructure, with emphasis on their electronic conductivity, electrocatalytic properties, chemical and thermal stability under reducing conditions and long-term performance in fuel cell conditions using practical fuels and biofuels. Particularly exciting results are being achieved through impregnation and exsolution techniques, whereby performance and durability can often be enhanced simultaneously. Furthermore, these types of surface chemistries seem to be inherent in many of the key processes relating to activation and ageing in SOFC electrodes and will provide a particularly fruitful area for further academic and technical study in this field.

References

[1] A. Atkinson, S. Barnett, R.J. Gorte, J.T.S. Irvine, A.J. McEvoy, M. Mogensen, S.C. Singhal, J. Vohs, Nat. Mater. 3 (2004) 17–27.

[2] J.-H. Lee, H. Moon, H.-W. Lee, J. Kim, J.-D. Kim, K.-H. Yoon, Solid State Ionics 148 (2002) 15–26.

[3] W. Wang, C. Su, Y. Wu, R. Ran, Z. Shao, Chem. Rev. 113 (10) (2013) 8104–8151.

[4] B.C.H. Steele, R.M. Dell, Philos. Trans. R. Soc. Lond. A 354 (1996) 1695–1710.

[5] E.S. Hecht, G.K. Gupta, H. Zhu, A.M. Dean, R.J. Kee, L. Maier, O. Deutschmann, Appl. Catal. A Gen. 295 (2005) 40–51.

[6] P. Holtappels, L.G.J. De Haart, U. Stimming, I.C. Vinke, M. Mogensen, J. Appl. Electrochem. 29 (1999) 561–568.

[7] T. Takeguchi, Y. Kania, T. Yano, R. Kikuchi, K. Eguchi, K. Tsujimoto, Y. Uchida, A. Ueno, K. Omoshiki, M. Aizawa, J. Power Sources 112 (2002) 588.

[8] S. McIntosh, R.J. Gorte, Chem. Rev. 104 (2004) 4845.

[9] P. Vernoux, J. Guindet, M.J. Kleitz, J. Electrochem. Soc. 145 (1998) 3487–3492.

[10] M. Boder, R. Dittmeyer, J. Power Sources 155 (2006) 13–22.

[11] R. Peters, E. Riesche, P. Cremer, J. Power Sources 86 (2000) 432.

[12] R. Peters, R. Dahl, U. Klüttgen, C. Palm, D. Stolten, J. Power Sources 106 (2002) 238.

[13] X.Y. Lu, P.W. Faguy, M.I. Liu, J. Electrochem. Soc. 149 (10) (2002) A1293–A1298.

[14] J. Sehested, Catal. Today 111 (2006) 103–110.

[15] S. Zha, Z. Cheng, M. Lui, J. Electrochem. Soc. 154 (2007) B201–B206.

[16] Z. Cheng, M. Lui, Solid State Ionics 178 (2007) 925–935.

[17] M. Cassidy, J.P. Ouweltjes, N. Dekker, Innovations in fuel cell technologies, in: R. Steinberger-Wilkens, W. Lehnert (Eds.), RSC Energy & Environment Series v2, 2010, pp. 153–189.

[18] N.Q. Minh, Solid State Ionics 174 (2004) 271–277.

[19] K. Huang, S.C. Singhal, J. Power Sources 237 (2013) 84–97.

[20] F. Tietz, H.-P. Buchkremer, D. Stöver, Solid State Ionics 152–153 (2002) 373.

[21] R.N. Basu, G. Blass, H.P. Buchkremer, D. Stover, F. Tietz, E. Wessel, I.C. Vinke, in: H. Yokokawa, S.C. Singhal (Eds.), Solid Oxide Fuel Cells VII, The Electrochemical Society Proceedings, Pennington, NJ, PV2001-16, 2001, p. 995.
[22] Q.X. Fu, F. Tietz, Fuel Cells 8 (2008) 283–293.
[23] M. Brown, S. Primdahl, M. Mogensen, J. Electrochem. Soc. 147 (2000) 475–485.
[24] M.D. Gross, J.M. Vohs, R.J. Gorte, Electrochem. Solid-State Lett. 10 (4) (2007) B65–B69.
[25] L. Miller, H. Taherparvar, N. Filkin, P. Slater, J. Yeomans, Solid State Ionics 179 (2008) 732–739.
[26] M. Cassidy, S. Boulfrad, J. Irvine, C. Chung, M. Jorger, C. Munnings, S. Pyke, Fuel Cells 9 (2009) 891–898.
[27] R. Leah, A. Bone, A. Selcuk, D. Corcoran, M. Lankin, Z. Dehaney-Steven, M. Selby, P. Whalen, ECS Trans. 35 (2011) 351–367.
[28] M.C. Tucker, J. Power Sources 195 (2010) 4570.
[29] M. Lang, T. Franco, G. Schiller, N. Wagner, J. Appl. Electrochem. 32 (2002) 871–874.
[30] P. Blennow, J. Hjelm, T. Klemensø, Å.H. Persson, S. Ramousse, M. Mogensen, Fuel Cells 11 (2011) 661–668.
[31] P. Blennowa, J. Hjelma, T. Klemensøa, S. Ramoussea, A. Krompb, A. Leonideb, A. Weber, J. Power Sources 196 (2011) 7117.
[32] H.J. Cho, Y.M. Park, G.M. Choi, ECS Trans. 25 (2009) 695–699.
[33] E. Baur, H. Preis, Z. Elektrochem. 43 (1937) 727.
[34] H.-H. Möbius, J. Solid State Electrochem. 1 (1997) 2.
[35] H.S. Spacil, US Patent 3,558,360; March, 1970.
[36] W.Z. Zhu, S.C. Deevi, Mater. Sci. Eng. A362 (2003) 228–239.
[37] T. Klemenso, C. Chung, P.H. Larsen, M. Mogensen, J. Electrochem. Soc. 152 (2005) A2186.
[38] B. Shri Prakash, S. Senthil Kumar, S.T. Aruna, Renew. Sust. Energ. Rev. 36 (2014) 149–179.
[39] M. Cassidy, K. Kendall, G. Lindsay, J. Power Sources 61 (1996) 189.
[40] D. Waldbillig, A. Wood, D.G. Ivey, J. Electrochem. Soc. 154 (2007) B133.
[41] A. Faes, A. Hessler-Wyser, A. Zryd, J. Van herle, Membranes 2 (2012) 585–664.
[42] A. Lanzini, P. Leone, C. Guerra, F. Smeacetto, N.P. Brandon, M. Santarelli, Chem. Eng. J. 220 (2013) 254–263.
[43] W.Y. Lee, J. Hanna, A.F. Ghoniem, J. Electrochem. Soc. 160 (2) (2013) F94–F105.
[44] M.L. Toebes, J.H. Bitter, A.J. van Dillen, K.P. de Jong, Catal. Today 76 (2002) 33–42.
[45] K. Sasaki, Y. Teraoka, J. Electrochem. Soc. 150 (2003) A878–A884.
[46] H.S. Bengaard, J.K. Nørskov, J. Sehested, B.S. Clausen, L.P. Nielsen, A.M. Molenbroek, J.R. Rostrup-Nielsen, J. Catal. 209 (2002) 365.
[47] W.L. Holstein, J. Catal. 152 (1995) 42.
[48] E.P. Murray, T. Tsai, S.A. Barnett, Nature 400 (1999) 649–651.
[49] C.A. Bernardo, I. Alstrup, J.R. Rostrupnielsen, J. Catal. 96 (1985) 517–534.
[50] W.S. Dong, H.S. Roh, K.W. Jun, S.E. Park, Y.S. Oh, Appl. Catal. A 226 (2002) 63–72.
[51] T. Bunluesin, R.J. Gorte, G.W. Graham, Appl. Catal. B 15 (1998) 107–114.
[52] S. Sharma, S. Hilaire, J.M. Vohs, R.J. Gorte, H.W. Jen, J. Catal. 190 (2000) 199–204.
[53] E. Ramirez-Cabrera, A. Atkinson, D. Chadwick, Solid State Ionics 136 (2000) 825–831.
[54] O.A. Marina, M. Mogensen, Appl. Catal. A 189 (1999) 117–126.
[55] S.D. Park, J.M. Vohs, R.J. Gorte, Nature 404 (2000) 265–267.
[56] N. Kiratzis, P. Holtappels, C.E. Hatchwell, M. Mogensen, J.T.S. Irvine, Fuel Cells 1 (2001) 211–218.
[57] H. Kim, S. Park, J.M. Vohs, R.J. Gorte, J. Electrochem. Soc. 148 (2001) A693–A695.

[58] J. Sfeir, P.A. Buffat, P. Mockli, N. Xanthopoulos, R. Vasquez, H.J. Mathieu, J. van Herle, K.R. Thampi, J. Catal. 202 (2001) 229–244.
[59] W. Kobsiriphat, B. Madsen, Y. Wang, M. Shah, L. Marks, S. Barnett, J. Electrochem. Soc. 157 (2010) B279–B284.
[60] S.W. Tao, J.T.S. Irvine, Nat. Mater. 2 (2003) 320–323.
[61] J.C. Ruiz-Morales, J. Canales-Vázquez, J. Peña-Matinez, D. Marrero-López, J.T.S. Irvine, P. Núñez, J. Mater. Chem. 16 (2006) 540–542.
[62] S. Tao, J.T.S. Irvine, S.M. Plint, J. Phys. Chem. 110 (2005) 21771–21776.
[63] S. Zha, P. Tsang, Z. Cheng, M. Liu, J. Solid State Chem. 178 (2005) 1844–1850.
[64] J.C. Ruiz-Morales, J. Canales-Vázquez, J. Peña-Martínez, D. Marrero-López, P. Núñez, Electrochim. Acta 52 (1) (2006) 278–284.
[65] S.P. Jiang, X.J. Chen, S.H. Chan, J.T. Kwok, K.A. Khor, Solid State Ionics 177 (2006) 149–157.
[66] S. Sengodan, S. Choi, A. Jun, T.H. Shin, Y.-W. Ju, H.Y. Jeong, J. Shin, J.T.S. Irvine, G. Kim, Nat. Mater. 14 (2015) 205–209.
[67] D. Neagu, J.T.S. Irvine, in: J. Reedijk, K. Poeppelmeier (Eds.), Comprehensive Inorganic Chemistry II, From Elements to Applications, vol. 4, Elsevier, Amsterdam, 2013, pp. 397–415.
[68] P.R. Slater, D.P. Fagg, J.T.S. Irvine, J. Mater. Chem. 7 (1997) 2495–2498.
[69] X. Li, H. Zhao, X. Zhou, N. Xu, Z. Xie, N. Chen, Int. J. Hydrog. Energy 35 (15) (2010) 7913–7918.
[70] J. Canales-Vázquez, J.C. Ruiz-Morales, J.T.S. Irvine, W. Zhou, J. Electrochem. Soc. 152 (7) (2005) 1458–1465.
[71] D. Neagu, J.T.S. Irvine, Chem. Mater. 23 (2011) 1607–1617.
[72] J.C. Ruiz-Morales, J. Canales-Vázquez, C. Savaniu, D. Marrero-López, W. Zhou, J.T.S. Irvine, Nature 439 (2006) 568–571.
[73] P. Blennow, A. Hagen, K.K. Hansen, L.R. Wallenberg, M. Mogensen, Solid State Ionics 179 (2008) 2047–2058.
[74] O.A. Marina, N.L. Canfield, J.W. Stevenson, Solid State Ionics 149 (2002) 21–28.
[75] S.Q. Hui, A.J. Petric, J. Electrochem. Soc. 149 (2002) J1–J10.
[76] Q. Ma, F. Tietz, D. Sebold, D. Stover, J. Power Sources 195 (2010) 1920–1925.
[77] O.A. Marina, L.R. Pederson, in: J. Huijsmans (Ed.), Proc. 5th European Solid Oxide Fuel Cell Forum, European SOFC Forum, Switzerland, 2002, pp. 481–489.
[78] M.R. Pillai, I. Kim, D.M. Bierschenk, S.A. Barnett, J. Power Sources 185 (2) (2008) 1086–1093.
[79] M.C. Verbraeken, T. Ramos, K. Agersted, Q. Ma, C.D. Savaniu, B.R. Sudireddy, J.T.S. Irvine, P. Holtappels, F. Tietz, RSC Adv. 5 (2) (2015) 1168–1180.
[80] S.Q. Hui, A. Petric, Solid State Ionics 143 (2001) 275–283.
[81] Y.-H. Huang, R.I. Dass, Z.-L. Xing, J.B. Goodenough, Science 312 (2006) 254–257.
[82] C. Bernuy-Lopez, M. Allix, C. Bridges, J. Claridge, J. Rosseinsky, Chem. Mater. 1 (2007) 1035–1038.
[83] Y. Ji, Y.-H. Huang, J.-R. Ying, J.B. Goodenough, Electrochem. Commun. 9 (8) (2007) 1881–1885.
[84] R. Martinez-Coronado, J.A. Alonso, M.T. Fernandez-Diaz, J. Power Sources 258 (2014) 76–82.
[85] B. Ma, U. Balachandran, Mater. Res. Bull. 33 (1998) 223–236.
[86] I. Fisher, C. James, S. Chuang, Catal. Commun. 10 (2009) 772–776.
[87] J.M. Haag, B.D. Madsen, S.A. Barnett, K.A. Poepellmeier, Electrochem. Solid-State Lett. 11 (2008) B51–B53.
[88] T.H. Shin, S. Ida, T. Ishihara, J. Am. Chem. Soc. 133 (2011) 19399–19407.

[89] P. Holtappels, F.W. Poulsen, M. Mogensen, Solid State Ionics 135 (2000) 675–679.
[90] D.M. Flot, J.T.S. Irvine, Solid State Ionics 135 (2000) 513–518.
[91] A. Kaiser, J.L. Bradley, P.R. Slater, J.T.S. Irvine, Solid State Ionics 135 (2000) 519–524.
[92] J.L. Rupp, A. Infortuna, L.J. Gauckler, Acta Mater. 54 (2006) 1721–1730.
[93] Z. Liu, B. Liu, D. Ding, M. Liu, F. Chen, C. Xia, J. Power Sources 237 (2013) 243–259.
[94] S.P. Jiang, Mater. Sci. Eng. A 418 (2006) 199–210.
[95] S.P. Jiang, S. Zhang, Y.D. Zhen, W. Wang, J. Am. Ceram. Soc. 88 (7) (2005) 1779–1785.
[96] L. Liu, K. Sun, X. Zhou, X. Li, M. Zhang, N. Zhang, Electrochem. Commun. 19 (2012) 63–66.
[97] S. McIntosh, J.M. Vohs, R.J. Gorte, J. Electrochem. Soc. 6 (11) (2003) A240–A243.
[98] A. Babaei, S.P. Jiang, J. Li, J. Electrochem. Soc. 156 (9) (2009) B1022–B1029.
[99] R.F. Martins, M.C. Brant, R.Z. Domingues, R.M. Paniago, K. Sapag, T. Matencio, Mater. Res. Bull. 44 (2) (2009) 451–456.
[100] R.J. Gorte, J.M. Vohs, S. McInstosh, Solid State Ionics 175 (2004) 1–6.
[101] G. Corre, G. Kim, M. Cassidy, J.M. Vohs, R.J. Gorte, J.T.S. Irvine, Chem. Mater. 21 (2009) 1077–1084.
[102] S.P. Jiang, Y.M. He, T.M. He, S.B. Ho, J. Power Sources 185 (1) (2008) 179–182.
[103] S. Barison, M. Fabrizio, C. Mortalo, P. Antponucci, V. Modafferi, R. Gerbasi, Solid State Ionics 181 (2010) 285–291.
[104] S. Lee, G. Kim, J.M. Vohs, R.J. Gorte, J. Electrochem. Soc. 155 (11) (2008) B1179–B1183.
[105] M.C. Verbraeken, B. Iwanschitz, A. Mai, J.T.S. Irvine, J. Electrochem. Soc. 159 (11) (2012) F757–F762.
[106] P. Holtappels, J.T.S. Irvine, B. Iwanschitz, L. Theil Kuhn, L.Y. Lu, Q. Ma, J. Malzbender, A. Mai, T. Ramos, J. Rass-Hansen, B.R. Sudireddy, F. Tietz, V. Vasechko, S. Veltz, M.C. Verbraeken, ECS Trans. 57 (1) (2013) 1175–1184.
[107] H. Kurokawa, L. Yang, C.P. Jacobson, L.C. De Jonghe, S.J. Visco, J. Power Sources 164 (2) (2007) 510–518.
[108] M.D. Grossz, K.M. Carver, M.A. Deighan, A. Schenkel, B.M. Smith, A.Z. Yee, J. Electrochem. Soc. 156 (4) (2009) B540–B545.
[109] Q. Ma, F. Tietz, A. Leonide, E. Ivers-Tiffee, J. Power Sources 196 (2011) 7308–7312.
[110] X.C. Lu, J.H. Zhu, Z. Yang, G. Xia, J.W. Stevenson, J. Power Sources 192 (2) (2009) 381–384.
[111] J.-S. Park, I.D. Hasson, M.D. Gross, C. Chen, J.M. Vohs, R.J. Gorte, J. Power Sources 196 (18) (2011) 7488–7494.
[112] E. Stefan, G. Tsekouras, J.T.S. Irvine, Adv. Energy Mater. 3 (2012) 1454–1462.
[113] H. Tanaka, M. Uenishi, M. Taniguchi, I. Tan, K. Narita, M. Kimura, K. Kaneko, Y. Nishihata, J. Mizuki, Catal. Today 117 (2006) 321–328.
[114] B.D. Madsen, W. Kobsiriphat, Y. Wang, L.D. Marks, S.A. Barnett, J. Power Sources 166 (2007) 64–67.
[115] D.M. Bierschenk, E. Potter-Nelson, C. Hoel, Y. Liao, L. Marks, K.R. Poeppelmeier, S. A. Barnett, J. Power Sources 196 (6) (2011) 3089–3094.
[116] T. Jardiel, M.T. Caldes, F. Moser, J. Hamon, G. Gauthier, O. Joubert, Solid State Ionics 181 (2010) 894–901.
[117] S.-H. Cui, J.-H. Li, X.-W. Zhou, G.-Y. Wang, J.-L. Luo, K.T. Chuang, Y. Bai, L.-J. Qiao, J. Mater. Chem. A 1 (2013) 9689–9696.
[118] D. Neagu, G. Tsekouras, D.N. Miller, H. Menard, J.T.S. Irvine, Nat. Chem. 5 (2013) 916–923.
[119] L. Adijanto, V.B. Padmanabhan, R. Kungas, R.J. Gorte, J.M. Vohs, J. Mater. Chem. 22 (2012) 11396–11402.

Cathodes

6

T. Kawada[1], *T. Horita*[2]
[1]Tohoku University, Sendai, Japan, [2]National Institute of Advanced Industrial Science and Technology (AIST) Tsukuba Central 5-2, 1-1-1 Higashi, Tsukuba, Ibaraki, 305-8565 JAPAN

6.1 Introduction

In fuel cells, oxygen dissociation reaction generally requires a large ovepotential and causes the major loss in efficiency. High-temperature operation of SOFC helps the electrode reaction to proceed without precious metal catalysis, which, however, does not mean any material can be used as the cathode. Properties required for cathode materials are:

(1) high catalytic activity,
(2) high electronic conductivity (preferably, electronic–ionic mixed conductivity),
(3) chemical stability and compatibility with the other cell components,
(4) morphological stability, and
(5) mechanical stability and compatibility.

They must be achieved in a cost-effective manner.

Historically, many types of oxides [1] including indium oxide, zinc oxide, perovskite-type and related oxides have been tested. The report by Tedmon et al. [2] in 1969 summarised the results of early-stage research, in which they concluded that a single fully satisfactory material was lacking. However, they had already introduced perovskite-type oxides with Co in the B-site, oxides which are still considered as promising candidates in current technologies of SOFCs. The high catalytic activity in oxygen reduction reactions partly comes from its mixed conductivity of electrons and oxide ions. When a mixed ionic electronic conductor (MIEC) is used as cathode, the electrochemical reaction site is not restricted at the boundaries with the electrolyte but can expand onto the surfaces of the electrode particles in a certain thickness of the porous electrode layer. The disadvantages of the cobaltite perovskite are chemical and mechanical instability on a YSZ electrolyte, that is, the high reactivity to YSZ [3] and much larger thermal expansion coefficient.

Thus, the focus of SOFC cathode research was shifted to lanthanum manganite perovskites such as (La, Sr or Ca) MnO_3, which are less reactive to YSZ and have closer thermal expansion coefficient. In the 1990's, the reactivity of lanthanum manganite to YSZ was intensively studied by experiments and thermodynamic calculations [4,5]. Introduction of A-site deficiency in $LaMnO_3$ was found to improve the chemical stability of A-site cations in the perovskite structure and to avoid reaction against YSZ [6,7]. Lanthanum manganite-based cathodes show sufficiently high electrochemical performance at temperatures higher than 800 °C and were utilised in tubular and planar cells fabricated by a slurry/sintering process [8].

Recently, SOFC has become applied to a smaller system such as an auxiliary power unit (APU) for vehicles [9,10] and for a residential CHP (combined heat and power)

High-temperature Solid Oxide Fuel Cells for the 21st Century. http://dx.doi.org/10.1016/B978-0-12-410453-2.00006-3

system of 1 kW class [11–13]. For such small systems, reduced temperature operation is preferable to reduce heat losses and achieve high energy-conversion efficiency. Moreover, use of metal parts for interconnects and casing requires the operating temperature to be lower than 800 °C. In an intermediate temperature SOFC (IT-SOFC), the performance of $LaMnO_3$-based cathodes was not enough because the activation energy of the oxygen incorporation reaction is rather high. At this stage, MIEC perovskites attracted attention again. Instead of pure cobaltite perovskite, solid solution with iron-based perovskite, that is, $(La,Sr)(Co,Fe)O_3$ (LSCF) was proposed [14] to have higher chemical and mechanical compatibility with other cell components. Although this composition is still reactive to YSZ, the introduction of a ceria-based protective interlayer enabled LSCF to be used on YSZ electrolyte without significant degradation [15]. It is now more and more popular among IT-SOFC developers.

Currently, both high-temperature SOFC and IT-SOFCs have succeeded in various demonstration stages and are now going into the penetration stage. Stability for more than 10 years is now the principal objective to be achieved. Since cathode overpotential is often a major source of efficiency loss, the chemical/morphological stability under operation conditions has become important. Understanding of mechanical behaviour of the cathode materials has developed strongly in the last decade [16]. Attempts to find a material with better performance also continued among perovskite and related oxides. $(Ba,Sr)(Co,Fe)O_3$ (BSCF) was found to show superior performance although the chemical stability has been a problem [17,18]. Recently, effects of adding second phase have shown enhancement of the catalytic activity not merely increasing ionic conduction path but also modifying the properties of the reaction site [19,20].

In this chapter, the physical and physicochemical properties of perovskites will be given briefly, and the relationship between defect concentration and the crystal structure, electrical properties, dimensional/morphological stability will be discussed.

6.2 Physical and physicochemical properties of perovskite cathode materials

6.2.1 Lattice structure

The lattice structure of perovskites, ABO_3, is shown in Figure 6.1. This oxide consists of three elements, namely, the large cations, A^{n+}; the small cations, $B^{(6-n)+}$ and the oxide ions, O^{2-}, where n is the positive charge on the A ions. Cations $B^{(6-n)+}$ are surrounded by 6 oxide ions, whereas cations A^{n+} have 12 oxide ion coordinates; the sites with 12 coordinates are often called the A sites, and the sites with 6 coordinates the B-sites. The geometrical fitting of cations and anions to this structure is measured with the Goldschmidt tolerance factor defined as follows:

$$t = (r_A + r_O)/\left\{\sqrt{2}(r_B + r_O)\right\} \quad (6.1)$$

where r_A, r_B and r_O are the effective ionic radii of A, B and O ions, respectively. Usually, this factor is evaluated from Shannon's ionic radii [21] for respective

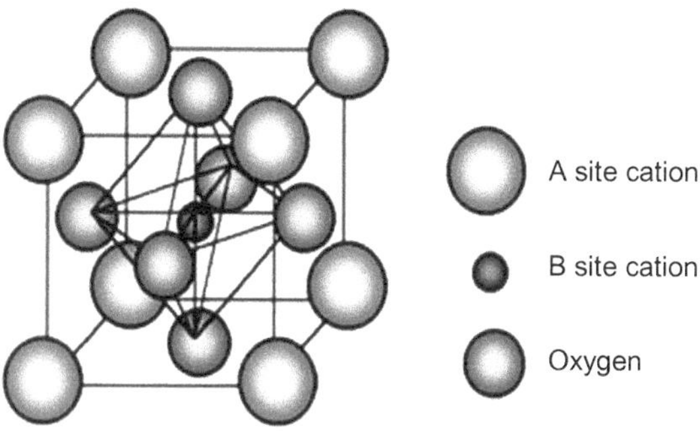

Figure 6.1 Schematic representation of lattice structure of perovskite, ABO_3.

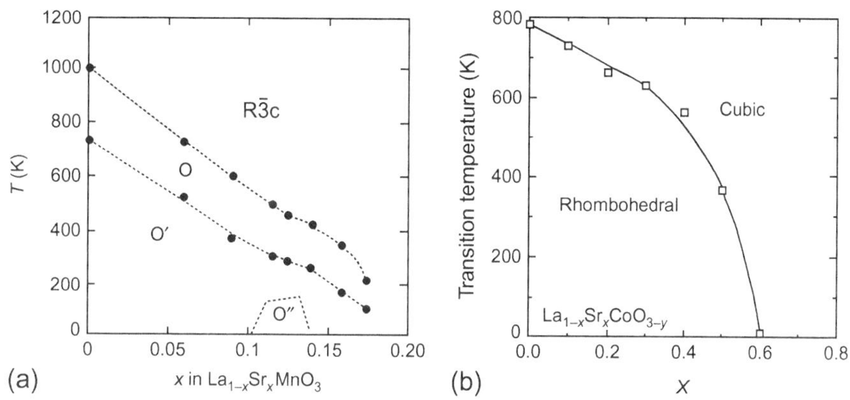

Figure 6.2 Variation of crystal structure with temperature and Sr content in (a) $La_{1-x}Sr_xMnO_{3+d}$ and (b) $La_{1-x}Sr_xCoO_{3-\delta}$.
(a) Reproduced from Ref. [22]. (b) Reprinted with permission from Ref. [23].

coordination numbers. When the tolerance factor is near unity, the structure is the ideal cubic. In other cases, orthorhombic or rhombohedral distortions appear. Undoped stoichiometric $LaMnO_3$ takes orthorhombic phases (Pbnm) at 25 °C and rhombohedral (R-3C) above 600 °C [22,23]. $(La,Sr)CoO_3$ and $(La,Sr)(Co,Fe)O_3$ are in rhombohedral (R-3C) at room temperature and show second-order phase transition to a cubic (or a pseudo cubic) phase at a certain temperature [23,24]. The transformation temperature strongly depends on the oxygen stoichiometry, and hence on the mean valence of the B-site cations (Figure 6.2).

6.2.2 Oxygen nonstoichiometry

In most perovskite-type oxides, the valence state of the transition metal in the B-site can be changed by doping aliovalent cations in the A-site or the B-site. In $LaMO_3$ (M = transition metal cation), doping Sr^{2+} in La^{3+} site increases the average valence number of the M ion over 3. The generated positive charge is known to behave as an itinerant hole in $La_{0.6}Sr_{0.4}CoO_{3-\delta}$ and as a small polaron in $(La,Sr)FeO_{3-\delta}$ [25,26].

The valence state of the B-site cation also changes when oxygen vacancy is formed under reduced oxygen partial pressure as

$$O_O + 2h \rightarrow V_O^{\bullet\bullet} + \frac{1}{2}O_2 \tag{6.2}$$

Since the perovskite structure allows a relatively large amount of oxygen vacancy to be formed in the crystal lattice, the equilibrium constant for oxygen vacancy formation is mainly determined by the property of B-site cations. Figure 6.3 compares oxygen nonstoichiometry concentration in $(La,Sr)CoO_3$, $(La,Sr)FeO_3$ and $(La,Sr)MnO_3$.

Among them, $(La,Sr)CoO_{3-\delta}$ (LSC) shows monotonic increase of oxygen vacancy concentration with decreasing oxygen partial pressure down to decomposition pressure. Mizusaki et al. [30] represented the oxygen partial pressure dependence of $La_{1-x}Sr_xCoO_{3-\delta}$ assuming random distribution of oxygen vacancies and delocalised electrons, where the relation between δ and oxygen partial pressure is written as

$$\begin{aligned} \frac{RT}{2}\ln p_{O_2} &= \mu_O - \mu_O{}^\circ - T(s_O - s_O{}^\circ) \\ &= [\Delta h_O{}^\circ(x) - a(x)\delta] - T\left[\Delta s_O{}^\circ(x) + R\ln\left(\frac{\delta}{3-\delta}\right)\right] \end{aligned} \tag{6.3}$$

The dependence of the entropy term on δ (in the latter bracket) is for the configuration of oxygen vacancies, and that of the enthalpy term (in the former bracket) is due to the change in vacancy formation energy. Lankhorst et al. attributed the energy change to the shift of Fermi energy by filling the 3d band of cobalt ions with the electrons generated by the vacancy formation [32].

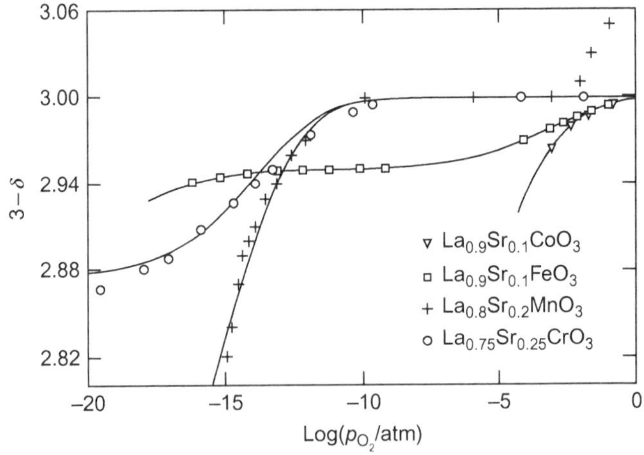

Figure 6.3 Comparison of oxygen nonstoichiometry in $La_{1-x}Sr_xMO_{3-\delta}$.
Reprinted with permission from Ref. [27]. The original data are taken from Refs. [28–31].

In contrast to cobaltite, (La,Sr)$FeO_{3-\delta}$ (LSF) shows a plateau in the medium partial pressure region where charge compensation is by Fe^{3+} ions. The defect equilibrium is thus represented assuming localised electrons and holes, Fe^{2+} and Fe^{4+}, respectively, which are equilibrated with Fe^{3+} through disproportionation reaction [29]. It can be written using the Kröger–Vink notation as

$$Fe_{Fe}^{\times} = Fe_{Fe}' + Fe_{Fe}^{\bullet} \tag{6.4}$$

With the redox equilibrium

$$O_O^{\times} + 2\left[Fe_{Fe}^{\bullet}\right] \rightarrow 2\left[Fe_{Fe}^{\times}\right] + V_O^{\bullet\bullet} + \frac{1}{2}O_2 \tag{6.2'}$$

and charge neutrality condition

$$\left[Fe_{Fe}^{\bullet}\right] - \left[Fe_{Fe}'\right] + 2\delta - \left[Sr_{La}'\right] = 0 \tag{6.5}$$

Solving the simultaneous equations,

$$p_{O_2}{}^{1/4} = \frac{1}{2}\left\{ -K_{\mathrm{red}}\frac{2\delta - x}{2\delta + 1 - x}\left(\frac{3-\delta}{\delta}\right)^{1/2} + \left(K_{\mathrm{red}}\left(\frac{2\delta - x}{2\delta + 1 - x}\right)^2\left(\frac{3-\delta}{\delta}\right) + 4K_{\mathrm{red}}K_i\left(\frac{-2\delta + 1 + x}{2\delta + 1 - x}\right)\left(\frac{3-\delta}{\delta}\right)\right)^{1/2}\right\} \tag{6.6}$$

is derived.

As is expected from the composition, oxygen nonstoichiometry of LSCF appears between those for LSC and for LSF [33]. With Fe concentration smaller than 0.4, the nonstoichiometry isotherm can be represented by the itinerant electron model, whereas the semiconductor feature appears when Fe concentration exceeds 0.6. They reported that LSCF showed "gradual decomposition" (hatched area in Figure 6.4) at oxygen partial pressures closer to the stability limit of LSC than that of LSF. Chemical stability may not be significantly improved by addition of Fe to LSC (Figure 6.5).

Among the transition metal perovskites shown in Figure 6.3, lanthanum manganite shows a notable feature of oxygen nonstoichiometry, that is, in addition to the oxygen deficiency, the "oxygen-excess" region [31,34] appears as illustrated in Figure 6.6a. The oxygen content exhibits two plateaus in its oxygen potential dependence: one is the saturation of oxygen excess (3 + d) at high oxygen partial pressures (Region I), and the other being the stoichiometric one (Region III in Figure 6.6a) at the lower oxygen partial pressures. Particular interest has focused on the oxygen-excess region. As will be discussed in Section 6.4, isothermal dilatometry [35] revealed that the oxygen incorporation in the oxygen-excess region accompanies the creation of a new crystal lattice. It means the oxygen excess is due to the formation of cation vacancies rather than oxygen interstitials.

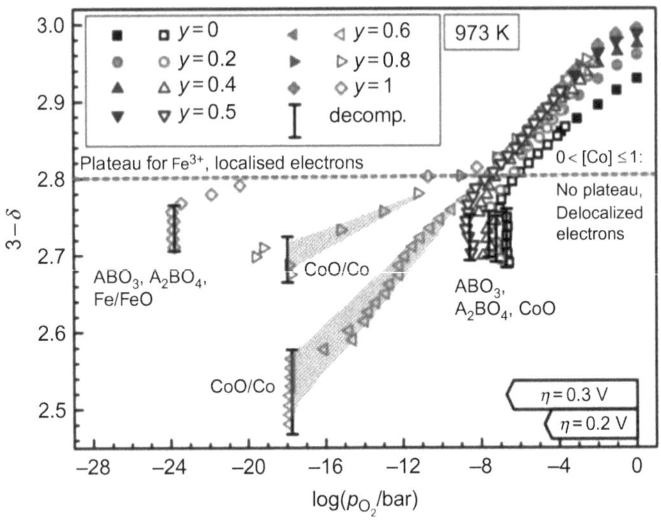

Figure 6.4 Oxygen nonstoichiometry and stability of $La_{0.6}Sr_{0.4}Co_{1-y}Fe_yO_{3-\delta}$. Reprinted with permission from Ref. [33].

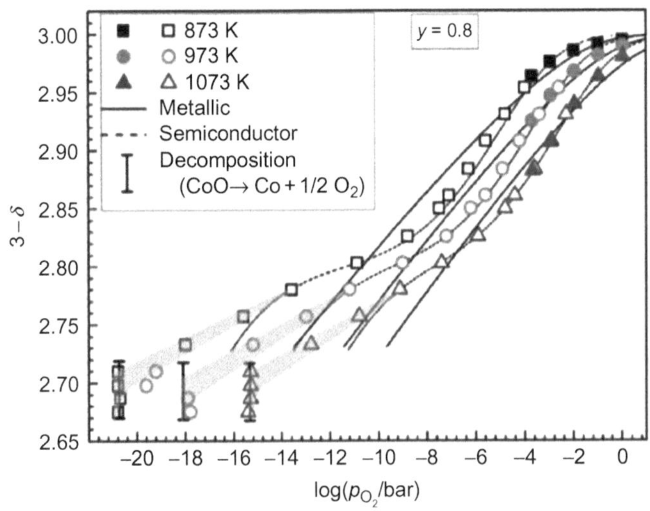

Figure 6.5 Temperature dependence of oxygen nonstoichiometry and stability of $La_{0.6}Sr_{0.4}Co_{0.2}Fe_{0.8}O_{3-\delta}$.
Reprinted with permission from Ref. [33].

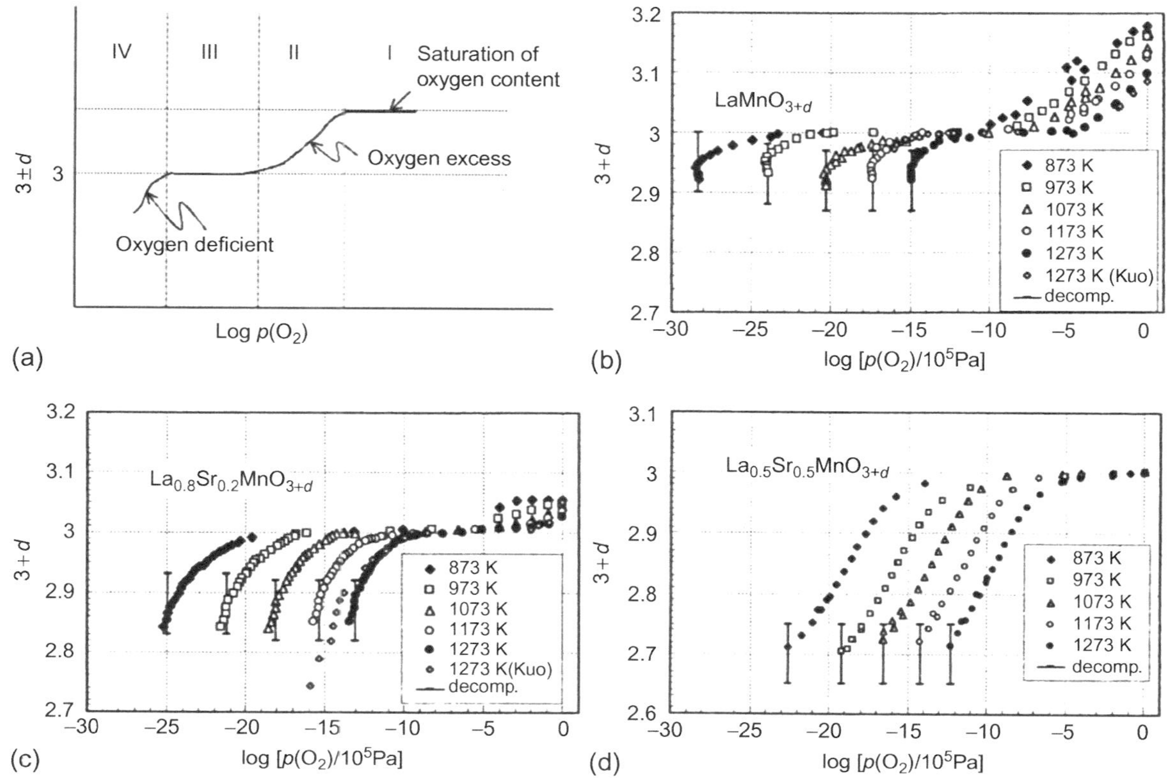

Figure 6.6 Oxygen nonstoichiometry of $LaMnO_3$ as a function of oxygen partial pressure. (a) Schematic representation of oxygen nonstoichiometry. (b) Oxygen content of $LaMnO_{3+d}$. (c) Oxygen content of $La_{0.8}Sr_{0.2}MnO_{3+d}$. (d) Oxygen content of $La_{0.5}Sr_{0.5}MnO_{3+d}$. Reprinted with permission from Ref. [34].

Attempts have been made to explain the thermogravimetric results by considering the formation of metal vacancies in the A- and the B-sites. Roosmalen and Cordfunke [36–39] tried to interpret the defect structure of the oxygen-excess composition by assuming the simultaneous formation of A- and B-site vacancies and charge disproportionation of Mn^{3+} to Mn^{2+} and Mn^{4+}. Results of their calculations can reproduce the thermogravimetric results fairly well. However, their assumption that $[Mn^{4+}]$ was constant implies the concentration of Mn^{3+} becomes lower than that of Mn^{4+} depending on the atmosphere. This does not happen in a system in which equilibration proceeds reasonably. Yokokawa et al. [40] made thermodynamic consideration to explain the oxygen nonstoichiometry of $La_{1-x}MnO_{3+d}$ with different lanthanum deficits as a function of oxygen potential and succeeded in reproducing the thermogravimetric results by a single model. The concentrations of the respective manganese ions derived from this model show a reasonable oxygen potential dependence as expected in accordance with normal chemical understanding. Mizusaki et al. [34] have proposed a defect clustering model to explain the fact that the oxygen-excess region disappears for $x > 0.4$ in $La_{1-x}Sr_xMnO_{3+d}$. In addition to metal vacancies, they propose a "vacancy excluding space" that is needed for each cation vacancy to exist stably without having the vacancy in its neighbouring space. The maximum oxygen excess can be determined from the maximum number of vacancy excluding spaces available in the lattice. For $x > 0.4$, there is no room for the vacancy excluding space around the metal vacancies, and this model can provide a good explanation for the disappearance of the oxygen excess.

6.2.3 Electrical conductivity

As expected from the Equations (6.2′) and (6.4), the electronic transport in (La,Sr) $FeO_{3-\delta}$ is characterised by hopping of localised charge carriers on Fe ions. Figure 6.7 shows oxygen partial pressure dependence of the electronic conductivity and Seebeck coefficient reported by Mizusaki et al. [41]. The electronic conductivity shows a minimum at the middle of the oxygen partial pressure range where the charge carrier changes from *p*- to *n*-type on reduction. It corresponds to the plateau region in oxygen nonstoichiometry isotherm in Figure 6.3. The conductivity can reach $100\ S\ cm^{-1}$ with heavy doping of Sr or Ca on the A-site.

Compared to the $LaFeO_3$-based perovskite, $La_{1-x}Sr_xCoO_3$ shows higher electronic conductivity even with a low concentration of Sr. For the composition range $x < 0.3$, the electron transport appears to be thermally activated semiconductor at lower temperatures but changes to metallic at higher temperatures. The transition temperature falls down below room temperature for the composition $x > 0.4$. The mobility of electron changes with Sr content, temperature and atmosphere. It is discussed in terms of the extent of crystal lattice distortion or the angle α of the rhombohedral crystal [42].

The electronic conduction characteristic of $(La,Sr)Co_{1-y}Fe_yO_3$ is in between $(La,Sr)CoO_3$ and $(La,Sr)FeO_3$. Similar to oxygen nonstoichiometry, the electronic conductivity is closer to $(La,Sr)FeO_3$ than $(La,Sr)FeO_3$ for $y = 0.5$. However, the effect of B-site composition decreases at a higher temperature range. Actually, the composition $y = 0.8$ is widely used for practical SOFCs for better thermomechanical

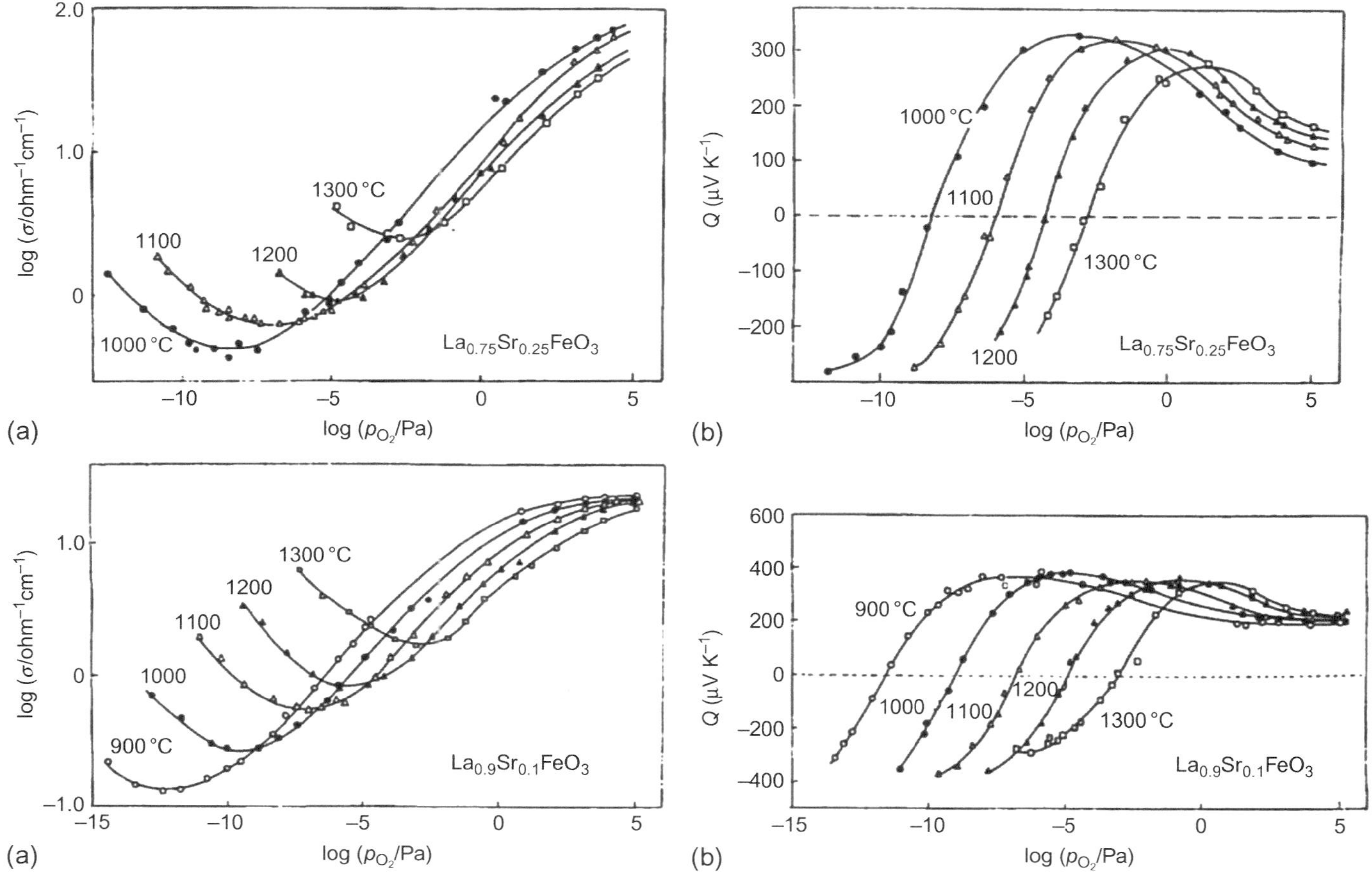

Figure 6.7 Electronic conductivity (a) and Seebeck coefficient (b) of $(La,Sr)FeO_{3-\delta}$. Reprinted with permission from Ref. [41].

compatibility as will be described in a later section. The electronic conductivity of $La_{1-x}Sr_xCo_{0.2}Fe_{0.8}O_{3-\delta}$ increases significantly with increasing Sr content x, reaching higher than 10^2 S cm^{-1} with $x = 0.4$ over a wide temperature range. The conduction mechanism is more like small polaron hopping for small y composition and changes to metallic with increasing y (Figures 6.8 and 6.9).

$LaMnO_3$-based perovskite also shows sufficient electrical conductivity, higher than 10 S cm^{-1}, at 700 °C. Figure 6.10 shows temperature dependence of the

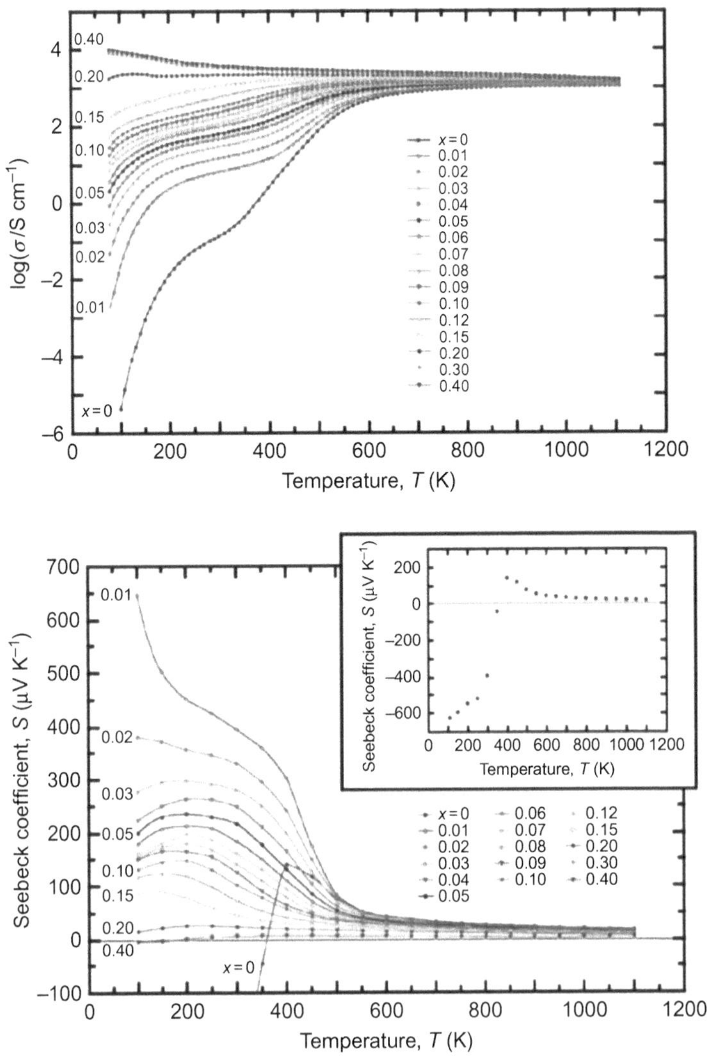

Figure 6.8 Electronic conductivity and Seebeck coefficient of $La_{1-x}Sr_xCoO_{3-\delta}$ in air. Reprinted with permission from Ref. [42].

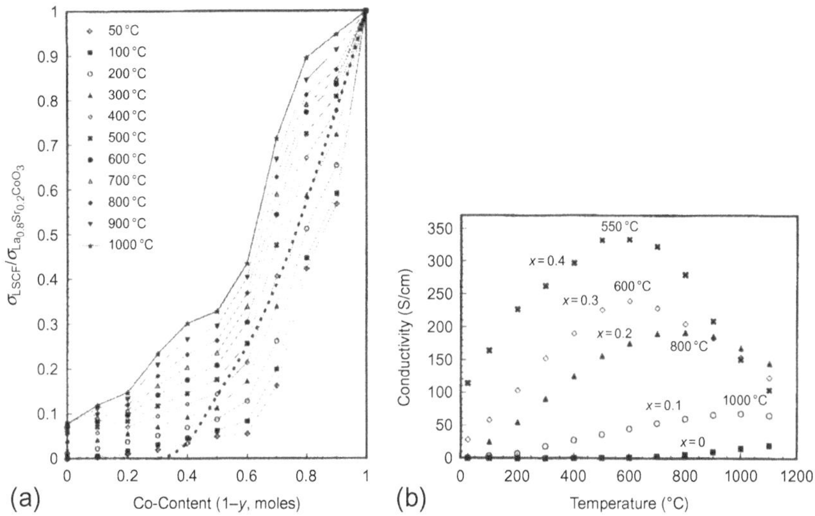

Figure 6.9 Electronic conductivity of $La_{0.8}Sr_{0.2}Co_{1-y}Fe_yO_{3-d}$ (a) and $La_{1-x}Sr_xCo_{0.2}Fe_{0.8}O_3$ (b).
Reprinted from Refs. [43,44] with permission.

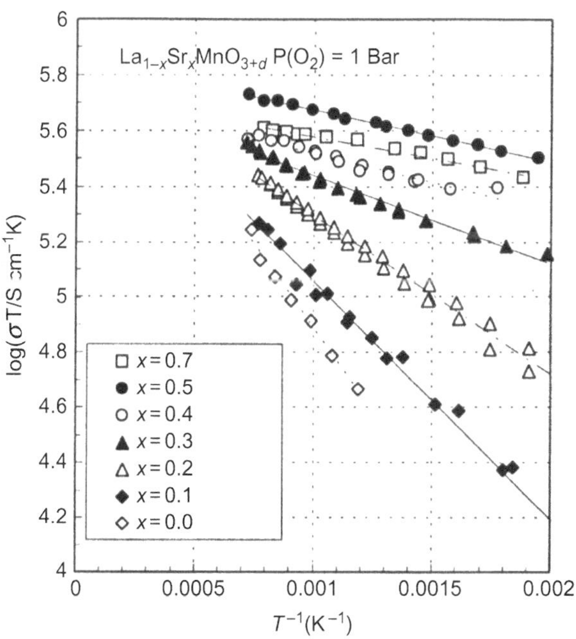

Figure 6.10 Electrical conductivity of Sr-doped $LaMnO_{3+d}$ as a function of inverse temperature.
Reprinted from Ref. [45] with permission.

electrical conductivity [45]. Straight lines in this plot suggest small polaron hopping conduction. In the oxygen-deficient $La_{1-x}Sr_xMnO_{3+d}$ ($d<0$), the conductivity is determined by the mean Mn valence and temperature. The conductivity decreases and the Seebeck coefficient increases with decreasing oxygen partial pressure, which suggests that the carrier concentration decreases with oxygen partial pressure. In contrast, the conductivity in the oxygen-excess region ($d>0$) is essentially constant irrespective of oxygen content or metal vacancies. Therefore, the concentration of carrier in the oxygen-excess region is about the same as that in the stoichiometric composition, $LaMnO_3$.

6.2.4 Oxygen transport

When a porous oxide is used as a cathode of SOFC, oxygen gas adsorbs on the surface of the particles, dissociates and diffuses in/on the particles to reach the electrolyte. A mixed electronic and oxide ionic conductor can serve a wider reaction area providing fast bulk diffusion paths. Higher diffusivity enables thicker electrode layer to participate in the oxygen incorporation reaction so that the overall electrode reaction resistance decreases proportionally to the square root of the diffusion resistance in one-dimensional approximation. Decreasing surface reaction resistance similarly reduces the electrode resistance. Surface reaction rate and oxygen transport rate are useful parameters for screening electrode materials.

In an oxygen sub-stoichiometric oxide, oxygen diffuses via vacancy mechanism where an oxide ion jumps into a neighbouring vacancy site leaving a vacancy behind. Thus, the self diffusion coefficient of oxygen, D_O, is characterised by the vacancy concentration c_V and vacancy diffusion coefficient D_V. The self diffusion coefficients are determined through the measurement of tracer diffusion coefficient D^*, chemical diffusion coefficient D, or partial oxide ionic conductivity σ_O, which are interrelated to each other by the following equation provided that the electronic conductivity is much higher than that of the oxide ion.

$$D_O = \frac{c_V}{c_O} D_V = \frac{1}{f} D_O^* = \frac{1}{W} \tilde{D}_O = \frac{RT}{4F^2 c_O} \sigma_O \left(W = \frac{1}{2} \frac{d \ln p_{O_2}}{d \ln c_O} \right) \tag{6.7}$$

where f is the correlation factor which is determined by the coordination of oxygen sublattice in the crystal. In a cubic perovskite-type oxide, f is equal to 0.69 [46]. The factor W is the so-called thermodynamic factor which is determined from the first-order derivative of the oxygen nonstoichiometry isotherm.

Tracer diffusion coefficient is measured by $^{16}O/^{18}O$ isotope exchange experiments. The sample is first equilibrated with the surrounding atmosphere of controlled oxygen partial pressure and then exposed to the $^{18}O_2$-enriched gas of the same total oxygen partial pressure. Tracer diffusion coefficient is evaluated either from the transient change of the isotope ratio in the gas phase or from the diffusion profile formed in the sample which can be measured by a secondary ion mass spectrometer (SIMS)

[47–49]. In these measurements, the diffusion coefficient is determined by fitting the solution of Fick's second law [50] under the transmitting boundary condition as

$$j_O^* = k^*\left(C_g^* - C_s^*\right) \tag{6.8}$$

where k^* is the surface exchange coefficient. $C_g{}^*$ and $C_s{}^*$ are the ^{18}O fractions in the gas and at the surface, respectively.

Kilner and co-workers have collected k^* and D^* data for several perovskites and derived the linear correlation between these two parameters [51,52]. They found a linear regression of the (logarithmic) data gives a slope near 0.5. Similarly, Wang et al. [53] reported that a similar relation between effective surface reaction constant and ionic conductivity is valid over a wide variety of perovskite materials. Since these materials are electronically conducting, there are many mobile electrons. On oxygen incorporation, therefore, the concentration of oxide ion vacancies plays an important role both for surface oxygen exchange and oxide ion diffusion (Figure 6.11).

Another popular method for determining the oxygen transport parameter is the electrical conductivity relaxation (ECR) method (Figure 6.12). It can be applied to a sample when it exhibits electronic conductivity that linearly changes with concentration of oxygen vacancy. The sample is first equilibrated with an atmosphere, and then the oxygen partial pressure of the surrounding gas is abruptly changed. The electronic conductivity changes following the relaxation of oxygen vacancy concentration towards the new equilibrium value. The electronic conductivity is measured using the four-terminal method with a bar-shaped sample or van der Pau method with a plane sheet sample. Since this ECR method is easy to carry out with a simple experimental setup, many works have been reported on various perovskite-type mixed conductors.

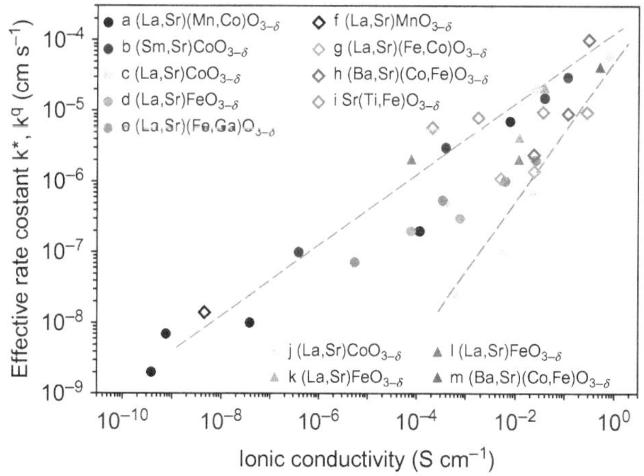

Figure 6.11 Relationship between surface reaction coefficient and ionic conductivity at 750 °C under 1 bar oxygen.
Reprinted from Ref. [53] with permission.

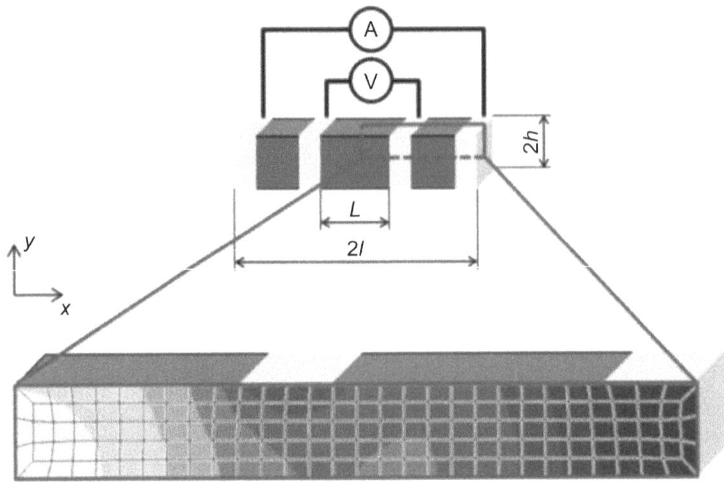

Figure 6.12 The oxygen concentration profile during electrical relaxation measurements using catalytic probes.
Reprinted from Ref. [54] with permission.

However, the k and $\widetilde{D}$ values obtained by this ECR method should be carefully evaluated. They might include uncertainties partly due to the intrinsic difficulty in extraction of k and $\widetilde{D}$ from a single relaxation curve [55] and due to the perturbation of the diffusion paths by the existence of current and potential probes. Kudo et al. [54] pointed out that the catalytic activity of Pt probes in four-terminal measurements can cause the oxygen exchange site to be mainly around the Pt probes. It results in the oxygen flux in the sample to be in the longitudinal direction rather than in the thickness direction, which makes the effective diffusion length longer, and underestimates $\widetilde{D}$.

As is shown in Equation (6.8), the oxygen diffusion coefficient D_O of perovskite-type oxides can be characterised by the availability of a vacancy site and its mobility. When vacancy concentration is small, the mobility of the vacancy would remain constant and the oxygen diffusivity would be proportional to the vacancy concentration. However, several researchers claim that the mobility of oxygen vacancy decreases with increasing vacancy concentration. Figure 6.13 summarises the oxygen vacancy concentration dependence of the vacancy diffusion coefficient D_V of $La_{0.6}Sr_{0.4}Co_{0.2}Fe_{0.8}O_{3-\delta}$, $La_{0.6}Sr_{0.4}CoO_{3-\delta}$ and $La_{0.6}Sr_{0.4}FeO_{3-\delta}$ collected from several papers. Large scatter is observed among the reports, and the difference is more significant in the larger vacancy concentration range. The data obtained by oxygen isotope exchange show less dependence on the vacancy concentration, while those obtained by ECR method tend to show significant decrease with increasing δ. A possible explanation for the discrepancy is that the oxygen diffusion path is significantly perturbed in ECR measurements as shown in Figure 6.13, especially for large δ, that is, under low oxygen partial pressures. This is plausible since oxygen exchange

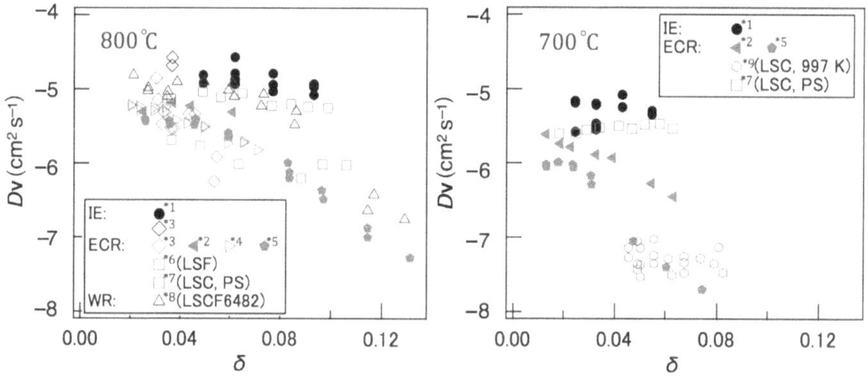

Figure 6.13 Oxygen vacancy diffusion coefficient, D_V, of $La_{0.6}Sr_{0.4}Co_{0.2}Fe_{0.8}O_{3-\delta}$ as a function of δ at (a) 1073 K and (b) 973 K; *1Kudo et al. [56], *2Yashiro et al. [57], *3Li et al. [58], *4Lane et al. [59], *5Bouwmeester et al. [60], *6Sogaard et al. ($La_{0.6}Sr_{0.4}FeO_{3-\delta}$) [61], *7Sogaard et al. ($La_{0.6}Sr_{0.4}CoO_{3-\delta}$ porous layer on the surface) [62], *8Katsuki et al. ($La_{0.6}Sr_{0.4}Co_{0.8}Fe_{0.2}O_{3-\delta}$) [63] and *9Sitte et al. ($La_{0.6}Sr_{0.4}CoO_{3-\delta}$) [64].

reaction rate on the surface of LSCF decreases with decreasing oxygen partial pressure, which makes the perturbation more prominent. In a practical consideration, it is reasonable to assume D_V of $La_{0.6}Sr_{0.4}Co_{0.2}Fe_{0.8}O_3$ to take constant values of $\sim 1 \times 10^{-5}$ cm^2 s^{-1} at 800 °C and $\sim 3 \times 10^{-6}$ cm^2 s^{-1} at 700 °C, and oxygen diffusivity D_O in air to be around $\sim 4 \times 10^{-7}$ at 800 °C and $\sim 6 \times 10^{-8}$ cm^2 s^{-1} at 700 °C, which corresponds to 0.1 S cm^{-1} and 0.02 S cm^{-1}, respectively, in conductivity.

Compared to the diffusion coefficients of cobalt or iron-containing perovskites, LSM-based perovskites show smaller oxygen diffusion coefficient due to the smaller vacancy concentration. Reported diffusion coefficients and surface oxygen exchange coefficients for selected LSM in comparison with LSC are shown in Table 6.1.

Table 6.1 Oxygen diffusion coefficients (D*) of perovskites

Composition	Temperature (°C)	pO_2 (atm)	D^* (cm^2 s^{-1})	Refs.
$La_{0.95}Sr_{0.05}MnO_3$	900	1	2.44×10^{-13}	[65]
$La_{0.90}Sr_{0.10}MnO_3$	1000	1	4.78×10^{-12}	[65]
$La_{0.80}Sr_{0.20}MnO_3$	900	1	1.27×10^{-12}	[65]
$La_{0.80}Sr_{0.20}MnO_3$	1000	1	1.33×10^{-11}	[65]
$La_{0.80}Sr_{0.20}MnO_3$	1000	1	6.60×10^{-13}	[66]
$La_{0.80}Sr_{0.20}MnO_3$	900	1	1.60×10^{-13}	[66]
$La_{0.80}Sr_{0.20}MnO_3$	800	1	4.00×10^{-15}	[66]
$La_{0.80}Sr_{0.20}MnO_3$	700	1	3.10×10^{-16}	[66]
$La_{0.80}Sr_{0.20}CoO_3$	1000	1	9.01×10^{-8}	[66]
$La_{0.80}Sr_{0.20}CoO_3$	900	1	1.03×10^{-8}	[66]
$La_{0.80}Sr_{0.20}CoO_3$	800	1	9.87×10^{-10}	[66]
$La_{0.80}Sr_{0.20}CoO_3$	700	1	1.04×10^{-10}	[66]

The oxygen diffusion coefficients are scattered over one or two orders of magnitude, and the values range in the orders of 10^{-13} to 10^{-11} $cm^2\ s^{-1}$ at 1000 °C at $p(O_2) = 1$ atm. These values for diffusivity are lower than required for mixed conducting oxide cathodes. This implies that when doped $LaMnO_3$ is used for a cathode, oxygen will be transported through the gaseous phase or on the surface of the $LaMnO_3$ to the major electrochemical reaction sites at the electrolyte/cathode/gas three-phase boundary. In the high overpotential region, however, some authors have pointed out that oxide ion diffusion can take place inside doped $LaMnO_3$ because the oxide ion vacancies can be formed [67–70].

6.3 Chemical stability and compatibility with the cell components

6.3.1 Thermodynamic stability of perovskite-type oxides

The perovskite lattice provides a strong stabilisation effect on the transition metal ions in B-sites. For example, the trivalent Co^{3+} ions are well stabilised in the perovskite structure although this valence state is not fully stable in other crystal structures. This stabilisation is due to the geometrical matching of the A-site and the B-site ions (see Figure 6.1). When the tolerance factor (Equation 6.1) derived from the ionic sizes is closer to unity, larger stabilisation energy is achieved [4,71]. For the rare earth transition metal perovskites, the tolerance factor is less than unity, suggesting that rare earth ions are small as A-site cations. Among rare earth ions, the largest ions, La^{3+}, therefore exhibit the largest stabilisation.

6.3.2 Reaction of perovskites with the zirconia component in YSZ

The La_2O_3 component in perovskite can react with the zirconia component in YSZ to form lanthanum zirconate, $La_2Zr_2O_7$. On reaction, the valence state of the transition metal ions may change as a result of the formation of other transition metal binary oxides:

$$LaMO_3 + ZrO_2 = 1/2La_2Zr_2O_7 + MO + 1/4O_2(g) \tag{6.9a}$$

$$LaMO_3 + ZrO_2 = 1/2La_2Zr_2O_7 + 1/2M_2O_3 \tag{6.9b}$$

$$LaMO_3 + ZrO_2 + 1/2O_2(g) = 1/2La_2Zr_2O_7 + MO_2 \tag{6.9c}$$

where M is the transition metal and MO_n is its binary oxide. During cell fabrication at high temperatures, reaction (6.9) becomes important because of its large entropy change. During cell operation, oxygen potential dependence becomes important. Since YSZ is an oxide ion conductor, the oxygen potential in the vicinity of the electrode/electrolyte interface is important in determining how electrolyte/electrode chemical reactions proceed under cell operation.

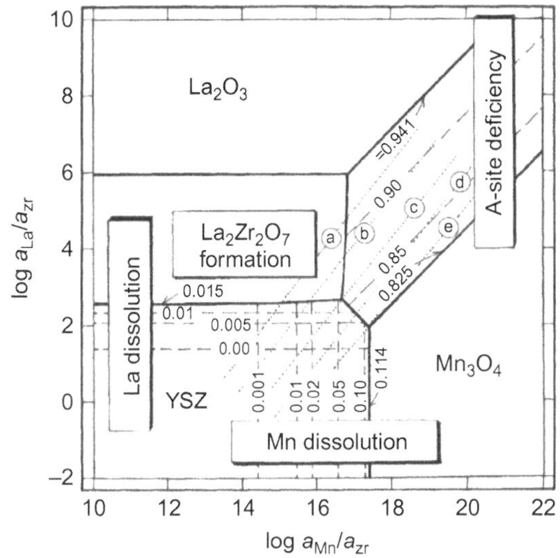

Figure 6.14 Chemical potential diagram for the La–Mn–Zr–O system.
Reprinted from Ref. [4] with permission.

The reactivity of $LaMnO_3$ with YSZ can be represented by the chemical potential diagram as given in Figure 6.14 [4]. It shows that the stoichiometric $LaMnO_3$ reacts with ZrO_2 to form the three-phase combination among ZrO_2, $La_2Zr_2O_7$ and an A-site-deficient $La_{1-x}MnO_3$, whereas the A-site-deficient $La_{1-x}MnO_3$ does not react with YSZ. In other words, the reaction of non-A-site-deficient $LaMnO_3$ with ZrO_2 can be written as follows:

$$LaMnO_3 + x(ZrO_2 + 1/4O_2) = x/2La_2Zr_2O_7 + La_{1-x}MnO_3 \tag{6.10}$$

In this reaction, manganese ions are oxidised and oxygen is involved as a reactant.

To minimise reactions between the cathode and the electrolyte, research efforts have been made on the A-site-deficient lanthanum manganite. In addition to A-site deficiency, improvement in cathode performance can also be obtained by using composite cathodes consisting of LSM and YSZ [72–75]. During the fabrication of the composite cathode, some $La_2Zr_2O_7$ may form, shifting the composition of LSM to where no further formation of $La_2Zr_2O_7$ occurs. Thus, such a composite cathode forms no more $La_2Zr_2O_7$ at the interface with the YSZ electrolyte.

The thermodynamic compatibility of A-site substituted $LaMnO_3$ with YSZ is predicted as shown in Figure 6.15 [76]. The zirconate formation can be avoided with $(La_{0.5}Ca_{0.5})MnO_3$, whereas some zirconate formation is expected for $(La_{0.84}Sr_{0.16})MnO_3$. In fuel cell operation, however, Sr substituted $LaMnO_3$ can work as a good cathode. As given in Equation (6.9), the $La_2Zr_2O_7$ formation reaction is accompanied by "oxidation" of manganese ions in the perovskites. This suggests that $La_2Zr_2O_7$ formed at the interface will disappear because the overpotential causes the local oxygen potential shift to the more reducing side. On the other hand, on anodic polarisation (e.g. in water electrolysis), the $La_2Zr_2O_7$ formation is enhanced. It is thus understandable that, in the early stage of development of zirconia cells, Westinghouse developed

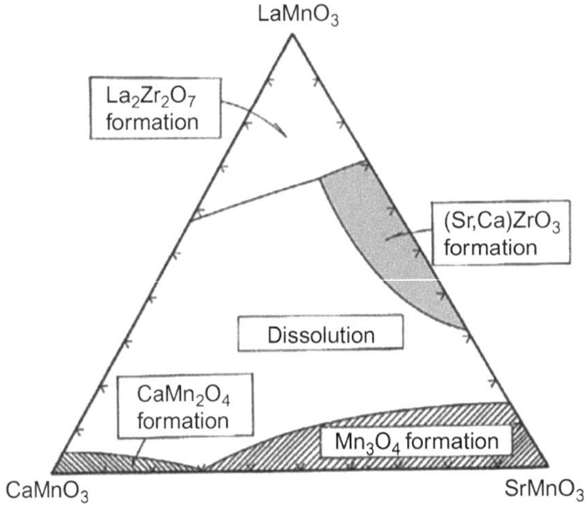

Figure 6.15 Composition diagram for (La,Sr,Ca)MnO_3 electrode. Westinghouse developed $(La_{0.84}Sr_{0.16})MnO_3$ for fuel cells, while Dörnier developed $(La_{0.5}Ca_{0.5})MnO_3$ for water electrolysers.
Reprinted from Ref. [76] with permission.

$(La_{0.84}Sr_{0.16})MnO_3$ for fuel cells, while Dörnier developed $(La_{0.5}Ca_{0.5})MnO_3$ for water electrolysers.

Thermodynamic calculation predicts that the driving force of reactions of $LaCoO_3$ with YSZ disappears below 1173 K. In reality, however, Sr substitution is necessary for a practical cathode, and it leads to high reactivity with YSZ, forming $SrZrO_3$. The reaction proceeds not only during the fabrication process but also in operation at reduced temperatures such as at 600 °C. Figure 6.16 shows the area-specific resistance at the interface between $La_{0.6}Sr_{0.4}CoO_3$ dense film electrode and YSZ single crystal, which was determined separately from the surface reaction resistance by using

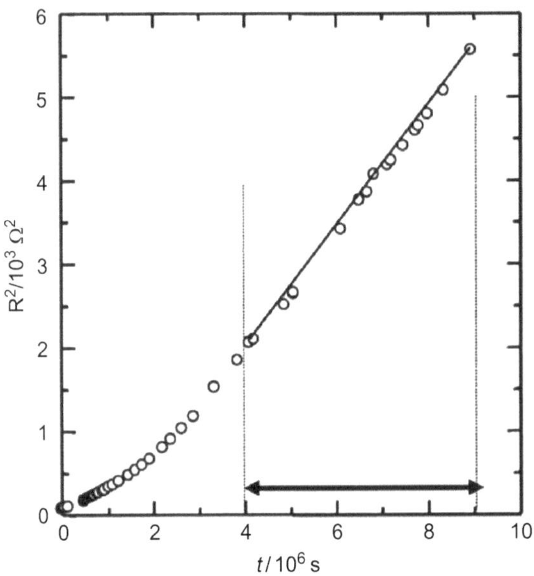

Figure 6.16 Change in the area-specific resistance at the interface between $La_{0.6}Sr_{0.4}CoO_3$ and $Ce_{0.9}Gd_{0.1}O_{1.95}$ operated at 973 K.
Reprinted from Ref. [77] with permission.

Figure 6.17 GDC interlayer (barrier layer) between (La,Sr)(Co,Fe)O_3 cathode and YSZ electrolyte.
SOFCPower, Reprinted from Ref. [80] with permission.

impedance spectroscopy [77]. The interface resistance was found to increase with time, accordingly to the parabolic rate law. The parabolic rate constant at 700 °C for the formation of $SrZrO_3$ was calculated to be about 2×10^{-18} cm^2 s^{-1}. Formation of $SrZrO_3$ significantly affected the electrode performance. The situation is similar with (La,Sr)FeO_3 and (La,Sr)(Co,Fe)O_3. It means those mixed conducting electrodes are not suitable to be used directly on YSZ electrolyte.

However, continuous attempts have been made to use cobaltite-based cathodes for reduced temperature SOFCs since a higher-performance cathode is necessary for reduced temperature operation. One of the practical solutions is the use of CeO_2-based electrolyte as an interlayer between YSZ and the cobaltite-based cathode. Compared with YSZ, ceria has less reactivity with perovskite cathode materials due to the less acidic nature of CeO_2 and less stabilisation energy in the perovskite lattice.

When ceria is heat treated on YSZ, the solid solution of ZrO_2 and CeO_2, which is less conductive than both oxides [78,79], is formed. Thus too high temperature should be avoided in the fabrication process. On the other hand, too low temperature sintering will leave many pores in the doped CeO_2 layer and the ability to suppress the formation of $SrZrO_3$ will be reduced. Recently, many developers successfully use $La_{0.6}Sr_{0.4}Co_{0.2}Fe_{0.8}O_{3-d}$ cathode on YSZ electrolyte with doped ceria interlayer. The degradation of the performance is minimised by optimising the fabrication process (Figure 6.17).

6.3.3 Chromia poisoning

In addition to compatibility with the electrolyte, compatibility of the cathode with the interconnect is also important. Even though many alloys have been investigated as metallic interconnects, almost all of them form chromia as a protective oxide scale [71].

Chromia containing alloys are also used for other parts in the cell stack assembly. The main issue associated with the use of such chromia-forming alloys is chromium poisoning both for the $LaMnO_3$-based cathode and $La(Co,Fe)O_3$-based cathode. However, the degradation mechanism is different for those two cases.

Taniguchi et al. [75] found that degradation by chromium poisoning occurs more severely at lower temperatures and that degradation measured in terms of the cathode life time is proportional to the logarithmic oxygen activity derived from the over-potential values, η, as follows:

$$\tau(\text{degradation}) \propto \Delta \log a_O (= 2F\eta/2.303RT) \tag{6.11}$$

Their observations on distribution of chromium in the electrolyte/cathode vicinity indicated that although the average Cr content in the cathode layer increases with increasing temperature, this quantity is not directly related to cathode degradation. Instead, concentrated chromium deposition on the three-phase boundaries can be directly related with the cathode lifetime.

It is well known that oxidation/corrosion of metallic materials is enhanced [81] in the presence of water vapour. Chromium poisoning can also be enhanced in the presence of water vapour. The vapour pressure of chromium-containing species increases with increasing water vapour pressure because $CrO_2(OH)_2$ is relatively stable in the presence of water vapour [82]. Its vapour pressure shows only a small temperature dependence so that even at low temperatures its vapour pressure is high. Taniguchi et al. found that the degradation increases with water vapour pressure consistent with the increasing vapour pressure of Cr-containing species.

Chromium poisoning was found to depend strongly on the particular combination of electrolyte and cathode by Matsuzaki and Yasuda [83]. In their investigations, the most severe degradation occurred in LSM/YSZ combination, whereas no apparent degradation occurred in LSCF/SDC (samaria-doped ceria) combination. These results appear to be consistent with Taniguchi's results [75]. As described previously, the cathode reaction mechanism is quite different in the LSM and the LSCF cathodes. In LSM, the oxide ion diffusivity is low and only the three-phase boundaries are the electrochemically active sites. The oxygen potential gradient appears near these sites. This gives rise to the driving force for chromium-containing species to attack the electrochemically active sites resulting in the deposition of chromia-containing oxides that are responsible for performance degradation. On the other hand, the electrochemically active sites in LSCF are less localised. They distribute on the surfaces of the electrode particles located in a certain thickness from the electrolyte. The poisoning on one site can be compensated by activation of another site.

To minimise/avoid Cr poisoning, several approaches have been tried. These include coating a dense electrically conducting oxide, such as cathode-related perovskites (LSM) or lanthanum chromite-based oxides, on metallic interconnects, and utilisation of Cr getters. Cr-containing species easily react with other oxides (particularly basic oxides) so that such basic oxides can act as Cr getters. La_2O_3 can be a strong Cr getter, but this material is hard to handle because of its hygroscopic nature. Lanthanum cobaltite can also be a Cr getter because the stabilisation energy of $LaCrO_3$ is larger than $LaCoO_3$.

6.3.4 Chemical and morphological instability under oxygen potential gradient

In operation of an SOFC, the local oxygen potential at the cathode/electrolyte interface is kept lower than that in equilibrium with the gas phase to provide the driving force for the oxygen incorporation reaction. Since the chemical potentials of the components in the perovskite are interrelated by Gibbs–Dühem relation as

$$d\mu_A + d\mu_B + 3d\mu_O = 0 \tag{6.12}$$

the oxygen potential gradient acts also on the cations to move in the inverse direction to oxygen. Under cathodic overpotential, the cations are forced to move from the interface to the bulk of the electrode layer. This may cause the electrode morphology to change around the interface. Furthermore, if the mobilities of the component cations are much different from each other, it can cause separation of the components, which is called "kinetic demixing" or "kinetic decomposition" [84,85]. Although cations are much less mobile than oxide ions, it may be an issue in a long-term operation.

As is discussed in the previous sections, $LaMnO_3$ forms cation vacancies under high oxygen partial pressures. This causes increase in the cation mobility. Miyoshi et al. [35] showed that oxidation and reduction cycles on $LaMnO_3$ lead creation and annihilation of the crystal lattices, resulting in a drastic change of the morphology. Taura et al. treated a dense $LaMnO_3$ plate under an oxygen potential gradient and found that grain growth took place on the surface of the higher partial pressure side, while pores were formed on the opposite surface (Figure 6.18) [86].

In a realistic cathode material, Sr is doped on the La site. It decreases the equilibrium concentration of cation vacancies and mitigates the morphological instability. In some cases, however, morphological change has been seen after long-term operations. Kuznecov et al. reported that porosity increased in a LSM cathode around the interface after 5000-h operation [87]. In contrast, recent report by Yamaji [88] and Kobayashi et al. [89] showed that $(La,Sr,Ca)MnO_3$ cathode densified at the interface with ceria-based electrolyte during operation at temperatures lower than 800 °C. The driving force and the transport mechanism in the realistic system may include mutual diffusion of the elements between the cathode and the surrounding components (Figures 6.19 and 6.20).

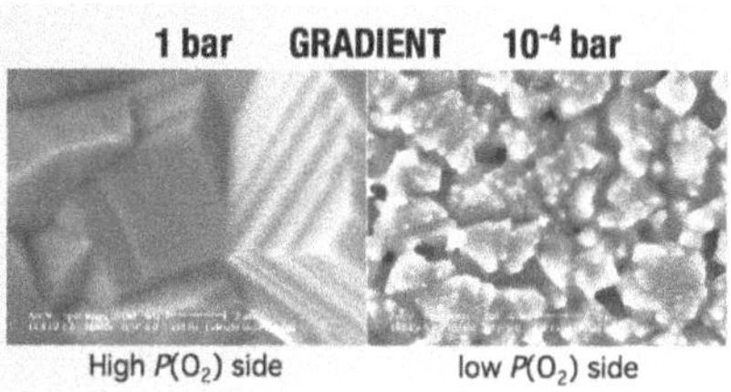

Figure 6.18 Morphological instability of $LaMnO_3$ under oxygen potential gradient. The sample was annealed between the gases of oxygen partial pressures 1 bar and 10^{-4} bar at 1273 K for 720 h. The left-hand side of the figure is the SEM image of the high oxygen partial pressure side, and the right-hand side, the low partial pressure side. [86]

Figure 6.19 SEM photograph at the interface between $(La_{0.5}Sr_{0.25}Ca_{0.25})MnO_3$ (LSCM) cathode and a $(Ce_{0.8}Sm_{0.2})O_2$ (SDC) interlayer after 5000 h operation at 760 °C. Reprinted from Ref. [89] with permission.

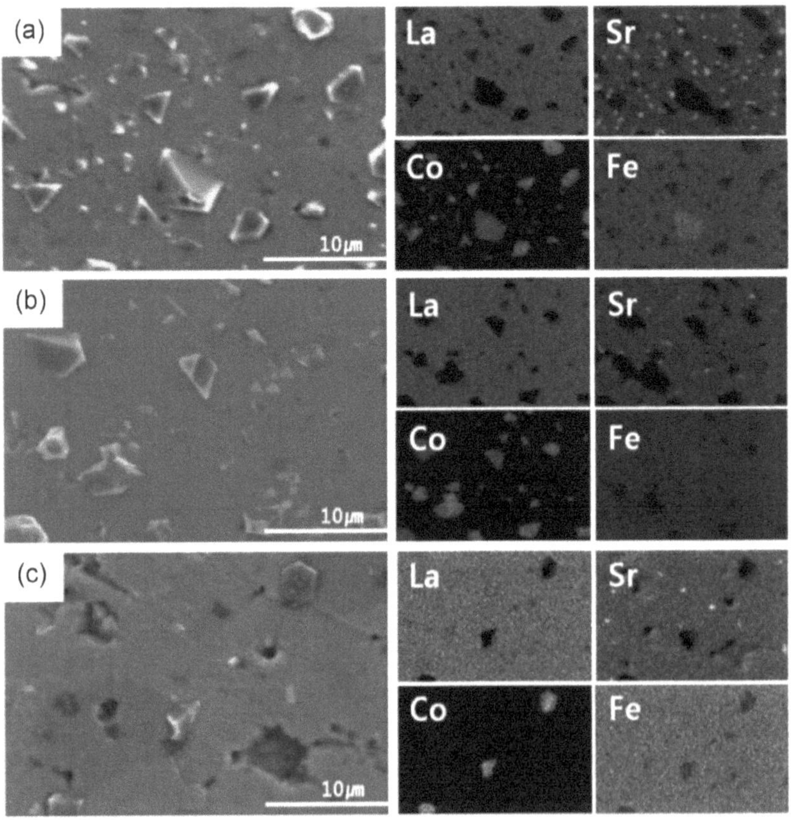

Figure 6.20 Microstructure and composition of the LSCF6428 top surfaces after the heat treatment at 1273 K for 1 week under the oxygen partial pressures of 1 bar on the top side and of 10^{-4} bar (a), 10^{-2} bar (b), 10^{-1} bar (c) on the bottom side.
Reprinted from Ref. [90] with permission.

While morphological instability is an issue of LSM cathode, LSCF electrode is claimed to have possible compositional instability for a long-term operation. Several reports can be found on the kinetic decomposition of LSCF under oxygen potential gradient. For example, Oh et al. [90] placed a dense $La_{0.6}Sr_{0.4}Co_{0.2}Fe_{0.8}O_{3-d}$ pellet between O_2–Ar mixed gases of different oxygen partial pressures for 1 or 2 weeks at elevated temperatures and found that Co oxide or Co–Fe oxide were formed on the surface of the high oxygen partial pressure side. The decomposition was more severe under higher-potential gradient but not observed under a uniform atmosphere, which indicates kinetic decomposition. Detailed discussion, however, has not been made since the mass transport pathway was not simple. Grain boundary diffusion and the resulting complex potential profile may have to be taken into account.

In a practical use of LSCF cathode, severe kinetic decomposition has not been reported. It is partly due to the reduced operation temperatures and also to the low overpotential that leads to low driving force for the kinetic decomposition.

6.4 Thermo-chemo-mechanical properties

6.4.1 Thermal and chemical strain

Since SOFC consists of a layered composite of multiple materials, the mismatch in the coefficient of thermal expansion (CTE) leads to the mechanical stress that may cause damage on the cell. The most serious trouble is fracture of the electrolyte. From the reliability point of view, the electrode materials should have slightly higher CTE than the electrolyte so that the electrolyte will be under compressive stress by cooling down the cell from the fabrication temperature. Too large mismatch, however, may cause the electrolyte to be broken by buckling.

In addition to the thermal strain, "chemical strain" caused by oxygen non-stoichiometry is also important for the cathode materials [91–93]. When oxygen vacancy is formed in the lattice, the average valence number of the B-site cation decreases. It leads to increase in the B-site ionic radius and hence to lattice expansion. When cation vacancy is formed as is the case of $LaMnO_3$, the lattice volume decreases, but the total volume increases due to the formation of new lattices.

The reported coefficients of thermal expansion (CTE) for $LaMnO_3$-based perovskites are listed in Table 6.2. They are in a tolerable range in terms of the compatibility with YSZ which has CTE of $10–11 \times 10^{-6}\ K^{-1}$. Furthermore, for practical cathodes with high dopant concentrations, the chemical strain is not necessary to be considered since cation vacancy concentration remains negligible as has been discussed in the previous section. Only purely "thermal" expansion is important unless the electrode is exposed to reducing conditions. With the dopant fraction less than 0.3, however, the chemical strain due to the cation vacancy formation must have non-negligible effect [35]. In the reported CTE values, chemical strain is often neglected since the cation vacancy formation/annihilation is a slow process so that equilibrium is not reached in

Table 6.2 Composition and thermal expansion coefficient of (La,M)MnO_3

Composition	CTE (10^{-6} K^{-1}) [94,95]
$La_{0.99}MnO_3$	11.2
$La_{0.69}Sr_{0.30}MnO_3$	12.8
$LaMnO_3$	12.5
$La_{0.7}Sr_{0.3}MnO_3$	11.8, 12.0
$La_{0.9}Ca_{0.1}MnO_3$	10.6
$La_{0.7}Ca_{0.3}MnO_3$	10.5

a normal temperature-scanning measurement. Mori et al. [95] reported that the sample size gradually changed during repeated scans of temperature. It is probably due to the chemical strain caused by cation vacancies.

The intermediate temperature cathode material (La,Sr)CoO_3 shows CTE of around 20×10^{-6} K^{-1} [96] which is too high to be used on YSZ. One of the reasons for the high CTE is attributed to the spin-state transition from intermediate to high spin. Another important reason is the chemical strain caused by the formation of oxygen vacancies. Since (La,Sr)$CoO_{3-\delta}$ shows high oxygen nonstoichiometry even in air, the vacancy concentration shows nontrivial increase with increasing temperature. Therefore, the apparent thermal expansion under isobaric condition is the sum of purely "thermal" and thermally induced "chemical" expansions. The apparent linear CTE is in the form;

$$\mathrm{CTE} = \frac{\Delta l}{l_0} = \frac{1}{3V_0}\frac{\partial V}{\partial T}\bigg|_{\delta} + \frac{1}{3V_0}\frac{\partial V}{\partial \delta}\bigg|_{T}\frac{\partial \delta}{\partial T}\bigg|_{\mu_O} = \alpha_{th} + \alpha_{chem}\frac{\partial \delta}{\partial T}\bigg|_{\mu_O} \tag{6.13}$$

In reality, δ is not a linear function of temperature, and the induced chemical expansion shows non-linear behaviour in the temperature range of SOFC operation. Thus, the thermochemical expansion is rather represented as a direct function of T and δ as,

$$\frac{\Delta l}{l_0} = \alpha_{th}(T - T_0) + \alpha_{chem}(\delta - \delta_0) \tag{6.14}$$

According to Chen et al. [97], purely thermal expansion coefficient α_{th} of $La_{0.6}Sr_{0.4}CoO_3$ is 15.8×10^{-6} K^{-1}. The remaining part of the apparent CTE is the contribution of the chemical expansion.

So far, attempts have been made to reduce CTE of $LaCoO_3$-based oxides by doping various cations on A-site and B-site. Especially, substitution of Co by Fe has been studied intensively [43,98,99]. Equation (6.13) implies that reduction of CTE is possible by the reduction of δ as well as the reduction of α_{th} and α_{chem}. From the viewpoint of the electrochemical performance, δ is preferred to remain high with reduced α_{chem}.

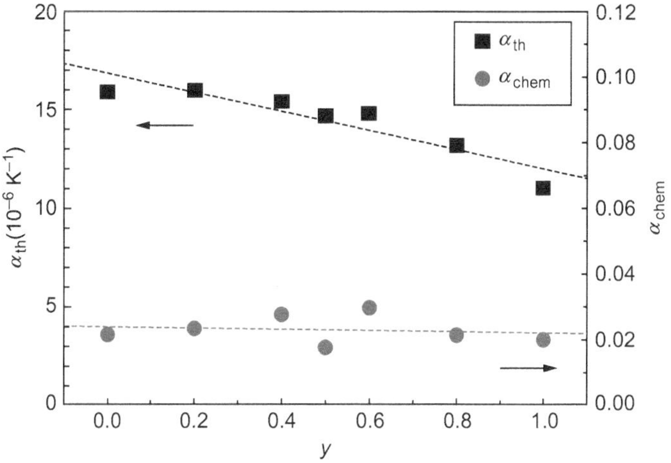

Figure 6.21 Thermal expansion coefficient α_{th} and chemical expansion coefficient α_{chem} as a function of y in $La_{0.6}Sr_{0.4}Co_{1-y}Fe_yO_{3-\delta}$.
Reprinted from Ref. [100] with permission.

However, α_{chem} is not sensitive to the composition as shown in Figure 6.21, and the reduction of chemical capacitance inevitably accompanies the reduction of δ in LSCF series. Practically the most widely adopted composition is $La_{0.6}Sr_{0.4}Co_{0.2}Fe_{0.8}O_{3-d}$ (often abbreviated as LSCF6428) which still shows high electrochemical performance while apparent CTE in air is reduced to $\sim 15 \times 10^{-6}\,K^{-1}$. According to Hashimoto et al. [101], the strain in LSCF6428 is written as

$$\frac{\Delta l}{l_0(T=300\mathrm{K}, \delta=0)} = 1.32 \times 10^{-5}T + 2.15 \times 10^{-2}\delta - 3.88 \times 10^{-3} \tag{6.15}$$

6.4.2 Mechanical properties of LSM, LSC, LSF and LSCF

The thermo-chemo-mechanical strain of the cathode materials may cause serious failures when the cell design or the operation sequences are not appropriate. To ensure the reliability of the cells and stacks, knowledge on mechanical properties, such as Young's modulus, Poisson's ratio, creep parameters and strength is necessary. Especially, the information related to the elastic and non-elastic deformation is essential for modelling the stress developed in the cells and stacks. The paper by Nakajo et al. in 2012 provides a useful list of the literature data on mechanical properties of the SOFC materials [102]. At that moment, data for the cathode materials were limited and only room temperature data were involved. More recently, dynamic or static measurements of elastic moduli were performed at elevated temperatures under controlled atmospheres [103–105]. They revealed the complicated behaviour of the elastic moduli of the cathode materials with temperature and oxygen partial pressure.

In a resonant method, bending and torsion vibrations are applied to a sample, and Young's modulus E, shear modulus G and Poisson's ratio ν are obtained from their resonant frequencies. Internal friction Q is also evaluated from the declining amplitude of the vibration after stopping the external vibration. Uniaxial pressing, four-point or three-point bending or small-punch bending tests give static Young's modulus from the stress–strain behaviour of the samples.

Figure 6.22 shows the temperature dependence of Young's modulus, Poisson's ratio and internal friction of $La_{0.6}Sr_{0.4}MnO_3$. Young's modulus gradually decreases with increasing temperature from the room temperature to 1000 K but shows abrupt jump at around 1000–1100 K. This corresponds to the temperature of second-order phase transition from the rhombohedral phase to the cubic phase. At the same temperature, the Poisson's ratio changes.

The similar tendency was also reported for LSC, LSF and LSCF [104,105]. According to the shift of the phase transition temperature with oxygen partial pressure, the temperature at which Young's modulus shows abrupt change shifted with the atmosphere. The increase in Young's modulus with decreasing oxygen partial pressure at 973 K (Figure 6.23b) is due to the phase transition. In contrast, Young's modulus decreased with decreasing oxygen partial pressure at 1173 K. It is because of the formation of oxygen vacancies and the resulting expansion of the lattice volume. Consequently, oxygen partial pressure dependence shows complicated behaviour as shown in Figure 6.23b.

The abrupt jump in the Young's modulus is explained by the Landau theory of the second-order phase transition as discussed for $LaAlO_3$ by Carpenter et al. [106]. Furthermore, rhombohedral perovskite is known to show ferroelastic deformation

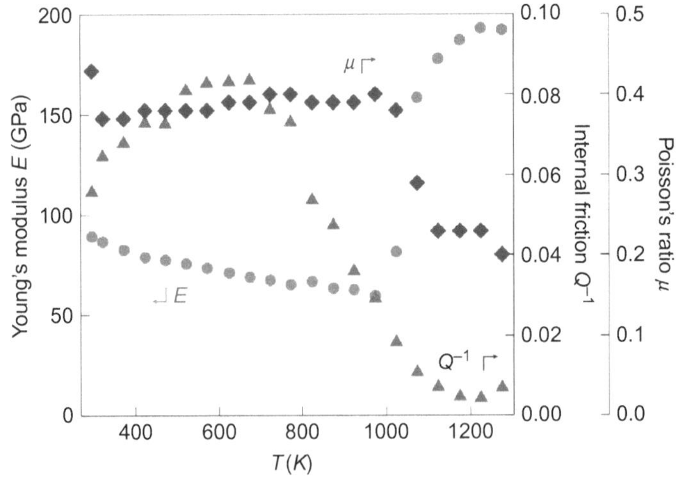

Figure 6.22 Young's modulus, Poisson's ratio and internal friction of $La_{0.6}Sr_{0.4}MnO_3$ as functions of temperature in $10\%O_2$–Ar mixed gas [103].

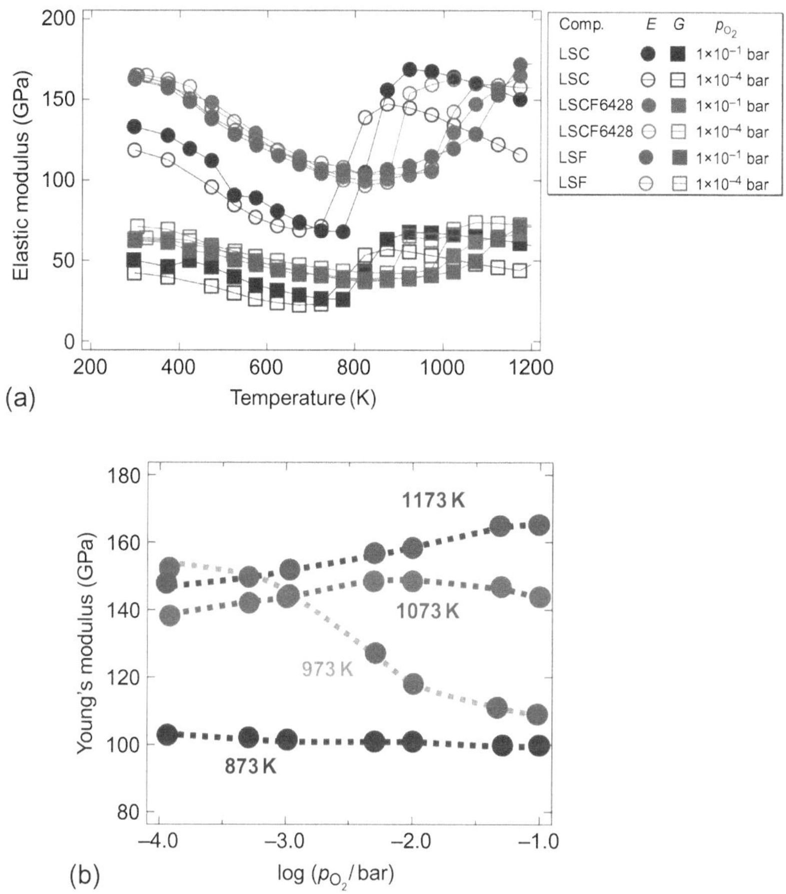

Figure 6.23 (a) Young's moduli and shear moduli of $La_{0.6}Sr_{0.4}CoO_3$, $La_{0.6}Sr_{0.4}FeO_3$ and $La_{0.6}Sr_{0.4}Co_{0.2}Fe_{0.8}O_3$ in 10% and 100 ppm oxygen. (b) Oxygen partial pressure dependence of Young's modulus of $La_{0.6}Sr_{0.4}Co_{0.2}Fe_{0.8}O_3$ [104].

resulted from the switching of twin domains [107–110]. Araki and Malzbender [110] showed clear ferroelastic behaviour of LSCF6428 by uniaxial pressing measurements as shown in Figure 6.24. The decrease in the apparent Young's modulus and the large internal friction shown in Figure 6.23 is explained by the contribution of the switching and back-switching of the twin domains in the resonance measurements.

In the SOFC operation, the abrupt change in the Young's modulus may affect the stress distribution in the cells. The modelling must be based on the precise data including temperature and partial pressure dependence. The effect of ferroelastic deformation should also be taken into account. Since the degree of the ferroelastic strain increases with decreasing temperature, it may mitigate the stress caused by the

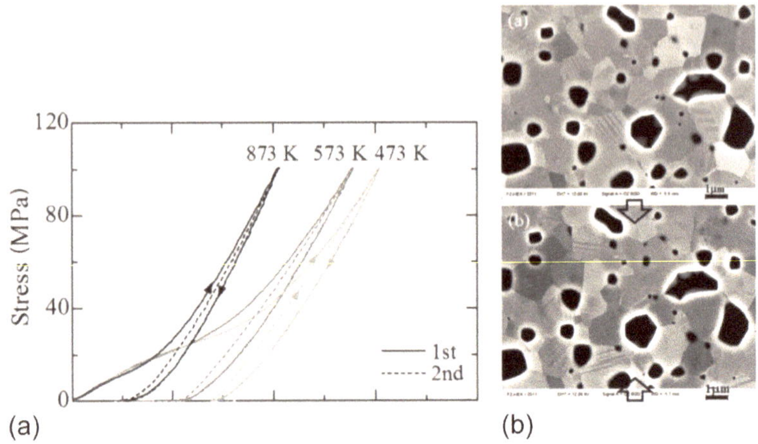

Figure 6.24 Ferroelastic deformation in air at various temperatures and domain switching at room temperature of $La_{0.6}Sr_{0.4}Co_{0.2}Fe_{0.8}O_3$.
Reprinted from Ref. [110] with permission.

mismatch in CTE when cooling the cell from fabrication temperature. It may be one of the reasons why the cathode materials with relatively large CTE are safely used in real cells and stacks.

6.5 Summary and further researches

Strontium-substituted lanthanum manganite, LSM, is a promising cathode material for high-temperature SOFCs, showing high chemical and thermomechanical compatibility with YSZ electrolyte. The oxygen-excess nonstoichiometry is due to the formation of cation vacancies, which causes the increased cation mobility and the morphological instability during long-term operation. Chromia poisoning is another issue for the durability. The effect of chromia is rather severe in LSM due to the restricted reaction site at triple-phase boundaries that is at the same time the preferential site for the chromia deposition.

The relatively high activation energy for oxygen reduction on LSM electrode motivated people to develop mixed conductor cathodes for reduced temperature SOFCs. $(La,Sr)CoO_3$ or LSC is a typical mixed conductor cathode and shows high electrochemical activity. However, the high thermal and chemical expansion made this material difficult to be used for a practical SOFC. Instead, $La_{0.6}Sr_{0.4}Co_{0.2}Fe_{0.8}O_3$, LSCF, with less CTE is widely used by the developers. The ferroelastic nature of the rhombohedral lattice mitigates the CTE mismatch with the electrolyte. Since both LSC and LSCF react with YSZ, ceria-based interlayer is often used to prevent reaction. In LSCF, a risk of kinetic decomposition should also be kept in mind.

In addition to the research for better understanding of LSM and LSC/LSCF, attempts are continued to find a new material with higher performance. $Ba_{0.5}Sr_{0.5}Co_{0.8}Fe_{0.2}O_3$ was found to show superior performance [17], and Wang et al. reported that the high surface reaction was caused by the high ionic conductivity [111]. Although this material is promising in terms of the performance, Wang et al. claimed that the thermochemical stability is not enough to be used in practical cells [18]. Ordered perovskite called "double perovskite" such as $PrBaCo_2O_{5+d}$ is also reported to show high catalytic activity for oxygen reduction [112]. Further understanding of the properties is necessary to utilise those materials. Compounds with perovskite-related layered structure, for example, K_2NiF_4-type oxides such as La_2NiO_4, are also interesting groups of materials showing interstitial oxygen [113,114]. The oxygen gas–solid exchange reaction on $(La,Sr)CoO_3$ was somehow fast at the boundary with K_2NiF_4-type $(La,Sr)_2CoO_4$ [20]. Change in the electron structure, elastic strain and local composition is discussed as the source for the enhancement [115,116]. Addition of ceria as the second phase is also reported to help improving electrode activity [19], which may not be simply due to the formation of fast ionic paths but modification of the catalytic properties [117].

The above-mentioned new materials and new phenomena may lead to the next-generation cathode material, when their properties are well understood by further research, and the total design of the cell is optimised for them to be integrated.

References

[1] J.B. Goodenough, J.M. Longo, in: Magnetic Oxides and Related Oxides, vol. 4a, 1970, pp. 126–314 (Chapter 3); Landolt-Bernstein. S. Nomura, Magnetic Oxides and Related Oxides, vol. 4a, 1978, pp. 368–520. Landolt-Berstein.

[2] C.S. Tedmon Jr., H.S. Spacil, S.P. Mito, J. Electrochem. Soc. 116 (1969) 1170.

[3] O. Yamamoto, Y. Takeda, R. Kanno, M. Noda, Solid State Ionics 22 (1987) 241–246.

[4] H. Yokokawa, N. Sakai, T. Kawada, M. Dokiya, J. Electrochem. Soc. 138 (9) (1991) 2719–2727.

[5] S.C. Singhal, Progress in tubular solid oxide fuel cell technology, in: S.C. Singhal, M. Dokiya (Eds.), Solid Oxide Fuel Cells VI, Electrochemical Society Proceedings, 1999, pp. 39–51, Pennington, NJ, PV 99-19.

[6] M. Dokiya, N. Sakai, T. Kawada, H. Yokokawa, T. Iwata, M. Mori, Overview of planar SOFC development at NCLI, in: S.C. Singhal (Ed.), Solid Oxide Fuel Cells I, Electrochemical Society Proceedings, 1995, pp. 325–336, Pennington, NJ, PV 89-11, 1989; see also T. Kawada, thesis.

[7] M. Suzuki, H. Sasaki, S. Otoshi, A. Kajimura, N. Sugiura, M. Ipponmatsu, J. Electrochem. Soc. 141 (7) (1994) 1928–1931.

[8] H. Takeuchi, H. Nishiyama, A. Ueno, S. Aikawa, M. Aizawa, H. Tajiri, T. Nakayama, S. Suehiro, K. Shukuri, Current status of SOFC development by wet process, in: S.C. Singhal, M. Dokiya (Eds.), Solid Oxide Fuel Cells VI, Electrochemical Society Proceedings, 1999, pp. 879–884, Pennington, NJ, PV 99-19.

[9] L. Petruzzi, S. Cocchi, F. Fineschi, J. Power Sources 118 (2003) 96–107.

[10] P. Lamp, J. Tachtler, O. Finkenwirth, S. Mukerjee, S. Shaffer, Fuel Cells 3 (3) (2003) 146–152.

[11] M. Suzuki, T. Takuwa, S. Inoue, K. Higaki, ECS Trans. 57 (1) (2013) 309–314.
[12] A. Mai, B. Iwanschitz, J.A. Schuler, R. Denzler, V. Nerlich, A. Schuler, ECS Trans. 57 (1) (2013) 73–80.
[13] R. Payne, J. Love, M. Kah, ECS Trans. 35 (1) (2011) 81–85.
[14] B.C.H. Steele, J.-M. Bae, Solid State Ionics 106 (1998) 255–261.
[15] S. Mukerjee, K. Haltiner, R. Kerr, J.Y. Kim, V. Sprenke, ECS Trans. 35 (1) (2011) 139–146.
[16] A. Nakajo, F. Mueller, J. Brouwer, J. Van herle, D. Favrat, Int. J. Hydrogen Energy 37 (11) (2012) 9249–9268.
[17] Z. Shao, S. Haile, Nature 431 (2004) 170–173.
[18] F. Wang, T. Nakamura, J. Mizusaki, K. Yashiro, K. Amezawa, Phys. Chem. Chem. Phys. 16 (2014) 7307–7314.
[19] V. Dusastre, J.A. Kilner, Solid State Ionics 126 (1999) 163–174.
[20] M. Sase, K. Yashiro, K. Sto, J. Mizusaki, T. Kawada, N. Sakai, K. Yamaji, T. Horita, H. Yokokawa, Solid State Ionics 178 (2008) 1843–1852.
[21] R.D. Shannon, Acta Cryst. A32 (1976) 751–768.
[22] B.C. Tofield, W.R. Scott, J. Solid State Chem. 10 (1974) 183.
[23] L. Pinsard, J. Rodríguez-Cavajal, A. Revcolevschi, J. Alloys Compd. 262/263 (1997) 152–156.
[24] S. Wang, M. Katsuki, M. Dokiya, T. Hashimoto, Solid State Ionics 159 (2003) 71–78.
[25] J. Mizusaki, J. Tabuchi, T. Matsuura, S. Yamauchi, K. Fueki, J. Electrochem. Soc. 136 (7) (1989) 2082–2088.
[26] J. Mizusaki, T. Sasamoto, W.R. Cannon, H.K. Bowen, J. Am. Ceram. Soc. 65 (8) (1982) 363–368.
[27] B.A. van Hassel, T. Kawada, N. Sakai, H. Yokokawa, M. Dokiya, Solid State Ionics 66 (1993) 295–305.
[28] J. Mizusaki, S. Yamauchi, K. Fueki, A. Ishikawa, Solid State Ionics 12 (1984) 119–124.
[29] J. Mizusaki, M. Yoshihiro, S. Yamauchi, K. Fueki, J. Solid State Chem. 58 (1985) 257–266.
[30] J. Mizusaki, Y. Mima, S. Yamauchi, K. Fueki, H. Tagawa, J. Solid State Chem. 80 (1989) 102–111.
[31] J.H. Kuo, H.U. Anderson, D.M. Sparlin, J. Solid State Chem. 83 (1989) 52–60.
[32] M.H.R. Lankhorst, H.J.M. Bouwmeester, H. Verweij, Solid State Ionics 96 (1997) 21–27.
[33] M. Kuhn, Y. Fukuda, S. Hashimoto, K. Sato, K. Yashiro, J. Mizusaki, J. Electrochem. Soc. 160 (1) (2013) F34–F42.
[34] J. Mizusaki, N. Mori, H. Takai, Y. Yonemura, H. Minamiue, H. Tagawa, M. Dokiya, H. Inaba, K. Naraya, T. Sasamoto, T. Hashimoto, Solid State Ionics 129 (2000) 163–177.
[35] S. Miyoshi, J.-O. Hong, K. Yashiro, A. Kaimai, Y. Nigara, K. Kawamura, T. Kawada, J. Mizusaki, Solid State Ionics 154 (155) (2002) 257–263.
[36] J.A.M. Van Roosmalen, E.H.P. Cordfunke, R.B. Helmholdt, J. Solid State Chem. 110 (1994) 100–105.
[37] J.A.M. Van Roosmalen, E.H.P. Cordfunke, J. Solid State Chem. 110 (1994) 106–108.
[38] J.A.M. Van Roosmalen, E.H.P. Cordfunke, J. Solid State Chem. 110 (1994) 109–112.
[39] J.A.M. Van Roosmalen, E.H.P. Cordfunke, J. Solid State Chem. 110 (1994) 113–117.
[40] H. Yokokawa, T. Horita, N. Sakai, M. Dokiya, T. Kawada, Solid State Ionics 86–88 (1996) 1161–1165.
[41] J. Mizusaki, T. Sasamoto, W.R. Cannon, H.K. Bowen, J. Am. Ceram. Soc. 66 (4) (1983) 247–252.

[42] K. Iwasaki, T. Ito, T. Nagasaki, Y. Arita, M. Yoshino, T. Matsui, J. Solid State Chem. 181 (2008) 3145–3150.
[43] L.-W. Tai, M.M. Nasrallah, H.U. Anderson, D.M. Sparlin, S.R. Sehlin, Solid State Ionics 76 (1995) 259–271.
[44] L.-W. Tai, M.M. Nasrallah, H.U. Anderson, D.M. Sparlin, S.R. Sehlin, Solid State Ionics 76 (1995) 273–283.
[45] J. Mizusaki, N. Mori, H. Takai, Y. Yonemura, H. Minamiue, H. Tagawa, M. Dokiya, H. Inaba, K. Naraya, T. Sasamoto, T. Hashimoto, Solid State Ionics 132 (2000) 167–180.
[46] T. Ishigaki, S. Yamauchi, K. Kishio, J. Mizusaki, K. Fueki, J. Solid State Chem. 73 (1988) 179–187.
[47] T. Ishigaki, S. Yamauchi, J. Mizusaki, K. Fueki, H. Tamura, J. Solid State Chem. 54 (1984) 100–107.
[48] J.A. Kilner, B.C.H. Steele, L. Ilkov, Solid State Ionics 12 (1984) 89–97.
[49] R.J. Chater, S. Carter, J.A. Kilner, B.C.H. Steele, Solid State Ionics 53–56 (1992) 859–867.
[50] J. Crank, The Mathematics of Diffusion, Oxford University Press, Oxford, 1975.
[51] J.A. Kilner, R.A. De Souza, I.C. Fullarton, Solid State Ionics 86–88 (1996) 703–709.
[52] R.A. De Souza, J.A. Kilner, Solid State Ionics 126 (1999) 153–161.
[53] L. Wang, R. Merkle, Y.A. Mastrikov, E.A. Kotomin, J. Maier, J. Mater. Res. 27 (15) (2012) 2000–2008.
[54] H. Kudo, K. Yashiro, S. Hashimoto, K. Amezawa, T. Kawada, Solid State Ionics 262 (2014) 696–700.
[55] Y. Li, K. Gerdes, H. Diamond, X. Liu, Solid State Ionics 204–205 (2011) 104–110.
[56] H. Kudo, K. Yashiro, S. Hashimoto, K. Amezawa, T. Kawada, J. Mizusaki, ECS Trans. 50 (44) (2013) 37–42.
[57] K. Yashiro, I. Nakano, M. Kuhn, S. Hashimoto, K. Sato, J. Mizusaki, ECS Trans. 35 (1) (2011) 1899.
[58] Y. Li, K. Gerdes, T. Horita, X. Liu, J. Electrochem. Soc. 160 (4) (2013) F343.
[59] J.A. Lane, J.A. Kilner, Solid State Ionics 136–137 (2000) 997.
[60] H.J.M. Bouwmeester, M.W. Den Otter, B.A. Boukamp, J. Solid State Electrochem. 8 (2004) 599.
[61] M. Sogaard, P.V. Hendriksen, M. Mogensen, J. Solid State Chem. 180 (2007) 1489.
[62] M. Sogaard, P.V. Hendriksen, M. Mogensen, F.W. Poulsen, E. Skou, Solid State Ionics 177 (2006) 3285.
[63] M. Katsuki, S. Wang, M. Dokiya, T. Hashimoto, Solid State Ionics 156 (2003) 453.
[64] W. Sitte, E. Bucher, W. Preis, Solid State Ionics 154–155 (2002) 517.
[65] S. Adler, J.A. Lane, B.C.H. Steele, J. Electrochem. Soc. 143 (1996) 3554.
[66] I. Yasuda, K. Ogasawara, M. Hishinuma, T. Kawada, M. Dokiya, Solid State Ionics 86–88 (1996) 1197–1201.
[67] A. Hammouche, E. Siebert, A. Hammou, M. Kleitz, J. Electrochem. Soc. 138 (1991) 1212.
[68] E. Siebert, A. Hammouche, M. Kleitz, Electrochim. Acta 40 (11) (1995) 1212.
[69] J. Van herle, A.J. McEvoy, K.R. Thampi, Electrochim. Acta 41 (1996) 1447.
[70] T. Horita, K. Yamaji, N. Sakai, H. Yokokawa, T. Kato, J. Electrochem. Soc. 148 (5) (2001) J25–J30.
[71] T. Kawada, H. Yokokawa, Materials and characterization of solid oxide fuel cells, in: J. Nowotny, C.C. Sorrell (Eds.), Electrical Properties of Ionic Solids, Trans Tech Publications, 1997, pp. 187–248.
[72] M. Juhl, S. Prindahl, C. Manon, M. Mogensen, J. Power Sources 61 (1996) 173.
[73] A.V. Virkar, J. Chen, C.W. Tanner, J.-W. Kim, Solid State Ionics 131 (2000) 189–198.

[74] D. Ghosh, E. Tang, M. Perry, D. Prediger, M. Pastula, R. Boersma, Status of SOFC developments at global thermoelectric, in: H. Yokokawa, S.C. Singhal (Eds.), Solid Oxide Fuel Cells VII, Electrochemical Society Proceedings, 2001, pp. 100–110, Pennington, NJ, PV 2001-16.
[75] S. Taniguchi, M. Kadowaki, H. Kawamura, T. Yasuo, Y. Akiyama, Y. Miyake, T. Saitoh, J. Power Sources 55 (1995) 73–79.
[76] H. Yokokawa, N. Sakai, T. Kawada, M. Dokiya, Solid State Ionics 40 (41) (1990) 398.
[77] M. Sase, D. Ueno, K. Yashiro, A. Kaimai, T. Kawada, J. Mizusaki, J. Phys. Chem. Solids 66 (2-4) (2005) 343–348.
[78] H. Naito, N. Sakai, T. Otake, H. Yugami, H. Yokokawa, Solid State Ionics 135 (2000) 669–673.
[79] A. Tsoga, A. Naomidis, D. Stöver, Solid State Ionics 135 (2000) 403–409.
[80] M. Bertoldi, O. Bucheli, A. Ravagni, N. Autissier, M. Signorini, S. Modena, ECS Trans. 25 (2) (2009) 105–114.
[81] P. Kofstad, R. Bredesen, Solid State Ionics 52 (1992) 69–75.
[82] K. Hilpert, D. Das, M. Miller, D.H. Peck, R. Weiß, J. Electrochem. Soc. 143 (1996) 4013.
[83] Y. Matsuzaki, I. Yasuda, J. Electrochem. Soc. 148 (2001) A126.
[84] H. Schmalzried, J. Chem. Soc. Faraday Trans. 86 (1990) 1273–1280.
[85] M. Martin, Pure Appl. Chem. 75 (7) (2003) 889–903.
[86] T. Taura, Ms. Thesis, Graduate School of Engineering, Tohoku University, 2004.
[87] M. Kuznecov, P. Otschik, N. Trofimenko, K. Eichler, Russ. J. Electrochem. 40 (11) (2004) 1162–1169. From Electrokhimiya, 40(11) (2004) 1355–1363.
[88] Yamaji et al. Private Communication, NEDO Project report on SOFC development (Japanese) (2013).
[89] Y. Kobayashi, K. Tomid, H. Tsukuda, Y. Shiratori, S. Taniguchi, K. Sasaki, J. Electrochem. Soc. 161 (3) (2014) F214–F225.
[90] M.-Y. Oh, A. Unemoto, K. Amezawa, T. Kawada, J. Electrochem. Soc. 159 (10) (2012) F659–F664.
[91] A. Atkinson, T.M.G.M. Ramos, Solid State Ionics 129 (2000) 259–269.
[92] S.B. Adler, J. Am. Ceram. Soc. 84 (9) (2001) 2117–2119.
[93] D. Marrocchelli, N.H. Perry, S.R. Bishop, Phys. Chem. Chem. Phys. 17 (2015) 10028–10039.
[94] S. Srilomsak, D.P. Schlling, H.U. Anderson, in: S.C. Singhal (Ed.), Solid Oxide Fuel Cell I, Electrochemical Society Proceedings, 1989, p. 129, Pennington, NJ, PV 89-11.
[95] M. Mori, Y. Hiei, in: S.C. Singhal, M. Dokiya (Eds.), Solid Oxide Fuel Cell VI, Electrochemical Society Proceedings, 1999, pp. 347–354, Pennington, NJ, PV 99-19.
[96] Y. Ohno, S. Nagata, H. Sato, Solid State Ionics 9–10 (1983) 1001–1008.
[97] X. Chen, J. Yu, S.B. Adler, Chem. Mater. 17 (17) (2005) 4537–4546.
[98] J.W. Stevenson, T.R. Armstrong, R.D. Carneim, L.R. Pederson, W.J. Weber, J. Electrochem. Soc. 143 (9) (1996) 2722–2729.
[99] F. Tietz, I. Arul Raj, M. Zahid, D. Stöver, Solid State Ionics 177 (2006) 1753–1756.
[100] M. Kuhn, S. Hashimoto, K. Sato, K. Yashiro, J. Mizusaki, Solid State Ionics 241 (2013) 12–16.
[101] S. Hashimoto, Y. Fukuda, M. Kuhn, K. Sato, K. Yashiro, J. Mizusaki, Solid State Ionics 186 (2011) 37–43.
[102] A. Nakajo, J. Kuebler, A. Faes, U.F. Vogt, H.J. Schindler, L.-K. Chiang, S. Modena, J. Van herle, T. Hocker, Ceram. Int. 38 (2012) 3907–3927.
[103] Y. Shirai, Ms. Thesis, Graduate School of Environmental Studies, Tohoku University 2011.

[104] Y. Kimura, T. Kushi, S. Hashimoto, K. Amezawa, T. Kawada, J. Am. Ceram. Soc. 95 (8) (2012) 2608–2613.
[105] Y. Kimura, K. Yashiro, S. Hashimoto, T. Kawada, J. Electrochem. Soc. 161 (11) (2014) F3079–F3083.
[106] M.A. Carpenter, S.V. Sinogeikin, J.D. Bass, J. Phys. Condens. Matter. 22 (2010) 035404.
[107] T. Matsuura, J. Mizusaki, S. Yamauchi, K. Fueki, Jpn. J. Appl. Phys. 23 (9) (1984) 1197–1201.
[108] K. Kleveland, N. Orlovskaya, T. Grande, A.M.M. Moe, M.-A. Einarsrud, K. Breder, G. Gogotsi, J. Am. Ceram. Soc. 84 (9) (2001) 2029–2033.
[109] S. Faaland, T. Grande, M.-A. Einarsrud, P.E. Vullum, R. Holmestad, J. Am. Ceram. Soc. 88 (3) (2005) 726–730.
[110] W. Araki, J. Malzbender, J. Eur. Ceram. Soc. 33 (2013) 805–812.
[111] L. Wang, R. Merkle, J. Maier, J. Electrochem. Soc. 157 (12) (2010) B1802–B1808.
[112] D. Chen, R. Ran, K. Zhang, J. Wang, Z. Shao, J. Power Sources 188 (2009) 96–105.
[113] S.J. Skinner, J.A. Kilner, Solid State Ionics 135 (2000) 709s–712s.
[114] F. Mauvy, J.-M. Bassat, E. Boehm, J.-P. Manuaud, P. Dordor, J.-C. Grenier, Solid State Ionics 158 (2003) 17–28.
[115] N. Tsvetkov, Y. Chen, B. Yildiz, J. Mater. Chem. A 2 (2014) 14690–14695.
[116] D. Lee, Y.-L. Lee, W.T. Hong, M.D. Biegalski, D. Morgan, Y. Shao-Horn, J. Mater. Chem. A 3 (2015) 2144–2157.
[117] R.A. Budiman, S. Hashimoto, K. Yashiro, K. Amezawa, T. Kawada, ECS Trans. 57 (1) (2013) 1873–1878.

Interconnects

7

L. Niewolak, F. Tietz, W.J. Quadakkers
Forschungszentrum Jülich, Jülich, Germany

7.1 Introduction

The most important properties required for the interconnect material of a solid oxide fuel cell (SOFC) are high electronic conductivity, thermal stability in the cathode and anode side gas at the high service temperatures (600–900 °C), and a thermal expansion coefficient similar to that of the ceramic, electro-active components. Most SOFC designs utilise yttria-stabilised zirconia (YSZ) as electrolyte, $(La,Sr)MnO_3$ as cathode and Ni/ZrO_2-cermet as anode [1–3]. Ceramic, perovskite-type materials on the basis of $LaCrO_3$ have been shown to possess the property combination required for interconnect materials [1,2,4–6]. Main drawbacks of the ceramic materials are their brittleness, high price and troublesome shaping technology. This was especially of concern in some planar stack designs where the interconnector acted as the mechanical support for the thin ceramic parts and provided the gas-proof separation between fuel gas and oxidant. Metallic construction materials offer a number of advantages over ceramics: they are easier, and therefore cheaper to fabricate, less brittle, easier to machine and can be joined with a number of standard welding or brazing techniques. Additionally, they possess higher electrical and thermal conductivities than most ceramics.

For these reasons, SOFC development in the recent 25 years was accompanied by the replacement of nearly exclusively ceramic interconnects by metallic ones. The alteration became possible due to the development of a number of new materials and manufacturing technologies. These allowed a decrease of the operating temperatures from previously used 1000 °C down to for example, 800 °C or even lower [7–9]. Consequently, a new exploration field related to development and property evaluation of metallic SOFC interconnect materials was born.

Extensive research programmes resulted in an enormous progress in the understanding of the degradation mechanisms and subsequently in improvements of SOFC relevant properties of metallic interconnect materials. This review focuses on the progress in interconnect development and understanding of their degradation processes under SOFC conditions. Additionally, basic principles for selection of metallic interconnect materials for different types of SOFC designs are described.

7.2 SOFC environments

The common operating temperatures for today's SOFC units are between 750 and 850 °C [8–10], and more recent developments towards a further temperature decrease

High-temperature Solid Oxide Fuel Cells for the 21st Century. http://dx.doi.org/10.1016/B978-0-12-410453-2.00007-5

down to, for example, 600 °C became apparent [10–12]. In parallel, there is a tendency to replace hydrogen fuel by, for example, reformed hydrocarbons, biogas, ammonia or even H_2S [13–17]. As a consequence, a number of additional, potential degradation mechanisms have to be considered in the selection of interconnect materials.

The most common gas species present in both anode and cathode environments is water vapour. Although its concentration on the cathode side usually does not exceed a few percent, water vapour may alter the air oxidation behaviour of the interconnect materials and enhance the volatilisation of chromium [18–20].

On the anode side water vapour is a product of fuel oxidation and may prevail as a main constituent of the gas, especially near the exhaust side of the cell. Due to the low prevailing equilibrium oxygen partial pressure, water vapour-enhanced Cr-evaporation is generally not a problem at the anode side of the cell. However, presence of water vapour may promote internal oxide formation thus hampering formation of a protective external chromia scale [20].

In SOFC's utilising hydrocarbons as fuel, carbon containing gas species (CO, CO_2, CH_4) will be major constituents prevailing at the anode side. Alloy degradation processes which may occur in high-carbon activity gases are predominantly carburisation and/or metal dusting [21–24].

Another gas species which is present in the cathode side and which may be present in the fuel gas and might affect the lifetime of the metallic interconnect is nitrogen. In case of chromium-based alloys, Cr_2N may form at the scale/alloy interface or beneath the interconnect contact with nickel current collector [25–27]. In steels, formation of Cr_2N precipitates may affect material ductility and result in depletion of the surrounding metallic matrix in chromium thus decreasing its oxidation resistance [28,29].

Sulphur may prevail in the fuel gas as an impurity, for example, as intentionally added species (so-called odorants) [30] or as a constituent of a fuel [13,14]. Using hydrocarbons-based fuels, especially biogas might be prone to sulphur contamination. Lindermeir [15] showed that H_2S contents in biogas may reach concentrations up to 2400 ppm.

7.3 Ceramic interconnects

Ceramic interconnects played a dominant role in the first 25 years of SOFC developments and are still essential in full ceramic designs today [2,4,31]. Contrary to metallic interconnect materials, ceramics were always tightly connected to a specific stack design and manufacturing route. Figure 7.1 shows the various concepts of full ceramic SOFCs, which can be generally divided into tubular and monolithic [32,33,43,44]. Both of them utilise an interconnect in form of relatively thin, usually up to 100 μm thick, compact layer. In such arrangements, the interconnect is not a load-bearing component but rather provides the path for the electrical current. However, since the interconnect layers are in direct contact with other cell components, the differences in coefficients of thermal expansion (CTE) between the various components

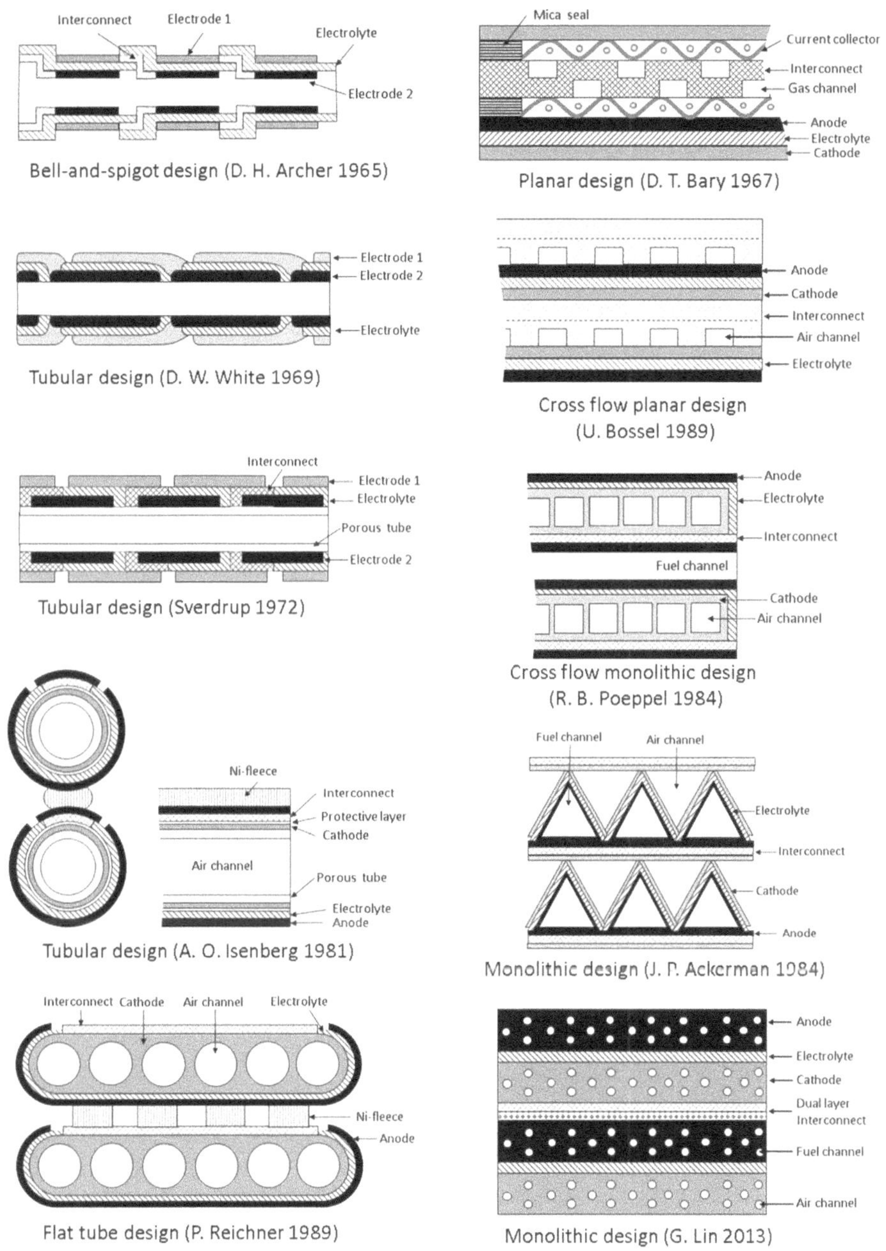

Figure 7.1 Various ceramic SOFC stack designs [32–42].

may result in initiation of substantial mechanical stresses during temperature changes. Additionally, mechanical stresses are generated in the interconnect by the large gradient of oxygen activity (about 20 orders of magnitude) prevailing within fuel and air side across the interconnect layer [45]. This problem can be avoided to a large extent by partial substitution of chromium with titanium in, for example, $(La,Ca)CrO_3$ [46,47]. This substitution leads to decreasing electrical conductivity [46], and complex tailoring of the composition is necessary to end up with a technical compromise for the various required properties.

The manufacturing of ceramic interconnects usually involves material synthesis, shaping procedure and assembling with other cell components. The manufacturing process may involve different steps like pressing, printing, wet spraying and sintering, or it consists of only a single step such as electrochemical vapour deposition (EVD) [48]. A detailed description of the various manufacturing methods falls outside the scope of this chapter. The reader is referred to a number of Refs. [3,4,31,49] in which the various manufacturing processes are described.

Virtually, all ceramic interconnects of present SOFC systems are based on $LaCrO_3$ with perovskite structure. There are also attempts to use titanates as interconnects [50,51], but they suffer from poor electronic conductivity on the air side. Therefore, a second perovskite layer is used [52] to maintain the conduction path, but this may induce mechanical stresses in this bilayer. Lanthanum chromite was selected as the most promising material for interconnect application in the late 1960s and since then continuously developed [1,2,4]. Pure $LaCrO_3$ is a refractory material with a melting temperature exceeding 2400 °C [2]. Consequently, in this form, it can hardly be used as an interconnect material due to its poor sinterability [6,53,54]. By modifying the stoichiometry with other elements such as Ca, Mg, Sr, Co, Ni and/or Ti, it is possible to modify the properties of this interconnect material with respect to sinterability, thermal expansion coefficient and electrical conductivity [4,53,54]. Detailed information regarding the influence of substituents on properties of $LaCrO_3$-based materials can be found in Refs. [4,31].

Today, substituted lanthanum chromites are state-of-the-art materials for ceramic interconnects, unless a new superior material would be found. In the recent years, competitive materials such as yttrium chromites [2,55], neodymium chromites or strontium titanates have been investigated, showing promising results [50,51,56,57]. However, the material costs of rare earth containing perovskites are rather high, and their application as ceramic interconnect seems to be only appropriate as long as the stack design requires only small amounts of the material [3]. Additionally, manufacturing technologies (shaping, sintering, etc.) for ceramic interconnects are usually cost- and labour intensive [3]. Therefore, the interest in development of ceramic interconnects decreased in recent years, and more and more attention was given to development of metallic interconnect materials. Since the year 2000, only a limited number of papers related to new developments of ceramic interconnect materials were published. Nonetheless, a number of research groups are working on full ceramic SOFC stacks [5,42,50,52,57]. Their activities seem to be, however, predominantly concentrating on improvements of design as well as manufacturing procedures and to a far less extent on material development.

7.4 High-temperature alloys for SOFC applications

The main drawback of metals and alloys for SOFC application is their reactivity with the anode and cathode side service environments at the high operating temperatures. The resulting high-temperature corrosion phenomena not only lead to dimensional changes and loss in load-bearing cross sections of the components, but also to formation of surface oxide scales which mostly possess poor electrical conductivities [1,27,58–61]. Therefore, noble metals have been considered as interconnector materials; however, because of limited availability and high cost, this solution has been abandoned for large-scale commercial application. Apart from obvious economic reasons, the main technical drawback of a noble metal such as silver is its ability to dissolve relatively large amounts of oxygen and hydrogen [62–64]. This results in formation of voids within the metallic interconnect during exposure under dual atmosphere conditions [64].

In early studies on metallic interconnects, conventional high-temperature alloys were considered as possible candidate materials for metallic interconnect application [65]. A major property of these high-temperature alloys is their ability to form protective surface oxide scales which possess sufficiently slow growth rates to keep the oxidation attack below a technologically acceptable level from the viewpoint of dimensional changes and loss of load-bearing cross section. The growth rate of oxide scales is commonly characterised by the parabolic rate constant k_p [60] if classical parabolic oxidation prevails. The scale growth kinetics is then written as

$$x^2 = k_p \times t \tag{7.1}$$

where x is the scale thickness, t the time and k_p the parabolic rate constant.

The scale growth kinetics is frequently expressed in terms of weight gain as a result of oxygen uptake:

$$(\Delta m)^2 = k_w \times t \tag{7.2}$$

where Δm is the area-specific weight gain. In many cases, chromia-based surface scales exhibit deviations from the ideal parabolic behaviour. The oxidation behaviour may then be expressed as [11]:

$$\Delta m = k \times t^n \tag{7.3}$$

in which k is the rate constant and n the oxidation exponent. Depending on the exact oxide scale growth mechanisms, more complex scaling kinetics may prevail [58].

Among the surface oxides which may form on commercially available construction materials exposed to high temperatures, only Al_2O_3, SiO_2, Cr_2O_3 and, in some special cases, NiO offer the possibility to obtain slow material degradation rates [60,66]. If one would consider oxidation rates at high temperatures as the first selection criterion for interconnector application, the so-called alumina formers of the type NiCrAl,

CoCrAl and especially FeCrAl would be the materials of choice [61]. These types of alloys provide excellent resistance against oxidation/corrosion attack due to the formation of an extremely slowly growing alumina scale which forms on the material surface upon high-temperature exposure. The low growth rate of the alumina scales is, however, accompanied by an extremely low electronic conductivity [27,63,67], making the mentioned alloys unsuitable for interconnect applications. A similar low electronic conductivity excludes silica forming alloys from application as interconnect materials. Therefore, most studies on qualification and/or development of metallic high-temperature materials for interconnector application relate to chromia forming alloys. Chromia provides less oxidation/corrosion protection than alumina; however, its electronic conductivity in the SOFC operation temperature range of approximately 600–900 °C is orders of magnitude larger than that of alumina or silica [29,67,68].

7.4.1 Chromia forming alloys

Table 7.1 shows a selection of chromia forming alloys evaluated by various researchers for application as SOFC interconnect materials. The vast majority of the commercially available chromia forming alloys is based on the systems NiCr, NiFeCr or FeCr. The two first mentioned alloy systems consist mainly of γ-phase with FCC structure. Nickel-based alloys, such as Haynes 230, have been studied as a possible interconnect material; however, until now they were hardly used in real SOFC stacks.

Commercially available alloys as well as materials specially developed for SOFC interconnector application (Table 7.2) are nearly exclusively based on the binary alloy system Fe–Cr. As the thermal expansion coefficient in Fe–Cr alloys decreases with increasing chromium content [130,148,149], chromium-based materials have, in most cases, been considered for application in zirconia electrolyte-based SOFC concepts [21,58,89,127,150,151]. If a slightly higher thermal expansion coefficient is required, as is, for example, the case in nickel cermet anode-based cell concepts or in cells using ceria as electrolyte, iron-rich alloys, that is, ferritic steels, are being considered [142,148,152–155]. Therefore, the following discussion will concentrate mainly on ferritic steels and chromium-based materials. Austenitic steels and nickel-based alloys will only briefly be addressed.

7.4.2 Chromium-based alloys

Extensive research work at Plansee and Siemens led to the development of an oxide dispersion strengthened (ODS) chromium-based alloy commercialised later as $Cr_5Fe_1Y_2O_3$ (EP0578855 Table 7.2) [18,21,58,67,127,131,156,157]. The material showed substantially improved oxidation and mechanical properties, for example, creep strength, compared to conventional chromium alloys [18,156,157]. The ODS materials are manufactured by elemental mixing of the starting powders (abbreviated as MIX alloy) or by mechanical alloying (MA), that is, by high energy ball milling of the chromium-iron alloy and oxide powders in a non-oxidising environment

Table 7.1 Nominal chemical compositions of potentially suitable metallic materials studied for interconnect applications

Type	Commercial name	Nominal composition in wt%															References
		Fe	Cr	Ni	Mn	Co	Si	Al	Ti	Nb	Mo	W	Y	La	Ce	Zr	
Ferritic steels	SUS430	Balance	17		0.5												[26,69–77]
	F17T		16.5		0.4		0.35		0.4								[78]
	AL 453		22	0.3	0.3		0.3	0.6	0.02					<0.1			[69,74,79–84]
	Crofer 22 A		22		0.5		0.1	0.1	0.1					0.1			[9,81,82,85–88]
	Crofer 22 APU		22		0.5		0.02	0.02	0.1					0.1			[9,11,12,16,23,28, 56,70,71,75,76,79, 81–83,87,89–110]
	Crofer 22 H		22		0.5		0.25	0.02	0.1	0.5		2		0.1			[83,91,93,96,100, 101,111–116]
	ZMG232		22	0.26	0.5		0.4	0.21						0.04		0.22	[25,70,72,81,82, 92,93,117–123]
	ZMG232L		22	0.33	0.48									0.08		0.2	[74,117,123]
	Sanergy HT		22.3		0.24		0.02	0.03	0.03	0.6	077					0.1	[56,70,83,92,93, 112,124,125]
	RA 446		25		<1.5		<1										[27,74,75,80, 122,126]
	ITM		25.7		0.04		0.03	0.04	0.21		1.9		0.24*				[89,107,108, 127–129]
	E-Brite		26	0.15	0.05		0.2			0.1	1						[70,74–76,81–83, 88,93,97,99,102, 107,108,110,126]
	AL 29-4C		29	0.3	0.5		0.35				4						[83,113]
Chromium-based materials	$Cr5Fe1Y_2O_3$	5.5	92.9										0.48*				[11,21,28,86,108, 127,129–132]
	CFY	5	94.6				0.12	0.12					0.1				[133]

*Yttrium added into the material in form of Y_2O_3 dispersion.

Continued

Table 7.1 **Continued**

Type	Commercial name	Nominal composition in wt%															References
		Fe	Cr	Ni	Mn	Co	Si	Al	Ti	Nb	Mo	W	Y	La	Ce	Zr	
Austenitic steels	304L	Balance	18	9.5	1												[81,82,134]
	316		17.5	11.5	1												[9,23,80,135]
	310		25	20	1		2										[9,24,136,137]
Ni-based alloys	Haynes 230	3	22	Balance	0.5	5	0.4	0.3			2	14		0.02			[69,84,88,94,99, 110,130,134, 137–140]
	Haynes 242	2	8		0.4	2.5	0.4	0.25			25						[88,99,110,130]
	HR-160	2	28		0.5	29	2.75		0.5	1	1	1					[80]
	Inconel 600	8	15		0.5		0.25										[24,140,141]
	Inconel 625	2.5	21		0.25	0.5	0.25	0.2	0.2	3.5	9						[49,102,110,134, 142–147]
	HastelloyX	18	22		0.5	1.5	0.5				9	0.6					[49,102,142–145, 147]
	HasteloyS	1.5	16		0.7	1		0.4						0.05			[49,139,147]

Table 7.2 **Patents related to metallic SOFC interconnect materials**

1st inventor	Affiliation	Title	Pat./app. number	Application date
C.S. Tedmon	General Electric	Silver–palladium cathodic current collector for high-temperature fuel cell	US3522097	12.06.1967
C.S. Tedmon	General Electric	Electrochemical cell including iron–chromium alloy conductor connected to cathode	US3516865	30.08.1967
N. Nazmy	Asea Brown Boveri AG	Current conduction element for stacked high-temperature fuel cells and method of manufacture	EP0424732	10.10.1990
H. Greiner	Siemens AG Plansee SE	Material für die metallischen Komponenten von Hochtemperatur-Brennstoffzellen-Anlagen	EP0578855	16.07.1992
B. Brückner	Siemens AG	Hochtemperatur-Brennstoffzellen-stapel und Verfahren zu seiner Herstellung	DE4237602	06.11.1992
T. Hitoshi	Sumitomo Corp.	Oxidation resistant metallic materials	US5800152	10.10.1996
U. Diekmann	Forschungszentrum Jülich	Werkstoff für Brennstoffzellen Interkonnektoren	DE19650704	06.12.1996
S. Rak-Hyun	Korea Energy Research Institute	Metallic interconnection material for slid oxide fuel cell and method for preparing the same	US6150048	30.07.1999
W.J. Quadakkers	Forschungszentrum Jülich	Material used at high temperatures	DE01/01415	20.05.2000
J.F. Grubb	ATI Inc.	Ferritic stainless steel having high-temperature creep resistance	US6641780	30.11.2001
M. Krumpelt	The University of Chicago	Compositionally graded metallic plates for planar SOFC	US6843960	12.01.2002
T. Uehara	Hitachi Metals Ltd.	Steel for separators of solid-oxide type fuel cells	JP2001296528	27.09.2001
I. Shinsuke	JFE Steel Corp.	Metal material for FC, FC using the same and method for producing the material	JP2002232367	09.08.2002

Continued

Table 7.2 Continued

1st inventor	Affiliation	Title	Pat./app. number	Application date
J.M. Rakowski	ATI Inc.	Oxidation resistant ferritic stainless steels	US8518234	03.09.2003
M. Schuisky	Sandvik AB	Heat-resistant steel	EP1600520	17.05.2005
J.M. Rakowski	ATI Inc.	Interconnects for solid oxide fuel cells and ferritic stainless steels adapted for use with solid oxide fuel cells	US7981561	28.06.2005
W.J. Quadakkers	Forschungszentrum Jülich	Creep resistant ferritic steel	EP1984533	18.02.2006
M. Schuisky	Sandvik AB Haldor Topsoe A/S	Ferritic chromium steel	EP1882756	26.07.2006
N. Yasuda	Hitachi Metals Ltd.	Steel for SOFC having excellent oxidation resistance	JP2009214525	16.09.2009

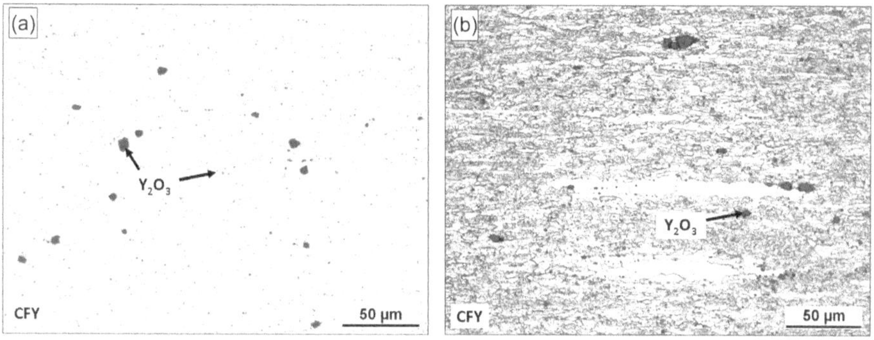

Figure 7.2 Microstructure of $Cr_5Fe_1Y_2O_3$ material, optical metallography (a) sample polished; (b) sample etched with 10% H_2SO_4.

[18,127,156,157]. The latter process can produce a very finely distributed oxide dispersion (typical particle diameter ≈ 15 nm) embedded in the chromium-based matrix, compared to the substantially coarser dispersion resulting from the elemental mixing process [18,127,156,157]. The commercial dispersion strengthened chromium-based alloys specifically developed for SOFC application mostly contain a few percent of iron to adjust the thermal expansion coefficient as close as possible to that of the electrolyte material YSZ [18,127,156,157]. As a dispersion, mostly rare earth oxides such as Y_2O_3, La_2O_3 or mixed oxides such as $LaCrO_3$ or $LaSrCrO_3$ are being used [158,159]. A major advantage of the powder metallurgical manufacturing route is that it offers the possibility to add far more reactive elements (REs) (in form of their oxides) than possible in conventional metallurgy [127–129]. Figure 7.2 shows microstructures of the ODS material $Cr5Fe1Y_2O_3$ designed for SOFC interconnect applications.

A further chromium-based alloy developed by Plansee is CFY [133]. Components of this alloy are manufactured using net-shape methods based on a powder-pressing process in order to keep unit costs lower than when using mechanically alloyed material. Final strength and physical properties are obtained by suitable heat treatment processes.

Chromia scales forming on pure chromium or chromium-based alloys at high temperatures exhibit slow growth rates; however, when formed during exposure in air or oxygen, the scales tend to buckle and possess poor adherence to the metallic substrate. Addition of the RE oxide dispersion leads to an improvement in scale adherence and reduction in scale growth rate [18,160]. Similar effects can be obtained by addition of REs such as Y, La, Ce and Zr in metallic form or by ion implantation [161–164]. Several reviews on this subject are available, and substantial information can be found in text books [60,61]. This RE imparted change of the most important oxide scale properties is frequently claimed to be related to a suppression of the outward transport of chromium ions in the oxide scale [161–164].

When using chromium-based alloys in air or oxygen, it has to be considered that the chromia scales may form volatile species [156,165] such as CrO_3 and/or $CrO_2(OH)_2$ according to the following reactions:

$$Cr_2O_3 + 1.5O_2 \rightarrow 2CrO_3 \quad (7.4)$$

$$Cr_2O_3 + 1.5O_2 + H_2O \rightarrow 2CrO_2(OH)_2 \quad (7.5)$$

From these reactions, it is apparent that substantial volatility prevails in environments containing a high pO_2. Formation of volatile chromium species is thus hardly significant in anode side environments [6].

Several authors have demonstrated that the existence of volatile chromium species in the gas phase can cause serious cell degradation [6,141]. Therefore, a number of protection methods have been proposed to minimise evaporation of volatile chromium species, such as coating of the interconnector with metallic or oxide layers [92,107,108,166–169], as will be discussed in a later section.

Chromia scales formed on chromium-based ODS alloys in $Ar/H_2O/H_2$, that is, simulated anode gas, exhibit generally better adherence to the metallic substrate than those formed in air or oxygen [58,151,170]. Hänsel et al. [151] found the oxidation rate of chromium-based ODS alloys at 950–1050 °C in an $Ar/H_2/H_2O$-mixture (equilibrium oxygen partial pressure at 1000 °C approximately 10^{-15} bar) to be higher than that in high-pO_2 environments (Air, Ar/O_2), whereby the difference could only partly be explained by formation of volatile chromium species in the high-pO_2 gases. The actual oxidation rate strongly depended on the alloy manufacturing method [150]. In the $Ar/H_2/H_2O$-environment, the alloys tended to form whisker type oxide morphologies, the extent of whisker formation being decreased by an "optimum" addition of a RE (oxide dispersion).

As already mentioned in Section 7.2, oxygen in SOFC relevant environments will not be the only gaseous specie which may react with the chromium based interconnector alloys. The second reactive gaseous specie which is of importance during service of chromium-based materials is nitrogen. Especially, for non-RE-doped chromium-based alloys, reaction with nitrogen is of major concern because it leads to subscale nitride formation and environmentally induced embrittlement during high-temperature exposure [29,150]. The sensitivity for nitrogen uptake depends on the gas-tightness of the chromia layer because nitrogen transport between the gas atmosphere and the metal or alloy is known to occur via gas molecules rather than via solid-state diffusion through the scale [29,150]. Generally, it can be said that the thick chromium oxide scales formed on non-RE-doped alloys are more permeable for molecules of nitrogen than the thinner, that is, more protective, scales on RE or RE-oxide-doped materials. This observation in fact indicates that the protective properties of the chromium oxide scales are to a large extend determined by the ability to prevent molecular transport of oxidising species rather than by the detailed type and concentration of point defects in the chromia lattice [160,171,172].

If carbon-containing gas species such as CO and/or CH_4 are present at the fuel side of the cell, all chromium based alloys tend to form a sub-scale layer containing chromium carbide, that is, mainly $Cr_{23}C_6$ sometimes in combination with Cr_7C_3 [21]. In case of a CH_4-based test gas, the carbides exist as a near-continuous layer [21], whereas in the case of a CO-based gas, the sub-scale layer tends to prevail as an oxide/carbide mixture [173,174]. In a similar manner as discussed for nitridation, the extent of carbon attack, which occurs by molecular gas transport through the scale, tends to decrease with decreasing growth rate of the oxide layer [173]. This again illustrates that in the case of chromia scales, protectiveness in mixed gas environments means to a large extent resistance against molecular transport from the gaseous atmosphere to the metal or alloy [21,160,171,172].

7.4.3 Ferritic steels

Formation of a protective chromia surface layer during high-temperature exposure of a ferritic steel requires a chromium content of approximately 17–20%, depending on temperature, surface treatment, minor alloying additions and impurities (e.g. C, S ,P) [9,175]. Thus, in commercially available ferritic steels considered for SOFC application, the chromium content varies between approximately 18 and 28 mass% (Tables 7.1 and 7.2), whereby the oxidation resistance generally increases with increasing chromium content [9,153,175]. Although the oxide scales formed on such materials in the temperature range of 600–900 °C consist predominantly of chromia, the detailed scale composition and growth rate are substantially affected by presence of minor alloying additions [11,58,103,153].

With respect to oxidation behaviour, the most important minor alloying additions in commercially available ferritic steels are titanium, manganese, aluminium and silicon, their concentrations commonly being a few tenths of a percent. Upon high-temperature oxidation, the first two mentioned elements tend to become incorporated in the scale [29,91,105,176,177] although titanium can also be present in the form of internal titanium oxide precipitates in the sub-surface region (Figure 7.3). Titanium tends to become dissolved in chromia at low oxygen partial pressures, that is, in the inner part of the chromia-based scale [101,178,179]. However, at higher oxygen partial pressures, thus in the outer part of the scale, it tends to become re-precipitated because of its decreasing solubility in chromia with increasing pO_2 [101,178,179]. Silicon and aluminium form oxides which are thermodynamically far more stable than chromia. Therefore, these elements mostly prevail in internal oxides rather than being incorporated into the external scale. Internal oxide precipitates of the mentioned elements may exhibit different morphologies (Figure 7.3). Internal titanium oxide frequently forms fine spherically shaped precipitates (Figure 7.3) which results in strengthening of the surface-near alloy region thus reducing the tendency for occurrence of surface wrinkling caused by relaxation of oxidation-induced stresses [58,131]. Aluminium oxide tends to form needle- or plate-like precipitates [11,58,103,153], and its formation resulting in localised outward flow of surrounding metal thereby forming protrusions in the outer oxide scale. Consequently, higher scaling rates for the steels containing small amounts of aluminium, for example, 0.1% Al, are observed. Silica forms mostly lens-shaped precipitates at or near the oxide/metal interface. Its formation decreases the electrical conductivity and, in some cases, deteriorates the adherence of the outer chromia scale [29,58,176,177].

Considering pure metals and oxides, the dissociation pressure of MnO is slightly lower than that of Cr_2O_3. Whether manganese may oxidise beneath an external chromia scale depends on the concentration, or better activity of chromium and

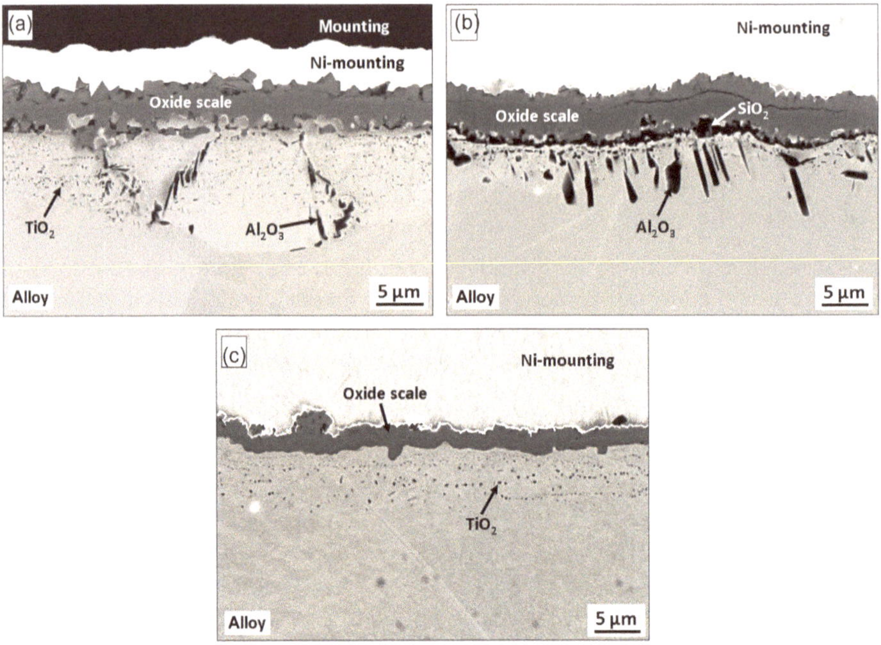

Figure 7.3 SEM/BSE pictures showing microstructures of oxide scales formed on various ferritic steels during 1000 h discontinuous oxidation at 800 °C in air (a) Crofer 22 A; (b) ZMG 232; (c) Crofer 22 APU (measured for specimens of 2 mm thickness).

manganese, the latter being generally substantially smaller than that of chromium in high chromium ferritic steels. Manganese possesses only a very limited solubility in chromia, but its diffusivity in the scale is relatively high. Therefore, it is mostly found in the outer part of the scale in the form of a Cr,Mn-spinel layer [180,181]. This spinel phase may exist in two forms, that is, the so-called normal spinel $MnCr_2O_4$ and inversed one $(Mn,Cr)_3O_4$ [180,181]. Both forms differ in their chemical composition, a number of physical properties and occurrence conditions. The normal spinel $MnCr_2O_4$ is stable at low pO_2, such as prevailing in SOFC anode gas. The inverse spinel $(Mn,Cr)_3O_4$ forms in gases with high pO_2, such as in air [101,180–182]. Covering of the oxide scale surface by the Cr,Mn-spinel alters remarkably the growth rate of the inner chromia layer. The observed differences in oxidation rates of various chromia forming alloys during oxidation in low or high pO_2 gas can frequently be explained by presence of minor manganese additions [101]. Generally, formation of Cr,Mn-spinel decreases the growth rate of chromia and the overall oxidation rate in low pO_2 gases, whereas at high pO_2, the opposite effect is found [101,182]. In cathode, side gas (high pO_2) formation of an outer $(Mn,Cr)_3O_4$ layer is considered to be beneficial since it reduces chromium evaporation and decreases the electrical resistivity of the oxide scale [29,59]. Due to these beneficial effects, minor amounts (a few tenths of percent) of manganese are present in most of the recently developed interconnect steels. Higher concentrations lead to an undesired high-scale growth rate in cathode side environments [122].

Other alloying elements which may be found in high chromium steels are tungsten, molybdenum and niobium (Table 7.1). Tungsten and molybdenum have a substantial solubility in the ferrite matrix and are added to reduce the CTE and/or improve the steel creep strength [91,96,111,149] by solid solution strengthening [91,149]. Niobium exhibits a relatively small, temperature-dependent, solubility in ferrite. Addition of a few tenths of a percent into the steel results in formation of precipitates of Laves phase $(Fe,Cr)_2Nb$, resulting in improved creep resistance [91,96,111].

Tungsten and molybdenum oxides are thermodynamically less stable than Cr_2O_3. Thus, both elements cannot oxidise beneath the external chromia scale. Consequently, they do not directly influence the growth of the surface oxide scale on the steels. Their indirect effect on the oxidation rate, resulting from strengthening of the metallic substrate, will be discussed in detail in Section 7.8. When considering pure metals, niobium oxide has a lower dissociation pressure than Cr_2O_3, thus the element may oxidise beneath the chromia surface scale. Formation of niobium oxide is undesired because it increases the growth rate of the chromia surface scale (doping effect) and depletes/removes Laves phase precipitates from the metallic matrix [91] (Figure 7.4). Niobium oxidation may be prevented by proper addition of silicon into niobium-alloyed steels [91], as will be discussed later.

The presented findings illustrate that only slight changes in alloy composition, service conditions or design-related parameters can fundamentally change the oxidation properties of high-chromium ferritic steels. Additionally, a change from one oxide scale type into another may occur during long-term service of a given steel due to depletion of minor alloying element additions in the alloy matrix because of their low initial concentrations [101,177,183,184].

The optimum steel composition for an SOFC interconnector will depend on the used cell design (e.g. electrolyte or anode-supported concept) and the service conditions (temperature, thermal cycling, fuel gas composition, required service life)

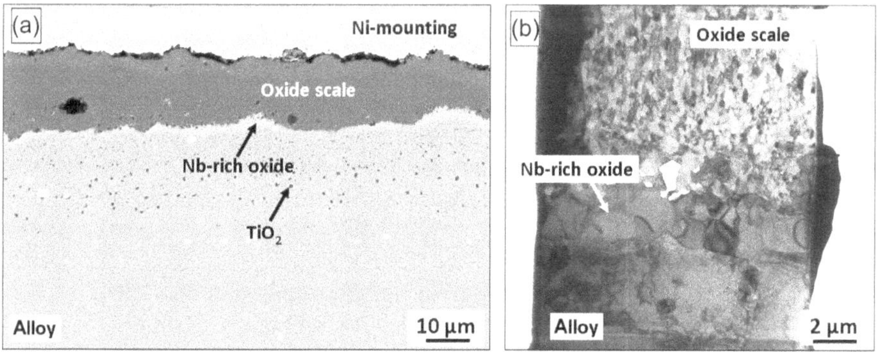

Figure 7.4 SEM/BSE (a) and TEM (b) pictures showing microstructure of oxide scales formed on niobium-alloyed high chromium ferritic steel during 1000 h discontinuous oxidation at 900 °C in air [11,91].

[1,3,7,43,58,145,146,165]. If an YSZ electrolyte-supported design is considered, a crucial requirement for the interconnector is a low thermal expansion coefficient of approximately $8–10 \times 10^{-6}\ K^{-1}$. Using binary Fe–Cr alloys with chromium contents commonly present in ferritic steels (see Table 7.1), such a low value cannot be obtained [148,153]. Moreover, the electrolyte-supported cells are, due to their high electrical resistivity, mostly considered for operating temperatures above 900 °C [185]. At such high operating temperatures, the oxidation limited lifetime of ferritic steel components will be quite short and therefore, chromium-based alloys are considered as interconnects [133].

At substantially lower service temperatures between 600 and 800 °C, ferritic steels are potentially suitable construction materials for interconnects. Such low temperatures are commonly applied when using an anode-supported cell design which allows applying an extremely thin YSZ electrolyte [98,186]. The choice of such a concepts requires interconnectors with CTEs similar to that of the Ni/ZrO_2 anode, that is, in the range of $11–12 \times 10^{-6}\ K^{-1}$. Such CTE values can be achieved by ferritic steels with Cr contents of approximately 20%.

7.4.4 Optimised ferritic steels for SOFC applications

Quadakkers et al. [153,155] studied the suitability of model steels with chromium contents of 16–25% and evaluated the effect of various RE additions to achieve an optimum combination of low scale growth rate and excellent scale adherence. Lanthanum appeared to be the preferred RE addition, mainly because it does, contrary to other commonly used REs, not form intermetallic compounds with iron [58,104,153]. This allows lanthanum to become evenly distributed in the alloy rather than being precipitated in the form of intermetallics on the alloy grain boundaries, as is, for example, observed if yttrium or cerium additions [63] are being used. Although in RE-doped steels, protective behaviour can already be obtained if the chromium content is only 16–18%, a substantially higher level of 22% was chosen to obtain a larger "chromium reservoir" to counteract the danger for occurrence of breakaway oxidation due to presence of high water vapour contents in the anode gas [20,175], chromium loss because of interaction with contacting materials [152] or rapid exhaustion of the chromium reservoir in very thin interconnects [177,183,184]. Small amounts of manganese and titanium were added to obtain external spinel formation which is known to decrease the formation of volatile chromium species [106,107,122,187–189]. Additionally, aluminium and silicon levels below 0.01% were defined to prevent extensive internal oxide formation [29,184]. This approach leads to the development of the now commercially available steel Crofer 22 APU [11,190]. The defined low levels of aluminium and silicon require the application of high purity raw materials and special steel manufacturing technologies such as vacuum induction melting (VIM).

Further development of ferritic steels which are especially designed for SOFC application was driven by the need for improvement of creep resistance and reducing manufacturing costs [91,115]. Recently, developments on the basis of 22% Cr ferritic steels with systematic variations of minor alloying additions such as tungsten, molybdenum, niobium and tantalum were described [191]. Froitzheim et al. [91,192] showed that using optimised, combined additions of niobium, tungsten and silicon, the creep strength of high-chromium ferritic steels can be substantially improved. This resulted in the development of the steel Crofer 22 H [91,96,101]. The room temperature tensile properties of the steel were similar to those of Crofer 22 APU. The addition of minor amounts of silicon largely suppressed the adverse effect of niobium on the oxidation behaviour, resulting in slowly growing, well-adhering surface oxide scales, which possess a high electronic conductivity at SOFC operation temperatures [193]. A major finding of these investigations was that an adjusted Si/Nb ratio had as result, that the undesired internal oxidation of silicon could be prevented. Allowing silicon additions of a few tenths of a percent has an important consequence, that the steel can be produced by conventional manufacturing methods thus reducing material costs. Thus, the combined, carefully adjusted additions of niobium and tungsten along with silicon, resulted in an alloy with substantially higher creep strength than Crofer 22 APU without deteriorating the oxidation resistance and the electrical conductivity of the oxide scales [192].

An alternative approach for obtaining high creep strength was applied by Honegger et al. [154] who developed a high-chromium ferritic steel produced by powder metallurgical techniques, thus allowing incorporation of oxide dispersions to obtain dispersion strengthening. The MA process allows producing materials with substantially higher concentrations of REs (here yttrium) than can be obtained by conventional alloy manufacturing methods. Therefore, the ferritic ODS steels generally exhibit lower oxidation rates than RE containing conventionally manufactured wrought alloys of similar composition. A high RE content in the surface regions can also be obtained by applying a RE containing coating or by RE ion implantation [58,162,164]. In this way, the oxidation rate of commercial, ferritic 18–26% chromium steels may substantially be decreased.

A major drawback of high-chromium ferritic steels is their tendency to form σ-phase (Figures 7.5 and 7.6) at lower temperatures. Additionally, the so-called α' precipitates may form below approximately 545 °C. As both processes lead to embrittlement of the steels, the exact steel composition has to be carefully adapted to the actually envisaged application temperature. For instance, steels with a chromium content of more than 25% are prone to σ-phase formation even at temperatures as high as 800 °C, whereby the tendency to its formation increases with decreasing temperature (Figures 7.5 and 7.6). Moreover, the σ-phase formation is known to be promoted by alloying additions, for example, molybdenum. Hammer et al. [82] illustrated that the σ-phase nucleation is not only affected by the alloy composition but also by the actually prevailing service environment. Steels with chromium contents substantially lower than 20% are less susceptible to σ-phase embrittlement. However, they are prone to breakaway oxidation in high water vapour containing anode side gases [20,175]. They also may run into early breakaway oxidation in cathode side gases in case of very thin components [184].

7.4.5 Austenitic steels and nickel-based alloys

Austenitic steels and nickel-based alloys (Table 7.1) are commonly preferred over ferritic steels for high-temperature applications if a high creep resistance is a primary design requirement. Two main drawbacks of these alloys as construction materials for SOFC interconnects are their high price and high coefficient of thermal expansion.

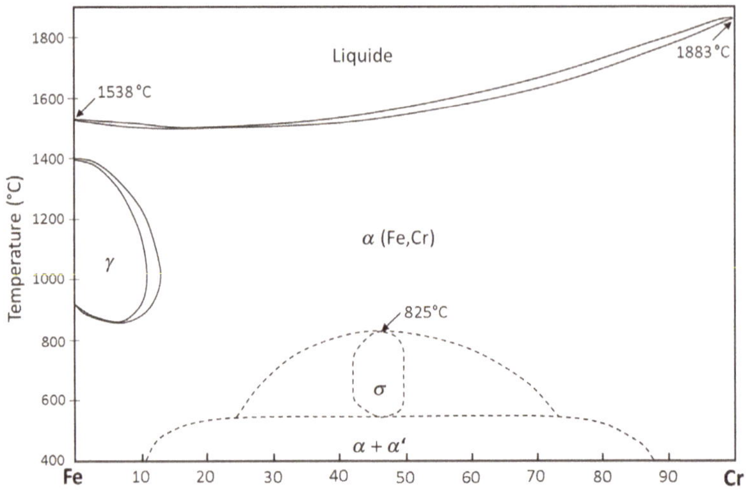

Figure 7.5 Binary Fe–Cr system derived from various sources [194–196]. Dashed lines correspond to fields which are still not unequivocally established.

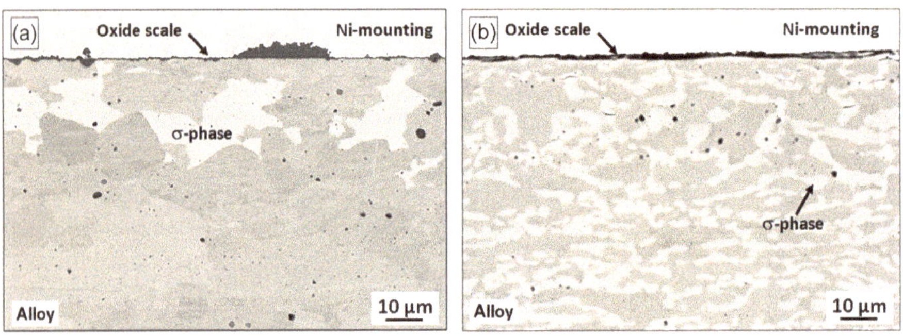

Figure 7.6 SEM/BSE pictures showing microstructures of alloy ITM (Table 7.1) after 1000 h oxidation in simulated anode gases at (a) 800 °C; (b) 650 °C [191].

Among the nickel-based alloys, Haynes 230 is the most frequently evaluated material for SOFC applications [69,84,88,94,99,110,130,134,137–140]. The relatively low CTE in combination with good oxidation resistance and high electrical conductivity of the oxide scale makes alloy Haynes 230 potentially suitable for interconnect applications. However, its commercial application was mainly hampered by the high material costs. Thus, austenitic steels may be considered as alternatives for nickel-based alloys. It was anticipated [8,9] that limitations set by the intrinsically high coefficient of thermal expansion of austenitic steels could be overcome by lowering the SOFC operation temperatures resulting in less stringent requirements to the CTE mismatch [9] and expected lower oxidation rates [8,9]. However, Hammer et al. [82] and later Niewolak et al. [9,23] showed that 18% Cr austenitic steels such as 304 or 316 exhibit relatively high oxidation rates at 600–700 °C, especially in anode side gas [9,81]. Increasing the chromium content up to 25%Cr such as prevailing in 310 stainless steel, substantially improved the oxidation resistance in all SOFC relevant atmospheres [9,24]; however, the high chromium contents make the steels prone to σ-phase formation, especially at service temperatures at and below 700 °C [23].

Summarising, it can be said that some of the commercially available austenitic steels and nickel-based alloys may fulfil part of the requirements for interconnect construction materials. The alloys are, however, hardly used in presently designed SOFCs because of the major drawbacks, that is, high CTE and high material cost.

7.5 Growth rates of chromia base surface scales

The growth rate of the oxide scales formed on the interconnector surfaces is an important parameter to be considered in SOFC applications because of dimensional constraints in a multilayered cell stack. Additionally, fracture mechanical concepts describing scale integrity during thermal cycling [121,162,184,197] predict that the onset for scale spalling depends on the oxide scale thickness.

The growth rates of scales on chromium-based ODS alloys are substantially lower than those formed on pure chromium or chromium-based alloys [11,21,58, 170,198,199] without addition of a RE (oxide). It is difficult to give a quantitative figure of this reduction in oxidation rate by the presence of the RE because chromium as well as chromium-based alloys do frequently not exhibit the classical parabolic time dependence for oxide scale growth [60,66]. Classical Wagner oxidation theory predicts that the scale growth will obey parabolic time dependence (Equation 7.1) if oxidation is controlled by diffusion of metal and/or oxygen ions through the oxide lattice [200]. For chromium, chromium-alloys and especially chromium-based ODS alloys, Equation (7.1) is hardly ever obeyed [173] because the assumptions made to describe scale growth by Equation (7.1) are mostly not completely fulfilled [171,198]. Chromia scale growth proceeds to a large extent via rapid diffusion paths such as oxide grain boundaries and is affected by micro-voids or micro-cracks [150,201]. Furthermore, in oxygen-rich environments (e.g. air), formation of volatile oxides and/or hydroxides may affect chromia growth [60]. The combination of all these factors makes the description of the scaling kinetics quite complex and a (near) fit of a measured curve to a parabolic rate law is frequently more or less accidental [11].

In the case of ODS alloys, it has to be considered that the desired low oxide growth rate of the scales requires the RE to become incorporated in the surface oxide layer [164,172,202]. This occurs by embedding of the dispersions in the inwardly growing oxide scale and subsequent diffusion of the RE to the oxide grain boundaries [172]. Consequently, the beneficial effect does not immediately become apparent but only appears after a certain oxidation time, with the consequence that initially a rapid oxide scale growth occurs which strongly decreases after longer time [11,58]. Several authors have tried to explain the RE-induced reduction in scale growth rate and improvement in scale adherence by a single mechanism. However, implantation experiments by Hou et al. [202] revealed that the presence of the RE yttrium in the oxide scale decreased scale growth rate, but it did not necessarily increase scale adherence. Thus, the improvement of the two major oxide scale properties seems to be imparted by two fundamentally different mechanisms. Similar conclusions can be drawn from TG studies on chromium-based ODS alloys: the reduction in scale growth rate strongly depended on dispersion size and distribution, whereas scale adherence did not [21,151,198].

The oxidation rates for chromium-ODS are generally lower than those observed for RE containing conventional wrought alloys such as high chromium steels or nickel-based alloys. This is related to the much lower RE contents and the presence of minor alloying elements such as manganese, titanium or silicon in the latter. These elements may form their own oxides or mixed oxides with chromia, thus affecting the growth rate of the oxide scales. The formation of mixed oxides such as Cr,Mn-spinels may substantially vary with time [101]. Huczkowski et al. [184,190] showed that growth rates of chromia/Cr,Mn-spinel scales on Crofer 22 APU type steel are affected by depletion of minor alloying elements. This depletion, and in consequence the oxidation rate, is affected by the oxidation temperature and the component surface/volume ratio [184,190]. Horita et al. [72,117,120] investigating oxidation of the steel ZMG232 and a model Laves phase containing ferritic steel found that the alloy grain boundaries play an important role in the formation of Cr,Mn-spinel because the alloy grain boundaries act as fast diffusion paths for manganese. Formation of Laves phase precipitates on the alloy grain boundaries hampers manganese transport resulting in reduction of the oxidation rate [72]. Zurek et al. [203] investigated oxidation of model nickel-based alloys and found that chromia scale growth is faster in Ar–H_2–H_2O than in oxygen. This was also demonstrated for the oxidation of pure chromium [204,205], for the oxidation of Ni–Cr under conditions where no spinel layer formed [203], and for the oxidation of Fe–Cr alloys under conditions where only Cr_2O_3 formed [206–208]. If, however, the chromia scale is covered by a Mn,Cr-spinel layer, the scaling rate is slower in Ar–4%H_2–2%H_2O than in air. The effect was later investigated by Niewolak et al. [101] who found that formation of the spinel layer on the steel is a time- and pO_2-dependent process. The authors found that the oxidation rates of Crofer 22 H at 800 °C in air were initially (up to approximately 250 h) smaller than those in Ar–H_2–H_2O; however, after longer exposure times, the opposite effect was observed. This was shown to be correlated with partial or complete coverage of the chromia by the spinel.

Generally, a change of the test atmosphere from air to an anode gas-simulating Ar–H_2–H_2O mixture has similar effects for high-chromium ferritic steels as described for chromium-based alloys. In high-chromium steels, the morphology of the chromia- and spinel-rich surface layers [101] is modified and the adherence of the scale is improved [151,154]. In Ar–H_2–H_2O gases, chromia scales and especially Cr,Mn-spinel oxides tend to exhibit a whisker-type surface morphology [101,174,205].

Comparing the equilibrium oxygen activities prevailing in typical anode gases (Figure 7.7) with the dissociation pressures of the oxides of the main alloying elements in ferritic steels reveals that under SOFC conditions only chromium, manganese, niobium and partially iron (depending on the service conditions) can oxidise, whereas nickel remains in metallic form.

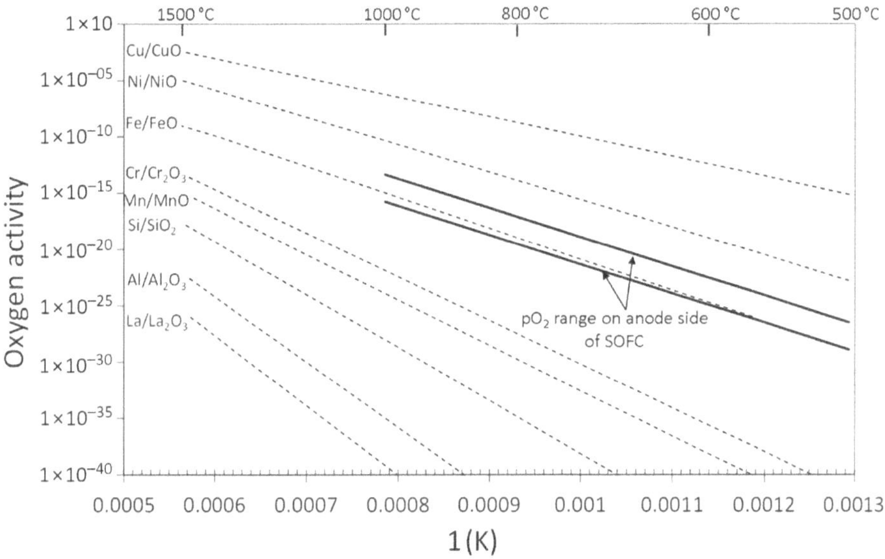

Figure 7.7 Dissociation pressures for selected oxides in equilibrium with respective pure metals compared with indicated typical range of oxygen activities in SOFC anode side gases. Calculated with FactSage software.

As iron can oxidise in many anode side gases, sometimes high oxidation rates are observed for ferritic steels in anode side environments. For ferritic steels with intermediate chromium contents (e.g. 16–18%) [9,87,175], it is well established that in environments in which water vapour is the main oxidising specie the tendency to form a protective external chromia base scale is substantially suppressed compared to that during air or oxygen exposure [206,209]. Presence of high amounts of water vapour thus shifts the critical chromium content for obtaining protective chromia scale formation towards higher concentrations. This has as a result that steels with an intermediate chromium content, which may form a protective chromia base scale during air or oxygen exposure, may form a poorly protective, iron-based oxide scale during exposure in a high water vapour containing environment [20]. The tendency to non-protective scale formation is suppressed by small alloy grain size and surface deformation of the steel [9,175].

The formation of poorly protective, iron-rich oxides will, of course, only occur if the equilibrium oxygen partial pressure in the gas is higher than the dissociation pressures of the iron oxides. In an H_2/H_2O gas mixture at typical SOFC temperatures in the range of 600–900 °C, this is the case if the H_2/H_2O is smaller than approximately 1. Considering a typical fuel utilisation of 80%, the poorly protective oxides may thus form in gas flow direction near to the gas outlet, but not in areas near the stack entrance.

A discussion of the mechanisms proposed for explaining the detrimental effect of water vapour on protective chromia scale formation can, for example, be found in reference [20]. It was shown that for a given ferritic steel, a temperature window exists in which chromia-based scales are being formed when exposed to H_2O/H_2-based gases

[175]. Non-protective oxide formation occurs at higher as well as at lower temperatures. The latter has far reaching implications for selection of interconnect materials in SOFCs with low operating temperatures around 600 °C [9,87].

7.6 Degradation in carbon containing anode gases

Compared to other fuel cell systems, a main advantage of the SOFC is its ability to use not only hydrogen but also presently available fossil fuels (such as methane, butane or even reformed gasoline or diesel), thus reducing operating costs and increasing flexibility [10]. As a consequence, the anode gas will then contain gas species such as CO, CO_2 and/or CH_4 in addition to H_2 and H_2O.

Degradation processes which may occur in high-carbon activity gases are predominantly carburisation, carbon deposition and/or metal dusting [21–24,210,211]. Carburisation may occur if the carbon activity which establishes at the metal surface is higher than the equilibrium carbon activity required to form carbides of at least one of the alloying elements [210]. In case of chromium-based alloys or ferritic steels, the main carbide forming element will be chromium. Carbon deposition or metal dusting can only occur if the carbon activity which establishes at the metal surface is larger than unity [210,211]. Carbon deposition may hamper the cell performance, for example, by blocking narrow gas passages, but it does not necessarily adversely affect the SOFC relevant properties of the interconnect. Metal dusting, however, is a process which commonly results in very rapid damage of metallic materials [210,211]. The mechanisms of the metal dusting process are still subject of scientific debate and a detailed description falls outside the scope of this chapter. The reader is referred to a number of references [21–24,210,211] in which the various steps in the degradation process are described.

In CO/H_2 – base anode side environments, the carbon activity will generally increase with decreasing temperature [23]. In real SOFC systems, the interconnect may be subjected to environments in which the oxygen partial pressure and the carbon activity substantially change as function of service time. During heating up the stack to the operation temperature, the interconnect is subjected to a gas with a high carbon activity and thus a substantial danger for occurrence of metal dusting exists, especially in the temperature range of 400–650 °C. When reaching the SOFC operation temperature of 800 °C, the carbon activity is lower than unity and metal dusting attack is no longer possible.

Due to the electrochemical reactions occurring at the anode side of the cell, the concentrations of CO, CH_4 and H_2 decrease in gas flow reaction, whereas the concentrations of CO_2 and H_2O increase. As shown in the Figure 7.7, chromia is the stable phase in equilibrium with commonly used SOFC gas atmospheres at 800 °C. It is commonly accepted that presence of a gas tight chromia-based surface scale will suppress or even completely stop carburisation and metal dusting [210,211]. However, it is well established that formation of internal chromium carbide, for example, $Cr_{23}C_6$, may occur even if the equilibrium carbon activity in the gas would be much smaller than the activity required to form the carbide [21]. This is related to a decrease in oxygen partial pressure and a parallel increase in carbon activity upon molecular transport of the gaseous carbonaceous species through the surface oxide layer [210].

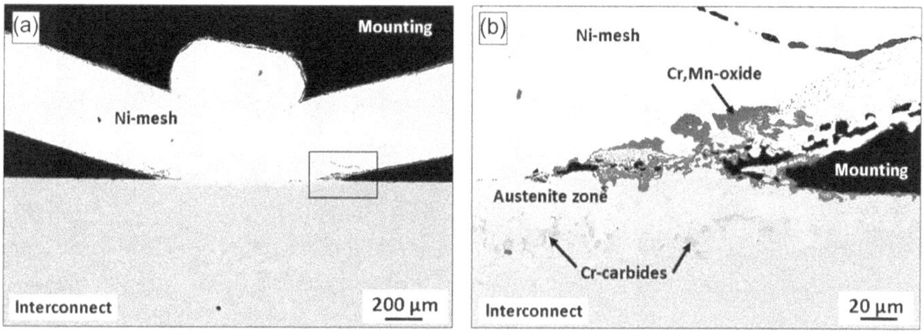

Figure 7.8 SEM/BSE images showing cross sections of contact between nickel-mesh and Crofer 22 APU interconnect in SOFC stack; (a) low magnification overview; (b) magnified area marked with frame in Fig. (a). The SOFC stack was operated for 4000 h at 800 °C with reformed methane. Chromium carbide formed in steel Crofer 22 APU as result of carbon uptake trough the nickel-mesh. Results from long-term SOFC stack test at Forschungszentrum Jülich (Stack F2036-04).

In case of ferritic steels, internal carburization does generally not result in formation of a continuous sub-scale carbide layer, but in formation of finely distributed carbide precipitates in the alloy matrix. Formation of chromium carbides such as $Cr_{23}C_6$ and/or Cr_7C_3 leads to material embrittlement and additionally depletes chromium from the alloy matrix in the surface near region of the alloy. The latter may result in a reduced chromium availability for transport to the surface required for protective scale formation and thus in a deterioration of the steel oxidation resistance.

Even if carbon uptake is prevented by a protective chromia scale formed on an interconnect, surface carburisation may occur in the same gas in a real stack. Niewolak et al. [23] showed that contacting the SOFC anode with the interconnect via a nickel wire mesh, as done in many SOFC designs, may promote carbon uptake of the ferritic steel interconnect. Figure 7.8 illustrates that presence of the nickel contact may result in transport of carbon from the gas into the steel, which is not commonly found when the bare steel is exposed to the anode side service environment [8].

7.7 Dual atmosphere exposures

In SOFC stacks, the metallic interconnect is simultaneously exposed to air at the cathode side and fuel at the anode side and thus experiences a chemical potential gradient from the fuel side to the air side. Nonetheless, the majority of laboratory oxidation tests of candidate materials for interconnect application are based on single atmosphere exposures in air or in various types of (simulated) anode gases.

For obtaining more realistic test conditions, a number of investigators introduced dual atmosphere testing [64,73,76,81,82,124,125,135,212]. Thereby, a piece of metallic material is on one side exposed to anode gas, whereas the other is subjected to air. A number of investigations have shown that under dual atmosphere conditions the oxidation behaviour of metallic materials on the cathode side may differ from that observed during single atmosphere exposures [76,135]. Dual atmosphere effects were mostly observed for a number of austenitic stainless steels [135] and for pure FCC metals such as nickel or silver [64,109]. In case of pure nickel subjected to dual gas condition (air/H_2–H_2O) exposure at 800 °C, diffusing hydrogen clearly changes the growth rate and morphology of the NiO scale at the air side [109]. The observations are in qualitative agreement with observations of Nakagawa et al. [213] who studied the oxidation behaviour of 2–12% Cr martensitic steels under conditions in which the inner surface of a tube was exposed at 600 and 650 °C to steam and the other side to air. The authors found clear evidences that hydrogen diffuses through the steel and changes the growth mechanism of the oxide scale on the air side [213].

A frequently reported phenomenon related to dual atmosphere exposures of high chromium ferritic steels is enhanced formation of iron-rich oxides on the sample surface on the air side [70,109,124,135]. The modified air side oxidation behaviour of ferritic steels under dual atmosphere conditions is similar to that found when such steels are exposed to wet air [81,207,212] or wet oxygen [97,151,175,205,214]. Kurokawa et al. [215] performed dual atmosphere exposures (air/Ar–H_2–H_2O) of the ferritic steel 430 at 800 °C and found virtually no influence of the hydrogen gradient on the air side oxide formation. A negligible influence of dual gas exposure was also reported for ZMG232 by Horita et al. [118]. Contrary to these findings are results from Yang et al. [76] who found significant differences in air side oxide scale morphologies formed on steels 430 and Crofer 22 APU during dual gas exposures (air/H_2–H_2O) at 800 °C in comparison to those formed in air. The authors pointed out that the mentioned effect decreased with increasing chromium contents in the steel and virtually disappeared for the 27%Cr steel E-Brite [76]. Rufner et al. [73] investigated steels 430 and 441 in dual gas exposure (air/H_2–H_2O) at 800 °C and reported a significant effect on the air side oxidation for the 18%Cr steel 441 and virtually no effect for the 17%Cr steel 430. Dual atmosphere exposures with Crofer 22 APU at 750 °C performed by Ziomek-Moroz et al. [216] on the same experimental setup showed significantly accelerated oxidation on the air side. The oxide scale formed during 200 h oxidation at 750 °C on the air side of the Crofer 22 APU had a thickness of about 10 μm. Such a thick oxide scale is comparable to that formed during more than 10,000 h single air exposure at 800 °C [91,101].

This evaluation illustrates that the literature data on the significance of dual atmosphere effects for various types of steel under simulated SOFC conditions are, at least to a certain extent, contradictory.

Dual exposure experiments are significantly more complex than single gas testing. They require a sophisticated experimental setup, and the results may be affected by sample geometry (thickness), gas leakage, undesired interactions of the tested material with system components (e.g. seals) and by the influence of the sample manufacturing procedure on oxidation properties. For instance, microstructural changes of the steel may occur as a result of a 3 h heat treatment at 1050 °C during soldering/sealing the samples. Effects of this parameter variation on oxidation properties were previously observed by Niewolak et al. [9,217]. Additional effects may occur due to the used sealing material. Wiener et al. [218] and Bram et al. [87] showed that volatile alkali elements present in a mica sealing can significantly accelerate oxidation of SOFC steels during dual gas exposure [87,218].

It is important to note that for high chromium ferritic steel interconnects dual atmosphere effects are, as far as known to the authors, not commonly observed in real stack experiments. In recent years, a substantial number of data became available for SOFC stacks with operation times up to 50,000 h [12]. According to the authors' knowledge, no clear experimental evidence for appearance of significantly enhanced oxidation as a result of the dual atmosphere conditions in these real SOFC stacks became clearly apparent.

7.8 Specimens thickness dependence of oxidation behaviour

Reduction of fuel cell stack weight is a major design requirement, especially if mobile applications of SOFCs are considered. Weight reduction can be achieved by decreasing the interconnect thickness down to a few tenths of a millimetre. When using such thin components of high chromium ferritic steels, it has to be considered that the long-term oxidation resistance is not only governed by the alloy chromium content but also by the chromium reservoir in the prevailing interconnect.

The growth of the chromia base surface scale consumes chromium from the steel with the result that after long-term service the initial chromium concentration in the component (C_0) may decrease to such a low level (C_B) that the growth of the protective scale can no longer be sustained. At this time (t_B), poorly protective iron-based oxides are being formed (breakaway oxidation). The time until the chromium content in a given steel component decreases to this critical concentration will obviously be shorter for thin than for thick components. Huczkowski et al. [177,183] used a simplified equation derived from studies on FeCrAl alloys to predict the time to breakaway (t_B) for selected high chromium ferritic steels as function of temperature and component thickness.

$$t_B = \left[2.3 \times 10^{-3}(C_0 - C_B) \cdot \frac{\rho d}{k}\right]^{1/n} \tag{7.6}$$

Here k and n are the oxidation parameters (Equation 7.3), d the component thickness and ρ the alloy density. In deriving this equation, it was assumed that the chromium depletion profiles beneath the oxide scales are virtually ideally flat. Validity of this assumption was experimentally proven for air oxidation in the temperature range of 800–900 °C. The reason for this observation is that in the given temperature range the ratio between oxidation rate k_p (in $cm^2\ s^{-1}$, Equation 7.1) and the diffusion coefficient (D) of chromium in the ferritic steel is quite small [183].

A major finding of Huczkowski et al. [177] was that not only the time to the start of breakaway oxidation (t_B) but also the intrinsic oxidation rate of the alloy depended on component thickness. Figure 7.9 shows k_w data for the high-chromium ferritic steels Crofer 22 A and Crofer 22 H oxidised discontinuously for 1000 h at 800 and 900 °C for specimens of different thickness [177,183]. The thickening rate of the two-phase oxide layer consisting of Cr_2O_3 and $(Mn,Cr)_3O_4$ formed on steel Crofer 22 APU appears to increase with decreasing specimen thickness. The difference in overall

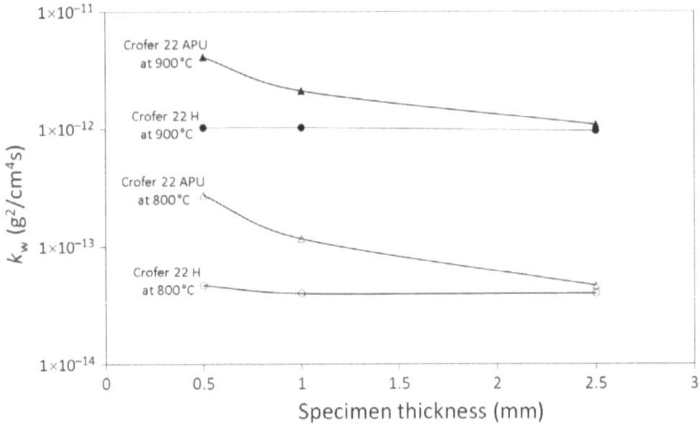

Figure 7.9 Instantaneous apparent oxidation rate k_w for the high-chromium ferritic steels Crofer 22 APU and Crofer 22 H oxidised discontinuously for 1000 h at 800 °C or 900 °C for specimens of different thickness.

scale thickness on high-chromium ferritic steels is mainly related to the difference in thickness of the inner chromia layer, whereas the thickness of the outer spinel layer varies only marginally [177,183]. The specimen thickness effect is, however, significantly less distinct for the steel Crofer 22 H [115]. Huczkowski et al. [177,183] and Zurek et al. [203] considered a number of mechanisms to explain the specimen thickness effect, including differences in minor alloying elements depletion as well as difference in oxide growth stress relaxation in thick and thin specimens. Asensio et al. [115] investigated the specimen thickness dependence of the oxidation rate on solution annealed and aged samples of Crofer 22 H at 800 °C in air. The results of the tests strongly indicate that the specimen thickness dependence is at least to a large extent affected by the intrinsic alloy creep strength of the steel. This finding, in combination with data in reference [91,96], indicates that differences in relief of oxide growth stress by plastic deformation of the thick and thin metallic substrate are responsible for the differences in oxidation kinetics between thick and thin specimens.

For an oxide scale under a biaxial compressive growth stress σ_{xx}, the diffusivity may be written as [203]:

$$D = A\ \exp\left(\frac{\Delta S^*}{R}\right)\exp\left(\frac{-\Delta H^*}{RT}\right)\exp\left(\frac{\frac{2}{3}\Delta V^*\sigma_{xx}}{RT}\right) \tag{7.7}$$

where ΔH^* and ΔS^* are the activation enthalpy and entropy, respectively, ΔV^* is the activation volume and A is a pre-exponential constant. Additionally, there is the possibility of a stress gradient over the oxide scale, whereby both cations and anions are transported from regions characterised by higher compressive stresses to regions where the stresses are lower. This results in a net transport of oxide into the tensile regions. Consequently, the vacancy concentration can be written as

$$C_V = C_V^0\ \exp\left(\frac{\sigma_h \Delta V^F}{RT}\right) \tag{7.8}$$

where C_V^o is the equilibrium concentration of vacancies in the unstressed state and ΔV^F the volume change associated with the formation of the relevant point defect. An estimate of the reduction caused by the effect of an assumed compressive stress [219] revealed for chromia that the effect of stress on the defect concentrations may be more significant than the direct effect on diffusivity [203]. This implies that the specimen thickness dependence is expected to differ from environment to environment in accordance with the observation that the effect is for ferritic steels far less pronounced in anode than in cathode gas [184]. The explanation also implies that an increase in creep strength of the ferritic steel would decrease the tendency for the oxidation rates to increase with decreasing component thickness, in agreement with the data in Ref. [184].

A further parameter which may affect the oxide scale formation is generation and relaxation of thermally induced stresses upon iterative heating and cooling. Cooling from SOFC operating temperature may results in initiation of a compressive stress in the oxide and a tensile stress in the metallic substrate. Consequently by a given oxide scale thickness, tensile stresses in the metallic substrate will increase with decreasing specimen thickness. It has been shown that creep of the metallic substrate may partly relax the stresses generated during cooling [219,220]. The creep relaxation is likely to be more substantial in a thin than in a thick specimen. Additionally, subsequent heating may generate a tensile stresses in the oxide scale resulting in microcrack formation in the oxide scale. This results in direct access of gas molecules to the metal surface leading to enhanced oxidation.

Important to mention that the enhanced oxidation would be expected to be more pronounced for thin components (thickness 0.5 mm or less) manufactured from an alloy with a poor resistance against creep [115,220].

7.9 Electronic conductivity of chromia-based scales

A crucial property of the interconnect materials is a high electronic conductivity. Since the electrical conductivity of metals is usually orders of magnitude higher than those for the respective oxides, the electrical conductivity of surface oxide scales is a major factor during selection of interconnect materials. Pure chromia is an electronic conductor and at temperatures around 1000 °C and above the electrical conductivity is claimed to be virtually independent of the oxygen partial pressure [221,222]. Below 1000 °C, the concentration of intrinsic point defects becomes so small that chromia changes into an extrinsic electronic conductor, the electronic conductivity being dominated by the presence of dopants or impurities [223–228]. The electrical conductivity of bulk chromia samples at 500–1000 °C has been reported to be in the range of $1 \times 10^{-2} - 1 \times 10^{2}$ S cm^{-1} [222] although lower values also have been found [227]. Substantially higher conductivity values were reported for chromia doped with other ions, for example, Ti, Mg, La or Li [223,226,227,229,230]. The doping also results in a complex dependence of the electrical conductivity on oxygen activity [223,226,227,229,230].

Beneath 1000 °C, magnesium-doped chromia was found to be a *p*-type conductor, whereby the electronic conductivity depends on oxygen partial pressure only at $pO_2 < 10^{-14}$ bar [222]. In contrast, chromia doped with M^{+4} ions such as Ti^{+4} is an *n*-type conductor at low oxygen partial pressures and a *p*-type conductor at near atmospheric oxygen partial pressure [222]. Nagai et al. [227,229] found that even doping with isovalent ions such as Y^{+3} or La^{+3} increases the electronic conductivity of chromia, whereby the conductivity decreases with increasing oxygen partial pressure, although this effect is not very pronounced, especially at high temperatures.

In H_2/H_2O mixtures, the conductivity could be affected by dissolved protons [222,224]. For MgO-doped chromia, protons dissolved in the oxide were claimed to decrease the conductivity, whereas in TiO_2-doped chromia, no effect of hydrogen on conductivity was detected [225]. The strongest increase of the conductivity was observed for doping by NiO, whereby, in this case, no oxygen partial pressure dependence was found. Also doping by CeO_2 or Fe_2O_3 increased the conductivity; however, the effect was much smaller than that observed for NiO-doping [227].

Most commercial ferritic steels as well as the steels specially developed for SOFC applications form scales containing Cr,Mn-spinel next to chromia. Naoumidis et al. [180] and later Garcia-Rosales et al. [181] showed that the exact structure and the chemical composition of Cr,Mn-spinel depends strongly on the oxygen partial pressure. At low oxygen partial pressures comparable to those prevailing on the anode side of the fuel cell, the stable form of spinel is tetragonal $MnCr_2O_4$, whereas at high oxygen partial pressures (e.g. in air), the spinel is cubic $(Mn,Cr)_3O_4$ [180,181]. Figure 7.10 shows electrical conductivity data for Mn,Cr-spinels derived from various literature sources. Generally, the electrical conductivity of the cubic spinel $(Mn,Cr)_3O_4$ measured in air is higher than that of pure chromia in air or the tetragonal spinel $MnCr_2O_4$ measured in H_2–H_2O atmospheres. Lu et al. [231] showed, however,

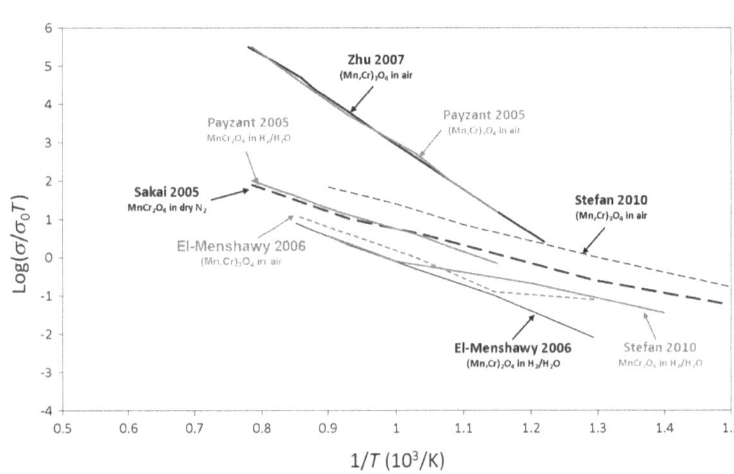

Figure 7.10 Electrical conductivity vs. reciprocal temperature for different Mn,Cr-spinels derived from various literature sources [68,231–234].

that the influence of the Mn:Cr ratio on the conductivity of the spinel is more significant than the influence of the oxygen partial pressure.

It is important to note that electrical conductivity data for oxides with defined chemical composition being in equilibrium with the surrounding oxygen partial pressure should be used with care when estimating properties of surface oxide scales. Strictly speaking, the surface scales formed on the various materials cannot be characterised by a single conductivity value because the scales do not consist of a one single oxide phase. Additionally, the scales contain impurities and they grow in an oxygen partial pressure gradient. According to the Wagner theory of metal oxidation [200], the oxygen activity at the metal/oxide interface equals the dissociation pressure of the oxide in equilibrium with the scale forming element in the alloy. Considering the case of a chromia scale grown in air, the corresponding oxygen activity (aO_2) at the metal/oxide interface varies between approximately 1×10^{-37} at 600 °C and 1×10^{-23} at 1000 °C, whereas the oxygen activity at the outer oxide/gas interface remains virtually constant at the level of 0.2. The latter may, however, not be true if a platinum contact layer, frequently used for conductivity measurements, is applied on the oxide surface [27,77,88,90,126,146,190,193]. In this case, the oxygen activity at the oxide/platinum layer interface remains undefined and depends on the oxygen permeability of the platinum layer. Under these conditions, the oxide scale may partially decompose and dissolve in the platinum layer [152]. Huczkowski et al. [190], Ertl [235] and Froitzheim [190,193,235] found that oxide scales formed on Crofer 22 APU may dissolve in metallic layers of, for example, nickel or silver applied on pre-oxidised samples (Figure 7.11).

A further factor which affects oxide scale conductivity is that the detailed composition of the scales usually varies as function of distance from the oxide/gas interface. Additionally, the element distribution within the oxide scales growing on interconnect steels may change with time and environment [25,29,72,101,117,119]. Therefore, electrical conductivities of growing scales can, strictly speaking, not directly be compared with the conductivities of bulk oxides.

If for a certain chromia forming alloy, the parabolic oxidation rate constant k_p (Equation 7.1) and the electrical conductivity σ of the oxide scale are known, and it is possible to estimate the contact resistance (ASR) of the oxide scale after a given oxidation time.

$$\mathrm{ASR} = \rho \times x \tag{7.9}$$

where $\rho = 1/\sigma$ and $x = (k_p \times t)^{1/2}$.

The area-specific contact resistance (ASR) is a more technologically relevant property and can easier be estimated than the conductivity since it does not require knowledge about the oxide scale thickness. Figure 7.12 shows a comparison of ASR values for selected interconnect steels measured during air exposure at 800 °C. The ASR values for oxide scales formed on materials especially designed for SOFC interconnect applications such as Crofer 22 APU are lower than those for other commercially available ferritic steels such as 1.4016. This is related to the precisely controlled chemical composition of the steel Crofer 22 APU which exhibits at 800 °C (see

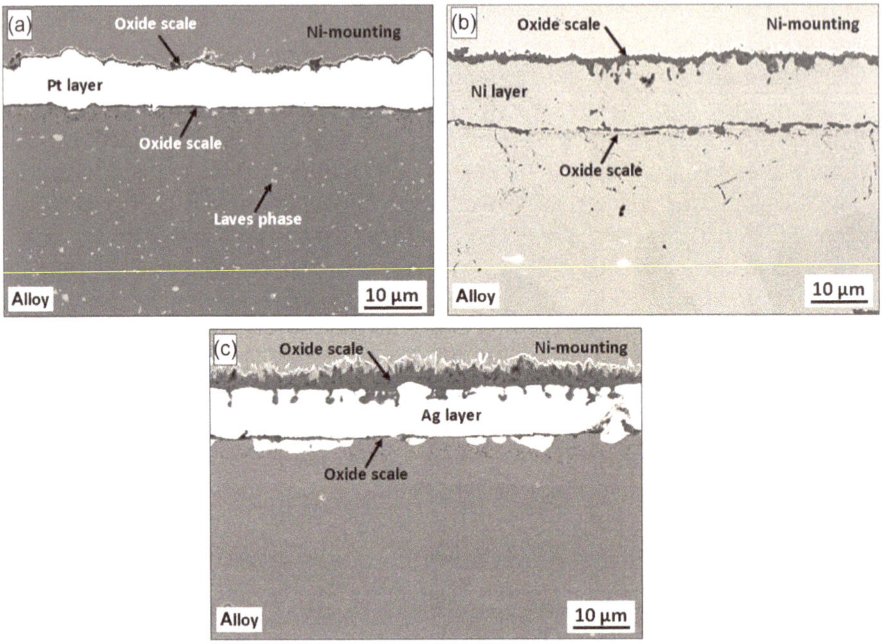

Figure 7.11 SEM/BSE images showing cross sections of oxides scales on Crofer 22 H (a) and Crofer 22 APU (b, c) coated with various metallic layers platinum (a), nickel (b) and silver (c). The samples were subsequently oxidised for 1000 h at 800 °C in simulated anode gas Ar–4% H_2–2%H_2O. Dissolution of oxide scales beneath the coating, and re-formation on the coating surface are clearly visible.

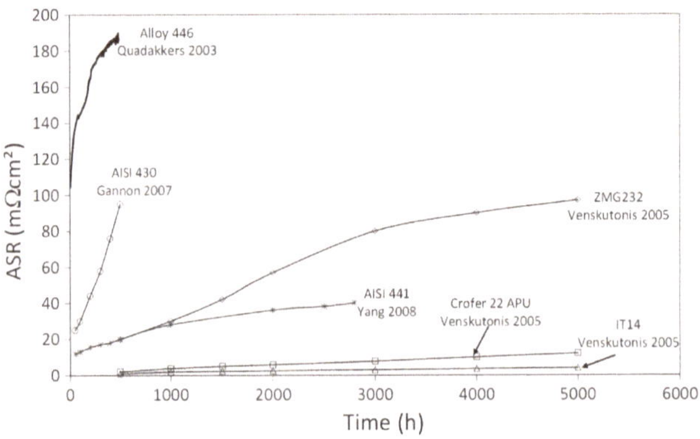

Figure 7.12 Area-specific resistance (ASR) of oxide scales on various interconnect steels at 800 °C in air derived from various literature sources [58,71,127,236].

Section 7.5) low ASR values between 10 and 40 mΩ cm^2, even during long-term testing (Figure 7.12).

It is important to mention that ASR values of oxide scales such as those shown in Figure 7.12 should be applied with care for real SOFC application because nearly all current stack designs utilise coated interconnects especially on the cathode side (see Section 7.14), and nickel mesh current collectors spot welded directly to the steel interconnects on the anode side. In the latter case, the electrical current thus flows from the cell via the nickel mesh into the interconnect, bypassing completely the oxide scale (Figure 7.8).

7.10 Volatile species and protection against chromium evaporation

During operation of SOFCs with ferritic steel or chromium-based alloy interconnects, rapid degradation of the cell performance has frequently been found [165,237]. As mentioned in Section 7.4.2, the effect was claimed to be caused mainly by chromium poisoning of the SOFC cathode and the cathode/electrolyte interface [25,95,145,237–240]. During operation at high temperature, volatile chromium-containing species are released from the interconnect surface (see Section 7.4.2); in the following, this process will be designated as “chromium evaporation”.

Hagel [241] found that the weight loss of Cr_2O_3 samples exposed at high temperature to oxygen is linearly dependent on time. Tedmon [242] applied these data and estimated that for oxidation of chromia forming alloys there is a point in time when the weight uptake resulting from chromium oxidation is equal the weight loss resulting from chromium evaporation. After that time, the thickness of the chromia scale remains virtually constant but the measured weight gain continuously decreases. Figure 7.13 shows the mentioned effect for the ferritic ODS steel ITM during discontinuous oxidation at 800 °C in laboratory air. The rate of material loss by volatilisation

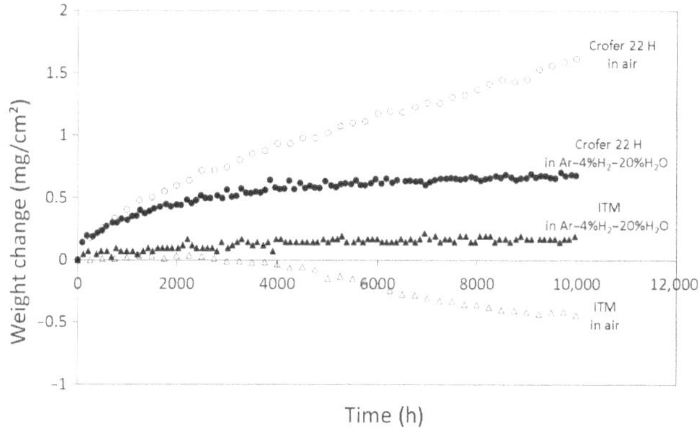

Figure 7.13 Weight change vs. time during long-term discontinuous oxidation of steels ITM and Crofer 22 H at 800 °C in air and in simulated anode gas Ar–4%H_2–20%H_2O [191].

is highly dependent on the partial pressures of the volatile species as well as the gas flow rate [106,108,189]. Caplan and Cohen [243] have shown that even small amounts of water vapour may substantially increase the chromium evaporation rate in high pO_2 environments. It has been experimentally proven unequivocally, that in wet air volatile $CrO_2(OH)_2$ is formed by reaction 5 in Section 4.2 [19,244–246].

Ebbinghaus [246] derived thermodynamic data for a large number of volatile chromium oxyhydroxides using the number of available experimental data available at that time. In recent years, several authors investigated chromium evaporation from metallic materials under SOFC relevant conditions, mostly in wet air (Table 7.3). The values derived from various sources cannot be directly compared due to the differences in experimental parameters such as flow rate and/or humidity. One may, however, conclude that the majority of the chromium evaporation rates measured at 800 °C are in the range of $2\text{–}9 \times 10^{-10}$ kg m^{-2} s^{-1} (Table 7.3).

Detailed analysis of data from Stanislowski [106,108] and Sachitanans [93] shows that chromium evaporation from ferritic steels decreases approximately linearly with increased manganese content in the material (Figure 7.14). The effect is in qualitative agreement with calculations from Holcomb [134], which show that chromium evaporation rates from Mn,Cr-spinel are 18–95 times lower than those for chromia. The effect results from a coverage of the chromia scale by an outer $(Mn,Cr)_3O_4$ layer [101,190], as assumed, for example, in the development of the steel Crofer 22 APU [192]. When comparing data for different sources, it should, however, be kept in mind that overgrowing of chromia by spinel is a time-dependent process which differs from alloy to alloy and depends on temperature as well as environment.

Several protection methods have been proposed to minimise evaporation of volatile chromium-species, as will be discussed in detail in Section 7.13.

Apart from chromium, three other common alloying elements (silicon, manganese and nickel) may form volatile species under SOFC relevant conditions [249–252]. Generally, increased humidity and/or decreased oxygen partial pressure favour formation of volatile silicon species. At low pO_2 such as those prevailing at the SOFC anode side, the predominant species are SiO and $Si(OH)_4$ [249]. Redeposition of electrically insulating silica on YSZ material was recently identified as an important degradation mechanism of the SOFC anode [251]. Ceramic refractory silicate materials or sealing glasses were identified as possible silicon sources [251]. There are, however, no data indicating that minor amounts of silicon present in most ferritic interconnect materials could be responsible for such a degradation mechanism.

Meschter et al. [250] reviewed available thermodynamic data for several technical important metals regarding their ability to form volatile species at high temperatures. The authors pointed out that currently available literature data do not allow to derive unequivocal conclusions regarding manganese volatilisation under SOFC relevant conditions. On the bases of presently available thermodynamic data, volatile MnH or MnO should be considered in simulated anode gas. However, as far as known to the authors virtually, no data are available in the literature regarding the significance of these volatile species for stack performance.

Table 7.3 Chromium evaporation rates for various alloys considered for SOFC interconnect

Source	Temp. (°C)	Humidity (vol.%)	Flow ($dm^3\ min^{-1}$)	Evaporation rate ($kg\ m^{-2}\ s^{-1}$) rate								
				Cr_2O_3	$Cr5Fe1Y_2O_3$	IT14	441	E-Brite	ZMG232	San.HT	22 H	22APU
[247]	800	7	5–12	4.7×10^{-10}	–	–	1.1×10^{-10}	3×10^{-10}	9.1×10^{-11}	–	–	3.5×10^{-10}
[108]	800	1.9	0.13	–	6×10^{-10}	1.6×10^{-10}	4.8×10^{-10}	6.8×10^{-10}	2×10^{-10}	–	–	2.5×10^{-10}
[248]	850	3	6	–	–	–	–	–	–	6.6×10^{-10}	–	–
[136]	850	3	0.3	2×10^{-3}	–	–	–	–	–	–	–	–
[112]	850	3	6	–	–	–	6.6×10^{-10}	–	–	8.8×10^{-10}	–	–
[93]	850	3	6	–	–	–	–	2×10^{-9}	9.4×10^{-10}	8×10^{-10}	6.6×10^{-10}	4.7×10^{-10}
[129]	850	–	–	–	7×10^{-10}	–	–	–	–	–	–	–
[86]	800	1.9	1.3	–	6×10^{-10}	–	–	–	–	–	–	2×10^{-10}

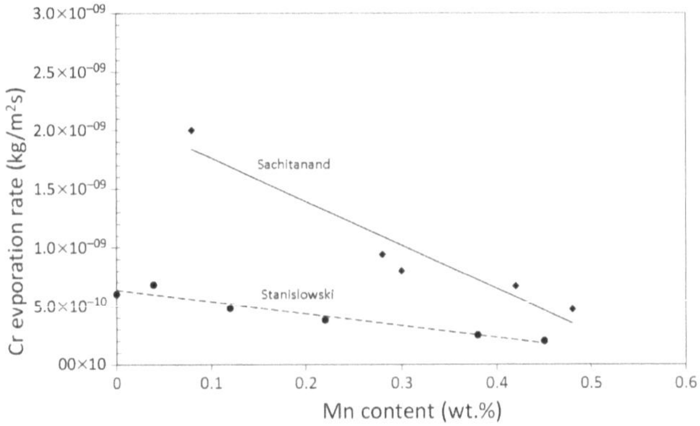

Figure 7.14 Chromium evaporation rates vs. Mn-content for various alloys considered for SOFC interconnect [93,108].

7.11 Interaction between interconnect and anode side contact materials

In currently used SOFC stacks, the interconnect on the anode side is in direct contact with the Ni/ZrO_2 cermet anode or, in most cases, with a nickel current collector [23,253,254]. The nickel current collector has usually the form of a wire mesh which provides the electrical contact between the interconnect and the anode. The elemental activity gradients between interconnect material and the nickel parts result in the diffusion of elements such as iron, chromium and manganese from the interconnect into the nickel. If, for instance, the nickel mesh is spot-welded to the interconnector, excellent contact resistance values are initially observed [253]. In case of a chromium-based interconnector, the interdiffusion can result in excessive Kirkendall void formation at the Cr/Ni-interface deteriorating the contact of both materials. Nickel plating of the interconnector has been applied to reduce this problem [255]; however, extensive outward diffusion of chromium into the nickel with subsequent chromia formation on the nickel layer during long-term exposure could not be prevented.

Interdiffusion between a ferritic steel interconnect and the spot-welded nickel wire mesh will lead to Ni transport into the steel resulting in local austenite formation (Figure 7.15) [22,193]. Niewolak et al. [23] found that in case of high chromium steels (Cr > 22%), interdiffusion with nickel may result in a further phase transformation. For example, the ferritic steel Crofer 22 APU exhibited formation of σ-phase near the interface between the original ferritic matrix and the austenite layer during anode gas exposure at 600 °C [23,116].

Transformation of ferrite into austenite and σ-phase as a result of interdiffusion with the nickel layer may have a number of important consequences for the overall material behaviour. The slower chromium diffusion in the austenite may affect the formation and growth of the protective chromia base surface scale [9,66].

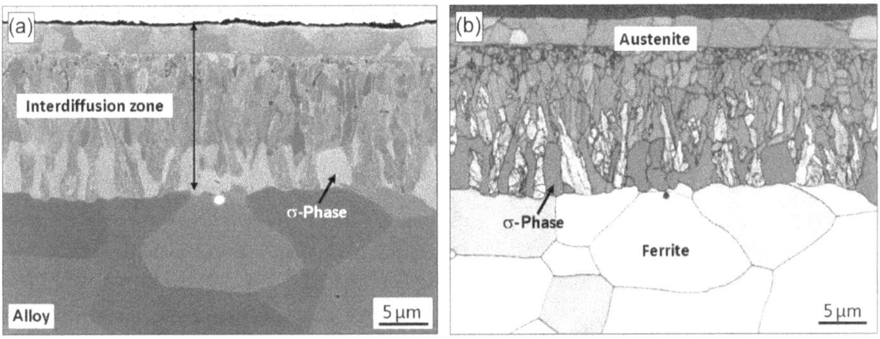

Figure 7.15 SEM/BSE (a) and EBSD phase map (b) pictures of nickel-coated Crofer 22 APU sample after 300 h oxidation in simulated anode gas at 600 °C.

Additionally, the austenite possesses a higher CTE than the ferrite which may result in generation of mechanical stresses during thermal cycling [58,132,142,148,149]. Presence of the brittle σ-phase may promote crack formation and thus decrease the overall performance of the interconnect. These property changes as a result of interdiffusion are expected to be of special concern if very thin interconnects are being used. Generally, the tendency to σ-phase formation increases with decreasing temperature and increasing chromium content (see Section 4.4).

Vice versa, transport of iron, chromium and other steel constituents into the wire mesh renders the latter into an alloy which will tend to form surface oxide layers in the anode side environment which may eventually even result in a decreased catalytic activity of the cermet anode, depending on the prevailing design [193,235].

If a carbon containing fuel is being used, the presence of the nickel contact may result in transport of carbon from the gas into the steel, which is not commonly found when the bare steel is exposed to the mentioned anode side service environment (see Section 7.6).

Pre-oxidation layers have been examined as possible barriers to minimise interdiffusion processes between the interconnect and the nickel current collector or the Ni/ZrO_2-anode. The pre-formed oxide scales tended, however, to become dissolved in the nickel [23,193,235] so that after longer exposure times the unwanted interdiffusion processes appeared again (see Section 7.9). A number of externally applied diffusion barrier layers have been proposed [56,128]. Quite promising results were obtained with sputtered ceria layers of a few micrometre thicknesses which virtually completely suppressed the interdiffusion phenomena [56]. During long-term operation, however, the outward growing Cr,Mn-spinel on the steel deteriorated the contact between oxide scale and ceria layer. An improvement of the electrical contact could be obtained by applying a duplex layer of ceria and copper on pre-oxidised ferritic steel. The high oxygen solubility in copper suppresses the dissociation of the oxide layer and the geometrical deterioration of the contact was reduced because the geometrical irregularities were compensated by plastic deformation of the copper layer [56].

7.12 Interaction of metallic interconnects with sealing materials

Planar SOFC sealings provide mechanical bonding of components and prevent mixing of fuel and oxidant. In most designs, they electrically insulate subsequent cells. In recent years, a number of sealing materials such as glasses, metallic brazing alloys or mica have been tested [25,87,218,253,256–258]. The state-of-the-art glass sealants used in SOFCs are based on the system barium-calcium-aluminosilicate (BCAS) [85,253]. The glasses frequently contain additionally minor amounts of other oxides, for instance to increase the wetting properties. Strictly speaking, the glass sealants used in SOFCs are glass-ceramic materials since they consist of glassy and crystalline phases [85,259].

The extent of reaction of high-chromium ferritic steels with the glass-ceramics depends on the exposure conditions and the distance of the interface of sealing glass and ferritic steel to the ambient air [85,257,259]. Near the three-phase boundary steel/glass/air, volatile chromium species may react with BaO thereby forming $BaCrO_4$ [146,253,258]. The presence of volatile earth alkaline or alkaline species could further enhance the tendency to breakaway oxidation. The effect may be especially significant at lower operating temperatures [87].

In the interior seal regions, hardly any access of oxygen from the air is possible and thus only a very minor reaction at the steel/glass interface occurs. Under these conditions, chromium tends to slightly dissolve into the glass [85,257,259]. The extent of reaction depends on the steel and glass composition. Crofer 22 APU was found to exhibit better chemical compatibility and bonding with the BCAS-based glass-ceramics than most other ferritic steels due to the growth of its specific oxide scale on the alloy surface during high-temperature exposure [258,260]. A further suppression of interface reactions and $BaCrO_4$ formation was obtained when alumina formation by, for example, local aluminising [261] or a (Mg,Al)-spinel coating were used [74,218].

More severe reactions may occur if the sealing material contains minor amounts of wetting agents such as PbO or ZnO. These oxides are thermodynamically less stable than the other glass constituents and could be easily reduced when exposed to anode gas. As confirmed by laboratory tests [260], the presence of lead vapour in equilibrium with liquid lead results in severe grain boundary attack in the high-chromium ferritic steels. This is accompanied by excessive internal oxidation of chromium and rapid growth of iron oxide on the ferritic steels near the triple-phase boundary air/glass/metal [85,260]. The high electronic conductivity of the iron-based oxide resulted in a short-circuiting of the stack. Haanappel et al. [85] showed that the trigger of this detrimental effect by lead vapour occurred near the triple-phase boundary at the anode side of the stack. Due to the low pO_2, no external iron oxide formation occurred in the anode side gas; however, the internal oxidation resulted in a large volume increase and thus in a local bulging of the steel [85,260]. The rate at which this lead vapour induced reaction occurred, appeared to depend on the amount of the minor alloying additions silicon and aluminium in the ferritic steel [85,260]. High purity steels (for instance,

Crofer 22APU) with very low aluminium and silicon contents in the range 0.01% were less prone to lead-vapour attack than steels with more common aluminium and/or silicon contents of approximately 0.1%.

In the recent years, a number of attempts have been made to use metallic brazing alloys of the Ag–Ti–Cu type as a sealing for SOFCs. The advantage offered by metallic sealings over glass-ceramics is their higher ductility which makes them less prone to failure by thermally induced stresses. The drawbacks of the silver-based metallic sealings are, however, their high electrical conductivity and high gas permeability [64,256,258]. Because of the high electrical conductivity, application of an additional locally applied electrically insulating coating on the interconnects is required. The high permeability for gases such as oxygen and especially hydrogen results in formation of steam voids within the brazing metal during service at high temperatures (see Section 7.4), resulting in deterioration of mechanical strength and gas tightness of the metallic sealing [64,256].

7.13 Protective coatings and contact materials

Besides all the suitable properties, metallic interconnects have two major disadvantages: Ohmic losses due to formation of surface oxide scale and release of volatile chromium species (see Section 7.10). The latter is nearly temperature-independent between 600 and 900 °C. In atmospheres containing water vapour, the most volatile compound is a chromium acid, $CrO_2(OH)_2$ [262], which is transported with the oxidant gas through the cathode to the cathode/electrolyte interface. In the case of (La,Sr) MnO_3 (LSM) cathodes, the Cr^{+6}-containing molecules compete with the oxygen molecules for the electrochemically active sites and block them after reduction to Cr^{+3} [262]. This results in an increase in cathode polarisation [263]. After the initial blocking of the electrochemically active sites, the ongoing transport and reduction of chromium species lead to decomposition of the perovskite material and the formation of $(Mn,Cr)_3O_4$ spinels [165,264]. A detailed study of the degradation of LSM cathodes has shown that the electrochemical voltage loss over time is strongly dependent on the applied current density and the kind of interconnect coating [240]. The frequently observed initial and quick voltage loss and the following slow linear voltage loss might be attributed to the different reaction processes of chromium poisoning, that is, the blocking of reactive sites and the perovskite decomposition, respectively.

Matsuzaki and Yasuda reported that $La_{0.6}Sr_{0.4}Fe_{0.8}Co_{0.2}O_3$ (LSCF) cathodes are much more stable against chromium poisoning than $La_{0.85}Sr_{0.15}MnO_3$ or $Pr_{0.6}Sr_{0.4}MnO_3$ cathodes [238], especially when ceria-based solid electrolytes are used, and no enrichment of chromium was found at the $La_{0.6}Sr_{0.4}Fe_{0.8}Co_{0.2}O_3$/ceria interface after 10 h. This has been proven in a long-term stack test lasting for about 2 years [265]. A possible explanation might be the different overvoltages of the cathodes for oxygen and $CrO_2(OH)_2$ reduction. While for the manganites, the reduction of the chromium oxyhydroxide is the energetically preferred reaction, in the presence of the ferrites the reduction of oxygen appears to require less activation energy. Instead, the LSCF cathode degraded by a steady release of strontium from the perovskite lattice leading to a continuous depletion of strontium by 50% at the outer surface after 2 years of operation. The released strontium, very likely as volatile $Sr(OH)_2$, reacted with the gaseous $CrO_2(OH)_2$ and formed $SrCrO_4$ crystals at the outer surface of the cathode [265], very similar to the results in model experiments where an LSCF cathode was attached to steel interconnects [239,266]. The overall chemical reactions can be written as

$$La_{1-x}Sr_x(Co,Fe)O_3 + yH_2O \rightarrow \left(La_{1-x}Sr_{x-y}, \bullet_y\right)(Co,Fe)O_{3-y} + ySr(OH)_2 \tag{7.10}$$

$$CrO_2(OH)_2 + Sr(OH)_2 \rightarrow SrCrO_4 + 2H_2O \tag{7.11}$$

in which • denotes a Sr-vacancy. Here, no redox reaction takes place as in the case of LSM cathodes, but a chemical transport reaction of two gaseous hydroxyl compounds forming a solid ($SrCrO_4$).

The extent of chromium evaporation from steels or other alloys and the effectiveness of protective coatings can be estimated before stack assembly by transpiration experiments [108]. Such investigations are useful in materials development of new interconnects and also for an understanding of degradation rates of cells and stacks.

A second major disadvantage of metallic interconnects is the formation of oxide scales leading to significant ohmic losses. The interaction of the metallic interconnect with the adjacent ceramic cell components and the resulting time-dependent resistance of these material combinations is highly important for stack performance. In the case of $Cr_5Fe_1Y_2O_3$, the best contact material for the cathode was found to be $LaCoO_3$ [267] before protective layers of $(La,Sr)CrO_3$ were applied to avoid chromium evaporation. $LaCoO_3$ was also successfully used in combination with ferritic steel [268], although the thermal expansion coefficient of the cobaltite is higher [268,269] than that of the other SOFC materials [132]. Besides the lower contact resistance (Figure 7.16), $LaCoO_3$ appears to react with the released chromium species to form a $La(Co,Cr)O_3$ perovskite and therefore retains some of the chromium vapour. During the last 10 years a wide variety of perovskites, spinels and binary oxides have been tested. An extended collection of coating materials and their properties in combination with metallic interconnects can be found in a review article on this subject [74].

From the aforementioned examples and technological needs, it is evident that coatings on an interconnect serve as protective coating against chromium poisoning as well as contact coating for minimising the area-specific resistance of the metal/ceramic interface. In addition, the coatings have a significant impact on the oxide scale growth due to interdiffusion of cations as well as on the adhesion of the ceramic (oxide scale) to the metal. However, on the one hand, not all coatings which show excellent low contact resistances are also applicable for chromium retention. And even the so-called protective coatings are not as protective against chromium poisoning as they should be when they are tested for thousands of hours [168]. On the other hand, an effective chromium retention is, in some cases, only achieved at the expense of an elevated contact resistance, for example, in the case of cobalt oxide [270].

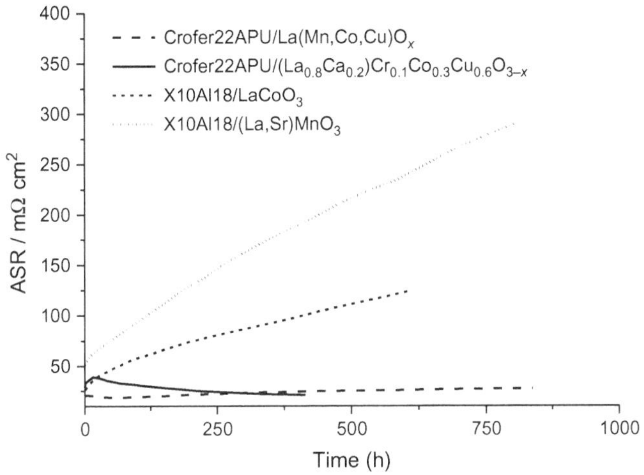

Figure 7.16 Change of electrical resistance of ferritic steel/perovskite ceramics combinations during exposure in air at 800 °C.

A chemical interaction between the chromium retention/contact layer and an electrode as well as the interconnect should not occur but cannot be avoided in most cases due to the reaction of the contact material with the chromia scale formed on the interconnect. In all cases where alkaline earth-containing perovskite materials have been used, the formation of chromates was observed [67,267,271] leading to progressive decomposition of the perovskite material. The change in contact resistance (Figure 7.16) is not only due to the scale formation on the surface of the interconnect but also driven by the reaction between the oxide scale and the volatile chromium species with the contact material, by the formation of alkaline-earth chromates and the steady depletion of material at the contact material/interconnect interface due to the volatility of these chromates. The latter process was demonstrated by Hou et al. [272] applying different cathode materials – (i) platinum, (ii) $La_{0.6}Sr_{0.4}CoO_3$, (iii) $La_{0.85}Sr_{0.15}MnO_3+La_{0.8}Sr_{0.2}Ga_{0.83}SMg_{0.17}O_3$ – onto an un-oxidised iron-based alloy with composition similar to X18CrN28 (1.4749). They found that the area-specific resistance of the cobaltite specimen increased at a greater rate than for the other two material combinations although the cobaltite is more conductive than the manganite/gallate mixture. For the contact material, it is important to have not only an initial low contact resistance but also a constant resistance with time (or even a decreasing resistance as shown in Figure 7.16).

At this point, several questions arise: Which kind of layer is a protective coating and which is rather a contact coating? Do the investigated materials serve for both requirements? What kind of layer shall be used for a specific application? Which layers are applicable and have been proven to be efficient for one or both purposes? For answering these questions, different cases with respect to cell design and application are necessary to be discussed.

7.13.1 Short-term applications with SOFCs having an LSCF cathode

Considering the fact that LSCF cathodes are less prone to chromium poisoning (see above) and stacks consisting of SOFCs with LSCF cathodes show voltage losses of a few percentages per 1000 h of operation [273], the interconnect coating can be preferentially optimised with respect to the contact resistance, because cell degradation is less important due to the limited period of operation. Here, the application of thin films of rare-earth oxides [79,274,275] or thick coatings of perovskites [276–278] has resulted in low contact resistances which are suitable for short-term automotive (auxiliary power units) or military applications. In the case of using rare earth oxides, the corresponding rare earth chromites are formed, but they do not show strong chromium retention. The ASR of units composed of cathode/perovskite coating/interconnect are far below 50 $m\Omega/cm^2$, but also do not show significant chromium retention, except when the perovskite coating can be applied as (nearly) dense layer [279].

7.13.2 Short-term applications with SOFCs having an LSM cathode

Since the LSM cathode is more sensitive to chromium poisoning and can show high voltage losses within few thousands of hours [240,273] without protective coating on the interconnect, short-term operations might be possible, but the application of an interconnect coating with measureable chromium retention [107] is very much recommended. Typically, binary oxides can be used in this case [107,280] and have shown sufficient long-term stability of stack operation [168]. The use of binary oxides like MnO_2, Co_3O_4 or CuO has three main technological advantages: rather low costs compared to complex oxides, wide availability and the possibility to apply them as metal coating using electro-deposition techniques. However, these oxides do not completely hinder the chromium to reach the cathode, and therefore, the chromium poisoning is retarded, up to several thousands of hours, but not eliminated. When the metals or corresponding oxides are applied as layers of only a few microns, an additional contact coating is useful to improve the contact resistance [77]. Contact materials are used in stack assembly not only for better electrical contact between cathode and interconnect as mentioned before, but also for compensation of dimensional tolerances of the construction parts. Such contact layers have no direct role in electrochemical reactions, but they can provide a homogeneous contact over the whole area of the fuel cell and minimise the ohmic losses within the stack.

Since there are no electrochemical requirements for the contact materials, they can be different from the electrode materials and be selected on the basis of their electrical conductivity and thermal expansion. Lanthanum cobaltites have high specific conductivities, up to 1700 S/cm [269]. However, as already mentioned, the thermal expansion of these cobaltites has a large mismatch with the other cell components as mentioned previously. For these ceramic contact materials, therefore, a compromise between acceptable conductivity and tolerable mismatch in thermal expansion is generally required. Therefore, the approach to use the cathode materials also as interconnect coating [277] is a consequent alternative as long as strong chromium retention is not the main target.

The maximum assembling temperature depends on the interconnect material used. In the case of $Cr_5Fe_1Y_2O_3$ interconnects, sintering can be utilised for stack assembly providing good contact without any contact material due to the high melting point (~1700 °C) of the alloy [157]. However, with ferritic steels as interconnects, the stack assembly temperature cannot be higher than 900–950 °C due to enhanced oxidation and thus, contact material with sintering ability at these temperatures is needed for good electrical contact.

7.13.3 Long-term applications with SOFCs having an LSCF or LSM cathode

For long-term operation, irrespective of application and type of cathode, very effective chromium retention is necessary to guarantee a stable performance and efficiency of the SOFC unit for several 10.000 h. Assuming an initial performance of 0.5 W/cm^2 and an end-of-life criterion of 30% performance loss, Figure 7.17 shows the impact of different degradation rates on the lifetime of an SOFC unit. As can be seen very clearly, a degradation rate of far less below 1%/1000 h is necessary to warrant a service life of about 5–10 years (45.000–90.000 h).

There exist only few examples of long-term stack or system tests in the open literature although several companies are commercialising small SOFC units for household applications or auxiliary power units for transportation [281–285]. The longest stack tests with planar SOFCs are currently ongoing at Forschungszentrum Jülich. One stack with manganese oxide as chromium protection layer and a perovskite as contact coating [12] is now in operation for more than 63,000 h showing an overall degradation rate of 0.8%/kh. The perovskite has been selected from the quasi-ternary system $LaMnO_3$–$LaCoO_3$–"$LaCuO_3$" [286,287]. A second stack has been started later

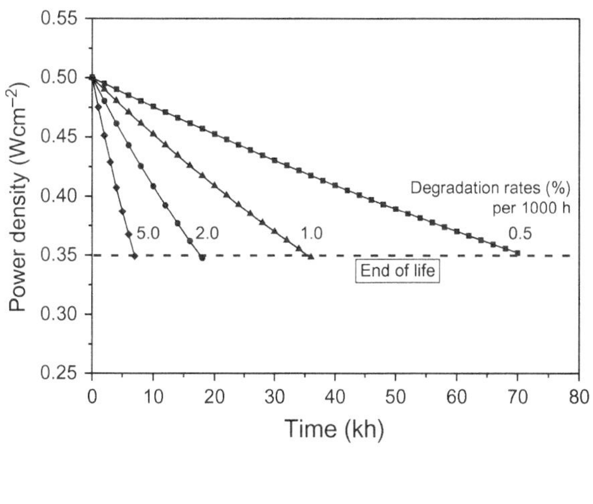

Figure 7.17 Dependence of different degradation rates on the power density and lifetime of an SOFC unit.

with a plasma-sprayed spinel coating as chromium protection layer [288] and the same perovskite as contact coating as in the forerunner. At the time of writing (status January, 2015), it has reached now 35,100 h with a degradation rate of 0.2%/kh [12]. Therefore, conductive spinels in combination with a nearly dense plasma-sprayed microstructure can comply with the everlasting target of long-term stability. On the one hand, the spinels have shown reasonable conductivity to achieve low ASR values, low or no diffusion of chromium into the spinel structure preserving the low ASR and effectively blocking the chromium diffusion into the solid. On the other hand, the plasma spraying creates a nearly dense layer directly at the start of operation and deposition-induced microcracks and pores are closed during crystallisation of the initially rather amorphous layer at operation temperature. This leads not only to an effective retention of the volatile chromium molecules but also to a reduced access of oxygen to the ferritic steel resulting in a reduced oxide scale growth.

In the recent years, much work has been spent on the characterisation and applicability of spinels. The most favourable compositions have been identified in the systems $(Mn,Co)_3O_4$ [288–290], $(Cu,Mn)_3O_4$ [290,291], $(Mn,Co,Fe)_3O_4$ and $(Cu,Ni,Mn)_3O_4$ [133].

During the recent two decades, a wide variety of deposition techniques have been used to apply ceramic coatings on interconnects. Many of these techniques are not of technological relevance, because of the limited size of area that can be coated, because of non-uniform coatings, especially when interconnects with gas channels or other structures are taken into account or because of the high costs related to starting materials or equipment. The following table (Table 7.4) gives a rough overview of the coating techniques and their relevance to SOFC technology. Since interconnects usually contain gas channels, a parameter tentatively indicating the uniformity of the coating on top and on the walls of a rectangular gas channel is given as the thickness ratio

$$t = t_{top}/t_{wall} \tag{7.12}$$

In the case $t = 1$, a very homogeneous coating with constant thickness on horizontal and vertical areas is achieved. If $t > 1$ or $t = \infty$, the walls are coated with smaller thickness than the horizontal faces or not coated at all, respectively.

Among these methods, electro-deposition, that is, electroplating, is very attractive due to the low cost of this process. Another important feature of electroplating is its capability to coat substrates with complex geometries, like the SOFC interconnect with rectangular-shaped gas channels [270]. Furthermore, the coated metallic layer can be *in situ* oxidised when the SOFC stack is started so that no extra processing steps are necessary to form spinel oxides on the interconnect surface. Recently Co, Co–Mn and Cu–Mn coatings have been successfully deposited by electroplating on ferritic stainless steels like UNS430 for SOFC interconnect applications [270,275,291,300]. Electroplating yields very homogeneous coatings with only little variation in thickness as long as the part that has to be coated is exposed to a homogeneous electrical field in the electroplating bath. Therefore, this method is not only very attractive for interconnects but also for other system components which are made from ferritic steels and are exposed to high temperatures.

Table 7.4 Coating techniques for chromium protection layers and their characteristics

Coating technique	Technological relevance	$t = t_{top}/t_{wall}$	Disadvantage	References
Electro-deposition	++	1	Difficulty for binary and ternary alloys, sequential coating laborious	[270,275,290–292]
Slurry coating, wet powder spraying	++	≈3–10	Porous coatings	[271,275,280,289]
Aerosol deposition	++	1–5		[279,293]
Screen printing	+	∞	Porous coatings	
Plasma spraying	+	≈2–8	Expensive process	[288,294,295]
Physical vapour deposition	–	≈4–10	Expensive equipment, time consuming for thick coatings	[167,275,296]
Sol–gel deposition	–	≈3–10	Very thin coatings, quality strongly dependent on interconnect surface	[169,297,298]
MOCVD, ALD	– –	≈2–5	Very thin coatings, expensive equipment	[79,166]
Pulsed laser deposition	– –	≈4–10	Expensive equipment, small coating area	[299]

Screen printing, wet powder spraying and slurry coating are conventional ceramic processing techniques, which are able to coat also large structural parts. The thickness ratio is typically $t > 1$ and not all surface structures can be coated properly. In addition, the ceramic coatings remain porous after stack assembly and therefore not useful for long-term operations. In contrast, aerosol depositions can deliver nearly dense ceramic coatings, and with proper movement systems, the process can also provide reasonable thickness ratios for three-dimensional interconnect parts [278].

Plasma spraying is very well known as a deposition technique which provides thick coatings between 20 and 200 μm in short processing time. When the spraying is carried out with a robot, the thickness ratio can be rather small. In addition, the various types of plasma spraying, including the new developments of liquid or suspension plasma spraying [294], offer new opportunities for adjusting the coating thickness and microstructure to the considered stack architecture. Depending on the processing parameters, the layers obtained by atmospheric plasma spraying can be rather dense and have shown so far the best chromium retention and lowest stack degradation rates. Already in the late 1990s, thermally sprayed manganite coatings have led to stable long-term performance lasting about 12,000 h with a degradation rate in cell voltage of less than 1%/1000 h.

Physical vapour deposition techniques, using rf magnetron sputtering, electron beam deposition or other, more sophisticated methods suffer from the large investment cost. Especially for large-sized parts and coatings thicker than 1–2 μm, the processes become very cost and time intensive and contribute significantly to the stack production costs. Therefore, it is unlikely that these methods will find their way into an SOFC production line. With the same arguments, also the chemical vapour deposition methods and pulsed laser deposition seem to remain an academic approach for chromium protection layers manufacturing.

Although sol–gel techniques are rather simple processing methods and do not require expensive equipment – but sometimes expensive precursors – the methods yield very thin layers which require a very smooth surface. Otherwise, the layers easily crack due to surface structures during heat treatment. Therefore, a multiple dip-coating process and sophisticated drying steps are necessary to obtain homogeneous and sufficiently thick layers. This can make coatings with a sol an elaborative process.

7.14 Summary

In the past 50 years, extensive research programmes brought enormous progress into SOFC technology. Nowadays, SOFCs are experiencing a breakthrough from research into manufacturing and commercialisation. One of the key components enabling this advancement is new interconnect material. In the first years of their development, SOFC designs were dominated by ceramic interconnects, whereas the recent years were accompanied by the successive replacement of ceramic by metallic interconnects. This alteration became possible through the development of a number of new materials and manufacturing technologies which allowed lower stack operating temperature from previously used 1000 °C down to, for example, 800 °C or even lower.

Since none of the conventional high-temperature alloys was able to satisfy the specific SOFC requirements for interconnect materials, there has emerged a need to develop metallic material especially designed for SOFC applications. Development of such alloys was challenging due to sometimes contradicting requirements such as low scaling rate and high electronic conductivity of the oxide scale. Finally, a compromise for these properties was found in a form of two types a chromia forming materials, for example, chromium-based ODS alloys and high chromium (22–26%Cr wt) ferritic steels.

Today, the high-chromium ferritic steels such as Crofer 22 APU or Crofer 22 H are state-of-the-art materials for manufacturing metallic SOFC interconnects. Due to a carefully controlled chemical composition, and/or sophisticated manufacturing technologies, it was possible to manufacture steel specially tailored for SOFC requirements. Since the interconnect material is subjected to a complex interactions with other SOFC components, it is virtually impossible to fulfil literally all requirements. This drawback can be, however, corrected by modifying the metal surface with appropriate coatings.

The interface between metallic interconnect and ceramic coating is very challenging because low corrosion and interaction, low contact resistance as well as low permeability of chromium species shall be realised, preferably, with only one functional layer. The recent results have shown that (a) optimised steels for SOFC applications are commercially available, (b) plasma-sprayed spinel coatings offer the best chromium retention and lowest degradation rates and (c) alkaline earth-free and cobalt-containing perovskites are the most suitable materials for contact layers. However, new solutions for stacking with permanently low resistances can only be reliably evaluated in long-term testing with test durations as long as several years.

References

[1] E. Ivers-Tiffée, W. Wersing, M. Schießl, H. Greiner, Ceramic and metallic components for a planar SOFC, Ber. Bunsenges. Phys. Chem. 94 (1990) 978–981.

[2] N.Q. Minh, Ceramic fuel-cells, J. Am. Ceram. Soc. 76 (1993) 563–588.

[3] F. Tietz, H.P. Buchkremer, D. Stover, Components manufacturing for solid oxide fuel cells, Solid State Ionics 152 (2002) 373–381.

[4] J.W. Fergus, Lanthanum chromite-based materials for solid oxide fuel cell interconnects, Solid State Ionics 171 (2004) 1–15.

[5] S. Wang, Y. Dong, B. Lin, J. Gao, X. Liu, G. Meng, Fabrication of dense $LaCrO_3$-based interconnect thin membrane on anode substrates by co-firing, Mater. Res. Bull. 44 (2009) 2127–2133.

[6] K. Hilpert, W.J. Quadakkers, L. Singheiser, Interconnects, in: W. Vielstich, A. Lamm, H.A. Gasteiger (Eds.), Handbook of Fuel Cells, vol. 4, John Wiley & Sons, Ltd., Chichester, UK, 2010.

[7] S.C. Singhal, Solid oxide fuel cells for stationary, mobile, and military applications, Solid State Ionics 152–153 (2002) 405–410.

[8] B. Rietveld, F. van Berkel, Y. Zhang-Steenwinkel, E. Bouyer, J. Irvine, M. Menon, L. Niewolak, S. Gross, A. Heel, P. Holtappels, S. Modena, The integrated project SOFC600 development of low-temperature SOFC, Solid Oxide Fuel Cells 11 (Sofc-Xi) 25 (2009) 29–34.

[9] L. Niewolak, E. Wessel, L. Singheiser, W.J. Quadakkers, Potential suitability of ferritic and austenitic steels as interconnect materials for solid oxide fuel cells operating at 600°C, J. Power Sources 195 (2010) 7600–7608.

[10] S.C. Singhal, Solid oxide fuel cells for power generation, Wiley Interdiscip. Rev. Energy Environ. 3 (2014) 179–194.

[11] L. Singheiser, P. Huczkowski, T. Markus, W.J. Quadakkers, 1.19 – High temperature corrosion issues for metallic materials in solid oxide fuel cells, in: B. Cottis, M. Graham, R. Lindsay, S. Lyon, T. Richardson, D. Scantlebury, H. Stott (Eds.), Shreir's Corrosion, Elsevier, Oxford, 2010, pp. 482–517.

[12] L. Blum, P. Batfalsky, L.G.J. de Haart, J. Malzbender, N.H. Menzler, R. Peters, W.J. Quadakkers, J. Remmel, F. Tietz, D. Stolten, Overview on the Jülich SOFC development status, ECS Trans. 57 (2013) 23–33.

[13] L. Aguilar, S. Zha, Z. Cheng, J. Winnick, M. Liu, A solid oxide fuel cell operating on hydrogen sulfide (H_2S) and sulfur-containing fuels, J. Power Sources 135 (2004) 17–24.
[14] A.I. Marquez, T.R. Ohrn, J.P. Trembly, D.C. Ingram, D.J. Bayless, Effects of coal syngas and H_2S on the performance of solid oxide fuel cells, J. Power Sources 164 (2007) 659–667.
[15] A. Lindermeir, R.-U. Dietrich, J. Oelze, in: Proc. International Symposium on Solid Oxide Fuel Cells: Materials, Science, and Technology, Daytona Beach, Florida, 2013.
[16] N. Christiansen, J.B. Hansen, H. Holm-Larsen, S. Linderoth, P.H. Larsen, P.V. Hendriksen, M. Mogensen, Solid oxide fuel cell development at Topsoe Fuel Cell and Risø, Fuel Cells Bull. 2006 (2006) 12–15.
[17] A. Fuerte, R.X. Valenzuela, M.J. Escudero, L. Daza, Ammonia as efficient fuel for SOFC, J. Power Sources 192 (2009) 170–174.
[18] W.J. Quadakkers, H. Greiner, W. Köck, H.P. Buchkremer, K. Hilpert, D. Stöver, in: Proc. 2nd EUROPEAN Solid Oxide Fuel Cell Forum, Oslo, 1996.
[19] H. Asteman, K. Segerdahl, J.-E. Svensson, L.-G. Johansson, M. Halvarsson, J.E. Tang, Oxidation of stainless steel in H_2O/O_2 environments – role of chromium evaporation, Mater. Sci. Forum 461–464 (2004) 775–782.
[20] W.J. Quadakkers, J. Żurek, 1.17 – Oxidation in steam and steam/hydrogen environments, in: B. Cottis, M. Graham, R. Lindsay, S. Lyon, T. Richardson, D. Scantlebury, H. Stott (Eds.), Shreir's Corrosion, Elsevier, Oxford, 2010, pp. 407–456.
[21] W.J. Quadakkers, M. Hänsel, T. Rieck, Carburization of Cr-based ODS alloys in SOFC relevant environments, Mater. Corros. 49 (1998) 252–257.
[22] L. Jian, J. Huezo, D.G. Ivey, Carburisation of interconnect materials in solid oxide fuel cells, J. Power Sources 123 (2003) 151–162.
[23] L. Niewolak, E. Wessel, T. Huttel, C. Asensio-Jimenez, L. Singheiser, W. J. Quadakkers, Behavior of interconnect steels in carbon containing simulated anode gas of solid oxide fuel cells, J. Electrochem. Soc. 159 (2012) F725–F732.
[24] Z. Zeng, K. Natesan, Corrosion of metallic interconnects for SOFC in fuel gases, Solid State Ionics 167 (2004) 9–16.
[25] N. Sakai, T. Horita, K. Yamaji, Y. Xiong, H. Kishimoto, M. Brito, H. Yokokawa, Material transport and degradation behavior of SOFC interconnects, Solid State Ionics 177 (2006) 1933–1939.
[26] C. Fu, K. Sun, X. Chen, N. Zhang, D. Zhou, Effects of the nickel-coated ferritic stainless steel for solid oxide fuel cells interconnects, Corros. Sci. 50 (2008) 1926–1931.
[27] W.A. Meulenberg, S. Uhlenbruck, E. Wessel, H.P. Buchkremer, D. Stöver, Oxidation behaviour of ferrous alloys used as interconnecting material in solid oxide fuel cells, J. Mater. Sci. 38 (2003) 507–513.
[28] M. Linder, T. Hocker, L. Holzer, K.A. Friedrich, B. Iwanschitz, A. Mai, J.A. Schuler, Cr_2O_3 scale growth rates on metallic interconnectors derived from 40,000 h solid oxide fuel cell stack operation, J. Power Sources 243 (2013) 508–518.
[29] P. Huczkowski, N. Christiansen, V. Shemet, L. Niewolak, J. Piron-Abellan, L. Singheiser, W.J. Quadakkers, Growth mechanisms and electrical conductivity of oxide scales on ferritic steels proposed as interconnect materials for SOFC's, Fuel Cells 6 (2006) 93–99.
[30] P.J. de Wild, R.G. Nyqvist, F.A. de Bruijn, E.R. Stobbe, Removal of sulphur-containing odorants from fuel gases for fuel cell-based combined heat and power applications, J. Power Sources 159 (2006) 995–1004.
[31] W.Z. Zhu, S.C. Deevi, Development of interconnect materials for solid oxide fuel cells, Mater. Sci. Eng. A 348 (2003) 227–243.
[32] D.W. White, US3460991, 1969.

[33] J.P. Ackerman, J.E. Young, US4476198, 1984.
[34] D.H. Archer, J.J. Alles, W.A. Englisch, L. Elikan, E.F. Sverdrup, R.L. Zahradnik, Westinghouse solid-electrolyte fuel cell, in: G.J. Young, H.R. Linden (Eds.), Fuel Cell Systems, vol. 47, American Chemical Society, 1969, pp. 332–342.
[35] D.T. Bary, L.D. La Grange, C.D. Park, M. Ulirich, US3300344 A, 1967.
[36] E.F. Sverdrup, C.J. Warde, R.L. Eback, Design of high-temperature solid-electrolyte fuel-cell batteries for maximum power output per unit volume, Energy Convers. 13 (1973) 129–141.
[37] A.O. Isenberg, US4490444, 1984.
[38] R.B. Poeppel, J.T. Dusek, US4476196, 1984.
[39] P. Reichner, P. Boro, US4874678, 1989.
[40] U.G. Bossel, EP0378812A1, 1989.
[41] G. Lin, US 2014/0099567 A1, 2013.
[42] S. Giles, G. Lin, A. Mohanram, Y. Narendar, J. Pietras, F.C. Qi, W.R. Robbins, R.J. Sliwoski, Saint-Gobains all ceramic SOFC stack: architecture and performance, ECS Trans. 57 (2013) 105–114.
[43] H. Yokokawa, N. Sakai, T. Horita, K. Yamaji, Recent developments in solid oxide fuel cell materials, Fuel Cells 1 (2001) 117–131.
[44] H.-H. Möbius, Chapter 2 – History, in: S.C. Singhal, K. Kendal (Eds.), High Temperature and Solid Oxide Fuel Cells, Elsevier Science, Amsterdam, 2003, pp. 23–51.
[45] T.L. Cable, S.W. Sofie, A symmetrical, planar SOFC design for NASA's high specific power density requirements, J. Power Sources 174 (2007) 221–227.
[46] G. Pudmich, B.A. Boukamp, M. Gonzalez-Cuenca, W. Jungen, W. Zipprich, F. Tietz, Chromite/titanate based perovskites for application as anodes in solid oxide fuel cells, Solid State Ionics 135 (2000) 433–438.
[47] M. Mori, Y. Hiei, Thermal expansion behavior of titanium-doped La(Sr)CrO_3 solid oxide fuel cell interconnects, J. Am. Ceram. Soc. 84 (2001) 2573–2578.
[48] U.B. Pal, S.C. Singhal, Electrochemical vapor-deposition of yttria-stabilized zirconia films, J. Electrochem. Soc. 137 (1990) 2937–2941.
[49] L. Jian, H. Bin, Z. Wenying, Metallic interconnect materials of solid oxide fuel cells, in: S.P. Jiang, Y. Yan, M. Lu (Eds.), Materials for High-Temperature Fuel Cells, Wiley-VCH Verlag GmbH & Co. KGaA, Weinheim, Germany, 2013, pp. 159–213.
[50] B.-K. Park, J.-W. Lee, S.-B. Lee, T.-H. Lim, S.-J. Park, R.-H. Song, W.B. Im, D.-R. Shin, La-doped $SrTiO_3$ interconnect materials for anode-supported flat-tubular solid oxide fuel cells, Int. J. Hydrogen Energy 37 (2012) 4319–4327.
[51] M. Mori, Z. Wang, N. Serizawa, T. Itoh, Evaluation of $SrTi_1{-}xCoxO_3$ perovskites ($0 \leq x \leq 0.2$) as interconnect materials for solid oxide fuel cells, J. Fuel Cell Sci. Technol. 8 (2011) 051010.
[52] B.-K. Park, J.-W. Lee, S.-B. Lee, T.-H. Lim, S.-J. Park, R.-H. Song, D.-R. Shin, A flat-tubular solid oxide fuel cell with a dense interconnect film coated on the porous anode support, J. Power Sources 213 (2012) 218–222.
[53] N. Sakai, H. Yokokawa, T. Horita, K. Yamaji, Lanthanum chromite-based interconnects as key materials for SOFC stack development, Int. J. Appl. Ceram. Technol. 1 (2004) 23–30.
[54] X.F. Ding, Y.J. Liu, L. Gao, L.C. Guo, Effects of cation substitution on thermal expansion and electrical properties of lanthanum chromites, J. Alloys Compd. 425 (2006) 318–322.
[55] S. Wang, B. Lin, Y. Dong, D. Fang, H. Ding, X. Liu, G. Meng, Stable, easily sintered Ca–Zn-doped $YCrO_3$ as novel interconnect materials for co-fired yttrium-stabilized zirconia-based solid oxide fuel cells, J. Power Sources 188 (2009) 483–488.

[56] J. Froitzheim, L. Niewolak, M. Brandner, L. Singheiser, W.J. Quadakkers, Anode side diffusion barrier coating for solid oxide fuel cells interconnects, J. Fuel Cell Sci. Technol. 7 (2010) 031020.
[57] N.R. Hosseini, N.M. Sammes, J.S. Chung, Manganese-doped lanthanum calcium titanate as an interconnect for flat-tubular solid oxide fuel cells, J. Power Sources 245 (2014) 599–608.
[58] W.J. Quadakkers, J. Piron-Abellan, V. Shemet, L. Singheiser, Metallic interconnectors for solid oxide fuel cells – a review, Mater. High Temp. 20 (2003) 115–127.
[59] P. Huczkowski, N. Christiansen, V. Shemet, L. Singheiser, W.J. Quadakkers, in: Proc. 6th European Solid Oxide Fuel Cell Forum, Luzern, CH, 2004.
[60] P. Kofstad, High Temperature Corrosion, Elsevier Applied Science, London/New York, 1988.
[61] G.Y. Lai, High-temperature corrosion of engineering alloys, ASM International, Metals Park, OH, 1990.
[62] C. Compson, S. Choi, H. Abemathy, Y. Choi, C. Liu, in: Proc. The American Ceramic Society's 31st International Conference on Advanced Ceramics and Composites, Daytona Beach, Florida, 2007.
[63] P. Kofstad, R. Bredesen, High-temperature corrosion in SOFC environments, Solid State Ionics 52 (1992) 69–75.
[64] R.W. Jackson, F.S. Pettit, G.H. Meier, The behavior of nickel and silver in a simulated solid oxide fuel cell environment, J. Power Sources 185 (2008) 1030–1039.
[65] H.S. Spacil, C.S. Tedmon, Electrochemical dissociation of water vapor in solid oxide electrolyte cells II. Materials, fabrication, and properties, J. Electrochem. Soc. 116 (1969) 1627–1633.
[66] N. Birks, G.H. Meier, F.S. Pettit, Introduction to the High Temperature Oxidation of Metals, Cambridge University Press, New York, 2006.
[67] W.J. Quadakkers, H. Greiner, M. Hansel, A. Pattanaik, A.S. Khanna, W. Mallener, Compatibility of perovskite contact layers between cathode and metallic interconnector plates of SOFCs, Solid State Ionics 91 (1996) 55–67.
[68] K. El-Menshawy, H.P. Buchkremer, F. Tietz, D. Stöver, Electrical Conductivity of sintered chromia mixed with TiO_2, CuO and Mn-oxides, J. Mater. Sci. Technol. 22 (2006) 245–251.
[69] L. Chen, Z. Yang, B. Jha, G. Xia, J.W. Stevenson, Clad metals, roll bonding and their applications for SOFC interconnects, J. Power Sources 152 (2005) 40–45.
[70] P. Gannon, R. Amendola, High-temperature, dual-atmosphere corrosion of solid-oxide fuel cell interconnects, JOM 64 (2012) 1470–1476.
[71] P. Gannon, V. Gorokhovsky, M. Deibert, R. Smith, A. Kayani, P. White, S. Sofie, Z. Yang, D. McCready, S. Visco, Enabling inexpensive metallic alloys as SOFC interconnects: an investigation into hybrid coating technologies to deposit nanocomposite functional coatings on ferritic stainless steels, Int. J. Hydrogen Energy 32 (2007) 3672–3681.
[72] T. Horita, H. Kishimoto, K. Yamaji, Y. Xiong, N. Sakai, M.E. Brito, H. Yokokawa, Effect of grain boundaries on the formation of oxide scale in Fe–Cr alloy for SOFCs, Solid State Ionics 179 (2008) 1320–1324.
[73] J. Rufner, Oxidation behavior of stainless steel 430 and 441 at 800°C in single (air/air) and dual atmosphere (air/hydrogen) exposures, Int. J. Hydrogen Energy 33 (2008) 1392–1398.
[74] N. Shaigan, W. Qu, D.G. Ivey, W. Chen, A review of recent progress in coatings, surface modifications and alloy developments for solid oxide fuel cell ferritic stainless steel interconnects, J. Power Sources 195 (2010) 1529–1542.
[75] M. Tucker, T. Sholklapper, G. Lau, L. DeJonghe, S. Visco, Progress in metal-supported SOFCs, ECS Trans. 25 (2009) 673–680.

[76] Z. Yang, M.S. Walker, P. Singh, J.W. Stevenson, T. Norby, Oxidation behavior of ferritic stainless steels under SOFC interconnect exposure conditions, J. Electrochem. Soc. 151 (2004) B669.
[77] Z. Yang, G.-G. Xia, P. Singh, J.W. Stevenson, Electrical contacts between cathodes and metallic interconnects in solid oxide fuel cells, J. Power Sources 155 (2006) 246–252.
[78] Q. Fu, F. Tietz, D. Sebold, E. Wessel, H.-P. Buchkremer, Magnetron-sputtered cobalt-based protective coatings on ferritic steels for solid oxide fuel cell interconnect applications, Corros. Sci. 54 (2012) 68–76.
[79] S. Fontana, R. Amendola, S. Chevalier, P. Piccardo, G. Caboche, M. Viviani, R. Molins, M. Sennour, Metallic interconnects for SOFC: characterisation of corrosion resistance and conductivity evaluation at operating temperature of differently coated alloys, J. Power Sources 171 (2007) 652–662.
[80] K. Gerdes, C. Johnson, Surface scale formation on solid oxide fuel cell proximal balance of plant components, J. Fuel Cell Sci. Technol. 6 (2009) 011018.
[81] J.E. Hammer, The Oxidation of Materials for Interconnects in Solid Oxide Fuel Cells, University of Pittsburgh, Pittsburgh, 2005, p. 108.
[82] J.E. Hammer, S.J. Laney, R.W. Jackson, K. Coyne, F.S. Pettit, G.H. Meier, The oxidation of ferritic stainless steels in simulated solid-oxide fuel-cell atmospheres, Oxid. Met. 67 (2007) 1–38.
[83] M. Palcut, L. Mikkelsen, K. Neufeld, M. Chen, R. Knibbe, P.V. Hendriksen, Corrosion stability of ferritic stainless steels for solid oxide electrolyser cell interconnects, Corros. Sci. 52 (2010) 3309–3320.
[84] P. Piccardo, R. Amendola, S. Fontana, S. Chevalier, G. Caboches, P. Gannon, Interconnect materials for next-generation solid oxide fuel cells, J. Appl. Electrochem. 39 (2009) 545–551.
[85] V.A.C. Haanappel, V. Shemet, S.M. Gross, T. Koppitz, N.H. Menzler, M. Zahid, W.J. Quadakkers, Behaviour of various glass–ceramic sealants with ferritic steels under simulated SOFC stack conditions, J. Power Sources 150 (2005) 86–100.
[86] E. Konysheva, U. Seeling, A. Besmehn, L. Singheiser, K. Hilpert, >Chromium vaporization of the ferritic steel Crofer 22 APU and ODS $Cr_5Fe_1Y_2O_3$ alloy, J. Mater. Sci. 42 (2007) 5778–5784.
[87] M. Bram, L. Niewolak, N. Shah, D. Sebold, H.P. Buchkremer, Interaction of sealing material mica with interconnect steel for solid oxide fuel cells application at 600°C, J. Power Sources 196 (2011) 5889–5896.
[88] S.J. Geng, J.H. Zhu, Z.G. Lu, Evaluation of several alloys for solid oxide fuel cell interconnect application, Scr. Mater. 55 (2006) 239–242.
[89] G. Kunschert, Hochtemperatureigenschaften einer ODS-Fe26Cr Legierung zum Einsatz als Interkonnektor in einer SOFC-APU, Material- und Geowissenschaften, Technischen Universität, Darmstadt, 2009, p. 156.
[90] M.P. Brady, B.A. Pint, Z.G. Lu, J.H. Zhu, C.E. Milliken, E.D. Kreidler, L. Miller, T.R. Armstrong, L.R. Walker, Comparison of oxidation behavior and electrical properties of doped NiO- and Cr_2O_3-forming alloys for solid-oxide, fuel-cell metallic interconnects, Oxid. Met. 65 (2006) 237–261.
[91] J. Froitzheim, G.H. Meier, L. Niewolak, P.J. Ennis, H. Hattendorf, L. Singheiser, W.J. Quadakkers, Development of high strength ferritic steel for interconnect application in SOFCs, J. Power Sources 178 (2008) 163–173.
[92] J. Froitzheim, H. Ravash, E. Larsson, L.G. Johansson, J.E. Svensson, Investigation of chromium volatilization from FeCr interconnects by a denuder technique, J. Electrochem. Soc. 157 (2010) B1295–B1300.

[93] R. Sachitanand, M. Sattari, J.-E. Svensson, J. Froitzheim, Evaluation of the oxidation and Cr evaporation properties of selected FeCr alloys used as SOFC interconnects, Int. J. Hydrogen Energy 38 (2013) 15328–15334.
[94] P.D. Jablonski, D.E. Alman, Oxidation resistance and mechanical properties of experimental low coefficient of thermal expansion (CTE) Ni-base alloys, Int. J. Hydrogen Energy 32 (2007) 3705–3712.
[95] E. Konysheva, J. Mertens, H. Penkalla, L. Singheiser, K. Hilpert, Chromium poisoning of the porous composite cathode, J. Electrochem. Soc. 154 (2007) B1252.
[96] B. Kuhn, C. Asensio-Jimenez, L. Niewolak, T. Hüttel, T. Beck, H. Hattendorf, L. Singheiser, W.J. Quadakkers, Effect of Laves phase strengthening on the mechanical properties of high Cr ferritic steels for solid oxide fuel cell interconnect application, Mater. Sci. Eng. A 528 (2011) 5888–5899.
[97] S.J. Laney, Modfying ferritic stainless steels for solid oxide fuel cell applications, Graduate Faculty of School of Engineering, University of Pittsburgh, Pittsburgh, 2007, p. 235.
[98] P. Leone, A. Lanzini, B. Delhomme, G.A. Ortigoza-Villalba, M. Santarelli, F. Smeacetto, M. Salvo, M. Ferraris, Experimental evaluation of planar SOFC single unit cell with Crofer22APU plate assembly, J. Fuel Cell Sci. Technol. 8 (2011) 031009.
[99] Y. Liu, Performance evaluation of several commercial alloys in a reducing environment, J. Power Sources 179 (2008) 286–291.
[100] S. Molin, M. Chen, J.J. Bentzen, P.V. Hendriksen, High temperature oxidation of ferritic steels for solid oxide electrolysis stacks, ECS Trans. 50 (2013) 11–20.
[101] L. Niewolak, D.J. Young, H. Hattendorf, L. Singheiser, W.J. Quadakkers, Mechanisms of oxide scale formation on ferritic interconnect steel in simulated low and high pO_2 service environments of solid oxide fuel cells, Oxid. Met. 82 (2014) 123–143.
[102] F.S. Pettit, G.H. Meier, Fundamental Studies of the Durability of Materials for Interconnects in Solid Oxide Fuel Cells, University of Pittsburgh, Pittsburgh, 2006, p. 125.
[103] J. Piron-Abellan, W.J. Quadakkers, Development of Ferritic Steels for Application as Interconnect Materials for Intermediate Temperature Solid Oxide Fuel Cells (SOFCs), Forschungszenrum Jülich, Jülich, 2005.
[104] W.J. Quadakkers, J. Piron-Abellan, P. Huczkowski, V. Shemet, L. Singheiser, Metallic construction materials for solid oxide fuel cell interconnectors, in: P. Lens, P. Westermann, M. Haberbauer, A. Moreno (Eds.), Biofuels for Fuel Cells, IWA Publishing, London, 2005.
[105] L. Singheiser, J. Piron-Abellan, P. Huczkowski, S.T. Ertl, V. Shemet, W.J. Quadakkers, Development of new ferritic steels for application as interconnector material in solid oxide fuel cells, in: B. Raj, S. Ranganathan, S.L. Mannan, K. Bhanu Sankara Rao, M.D. Mathew, P. Shankar (Eds.), Frontiers in the Design of Materials, CRC Press LLC, Taylor and Francis Group, Boca Raton, FL 33487, USA, 2007, pp. 175–184.
[106] M. Stanislowski, Verdampfung von Werkstoffen beim Betrieb von Hochtemperaturbrennstoffzellen (SOFC), RWTH Aachen, Jülich, Germany, 2006.
[107] M. Stanislowski, J. Froitzheim, L. Niewolak, W.J. Quadakkers, K. Hilpert, T. Markus, L. Singheiser, Reduction of chromium vaporization from SOFC interconnectors by highly effective coatings, J. Power Sources 164 (2007) 578–589.
[108] M. Stanislowski, E. Wessel, K. Hilpert, T. Markus, L. Singheiser, Chromium vaporization from high-temperature alloys, J. Electrochem. Soc. 154 (2007) A295.
[109] Z. Yang, G. Xia, M. Walker, C. Wang, J. Stevenson, P. Singh, High temperature oxidation/corrosion behavior of metals and alloys under a hydrogen gradient, Int. J. Hydrogen Energy 32 (2007) 3770–3777.

[110] Z.G. Yang, J.W. Stevenson, Durability of metallic interconnects and protective coatings, Handbook of Fuel Cells, John Wiley & Sons, Ltd., Hoboken, NJ, 2010.
[111] Y.-T. Chiu, C.-K. Lin, Effects of Nb and W additions on high-temperature creep properties of ferritic stainless steels for solid oxide fuel cell interconnect, J. Power Sources 198 (2012) 149–157.
[112] J.G. Grolig, H. Abdesselam, M. Gas, H.F. Windisch, J. Froitzheim, J.-E. Svensson, Copper based conversion coatings on ferritic stainless strip steel as solid oxide fuel cell interconnects: oxidation performance and chromium evaporation, ECS Trans. 57 (2013) 2339–2347.
[113] M. Palcut, L. Mikkelsen, K. Neufeld, M. Chen, R. Knibbe, P.V. Hendriksen, Efficient dual layer interconnect coating for high temperature electrochemical devices, Int. J. Hydrogen Energy 37 (2012) 14501–14510.
[114] C. Asensio-Jimenez, A. Chyrkin, L. Niewolak, V. Konoval, H. Hattendorf, B. Kuhn, L. Singheiser, W.J. Quadakkers, Subsurface depletion and enrichment processes during oxidation of a high chromium, Laves-phase strengthened ferritic steel, Electrochem. Solid-State Lett. 14 (2011) P17–P20.
[115] C. Asensio-Jimenez, L. Niewolak, H. Hattendorf, B. Kuhn, P. Huczkowski, L. Singheiser, W.J. Quadakkers, Effect of specimen thickness on the oxidation rate of high chromium ferritic steels: the significance of intrinsic alloy creep strength, Oxid. Met. 79 (2012) 15–28.
[116] L. Garcia-Fresnillo, V. Shemet, A. Chyrkin, L.G.J. de Haart, W.J. Quadakkers, Long-term behaviour of solid oxide fuel cell interconnect materials in contact with Ni-mesh during exposure in simulated anode gas at 700 and 800°C, J. Power Sources 271 (2014) 213–222.
[117] T. Horita, H. Kishimoto, K. Yamaji, N. Sakai, Y. Xiong, M.E. Brito, H. Yokokawa, Effects of silicon concentration in SOFC alloy interconnects on the formation of oxide scales in hydrocarbon fuels, J. Power Sources 157 (2006) 681–687.
[118] T. Horita, H. Kshimoto, K. Yamaji, N. Sakai, Y. Xiong, M. Brito, H. Yokokawa, Anomalous oxidation of ferritic interconnects in solid oxide fuel cells, Int. J. Hydrogen Energy 33 (2008) 3962–3969.
[119] T. Horita, K. Yamaji, H. Yokokawa, A. Toji, T. Uehara, K. Ogasawara, H. Kameda, Y. Matsuzaki, S. Yamashita, Effects of Si and Al concentrations in Fe–Cr alloy on the formation of oxide scales in H_2–H_2O, Int. J. Hydrogen Energy 33 (2008) 6308–6315.
[120] T. Horita, H. Kishimoto, K. Yamaji, M.E. Brito, Y. Xiong, H. Yokokawa, J. Power Sources 193 (2009) 180–184.
[121] T. Horita, M. Yoshinaga, H. Kishimoto, K. Yamaji, M.E. Brito, Y. Xiong, H. Yokokawa, Effects of thermal cycling on the formation of oxide scale of Fe-Cr alloy interconnects for solid oxide fuel cells, Int. J. Appl. Ceram. Technol. 8 (2011) 1374–1381.
[122] J. Piron-Abellan, F. Tietz, V. Shemet, A. Gil, T. Ladwein, L. Singheiser, W. J. Quadakkers, in: Proc. 5th European Silid Oxide Fuel Cell Forum, Luzerne, 2002.
[123] T. Uehara, N. Yasuda, M. Okamoto, C. Aoki, T. Ohno, A. Toji, Long-term oxidation behavior of Fe-Cr ferritic alloy ZMG232L for SOFC interconnects, ECS Trans. 25 (2009) 1455–1462.
[124] B.W.A. Skilbred, R. Haugsrud, The effect of dual atmosphere conditions on the corrosion of Sandvik Sanergy HT, Int. J. Hydrogen Energy 37 (2012) 8095–8101.
[125] A.W.B. Skilbred, R. Haugsrud, The effect of water vapour on the corrosion of Sandvik Sanergy HT under dual atmosphere conditions, Oxid. Met. 79 (2012) 639–654.

[126] K. Huang, Y.P. Hou, J.B. Goodenough, Reduced area specific resistance for iron-based metallic interconnects by surface oxide coatings, Mater. Res. Bull. 36 (2001) 81–95.
[127] A. Venskutonis, W. Glatz, G. Kunschert, High volume fabrication of ready-to-stack components for planar SOFC concepts, in: Proc. 16TH, Plansee Seminar: Powder Metallurgical High Performance Materials, 2005.
[128] T. Franco, M. Brandner, M. Rüttinger, G. Kunschert, A. Venskutonis, L. Sigl, Recent development aspects of metal supported thin-film SOFC, ECS Trans. 25 (2009) 681–688.
[129] A. Venskutonis, M. Brandner, W. Kraussler, L. Sigl, High volume fabrication of ready-to-stack components for planar SOFC concepts, ECS Trans. 25 (2009) 1353–1359.
[130] S. Linderoth, P.V. Hendriksen, M. Mogensen, Investigations of metallic alloys for use as interconnects in solid oxide fuel cell stacks, J. Mater. Sci. 31 (1996) 5077–5082.
[131] Y. Larring, R. Haugsrud, T. Norby, HT corrosion of a Cr-5 wt % Fe-1 wt % Y_2O_3 alloy and conductivity of the oxide scale, J. Electrochem. Soc. 150 (2003) B374.
[132] F. Tietz, Thermal expansion of SOFC materials, Ionics 5 (1999) 129–139.
[133] M. Brandner, C. Bienert, S. Megel, M. Kusnezoff, N. Trofimenko, V. Sauchuk, A. Venskutonis, W. Kraussler, A. Michaelis, L.S. Sigl, Long term performance of stacks with chromium-based interconnects (CFY), ECS Trans. 57 (2013) 2235–2244.
[134] G.R. Holcomb, Calculation of reactive-evaporation rates of chromia, Oxid. Met. 69 (2008) 163–180.
[135] G.R. Holcomb, M. Ziomek-Moroz, S.D. Cramer, B.S. Covino, S.J. Bullard, Dual-environment effects on the oxidation of metallic interconnects, J. Mater. Eng. Perform. 15 (2006) 404–409.
[136] L. Ge, A. Verma, R. Goettler, D. Lovett, R.K.S. Raman, P. Singh, Oxide scale morphology and chromium evaporation characteristics of alloys for balance of plant applications in solid oxide fuel cells, Metall. Mater. Trans. A 44 (2012) 193–206.
[137] L. Jian, P. Jian, X. Guangyuan, W. Shunxu, X. Jianzhong, Heat resistant alloys as interconnect materials of reduced temperature SOFCs, J. Power Sources 157 (2006) 368–376.
[138] Y. Li, Y. Jiang, J. Wu, R. Pineault, R. Gemmenn, X. Liu, Effect of electrical current on solid oxide fuel cells metallic interconnect oxidation in syngas, Int. J. Appl. Ceram. Technol. 7 (2010) 41–48.
[139] Z. Yang, G.-G. Xia, J.W. Stevenson, Evaluation of Ni–Cr-base alloys for SOFC interconnect applications, J. Power Sources 160 (2006) 1104–1110.
[140] W.Z. Zhu, S.C. Deevi, Opportunity of metallic interconnects for solid oxide fuel cells: a status on contact resistance, Mater. Res. Bull. 38 (2003) 957–972.
[141] S. Taniguchi, M. Kadowaki, H. Kawamura, T. Yasuo, Y. Akiyama, Y. Miyake, T. Saitoh, Degradation phenomena in the cathode of a solid oxide fuel-cell with an alloy separator, J. Power Sources 55 (1995) 73–79.
[142] J.W. Fergus, Metallic interconnects for solid oxide fuel cells, Mater. Sci. Eng. A 397 (2005) 271–283.
[143] D.M. England, A.V. Virkar, Oxidation kinetics of some nickel-based superalloy foils and electronic resistance of the oxide scale formed in air part I, J. Electrochem. Soc. 146 (1999) 3196–3202.
[144] D.M. England, A.V. Virkar, Oxidation kinetics of some nickel-based superalloy foils in humidified hydrogen and electronic resistance of the oxide scale formed part II, J. Electrochem. Soc. 148 (2001) A330–A338.
[145] H.U. Anderson, F. Tietz, Chapter 7 – Interconnects, in: S.C. Singhal, K. Kendal (Eds.), High Temperature and Solid Oxide Fuel Cells, Elsevier Science, Amsterdam, 2003, pp. 173–195.

[146] Z. Yang, K.S. Weil, D.M. Paxton, J.W. Stevenson, Selection and evaluation of heat-resistant alloys for SOFC interconnect applications, J. Electrochem. Soc. 150 (2003) A1188.
[147] J.W. Wu, X.B. Liu, Recent development of SOFC metallic interconnect, J. Mater. Sci. Technol. 26 (2010) 293–305.
[148] T. Malkow, U.v.d. Crone, A.M. Laptev, T. Koppitz, U. Breuer, in: Proc. SOFC-V, Aachen, 1997.
[149] M. Ueda, H. Taimatsu, Thermal expansivity and high-temperature oxidation resistance of Fe-Cr-W alloys, Mater. Trans. 40 (1999) 1390–1395.
[150] U.v.d. Crone, M. Hänsel, W.J. Quadakkers, R. Vaßen, Oxidation behavior of mechanically alloyed chromium based alloys, Fresenius J. Anal. Chem. 358 (1997) 230–232.
[151] M. Hänsel, W.J. Quadakkers, D.J. Young, Role of water vapor in chromia-scale growth at low oxygen partial pressure, Oxid. Met. 59 (2003) 285–301.
[152] K. Huang, P.Y. Hou, J.B. Goodenough, Characterization of iron-based alloy interconnects for reduced temperature solid oxide fuel cells, Solid State Ionics 129 (2000) 237–250.
[153] W.J. Quadakkers, T. Malkow, J. Piron-Abellan, U. Flesch, V. Shemet, L. Singheiser, in: Proc. 4th European SOFC Forum, Luzern, CH, 2000.
[154] K. Honegger, A. Plas, in: Proc. International Symposia on Solid Oxide Fuel Cells, Tsukuba, Japan, 2001.
[155] J. Piron-Abellan, V. Shemet, F. Tietz, L. Singheiser, W.J. Quadakkers, in: Proc. SOFC-VI, Tsukuba, Japan, 2001.
[156] W.J. Quadakkers, H. Greiner, W. Köck, in: Proc. European Solid Oxide Fuel Cell Forum, Lucerne, CH, 1994.
[157] W. Köck, H.P. Martinz, H. Greiner, M. Janousek, Proc. 4th Int. Symp. Solid Oxide Fuel Cells (SOFC-IV), Yokohama, Japan, 1995.
[158] R.-H. Song, D.-R. Shin, K.-S. Jeon, Y.-S. Han, D. Masayuki, US6150048, 2000.
[159] M. Krumpelt, T.A. Cruse, D.J. Carter, J.L. Routbort, R. Kumar, US6843960 B2, 2005.
[160] W.J. Quadakkers, H. Holzbrecher, K.G. Briefs, H. Beske, in: Proc. Europ. Coll. on the Role of Active Elements in the Oxidation Behaviour of High Temperature Alloys, Petten, NL, 1989.
[161] P. Hou, J. Stringer, Oxide scale adhesion and impurity segregation at the scale/metal interface, Oxid. Met. 38 (1992) 323–345.
[162] D.P. Whittle, J. Stringer, Improvements in high temperature oxidation resistance by additions of reactive elements or oxide dispersions, Philos. Trans. R. Soc. London, Ser. A 295 (1980) 309–329.
[163] F.A. Golightly, F.H. Stott, G.C. Wood, The influence of yttrium additions on the oxide-scale adhesion to an iron-chromium-aluminum alloy, Oxid. Met. 10 (1976) 163–187.
[164] B.A. Pint, in: Proc. John Stringer Symposium on High Temperature Corrosion, 2003.
[165] S.P.S. Badwal, R. Deller, K. Foger, Y. Ramprakash, J.P. Zhang, Interaction between chromia forming alloy interconnects and air electrode of solid oxide fuel cells, Solid State Ionics 99 (1997) 297–310.
[166] M. Burriel, G. Garcia, J. Santiso, A.N. Hansson, S. Linderoth, A. Figueras, Co_3O_4 protective coatings prepared by pulsed injection metal organic chemical vapour deposition, Thin Solid Films 473 (2005) 98–103.
[167] V.I. Gorokhovsky, P.E. Gannon, M.C. Deibert, R.J. Smith, A. Kayani, M. Kopczyk, D. VanVorous, Z. Yang, J.W. Stevenson, S. Visco, C. Jacobson, H. Kurokawa,

S.W. Sofie, Deposition and evaluation of protective PVD coatings on ferritic stainless steel SOFC interconnects, J. Electrochem. Soc. 153 (2006) A1886–A1893.

[168] F. Tietz, D. Sebold, Interface reactions between electrically conductive ceramics and ferritic steel-I. The system Cr–22Fe–0.5Mn/Mn_2O_3/(La,Ca)(Cr,Co,Cu)O_3, Mater. Sci. Eng. B 150 (2008) 135–140.

[169] J.H. Zhu, Y. Zhang, A. Basu, Z.G. Lu, M. Paranthaman, D.F. Lee, E.A. Payzant, $LaCrO_3$-based coatings on ferritic stainless steel for solid oxide fuel cell interconnect applications, Surf. Coat. Technol. 177–178 (2004) 65–72.

[170] W.J. Quadakkers, J.F. Norton, H.J. Penkalla, U. Breuer, A. Gil, T. Rieck, M. Hänsel, in: Proc. 3rd International conference on "Microscopy of Oxidation", Cambridge, UK, 1996.

[171] W.J. Quadakkers, Oxidation of ODS alloys, J. Phys. IV France 3 (1993) 177–186.

[172] W.J. Quadakkers, L. Singheiser, Practical aspects of the reactive element effect, Mater. Sci. Forum 369–372 (2001) 77–92.

[173] M. Hänsel, W.J. Quadakkers, L. Singheiser, H. Nickel, Korrosions- und Kompatibilitätsstudien an Cr-Basislegierungen für den metallischen Interkonnektor der Hochtemperaturbrennstoffzelle, Forschungszentrum Jülich, Jülich, 1998.

[174] X.G. Zheng, D.J. Young, High-temperature corrosion of Cr_2O_3-forming alloys in CO-CO_2-N_2 atmospheres, Oxid. Met. 42 (1994) 163–190.

[175] D.J. Young, J. Zurek, L. Singheiser, W.J. Quadakkers, Temperature dependence of oxide scale formation on high-Cr ferritic steels in Ar–H_2–H_2O, Corros. Sci. 53 (2011) 2131–2141.

[176] M. Schuisky, A. Rosberg, L. Mikkelsen, P.V. Hendriksen, N. Christiansen, J. Gutzton-Larsen, in: Proc. European Solid Oxide Fuel Cell Forum, Luzern, CH, 2006.

[177] P. Huczkowski, N. Christiansen, V. Shemet, J. Piron-Abellan, L. Singheiser, W.J. Quadakkers, Oxidation induced lifetime limits of chromia forming ferritic interconnector steels, J. Fuel Cell Sci. Technol. 1 (2004) 30.

[178] W.J. Quadakkers, High temperature corrosion in the service environments of a nuclear process heat plant, Mater. Sci. Eng. 87 (1987) 107–112.

[179] M. Shindo, W.J. Quadakkers, H. Schuster, Corrosion behaviour of high temperature alloys in impure helium environments, J. Nucl. Mater. 140 (1986) 94–105.

[180] A. Naoumidis, H.A. Schulze, W. Jungen, P. Lersch, Phase studies in the chromium-manganese-titanium oxide system at different oxygen partial pressures, J. Eur. Ceram. Soc. 7 (1991) 55–63.

[181] C. Garcia-Rosales, H.A. Schulze, A. Naoumidis, H. Nickel, Influence of oxygen partial pressure on the quasi-ternary system CrMnTi oxide, J. Am. Ceram. Soc. 76 (1993) 2869–2878.

[182] J. Zurek, D.J. Young, E. Essuman, M. Hänsel, H.J. Penkalla, L. Niewolak, W. J. Quadakkers, Growth and adherence of chromia based surface scales on Ni-base alloys in high- and low-pO_2 gases, Mater. Sci. Eng. A 477 (2008) 259–270.

[183] P. Huczkowski, V. Shemet, J. Piron-Abellan, L. Singheiser, W.J. Quadakkers, N. Christiansen, Oxidation limited life times of chromia forming ferritic steels, Mater. Corros. 55 (2004) 825–830.

[184] P. Huczkowski, S.T. Ertl, J. Piron-Abellan, N. Christiansen, T. Hofler, V. Shemet, L. Singheiser, W.J. Quadakkers, Effect of component thickness on lifetime and oxidation rate of chromia forming ferritic steels in low and high pO_2 environments, Mater. High Temp. 22 (2005) 253–262.

[185] M. Stelter, A. Reinert, B.E. Mai, M. Kuznecov, Engineering aspects and hardware verification of a volume producable solid oxide fuel cell stack design for diesel auxiliary power units, J. Power Sources 154 (2006) 448–455.

[186] I. Vinke, B. Röwekamp, F. Tietz, M. Zahid, W.J. Quadakkers, in: Proc. Solid Oxide Fuel Cells IX, Quebec, 2005.
[187] W.J. Quadakkers, J. Piron-Abellan, V. Shemet, Metallic materials in solid oxide fuel cells, Mater. Res. 7 (2004) 203–208.
[188] D.J. Young, B.A. Pint, Chromium volatilization rates from Cr_2O_3 scales into flowing gases containing water vapor, Oxid. Met. 66 (2006) 137–153.
[189] M. Stanislowski, D.H. Peck, S.K. Woo, L. Singheiser, K. Hilpert, O. Schulz, M. Martin, Vaporization and diffusion studies on the stability of doped lanthanum gallates, Fuel Cells 6 (2006) 270–283.
[190] P. Huczkowski, W.J. Quadakkers, Effect of Geometry and Composition of Cr Steels on Oxide Scale Properties Relevant for Interconnector Applications in Solid Oxide Fuel Cells (SOFCs), Forschungszentrums Jülich, Jülich, 2007.
[191] L. Niewolak, C. Asensio-Jimenez, W.J. Quadakkers, Final Report of Zeus III Project (WP Steel Development), Forschungszentrum Jülich, Jülich, 2011, p. 410.
[192] W.J. Quadakkers, V. Shemet, L. Singheiser, 2001.
[193] J. Froitzheim, Ferritic Steel Interconnectors and Their Interactions with Ni Base Anodes in Solid Oxide Fuel Cells (SOFC), Fakultät für Maschinenwesen, Rheinisch-Westfälischen Technischen Hochschule, Aachen, 2008, p. 181.
[194] T.B. Massalski, H. Okamoto, Binary Alloy Phase Diagrams, ASM International, Material Park, Ohio, 1990.
[195] J.O. Anderson, B. Sundman, Thermodynamic properties of the Cr-Fe system, Calphad 11 (1987) 83–89.
[196] W. Xiong, P. Hedström, M. Selleby, J. Odqvist, M. Thuvander, Q. Chen, An improved thermodynamic modeling of the Fe–Cr system down to zero kelvin coupled with key experiments, Calphad 35 (2011) 355–366.
[197] M. Schütze, M. Malessa, Standardisation of Thermal Cycling Exposure Testing, Woodhead Publishing, Cambridge, England, 2007.
[198] W.J. Quadakkers, Growth mechanisms of oxide scales on ODS alloys in the temperature range 1000–1100°C, Werkst. Korros. 41 (1990) 659–668.
[199] B.A. Pint, I.G. Wright, Oxidation behavior of ODS Fe–Cr alloys, Oxid. Met. 63 (2005) 193–213.
[200] C. Wagner, in: Proc. Thirty-Second National Metal Congress and Exposition, Chicago, October 21 to 27, 1950, Chicago, IL, 1950.
[201] W.J. Quadakkers, H. Holzbrecher, K.G. Briefs, H. Beske, Differences in growth mechanisms of oxide scales formed on ODS and conventional wrought alloys, Oxid. Met. 32 (1989) 67–88.
[202] P.Y. Hou, I.G. Brown, J. Stringer, Study of the effect of reactive-element addition by implanting metal ions in a preformed oxide layer, Nucl. Instrum. Methods Phys. Res. Sect. B 59–60 (Part 2) (1991) 1345–1349.
[203] J. Zurek, G.H. Meier, E. Essuman, M. Hänsel, L. Singheiser, W.J. Quadakkers, Effect of specimen thickness on the growth rate of chromia scales on Ni-base alloys in high- and low-pO_2 gases, J. Alloys Compd. 467 (2009) 450–458.
[204] M. Michalik, M. Hänsel, J. Zurek, L. Singheiser, W.J. Quadakkers, Effect of water vapour on growth and adherence of chromia scales in high and low pO_2-environments at 1000 and 1050°C, Mater. High Temp. 22 (2005) 213–221.
[205] M. Michalik, M. Hänsel, W.J. Quadakkers, Effect of water vapour on growth and adherence of chromia scales on pure chromium, Schriften des Forschungszentrums Jülich, Reihe Energietechnik/Energy Technology, RWTH Aachen, Jülich, 2007.
[206] E. Essuman, G.H. Meier, J. Żurek, M. Hänsel, W.J. Quadakkers, The effect of water vapor on selective oxidation of Fe–Cr alloys, Oxid. Met. 69 (2008) 143–162.

[207] E. Essuman, G.H. Meier, J. Żurek, M. Hänsel, L. Singheiser, W.J. Quadakkers, Enhanced internal oxidation as trigger for breakaway oxidation of Fe–Cr alloys in gases containing water vapor, Scr. Mater. 57 (2007) 845–848.
[208] J. Zurek, G.H. Meier, E. Wessel, L. Singheiser, W.J. Quadakkers, Temperature and gas composition dependence of internal oxidation kinetics of an Fe-10%Cr alloy in water vapour containing environments, Mater. Corros. 62 (2011) 504–513.
[209] T. Malkow, W.J. Quadakkers, L. Singheiser, H. Nickel, Untersuchungen zum Lanzeitverhalten von metallischen Interkonnektorwerkstoffen der Hochtemperatur-Brennstofzelle (SOFC) im Hinblick auf die Kompatibilität mit kathodenseitigen Kontaktschichten, Berichte des Forschungszentrum Jülich, Forschungszentrum Jülich, Jülich, 1998, p. 171.
[210] H.J. Grabke, Carburisation, carbide formation, metal dusting, cocikng, Mater. Technol. 36 (2002) 297–305.
[211] D.J. Young, High Temeperature Oxidation and Corrosion of Metals, Elsevier, Amsterdam, 2008.
[212] Y. Zhao, J. Fergus, High temperature oxidation behavior of stainless steel 441 in dual atmosphere – effects of flow rate and humidity, ECS Trans. 16 (2009) 57–64.
[213] K. Nakagawa, Y. Matsunaga, T. Yanagisawa, Corrosion behavior of ferritic steels on the air sides of boiler tubes in a steam/air dual environment, Mater. High Temp. 18 (2001) 51–56.
[214] J. Ehlers, D.J. Young, E.J. Smaardijk, A.K. Tyagi, H.J. Penkalla, L. Singheiser, W.J. Quadakkers, Enhanced oxidation of the 9%Cr steel P91 in water vapour containing environments, Corros. Sci. 48 (2006) 3428–3454.
[215] H. Kurokawa, Oxidation behavior of Fe–16Cr alloy interconnect for SOFC under hydrogen potential gradient, Solid State Ionics 168 (2004) 13–21.
[216] M. Ziomek-Moroz, G.R. Holcomb, B.S. Covino, S.J. Bullard, P.D. Jablonski, D. Alman, in: Proc. 30th International Conference on Advanced Ceramics and Composites, Cocoa Beach, Florida, 2006.
[217] L. Niewolak, M. Malessa, S.Y. Coleman, W.J. Quadakkers, M. Schütze, Influence of cycling parameter variation on thermal cyclic oxidation testing of high temperature materials (COTEST), Mater. Corros. 57 (2006) 31–42.
[218] F. Wiener, M. Bram, H.P. Buchkremer, D. Sebold, Chemical interaction between Crofer 22 APU and mica-based gaskets under simulated SOFC conditions, J. Mater. Sci. 42 (2007) 2643–2651.
[219] A.M. Huntz, A. Piant, J.L. Lebrun, S. Daghigh, Evidence of stress relaxation in thermally grown oxide layers—experiments and modelling, Mater. Sci. Eng. A 248 (1998) 44–55.
[220] C. Asensio-Jimenez, Effect of Composition, Microstructure and Component Thickness on the Oxidation Behaviour of Laves Phase Strengthened Interconnect Steel for Solid Oxide Fuel Cells (SOFC), RWTH Aachen, Jülich, 2014, 210 p.
[221] A. Holt, P. Kofstad, Electrical conductivity and defect structure of Cr_2O_3. I. High temperatures ($>$ ~1000°C), Solid State Ionics 69 (1994) 127–136.
[222] A. Holt, P. Kofstad, Electrical conductivity and defect structure of Cr_2O_3. II. Reduced temperatures ($<$ ~1000°C), Solid State Ionics 69 (1994) 137–143.
[223] J.S. Park, H.G. Kim, Electrical Conductivity and Defect Models of MgO-Doped Cr_2O_3, J. Am. Ceram. Soc. 71 (1988) 173–176.
[224] A. Holt, P. Kofstad, Electrical conductivity and defect structure of Mg-doped Cr_2O_3, Solid State Ionics 100 (1997) 201–209.
[225] A. Holt, P. Kofstad, Electrical conductivity of Cr_2O_3 doped with TiO_2, Solid State Ionics 117 (1999) 21–25.

[226] R.F. Huang, A.K. Agarwal, H.U. Anderson, Oxygen activity dependence of the electrical conductivity of Li doped Cr_2O_3, J. Am. Ceram. Soc. 67 (1984) 146–150.
[227] H. Nagai, S. Ishikawa, N. Amano, K. Shoji, Electrical conductivity of sintered Cr_2O_3 simultaneously doped with NiO and rare earth oxide at low oxygen potential, Trans. Jpn. Inst. Metals 26 (1985) 753–762.
[228] A. Atkinson, M. Levy, S. Roche, R. Rudkin, Defect properties of Ti-doped Cr_2O_3, Solid State Ionics 177 (2006) 1767–1770.
[229] H. Nagai, S. Ishikawa, K. Shoji, Electrical conductivity of sintered Cr_2O_3 with Fe_2O_3, Trans. Jpn. Inst. Metals 26 (1985) 44–51.
[230] W.C. Hagel, Electrical conductivity of Li-substituted Cr_2O_3, J. Appl. Phys. 36 (1965) 2586.
[231] Z. Lu, J. Zhu, E. Andrew Payzant, M.P. Paranthaman, Electrical conductivity of the manganese chromite spinel solid solution, J. Am. Ceram. Soc. 88 (2005) 1050–1053.
[232] J.H. Zhu, S.J. Geng, Z.G. Lu, W.D. Porter, Evaluation of binary Fe–Ni alloys as intermediate-temperature SOFC interconnect, J. Electrochem. Soc. 154 (2007) B1288.
[233] N. Sakai, T. Horita, Y.P. Xiong, K. Yamaji, H. Kishimoto, M.E. Brito, H. Yokokawa, T. Maruyama, Structure and transport property of manganese–chromium–iron oxide as a main compound in oxide scales of alloy interconnects for SOFCs, Solid State Ionics 176 (2005) 681–686.
[234] E. Stefan, J.T.S. Irvine, Synthesis and characterization of chromium spinels as potential electrode support materials for intermediate temperature solid oxide fuel cells, J. Mater. Sci. 46 (2010) 7191–7197.
[235] S.T. Ertl, Untersuchungen zur oxidationsbedingten Lebensdauer von Chromstählen für die Anwendung in der Hochtemperaturbrennstoffzelle (SOFC), Fakultät für Maschinenwesen, Rheinisch-Westfälischen Technischen Hochschule, Aachen, 2006, p. 146.
[236] Z. Yang, G.-G. Xia, C.-M. Wang, Z. Nie, J. Templeton, J.W. Stevenson, P. Singh, Investigation of iron–chromium–niobium–titanium ferritic stainless steel for solid oxide fuel cell interconnect applications, J. Power Sources 183 (2008) 660–667.
[237] F. Tietz, D. Sebold, A. Brisse, J. Schefold, Degradation phenomena in a solid oxide electrolysis cell after 9000 h of operation, J. Power Sources 223 (2013) 129–135.
[238] Y. Matsuzaki, I. Yasuda, Dependence of SOFC cathode degradation by chromium-containing alloy on compositions of electrodes and electrolytes, J. Electrochem. Soc. 148 (2001) A126–A131.
[239] E. Konysheva, H. Penkalla, E. Wessel, J. Mertens, U. Seeling, L. Singheiser, K. Hilpert, Chromium poisoning of perovskite cathodes by the ODS alloy Cr5Fe1Y[sub 2]O[sub 3] and the high chromium ferritic steel Crofer22APU, J. Electrochem. Soc. 153 (2006) A765.
[240] A. Neumann, N.H. Menzler, I. Vinke, H. Lippert, Systematic study of chromium poisoning of LSM cathodes – single cell tests, ECS Trans. 25 (2009) 2889–2898.
[241] W.C. Hagel, Factors controlling the high-temperature oxidation of chromium, Trans. ASME 56 (1963) 583–599.
[242] C.S. Tedmon, The effect of oxide volatilization on the oxidation kinetics of Cr and Fe-Cr alloys, J. Electrochem. Soc. 113 (1966) 766.
[243] D. Caplan, M. Cohen, The volatilization of chromium oxide, J. Electrochem. Soc. 108 (1961) 438–442.
[244] B.B. Ebbinghaus, Thermodynamics of gas phase chromium species: the chromium chlorides, oxychlorides, fluorides, oxyfluorides, hydroxides, oxyhydroxides, mixed oxyfluorochlorohydroxides, and volatility calculations in waste incineration processes, Combust. Flame 101 (1995) 311–338.

[245] C. Gindorf, L. Singheiser, K. Hilpert, Vaporisation of chromia in humid air, J. Phys. Chem. Solids 66 (2005) 384–387.
[246] B.B. Ebbinghaus, Thermodynamics of gas phase chromium species: the chromium oxides, the chromium oxyhydroxides, and volatility calculations in waste incineration processes, Combust. Flame 93 (1993) 119–137.
[247] M. Casteel, D.J. Lewis, A. Renko, P. Willette, Characterization of vaporization rates on SOFC interconnect alloys, ECS Trans. 35 (2011) 2601–2607.
[248] J. Froitzheim, E. Larsson, L.-G. Johansson, J.-E. Svensson, Cr evaporation of metallic interconnects: a novel method for quantification, ECS Trans. 25 (2009) 1423–1428.
[249] A.H. Heuer, V.L.K. Lou, Volatility diagrams for silica, silicon-nitride, and silicon-carbide and their application to high-temperature decomposition and oxidation, J. Am. Ceram. Soc. 73 (1990) 2789–2803.
[250] P.J. Meschter, E.J. Opila, N.S. Jacobson, Water vapor–mediated volatilization of high-temperature materials, Annu. Rev. Mater. Res. 43 (2013) 559–588.
[251] P.S. Gentile, S.W. Sofie, C.F. Key, R.J. Smith, Silicon volatility from alumina and aluminosilicates under solid oxide fuel cell operating conditions, Int. J. Appl. Ceram. Technol. 9 (2012) 1035–1048.
[252] O.A. Marina, L.R. Pederson, E.C. Thomsen, C.A. Coyle, K.J. Yoon, Reversible poisoning of nickel/zirconia solid oxide fuel cell anodes by hydrogen chloride in coal gas, J. Power Sources 195 (2010) 7033–7037.
[253] N.H. Menzler, F. Tietz, S. Uhlenbruck, H.P. Buchkremer, D. Stöver, Materials and manufacturing technologies for solid oxide fuel cells, J. Mater. Sci. 45 (2010) 3109–3135.
[254] B.C.H. Steele, Materials for IT-SOFC stacks 35 years R&D: the inevitability of gradualness? Solid State Ionics 134 (2000) 3–20.
[255] O. Teller, W.A. Meulenberg, F. Tietz, E. Wessel, W.J. Quadakkers, in: Proc. 7th Int. Symp. Solid Oxide Fuel Cells (SOFC-VII), Tsukuba, Japan, 2001.
[256] W. Kobsiriphat, S. Barnett, Ag–Cu–Ti braze materials for sealing SOFCs, J. Fuel Cell Sci. Technol. 5 (2008) 011002.
[257] J. Schilm, A. Rost, M. Kusnezoff, A. Michaelis, Sealing glasses for SOFC – degradation behaviour, Advances in Solid Oxide Fuel Cells V, John Wiley & Sons, Inc., 2010, pp. 183–193.
[258] J.W. Fergus, Sealants for solid oxide fuel cells, J. Power Sources 147 (2005) 46–57.
[259] S.M. Gross, R. Conradt, P. Geasee, V. Shemet, W.J. Quadakkers, J. Remmel, U. Reisgen, in: Proc. 6th European Solid Oxide Fuel Cell Forum, Luzern, CH, 2004.
[260] P. Batfalsky, V.A.C. Haanappel, J. Malzbender, N.H. Menzler, V. Shemet, I.C. Vinke, R. W. Steinbrech, Chemical interaction between glass–ceramic sealants and interconnect steels in SOFC stacks, J. Power Sources 155 (2006) 128–137.
[261] Y.-S. Chou, J.W. Stevenson, P. Singh, Effect of aluminizing of Cr-containing ferritic alloys on the seal strength of a novel high-temperature solid oxide fuel cell sealing glass, J. Power Sources 185 (2008) 1001–1008.
[262] D.H. Peck, M. Miller, H. Nickel, D. Das, K. Hilpert, in: Proc. 4th Int. Symp. Solid Oxide Fuel Cells (SOFC-IV), Yokohama, 1995.
[263] Y. Matsuzaki, I. Hishinuma, I. Yasuda, in: Proc. 6th Int. Symp. Solid Oxide Fuel Cells (SOFC-VI), Honolulu, Hawaii, 1999.
[264] S.P. Jiang, J.P. Zhang, L. Apateanu, K. Foger, Deposition of chromium species at Sr-doped $LaMnO_3$ electrodes in solid oxide fuel cells. I. Mechanism and kinetics, J. Electrochem. Soc. 147 (2000) 4013–4022.
[265] N.H. Menzler, P. Batfalsky, S. Groß, V. Shemet, F. Tietz, Post-test characterization of an SOFC short-stack after 17,000 hours of steady operation, ECS Trans. 35 (2011) 195–206.

[266] S.P. Jiang, S. Zhang, Y.D. Zhen, Deposition of Cr species at (La, Sr) (Co, Fe) O_3 cathodes of solid oxide fuel cells, J. Electrochem. Soc. 153 (2006) A127–A134.

[267] H. Schmidt, B. Brückner, K. Fischer, in: Proc. 4th Int. Symp. Solid Oxide Fuel Cells (SOFC-IV), Yokohama, Japan, 1995.

[268] H.P. Buchkremer, U. Diekmann, L.G.J. de Haart, H. Kabs, U. Stimming, D. Stöver, in: Proc. 5th Int. Symp. Solid Oxide Fuel Cells (SOFC-V), Aachen, 1997.

[269] A. Petric, P. Huang, F. Tietz, Evaluation of La–Sr–Co–Fe–O perovskites for solid oxide fuel cells and gas separation membranes, Solid State Ionics 135 (2000) 719–725.

[270] Q.X. Fu, D. Sebold, F. Tietz, H.P. Buchkremer, Electrodeposited cobalt coating on Crofer22APU steels for interconnect applications in solid oxide fuel cells, Solid State Ionics 192 (2011) 376–382.

[271] Y. Larring, T. Norby, Spinel and perovskite functional layers between plansee metallic interconnect (Cr-5 wt % Fe-1 wt % Y_2O_3) and ceramic (La0.85Sr0.15)0.91MnO_3 cathode materials for solid oxide fuel cells, J. Electrochem. Soc. 147 (2000) 3251–3256.

[272] P. Hou, K. Huang, W.T. Bakker, in: Proc. 6th Int. Symp. Solid Oxide Fuel Cells (SOFC-VI), Honolulu, Hawaii, 1999.

[273] L.G.J. de Haart, J. Mougin, O. Posdziech, J. Kiviaho, N.H. Menzler, Stack degradation in dependence of operation parameters; the real-SOFC sensitivity analysis, Fuel Cells 9 (2009) 794–804.

[274] K. Fujita, K. Ogasawara, Y. Matsuzaki, T. Sakurai, Prevention of SOFC cathode degradation in contact with Cr-containing alloy, J. Power Sources 131 (2004) 261–269.

[275] X. Montero, N. Jordán, J. Pirón-Abellán, F. Tietz, D. Stöver, M. Cassir, I. Villarreal, Spinel and perovskite protection layers between Crofer22APU and La0.8Sr0.2FeO_3 cathode materials for SOFC interconnects, J. Electrochem. Soc. 156 (2009) B188–B196.

[276] X. Montero, F. Tietz, D. Stöver, M. Cassir, I. Villarreal, Comparative study of perovskites as cathode contact materials between an La0.8Sr0.2FeO_3 cathode and a Crofer22APU interconnect in solid oxide fuel cells, J. Power Sources 188 (2009) 148–155.

[277] J.-J. Choi, J.-H. Lee, D.-S. Park, B.-D. Hahn, W.-H. Yoon, H.-T. Lin, Oxidation resistance coating of LSM and LSCF on SOFC metallic interconnects by the aerosol deposition process, J. Am. Ceram. Soc. 90 (2007) 1926–1929.

[278] Z. Yang, G.-G. Xia, G.D. Maupin, J.W. Stevenson, Evaluation of perovskite overlay coatings on ferritic stainless steels for SOFC interconnect applications, J. Electrochem. Soc. 153 (2006) A1852–A1858.

[279] J.-J. Choi, D.-S. Park, B.-D. Hahn, J. Ryu, W.-H. Yoon, Oxidation behavior of ferritic steel alloy coated with highly dense conducting ceramics by aerosol deposition, J. Am. Ceram. Soc. 91 (2008) 2601–2606.

[280] M. Zahid, F. Tietz, D. Sebold, H.P. Buchkremer, in: Proc. 6th European SOFC Forum, Luzerne, 2004.

[281] B. Borglum, E. Tang, M. Pastula, Development of solid oxide fuel cells at versa power systems, ECS Trans. 35 (2011) 63–69.

[282] A. Mai, B. Iwanschitz, U. Weissen, R. Denzler, D. Haberstock, V. Nerlich, A. Schuler, Status of Hexis' SOFC stack development and the Galileo 1000 N micro-CHP system, ECS Trans. 35 (2011) 87–95.

[283] S. Mukerjee, K. Haltiner, R. Kerr, J.-Y. Kim, V. Sprenkle, Latest update on Delphi's solid oxide fuel cell stack for transportation and stationary applications, ECS Trans. 35 (2011) 139–146.

[284] R.J. Payne, J. Love, M. Kah, CFCL's BlueGen product, ECS Trans. 35 (2011) 81–85.

[285] H. Yoshida, T. Seyama, T. Sobue, S. Yamashita, Development of residential SOFC CHP system with flatten tubular segmented-in-series cells stack, ECS Trans. 35 (2011) 97–103.
[286] F. Tietz, A. Schmidt, M. Zahid, Investigation of the quasi-ternary system $LaMnO_3$–$LaCoO_3$–"$LaCuO_3$"—I: the series La(Mn0.5Co0.5)1−xCuxO_3−δ, J. Solid State Chem. 177 (2004) 745–751.
[287] F. Tietz, I. Arul Raj, Q.X. Fu, M. Zahid, Investigation of the quasi-ternary system $LaMnO_3$–$LaCoO_3$–"$LaCuO_3$". II: The series LaMn0.25−x Co0.75−x Cu_2x O_3−δ and LaMn0.75−x Co0.25−x Cu_2x O_3−δ, J. Mater. Sci. 44 (2009) 4883–4891.
[288] M.J. Garcia-Vargas, M. Zahid, F. Tietz, A. Aslanides, ECS Trans. 7 (2007) 2399–2405.
[289] M.R. Bateni, P. Wei, X. Deng, A. Petric, Spinel coatings for UNS 430 stainless steel interconnects, Surf. Coat. Technol. 201 (2007) 4677–4684.
[290] M. Bertoldi, T. Zandonella, D. Montinaro, V.M. Sglavo, A. Fossati, A. Lavacchi, C. Giolli, U. Bardi, Protective coatings of metallic interconnects for IT-SOFC application, J. Fuel Cell Sci. Technol. 5 (2008) 011001.
[291] X. Deng, P. Wei, M.R. Bateni, A. Petric, Cobalt plating of high temperature stainless steel interconnects, J. Power Sources 160 (2006) 1225–1229.
[292] W. Wei, W. Chen, D.G. Ivey, Anodic electrodeposition of nanocrystalline coatings in the Mn−Co−O system, Chem. Mater. 19 (2007) 2816–2822.
[293] J.-J. Choi, J. Ryu, B.-D. Hahn, W.-H. Yoon, B.-K. Lee, D.-S. Park, Dense spinel $MnCo_2O_4$ film coating by aerosol deposition on ferritic steel alloy for protection of chromic evaporation and low-conductivity scale formation, J. Mater. Sci. 44 (2009) 843–848.
[294] M. Gindrat, A. Refke, O. Sabouni, R. Damani, H.-M. Höle, Thermal spray processes and materials for coating applications of SOFC components, ECS Trans. 25 (2009) 1369–1377.
[295] D.P. Lim, D.S. Lim, J.S. Oh, I.W. Lyo, Influence of post-treatments on the contact resistance of plasma-sprayed La0.8Sr0.2MnO_3 coating on SOFC metallic interconnector, Surf. Coat. Technol. 200 (2005) 1248–1251.
[296] D.-J. Jan, C.-T. Lin, C.-F. Ai, Structural characterization of La0.67Sr0.33MnO3 protective coatings for solid oxide fuel cell interconnect deposited by pulsed magnetron sputtering, Thin Solid Films 516 (2008) 6300–6304.
[297] W. Qu, J. Li, D.G. Ivey, Sol–gel coatings to reduce oxide growth in interconnects used for solid oxide fuel cells, J. Power Sources 138 (2004) 162–173.
[298] J.S. Yoon, J. Lee, H.J. Hwang, C.M. Whang, J.-W. Moon, D.-H. Kim, Lanthanum oxide-coated stainless steel for bipolar plates in solid oxide fuel cells (SOFCs), J. Power Sources 181 (2008) 281–286.
[299] L. Mikkelsen, M. Chen, P.V. Hendriksen, Å. Persson, N. Pryds, K. Rodrigo, Deposition of La0.8Sr0.2Cr0.97V0.03O_3 and $MnCr_2O_4$ thin films on ferritic alloy for solid oxide fuel cell application, Surf. Coat. Technol. 202 (2007) 1262–1266.
[300] W. Wei, X. Cui, W. Chen, D.G. Ivey, Electrochemical cyclability mechanism for MnO2 electrodes utilized as electrochemical supercapacitors, J. Power Sources 186 (2009) 543–550.

Cell and stack design, fabrication and performance

8

N.Q. Minh
University of California, San Diego, CA, USA

8.1 Introduction

A solid oxide fuel cell (SOFC) cell or single cell is a fully functional unit composed of an oxide electrolyte sandwiched between an anode and a cathode. The electrolyte, anode and cathode can be a single layer or a multi layer, each made of one or more than one material. Like any fuel cell technology, SOFC cells are stacked and connected in electrical series (via interconnects) to build voltage. A multi-cell stack is thus the fundamental building block of any practical fuel cell-based power systems. An SOFC stack also incorporates gas manifolds for ducting fuel and oxidant in and out of the stack structure and, depending on the design, may need sealing (commonly, edge sealing) to prevent gas cross-leakage. The main features of SOFC cell/stack technology are its solid-state construction (ceramics and metals) and high operating temperature (typically 600–1000 °C). The combination of these features not only leads to a number of advantages for the technology (such as cell/stack design flexibility, multiple cell fabrication options, multi-fuel capability and operating temperature choices) but also presents several technical challenges (such as limited material selections, increased undesirable chemical interactions between components and certain operation restrictions). This chapter provides an overview of SOFC cell and stack designs and fabrication processes and discusses key design factors and operating parameters impacting cell and stack performance.

8.2 Requirements

8.2.1 Requirements for single cells

An SOFC single cell designed for and fabricated from a selected set of materials must have desired physical, mechanical, chemical, electrical and electrochemical properties and characteristics set by the operating requirements of the intended application [1–3]. The discussion on requirements given in this section is qualitative as any specifics very much depend on selected materials and particular cell designs.

- *Physical requirements*: A single cell must have required shapes and dimensions with dimensional tolerances specified by its design. An example is required cell flatness for the planar design; planar cells must be as flat as possible to permit effective stacking. Another example is cell component thickness; thickness variations, especially for the electrolyte (typically, the

High-temperature Solid Oxide Fuel Cells for the 21st Century. http://dx.doi.org/10.1016/B978-0-12-410453-2.00008-7

least conducting), must be minimised to promote uniform current, thus temperature, distribution within the cell during operation. The electrolyte must be fully dense without pinholes (pinholes tend to create hot spots due to local gas cross-leakage, resulting in undesirable thermal stresses with potential crack propagation in cell operation for extended periods of time). The electrodes must be sufficiently porous with appropriate microstructures to facilitate gas transport of reactants and products to and from reaction sites.

- *Mechanical requirements*: A single cell must have sufficient strength and toughness for handling as required in processing and stacking. The cell must have adequate bonding between the electrolyte and the electrodes and must be designed to accommodate possible mismatches in coefficients of thermal expansion between the different cell components.
- *Chemical requirements*: A single cell must have the required chemical stability. This requirement means that a single cell must be designed to minimise deleterious chemical interaction, elemental migration/interdiffusion and phase transformation within the cell. For example, single cells with certain perovskites such as lanthanum strontium cobalt iron perovskite (LSCF) as the cathode and yttria stabilised zirconia (YSZ) as the electrolyte tend to interact at the operating temperature. Thus, doped ceria as an interlayer is commonly incorporated to prevent reactions between the two components (e.g. Refs. [4,5]). This chemical stability is required not only at operating conditions, especially the operating temperature, but also at fabrication conditions, especially the much higher temperatures at which the cell is generally fabricated.
- *Electrical requirements*: A single cell must have ohmic losses as low as possible. This requirement means that the components (especially those having low electrical conductivity) must be as thin as feasible. In SOFCs, due to the conductivity/electrical requirements, the oxide electrolyte as the least conducting component defines the cell operating temperature. Thus, SOFC cells can be designed to operate at different temperatures depending on electrolyte thickness [6].
- *Electrochemical requirements*: A single cell must have the desired electrochemical performance. This requirement means that the cell design must provide for full open circuit voltages (OCVs) and minimal polarisation losses. Thus, any significant gas cross-leakage and electrical short in the electrolyte must be avoided. The electrode microstructure must be designed to provide appropriate and sufficient reactive sites (to promote the electrochemical reactions) and allow fuel and oxidant to quickly reach and reaction products to remove from the reactive sites (to avoid gas diffusion limitation).

Table 8.1 summarises the requirements for SOFC single cells.

8.2.2 Requirements for multi-cell stacks

An SOFC stack formed from selected single cells and interconnects must be designed to have the desired electrical and electrochemical performance along with required chemical stability, thermal management, gas flow distribution and pressure drop and mechanical/structural integrity to meet operating requirements of specified power generation applications. Similar to Section 8.2.1, the discussion here is qualitative because the specific requirements depend on specific materials and stack designs.

- *Electrical performance*: This requirement means that the design must minimise ohmic losses in the stack. Thus, the current path in the components must be designed to be as short as possible. There must be good electrical contact and sufficient contact area between cells and interconnects. The current collector must also be designed to facilitate current distribution and flow in the stack.

Table 8.1 **Requirements for single cells**

Requirement	Property requirement	Target
Physical	Shape, dimension, density/ porosity	Required shape and dimensional tolerances for cells and cell components Fully dense electrolyte without pinholes and sufficiently porous electrodes
Mechanical	Strength and toughness	Adequate strength and toughness for handling Good bonding between components Design to accommodate thermal expansion mismatches
Chemical	Chemical stability	Insignificant deleterious chemical interaction, elemental migration/interdiffusion and phase transformation within cell during fabrication and operation
Electrical	Ohmic loss	Cell components as thin as feasible
Electrochemical	OCV and polarisation loss	Insignificant gas cross-leakage and no electrical short in electrolyte Electrode microstructure to provide appropriate and sufficient reactive sites and fast gas diffusion to and from reactive sites

- *Electrochemical performance*: This requirement means that the design must provide for full OCVs and minimal cell-to-cell performance variations. Thus, any significant gas leakage or cross-leakage and electrical short must be avoided. Cell-to-cell electrochemical performance must be as uniform as required to achieve optimal stack performance.
- *Chemical stability*: This requirement means that the design must minimise undesirable chemical interaction and elemental migration/interdiffusion between various stack components, especially between the cell and interconnect and between the cell and sealant.
- *Thermal management*: This requirement means that the design must provide means for stack cooling and more uniform temperature operation during operation. The design must also permit the highest possible gas temperature gradient across the stack [7].
- *Gas flow distribution and pressure drop*: This requirement means that the design must allow uniform distribution of fuel and oxidant across the area of each cell and to each cell of the stack. Minimising stack pressure drop is desired to lower power requirements for cathode air blower/compressor, thus reducing parasitic power consumption in the system [7].
- *Mechanical/structural integrity*: This requirement means that any SOFC stack must be designed to have adequate mechanical strength for assembly and handling. Thus, mechanical and thermal stresses must be kept to minimum to prevent cracking, delamination or detachment of the components under the variety of operating conditions the stack is expected to experience (e.g. normal operating temperature gradients, off-design temperature gradients, thermal shock conditions such as sudden power change and cold start-up and mechanical loading expected during installation, moving and vibration loading).

Table 8.2 summarises the requirements for SOFC multi-cell stacks.

Table 8.2 **Requirements for multi-cell stacks**

Requirement	Property requirement	Target
Electrical performance	Minimal ohmic loss	Short current path Good electrical contact and sufficient contact area Current collector design for uniform and short current path
Electrochemical performance	Full OCV, minimal cell-to-cell performance variation	Insignificant gas leakage or cross-leakage No electrical short Design for uniform cell-to-cell performance
Chemical stability	Required chemical stability	No detrimental chemical interaction and elemental interdiffusion between cell and interconnect and between cell and sealant
Thermal management	Required cooling, uniform stack temperature distribution and highest possible gas temperature gradient across stack	Simple and efficient means for cooling Appropriate gas flow configuration Design to permit highest gas temperature gradient
Gas flow distribution and pressure drop	Uniform gas distribution and reduced pressure drop	Design for uniform distribution of fuel and oxidant across cell area and to each cell Design to reduce stack pressure drop
Mechanical/ structural integrity	Adequate mechanical strength	Design for minimal mechanical and thermal stress

8.3 SOFC single cell

8.3.1 Cell design

An SOFC single cell can be configured into various shapes (tube, flat-plate or corrugated structure) because all the components are solid [2]. Each configuration may have several different versions. For example, the tubular cell configuration can be a closed-one-end tube [3], a microtube [8] or a flattened tube [9] (Figure 8.1). The flat-plate or planar cell configuration can be a rectangular plate, a circular disc or several other shapes. The three components of the cell (electrolyte, anode and cathode) can be a single layer or a multi layer, each made of one or more than one material, depending on selected materials and intended functions. For example, bilayer electrolytes consisting of YSZ/doped CeO_2 [10] and doped Bi_2O_3/doped CeO_2 [11] have

Closed-one-end tubular cells | Micro-tubular cells | Flattened tubular cells

Figure 8.1 Tubular cell configurations.

been considered and developed for the SOFC. The YSZ layer of the YSZ/doped CeO_2 bilayer electrolyte is exposed to the fuel side to protect ceria from reduction (with attendant electronic conduction) by the fuel reducing environment. The doped CeO_2 of the doped Bi_2O_3/doped CeO_2 bilayer electrolyte exposed to the fuel side protects the doped Bi_2O_3 from decomposing, while the Bi_2O_3 layer on the oxidant side blocks the electronic leakage current in the doped CeO_2 layer. Another example is a dual-layer anode consisting of a Cu–CeO_2 impregnated Ni/YSZ support outer layer and a Ni/YSZ electroactive inner layer for the direct operation on ethanol [12]. SOFC single cells may also include interlayers between the electrolyte and the electrodes. One example is the use of ceria barrier layers to prevent interactions between LSCF cathodes and YSZ electrolytes [4,5].

Various options are available for designing SOFC cells. Single cells can be configured using different components for structural support. Thus, cells can be self-supporting or external supporting (built on a separate support material). In the cell-supporting design, one of the cell components, often thickest layer, acts as the cell structural support. Thus, single cells can be designed as electrolyte-supported, anode-supported or cathode-supported. In the external supporting design, the single cell is configured as thin layers on the interconnect (interconnect-supported) or a substrate (substrate-supported). The various cell designs are schematically shown in Figure 8.2. Figure 8.3 shows examples of micrographs of cells of different designs [3,13]. The key features of each cell design are summarised in Table 8.3.

8.3.2 Cell fabrication

Fabrication of SOFC single cells generally involves various process steps to incorporate starting materials into the desired cell configuration. The selection of processes and process sequences for cell fabrication thus depends on cell materials and cell designs. Any selected process should produce single cells that meet the design specifications and the requirements discussed in Section 8.2.1. Considerations should be given to process capability to manage possible mismatches in coefficient of thermal expansion and potential chemical interactions between different cell materials. Other

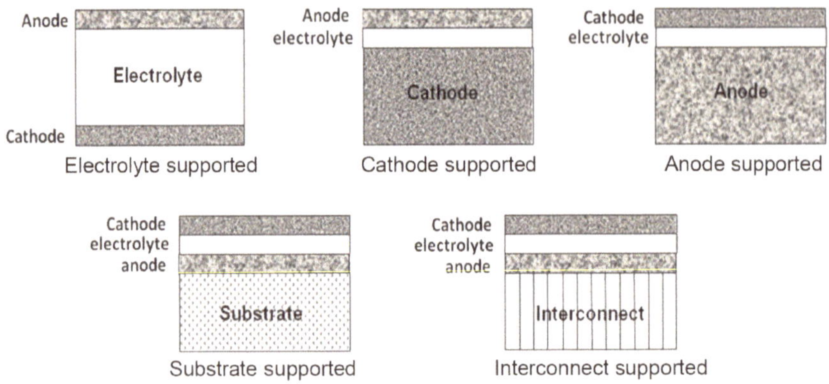

Figure 8.2 SOFC cell designs.

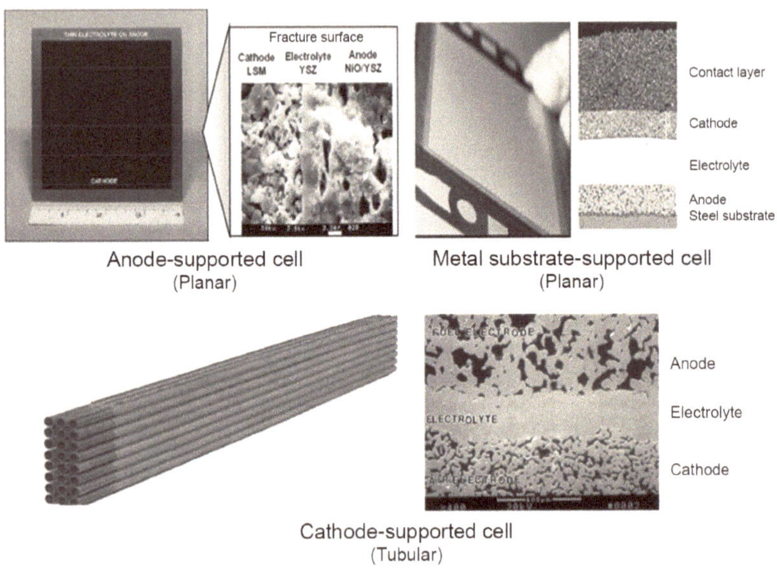

Figure 8.3 Examples of cell designs.

considerations for practical applications include process scalability and fabrication cost. The general issues in cell fabrication mainly relate to cell component microstructures (pinholes in electrolyte, electrode porous structures), thickness uniformity, adhesion/bonding, cell flatness/shape tolerances and microcracking.

The key step in any selected process sequence is the fabrication of fully dense electrolytes. Fabrication processes used to produce dense oxide electrolytes for the SOFC can be classified into two groups based on the fabrication approach: the particulate approach and the deposition approach. The particulate approach involves compaction of powders into layers and densification at elevated temperatures. An example of the

Table 8.3 **Features of cell designs**

Cell configuration	Advantage	Disadvantage
Self supporting		
Electrolyte-supported	Relatively strong structural support from dense electrolyte Less susceptible to failure due to anode re-oxidation (Ni/YSZ anodes)	Higher resistance due to low electrolyte conductivity Higher operating temperatures required to minimise electrolyte ohmic losses
Anode-supported	Highly conductive anode for reduced ohmic losses (Ni/YSZ anodes) Lower operating temperatures possible via use of thin electrolytes	Potential anode re-oxidation (Ni/YSZ anodes) Potential mass transport limitation due to thick anodes
Cathode-supported	No cathode oxidation issues Lower operating temperatures possible via use of thin electrolytes	Low cathode conductivity Potential mass transport limitation due to thick cathodes
External supporting		
Interconnect-supported	Thin cell components for lower operating temperatures Stronger structures from interconnects (especially metallic substrates)	Interconnect oxidation (metallic) Flowfield design limitation due to cell support requirement
Substrate-supported	Thin cell components for lower operating temperatures Potential for use of non-cell materials for support to improve properties	Increased complexity due to addition of non-cell materials Potential discontinuous layers on porous substrates

particulate process is tape casting. Tape casting involves making of a layer of slip (powder suspended in a liquid) using a doctor blade and drying this layer (or tape) on a temporary support. The dried layer (green tape) can be stripped from the support and fired to form a sintered layer. Multilayer tapes can be fabricated by sequentially casting one layer on top of another. The tape casting process is shown schematically in Figure 8.4. The deposition process involves formation of layers on a substrate by a chemical or physical process. An example of the deposition approach is plasma spraying. Plasma spraying is a thermal spraying technique in which melted (or heated) materials are sprayed onto a surface. Material in the form of powder is injected into a very high temperature plasma flame, where it is rapidly heated and accelerated to a high velocity. The hot material impacts on the substrate surface and rapidly cools

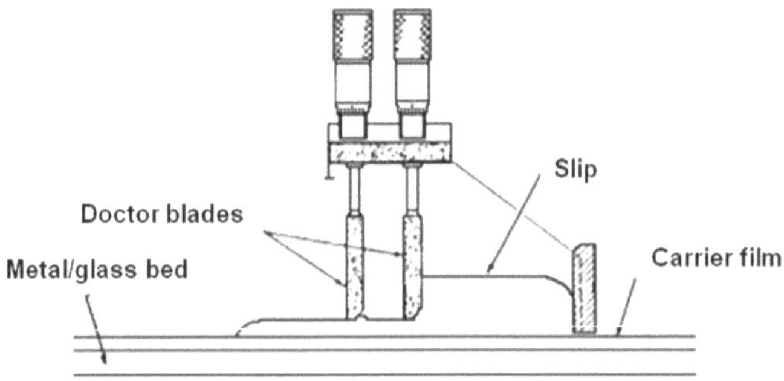

Figure 8.4 Tape casting.

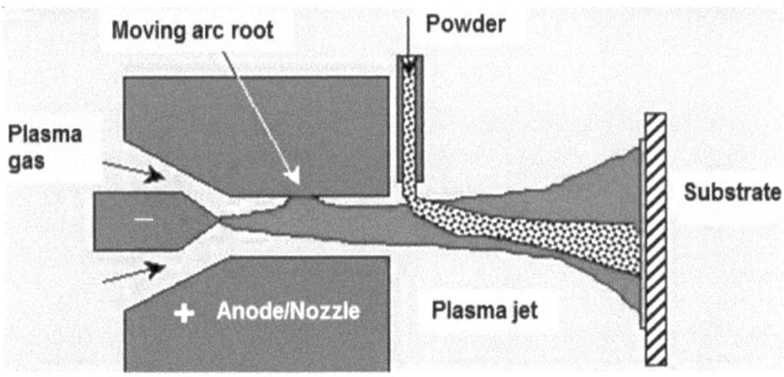

Figure 8.5 Plasma spraying.

forming a layer or coating. The substrate temperature can be kept low during processing avoiding damage, changes and distortion to the substrate material. The plasma spraying process is shown schematically in Figure 8.5. A wide range of fabrication processes have been considered and developed for the SOFC [2,14]. A list of selected processes is shown in Table 8.4.

A fabrication sequence based on tape calendering to make electrode-supported cells (e.g. anode-supported cells) is described here as an example of SOFC cell fabrication (Figure 8.6) [15]. The key steps of this sequence are tape forming, rolling, firing and electrode application. In the tape forming step, starting materials for electrolyte and anode tapes (powders, organic binders, plasticisers) are combined in a high shear mixer to create a homogeneous plastic mass. Each plastic mass is then rolled using a two-roll mill to produce a thin and flexible tape on the order of 0.25–2.5 mm thick. Green electrolyte and anode tapes are laminated and rolled using a two-roll mill in the rolling step. The rolling sequence can be devised to produce tapes with target thickness of the electrolyte (e.g. 1–10 μm) and electrolyte/electrode thickness ratio. The electrolyte and anode tapes of certain thickness ratio are laminated together to form a bilayer and subsequently rolled to reduce the thickness of each layer. Another anode tape is added to the bilayer and again rolled down. The addition of anode tapes has the effect of reducing the electrolyte thickness while maintaining

Table 8.4 **Selected SOFC cell fabrication processes**

Particulate process	Deposition process
Tape casting	Chemical vapour deposition/electrochemical vapour deposition
Tape calendering	Sputtering
Screen printing	Spray pyrolysis
Slurry coating	Pulsed laser deposition
Electrophoretic deposition	Plasma spraying

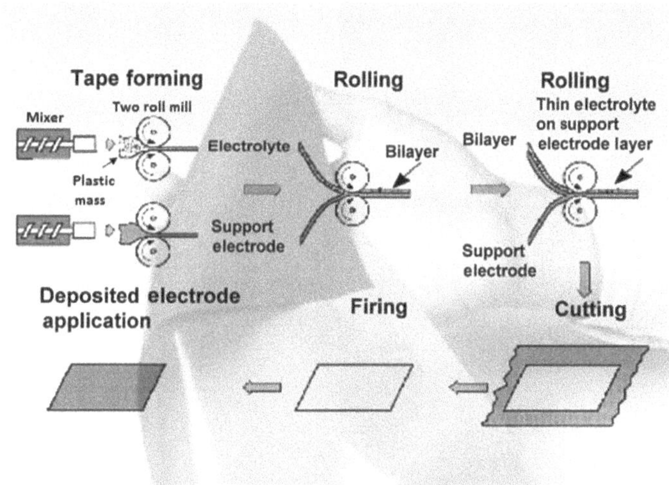

Figure 8.6 Cell fabrication sequence based on tape calendering.

the bilayer at a handleable thickness. Such process is repeated until the desired thickness of electrolyte is reached. An example of a final bilayer tape is shown in Figure 8.7a. The final green bilayer tape is cut to size and then fired in an air atmosphere to burn the organics and sinter the ceramic in the firing step. The cathode is applied (e.g. by printing) on the sintered bilayer and fired at an appropriate temperature to produce a single cell (Figure 8.7b). To produce cells meeting the required properties (Section 8.2.1), this tape calendering process requires definition of the processing/structure/property relationship for each layer. Thus, key relationships between starting powder characteristics, tape forming and firing parameters and sintered cell structure and properties need to be defined to achieve the required properties of each fabricated layer.

Fabrication of full-size cells is required for many practical applications. Tubular SOFC cells (typical diameters of >15 mm) have been fabricated in full active length (e.g. 150 cm) [3]. Planar SOFC cells, especially anode-supported cells, have been scaled up to sizes having active areas (e.g. 500–1000 cm^2) suitable for uses in large

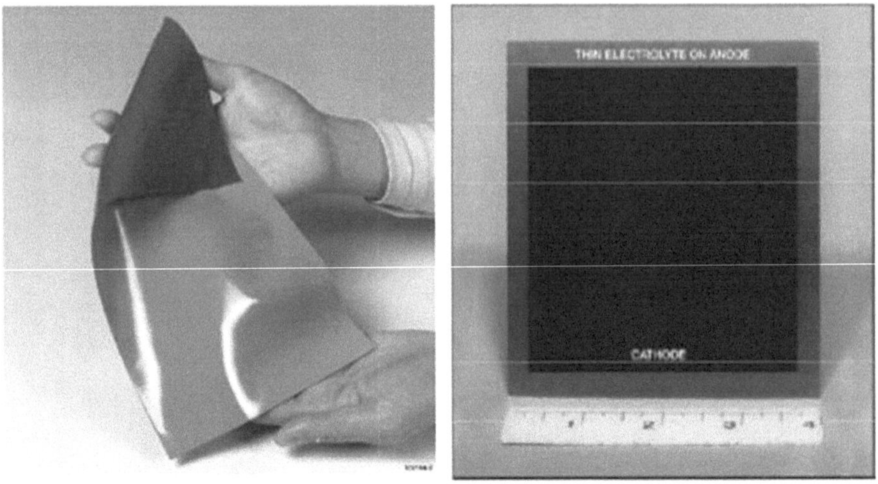

(a) (b)

Figure 8.7 (a) Green YSZ/NiO bilayer tape; (b) Anode-supported cell (anode not reduced).

Figure 8.8 Photograph of a large area (1200 cm^2) planar cell.

power systems. Manufacture of planar cells as large as 1200 cm^2 in total area has been demonstrated (Figure 8.8) [16]. In addition to fabrication scale-up for large power systems, SOFC cells have also been scaled down for certain applications such as consumer electronics devices and compact portable power supplies [17]. Examples of miniature SOFCs include micro-tubular (diameters of <5 mm) cells [17], thin-film cells micro-fabricated on silicon wafers [17] and single-chamber SOFCs [18].

8.3.3 Cell performance

SOFC single cells have exhibited peak power densities as high as 2 W/cm^2 (with pure hydrogen fuel and air oxidant, low fuel and air utilisations) and operated at temperatures as low as 400–500 °C, depending on electrolyte/electrode systems and cell structures and configurations [11,19,20]. An example of SOFC single cell

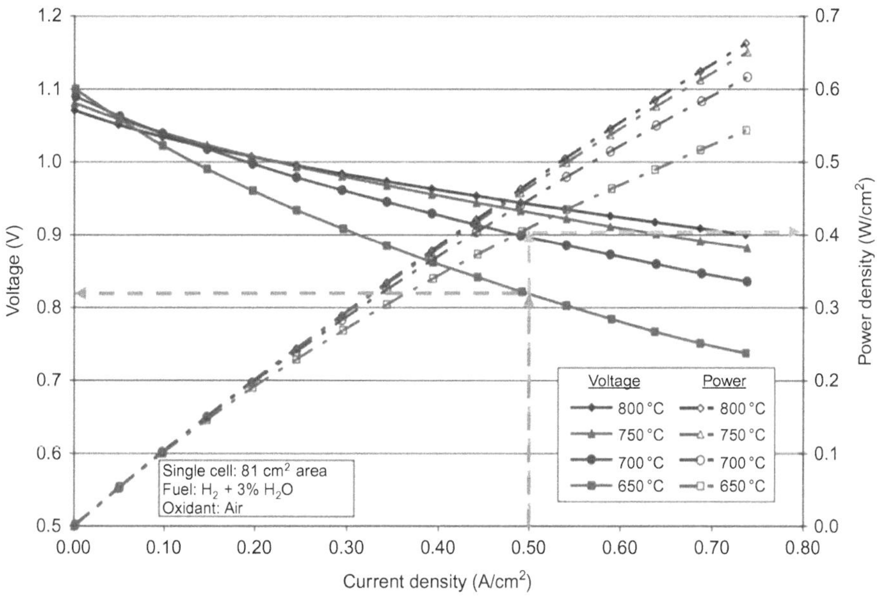

Figure 8.9 Example of SOFC single cell (anode-supported) performance.

performance is given in Figure 8.9 [21]. A power density of about 0.4 W/cm^2 at 0.82 V is achieved at 650 °C.

Cell performance generally increases with increasing temperatures and at a specified temperature varies depending on a number of operating parameters such as fuel and oxidant compositions, utilisations and pressures. Performance of SOFC cells is considerably higher when operated in oxygen instead of air [22]. Operation at higher fuel utilisations reduces cell power output. For practical applications, cells must be designed to support high fuel utilisations (>65–70%); fuel utilisations up to 90% have been demonstrated [23]. Pressurised operation, on the other hand, improves cell performance; testing up to 15 atm has been conducted [24]. Significant performance enhancement has been observed as pressure increases from 1 to 3 atm [25] and moderate improvement from 4 to 10 atm (Figure 8.10). When properly designed, SOFC cells have also been shown to have excellent thermal cyclability, but long startup and shutdown times are often required to prevent thermal shock.

For SOFC single cells having minimal ohmic resistance contributions from the components, cathode polarisation is generally the major contribution to cell performance losses (Figure 8.11) [26]. Thus, many cathode studies have been conducted and approaches developed to improve cathode performance [27]. One example is the use of infiltration as a potent means to form cathode nanostructures for electrode performance enhancement [28,29]. For example, infiltration of yttria-doped ceria (YDC) into LSM/YSZ cathode increased peak power density from 208 to 519 mW/cm^2 at 700 °C and power density at 0.7 V from 135 to 370 mW/cm^2 [28]. Infiltration of active components as dispersed particles or connected nanoparticulate

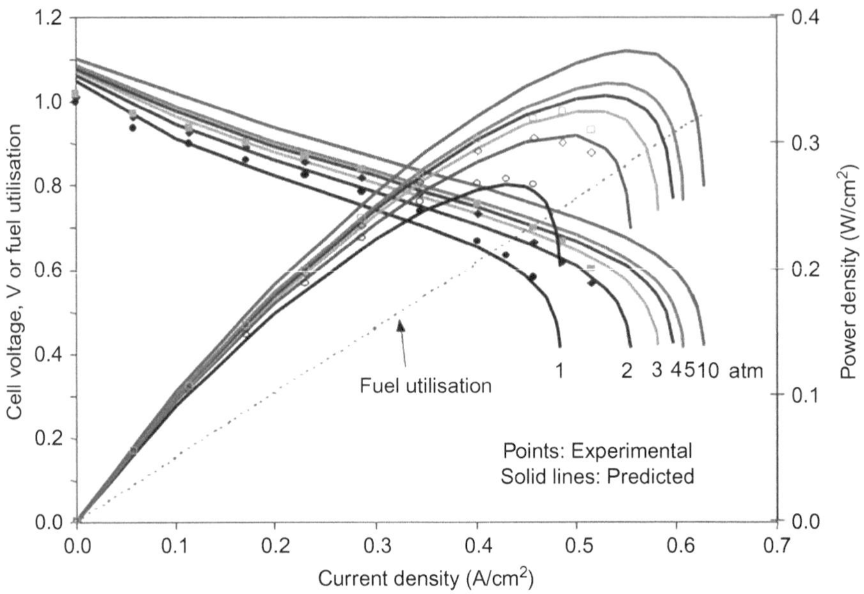

Figure 8.10 Cell performance at different operating pressures.

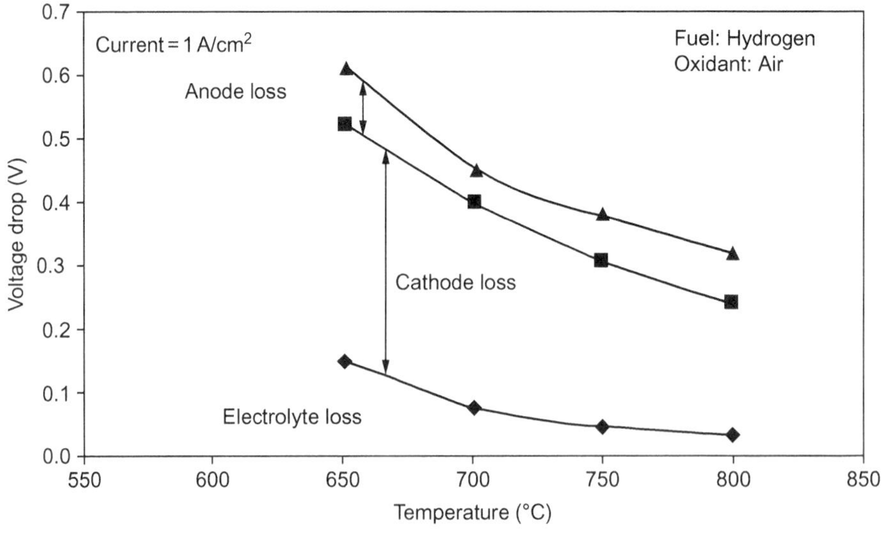

Figure 8.11 Contributions to cell performance loss.

networks to form nanostructures enhances cathode performance by modifying catalytic activities and/or conduction pathways of the electrode. Use of nanostructures has also been shown to improve anode performance [30]. The main issue is the stability of the nanostructure over extended periods of time at high operating temperatures. Operating the SOFC at reduced temperatures (e.g. <600 °C) or stabilising the nanostructure are potential approaches to maintain sufficient long-term stability [28,31].

SOFC cells can operate directly on fuels other than hydrogen (e.g. hydrocarbons, alcohols) via internal reforming (on fuel feeds with significant amounts of water) or direct utilisation (on fuel feeds with no water). Internal reforming using on-anode reformation is well known and has been demonstrated for the SOFC (Figure 8.12) [32]. Instead of complete (100%) internal reforming, it is possible to have a portion of the fuel reformed in an external reformer (referred to as a pre-reformer), and the resulting reformate plus the remaining fuel are fed to the SOFC where the fuel is internally reformed (via steam reforming) within the fuel cell. An example is the demonstration of operation of a 5 kW SOFC system with an auto-thermal reformation reformate containing about 7 volume% methane [33]. This pre-reforming/internal reforming option has been employed to use the endothermic reforming reactions to reduce cooling requirements in thermal management of the SOFC.

SOFC cells have been shown to have the capability for direct utilisation of different types of fuel [34]. For direct fuel utilisation operation, the anode material has been modified to address the carbon deposition issue associated with nickel commonly used in the anode composition (e.g. Cu/ceria instead of Ni/YSZ [35]). With modified anodes, high electrochemical performance can be achieved for direct SOFCs. For example, a peak power density of about 400 mW/cm^2 at 800 °C was obtained with 7.3% ethanol balance He fuel and air oxidant for an anode-supported SOFC with a

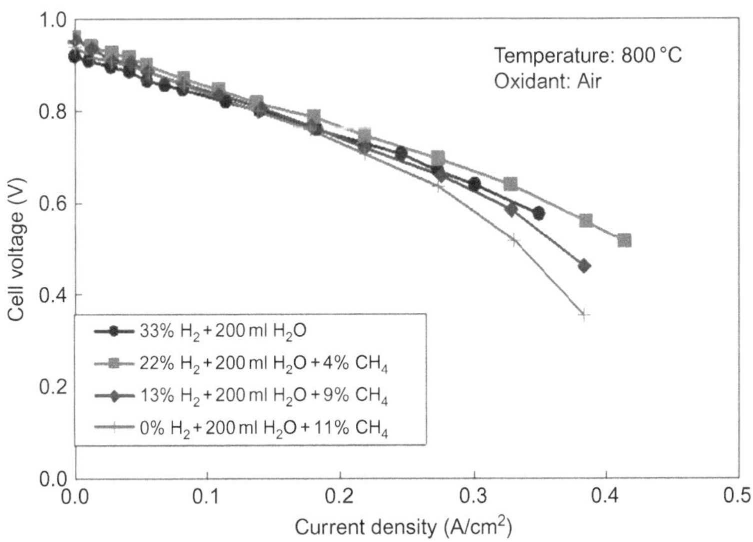

Figure 8.12 Cell performance with fuels containing water and methane.

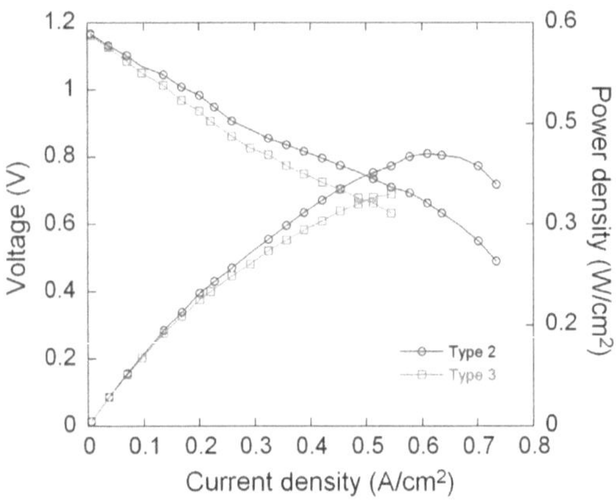

Figure 8.13 Performance at 800 °C with 7.3% ethanol balance He fuel and air oxidant for an anode-supported SOFC (with a dual-layer anode consisting of a Cu–CeO_2 impregnated Ni/YSZ support outer layer and a Ni/YSZ electroactive inner layer) (Type 2 and 3 indicates different thermal treatments of infiltrated anodes).

dual-layer anode consisting of a Cu–CeO_2 impregnated Ni/YSZ support outer layer and a Ni/YSZ electroactive inner layer [36] (Figure 8.13). Long-term performance stability of direct SOFCs without significant carbon deposition, however, remains to be demonstrated.

SOFC single cells, when properly prepared with conventional high-purity materials and operated on clean fuels and air, show minimal performance degradation for extended periods of time. For example, tubular cells were electrically tested for times as long as 8 years and showed satisfactory performance with less than 0.1% per 1000 h degradation [3]. SOFC cells, however, can experience significant performance degradation in realistic environments, depending on several factors such as gas input purities and component materials used in the stack/system [37]. Sulphur is the most prevalent fuel impurity in many practical fuels and its poisoning effects on the Ni/YSZ anode are well known [2]. It has also been shown that silicon impurities present in the fuel (originated from, for example, stack glass sealants or silica containing insulations in the system) can also poison the Ni/YSZ anode [38]. On the cathode side, presence of significant amounts of water or carbon dioxide in air can have deleterious effects on cell performance [39–41]. In long-term operation of cells in contact with metallic interconnects, cell performance can degrade due to chromium poisoning. Chromium present in the metallic interconnect can migrate to cathode reactive sites and interact with the cathode, poisoning the electrode thereby increasing cathode polarisation with time. The most common mitigating approach is to use conductive coatings (e.g. Co–Mn spinel) on the metallic interconnect to minimise the chromium transport and migration [42].

8.4 SOFC multi-cell stacks

8.4.1 Stack design

Four main stack designs have been proposed and developed for the SOFC: the tubular design, the segmented-cells-in-series design, the monolithic design and the planar (flat-plate) design [1,2]:

- *The tubular design* consists of tubular cells bundled in series and/or parallel electrical connection.
- *The segmented-cells-in-series design* consists of segmented cells (connected in electrical and gas flow series) either arranged as a thin banded structure on a porous support or fitted one into the other to form a self-supporting structure.
- *The monolithic design* consists of thin cell components and interconnects formed into a corrugated structure of either gas co-flow or cross-flow configurations.
- *The planar design* consists of cells and interconnects configured as planar plates and stacked in electrical series.

The designs differ in the extent of dissipative losses within the cells, in the manner of fuel and oxidant channels and in the cell-to-cell connection in a stack of cells. Each design may have several different versions. Figure 8.14 shows, as an example, the schematic diagrams of the various SOFC stack designs. Photographs of several stacks of different designs [24,43–45] are given in Figure 8.15. Table 8.5 compares the key characteristics of the four designs.

Three important features of the SOFC stack designs relate to gas flow configuration, electrical connection and gas manifolding which can be arranged in several ways:

(i) *Gas flow configurations*: Fuel and oxidant flows in SOFC stacks can be arranged to be cross-flow, co-flow or counter-flow. The selection of a particular flow configuration depends on the stack design and has significant effects on temperature and current distribution within the stack. Various flow patterns can be implemented in the different flow configurations, especially for the planar SOFC, including Z-flow, serpentine, radial and spiral patterns

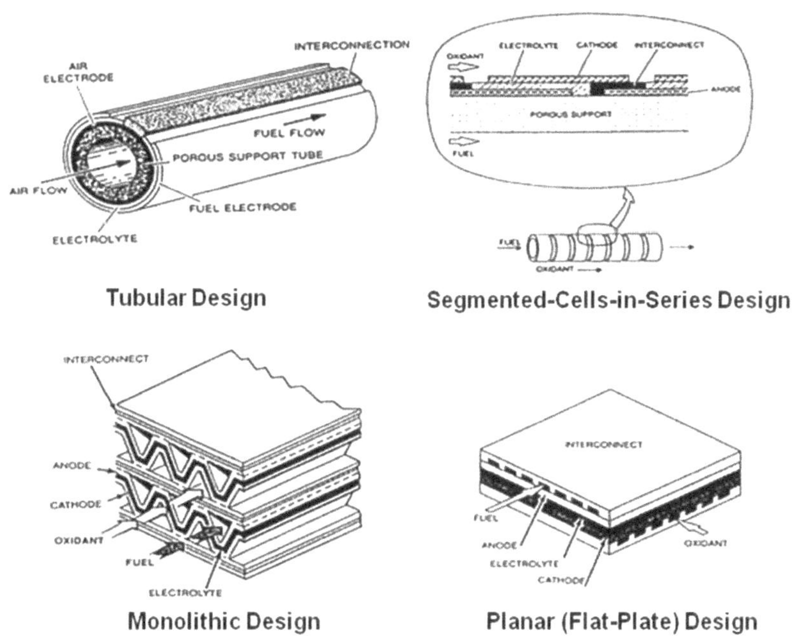

Figure 8.14 SOFC stack designs.

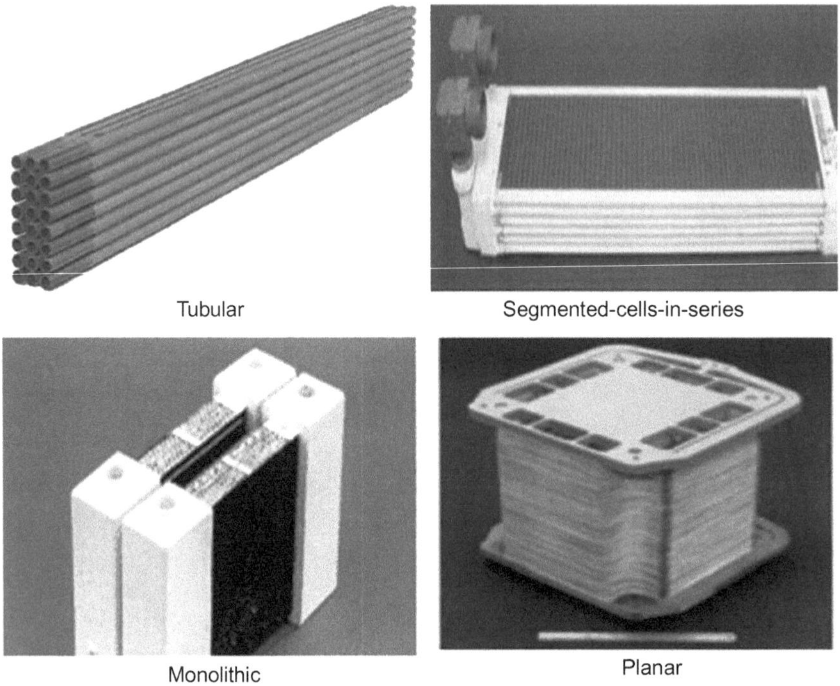

Figure 8.15 Examples of SOFC stacks of different designs.

Table 8.5 Features of stack designs

Design	Advantage	Disadvantage
Tubular	Less issues with thermally induced stresses Simpler sealing or sealless	Relatively long current path Relatively low power density
Segmented-cells-in-series	Assembly of a number of cells on a support Series gas flow improving performance	Seal issues due to many interconnects on an assembly Short cells to minimise current path
Monolithic	Compactness High power density	Difficulties to fabricate stack structures meeting design requirements Limits on selected materials and fabrication processes
Planar	Flexibility in terms of geometry, gas flow configuration and gas manifolding Flexibility in fabrication, especially with separate cells and interconnects	Potential scale-up limit to large sizes for self-supporting cells Stringent flatness and sealing requirements and potential contact issues (between cells and interconnects)

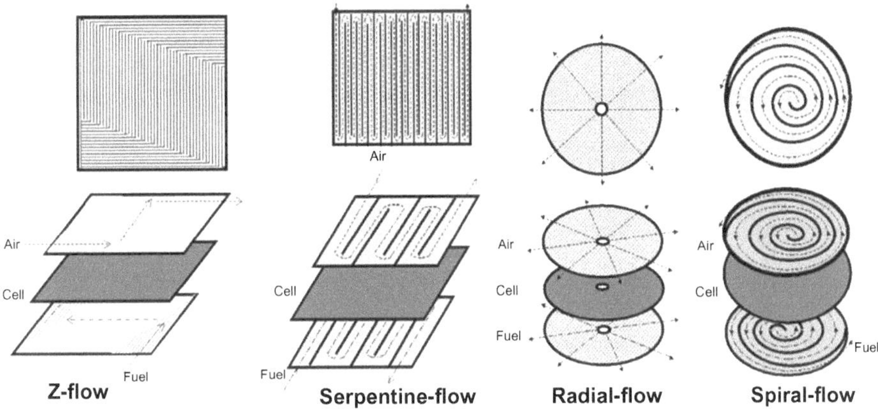

Figure 8.16 Gas flow patterns in planar SOFC stacks.

(Figure 8.16). Flowfields (flow channels) are used in planar stacks to increase uniformity of gas distribution and to promote heat and mass transport to each cell. In addition, the flowfield is often designed to have sufficient pressure drop through the cell to promote cell-to-cell flow uniformity within the stack. Thus, defining the flowfield for both fuel and oxidant flows is an important aspect in designing planar SOFCs. For a specific design, the shape and arrangement of the flowfield can be varied to improve/optimise stack design. Figure 8.17 shows two examples of flowfield design used in planar SOFCs [3]. Flowfields are commonly designed as part of the interconnect although certain planar designs include the flowfield in the electrodes [46]. Since the flowfield electrically connects the interconnect and the electrodes, contact area (between the flowfield and the electrodes) must be

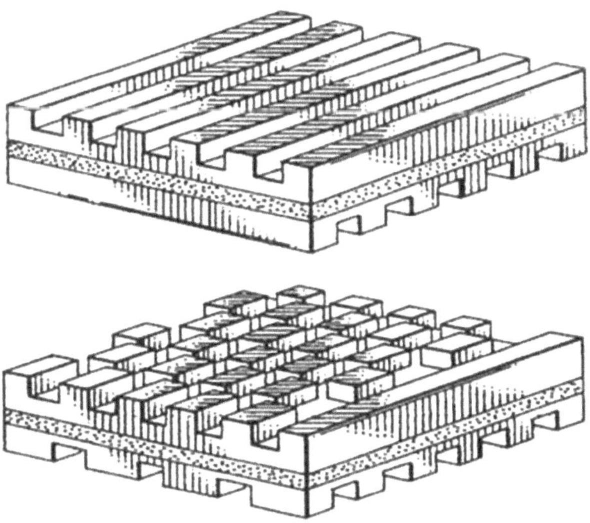

Figure 8.17 Examples of flowfield designs in planar stacks.

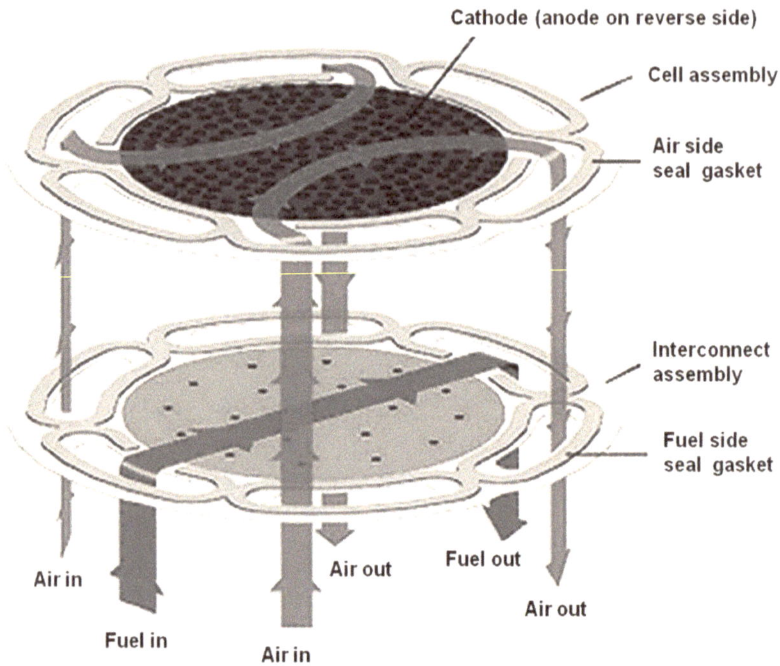

Figure 8.18 Example of a flow configuration design for a planar stack.

considered in the design to minimise contact resistance losses. An example of gas flow configuration design for a planar SOFC stack [47] is shown in Figure 8.18.

(ii) *Electrical connection*: In a stack, single cells are normally connected in series electrical connection via interconnects to build voltage. A stack design can also include parallel electrical connection [48–50]. The series and parallel electrical connection can be used to protect stack against complete failure if an individual cell fails. Examples of series and parallel electrical connection [48,49] are given in Figure 8.19. For tubular cells, electrical connections can be made in the reducing environment at the anode or in the oxidising atmosphere at the cathode side. The anode side connection allows the use of low-cost metals for cell-to-cell contacts and current collectors.

(iii) *Gas manifolding*: Any stack design must include gas manifolds for routing gases from a common supply point to each cell and removing unreacted gases and reaction products. Gas manifolds can be classified as external or integral. External manifolds are constructed separately from the cell or interconnect component of the stack. Figure 8.20 is an external manifold concept for cross-flow planar SOFCs [51]. Integral manifolds are designed and formed as part of the cell or interconnect (Figure 8.21) [52]. Figure 8.22 shows an example of a stack concept with integral manifolds [53]. Depending on the design, gas manifolds often require sealing to prevent gas leakage or crossover (although several sealless designs have been proposed [54,55]). The gas manifold seal is insulating to preclude cell-to-cell electrical shorts. In principle, the manifold must be designed to have low pressure drop (relative to individual cell pressure drop) to provide uniform flow distribution to the stack.

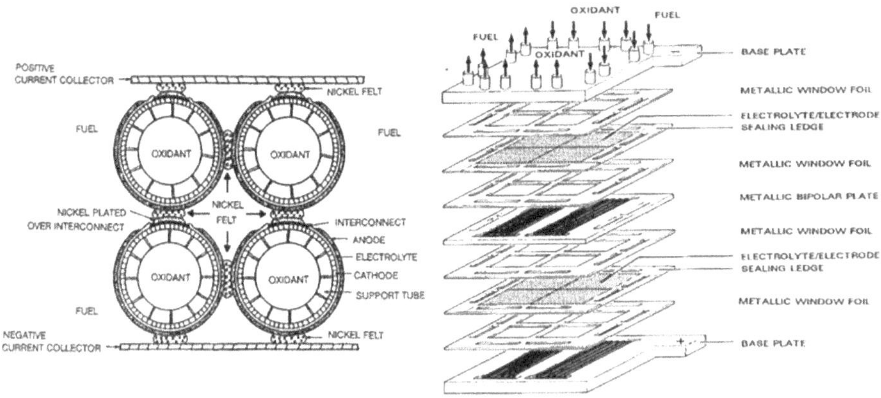

Figure 8.19 Examples of series and parallel electrical connection in SOFC stacks.

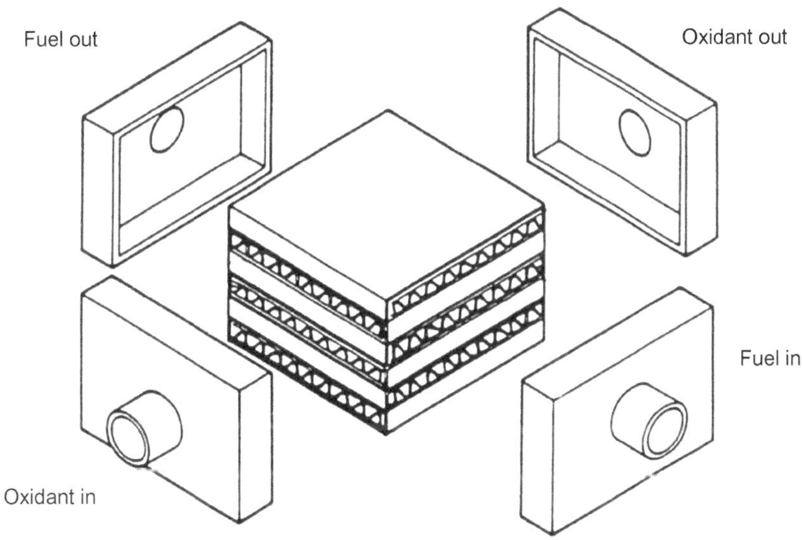

Figure 8.20 Example of external gas manifolds.

8.4.2 *Stack fabrication/assembly*

Fabrication/assembly of multi-cell stacks mainly involves incorporation of single cells and interconnects along with gas manifolds and seals (if any) into the architecture according to the design specifications. Depending on stack design and selected materials, cell components and interconnects can be fabricated together in a single-process step (e.g. co-firing of multilayer tapes) or sequentially in a progression of processing steps (e.g. deposition of one layer by one layer). In many cases, cells and interconnects (e.g. ceramic cells and metallic interconnects) are fabricated separately and then stacked to build the desired stack size.

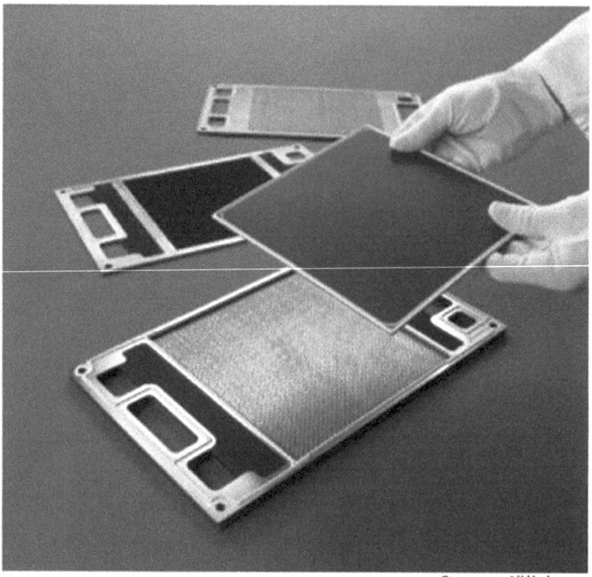

Source: Jülich

Figure 8.21 Gas manifolds as part of interconnect.

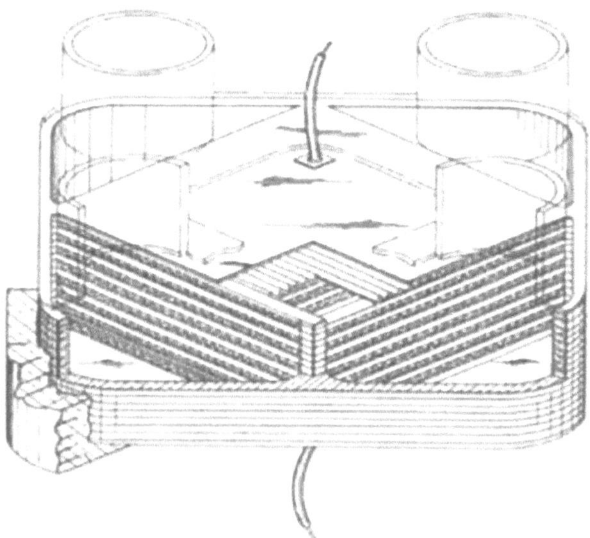

Figure 8.22 Example of integral gas manifolding concept.

SOFC multi-cell stacks have been built to sizes suitable for practical uses. For example, stacks up to 100 kW with tubular cells of full active length (e.g. 150 cm) have been assembled [3]. Planar stacks, especially those with anode-supported cells, have been scaled up to sizes having suitable active areas (e.g. 500–1000 cm^2) and stacks of up to 15 kW have been assembled as the stack building block for large power systems [56]. Segmented-cells-in-series SOFCs have been made into practical assembly/stack sizes (e.g. 60-cell assemblies of 60 W output [50]). These assemblies can be stacked and bundled to form modules of appropriate power levels. For example, 20 kW modules suitable for 1 MW systems have been constructed from 5 strips of 12 parallel bundles of six 60-cell assemblies [50]. SOFC stacks based on the monolithic design have been scaled up to 30 kW sizes [57].

8.4.3 Stack performance

SOFC multi-cell stacks have demonstrated electrochemical performance under operating conditions appropriate for practical uses, including operation with internal reforming and on gas products directly from external reformers. For example, a 96-cell planar stack shows a power density of about 0.3 W/cm^2 (voltage of about 0.82 V per cell at 0.364 A/cm^2), 715 °C on air (air utilisation Ua of 15%) and fuel containing 14.5% natural gas (NG) (fuel utilisation Uf of 68%) (Figure 8.23) [56]. An example of direct operation of a stack with reformates from catalytic partial oxidation (CPOX) of jet fuel JP-8 is given in Figure 8.24 [58].

Multi-cell stack performance is dependent on performance of individual cells in the stack. Thus, one important characteristic for stable and reliable stack performance is minimal cell-to-cell variation within the stack. Cell-to-cell variations in the stack can be due to cell performance reproducibility, stack design properties and other

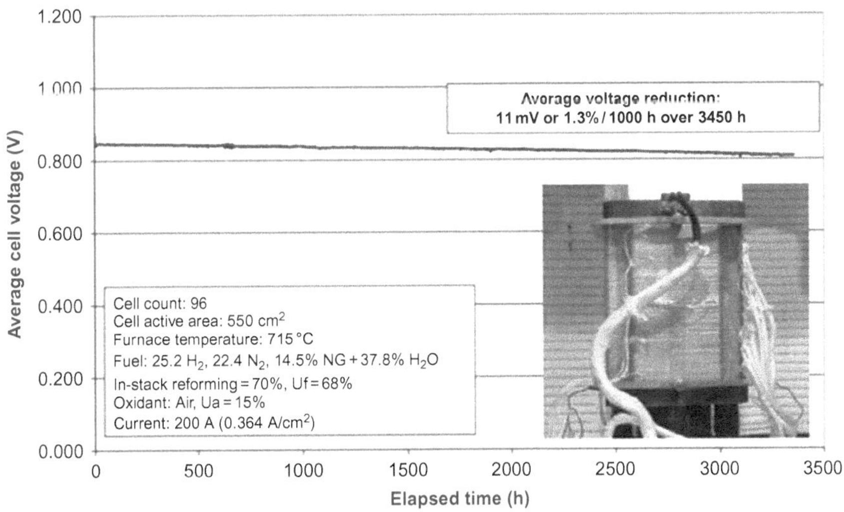

Figure 8.23 Performance of a 96-cell planar stack.

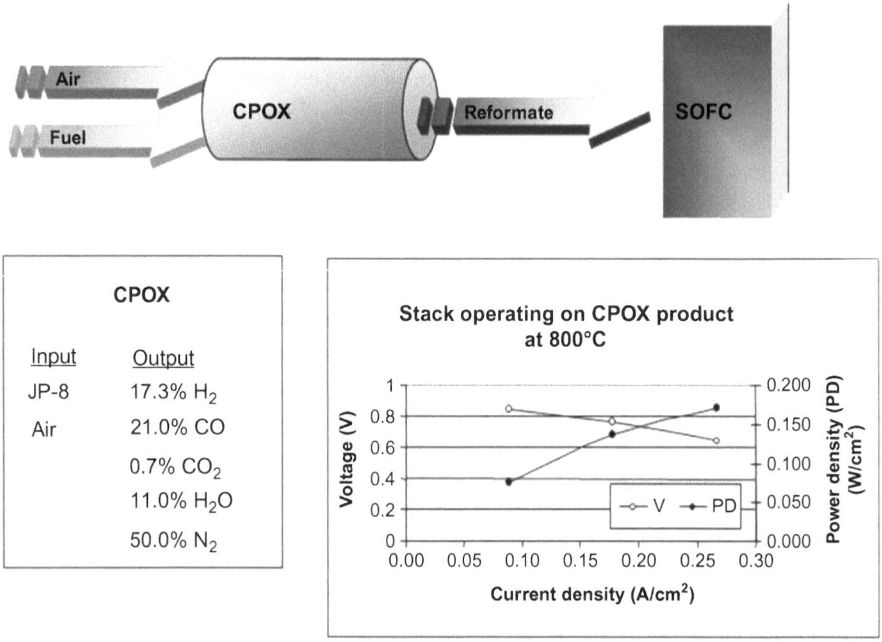

Figure 8.24 Performance of a stack on reformate from CPOX of JP-8.

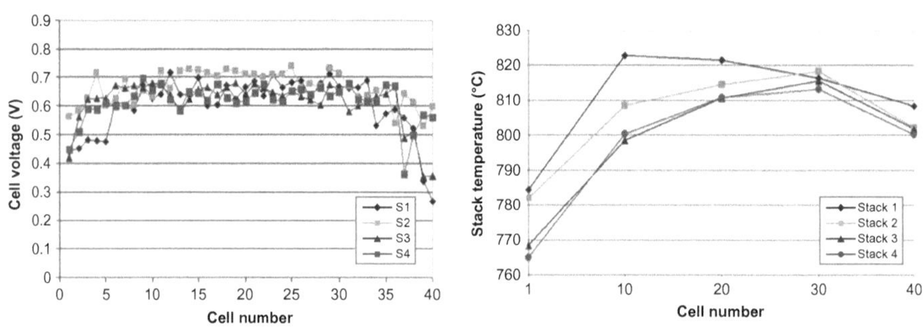

Figure 8.25 Voltage variations and temperature distribution in four 40-cell stacks operating at system peak power.

conditions during operation. For example, end effects can be seen in multi-cell stacks due to temperature distributions within the stack. Figure 8.25 shows such effects with lower cell voltages with corresponding lower temperatures at the top and the bottom of four 40-cell stacks [33]. Minimal cell-to-cell variations can be achieved via design and process control and an example is given in Figure 8.26 [59].

In SOFC stacks, especially planar stacks with metallic interconnects, contact resistance between the electrodes, especially the cathode, and the metallic interconnect is a major factor in stack performance losses (Figure 8.27) [60] and long-term

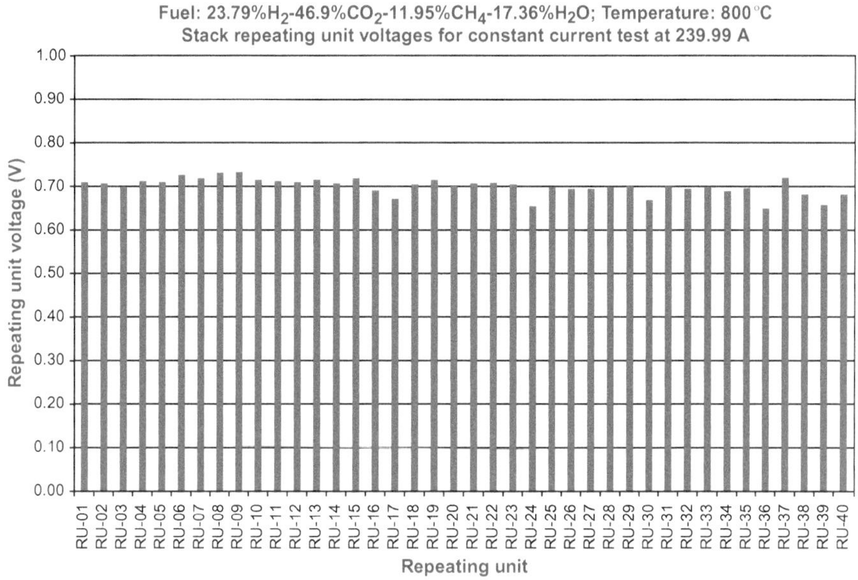

Figure 8.26 Example of cell-to-cell variation in a 40-cell planar stack.

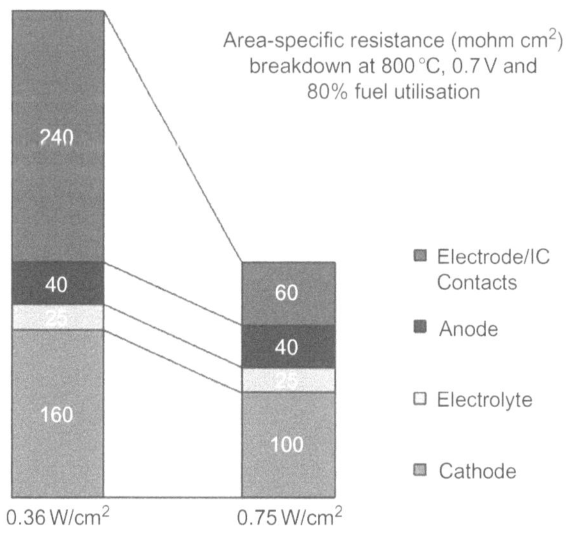

Figure 8.27 Area specific resistance breakdown for a planar stack (anode-supported cells, metallic interconnects).

performance degradation (discussed below). Conductive contact pastes have been used in planar stacks to minimise contact resistance; however, stability of such contact pastes over long duration is questionable.

An important performance degradation mode is chromium poisoning of the cathode in long-term operation of planar stacks having metallic interconnects. Chromium present in the metallic interconnect can migrate to cathode reactive sites and interact with the cathode (Figure 8.28), poisoning the electrode thereby increasing cathode polarisation with time. At present, the most common mitigating approach is to use conductive coatings (e.g. Co–Mn spinel) on the metallic interconnect to minimise the chromium transport and migration, thus reducing degradation rates [42]. Figure 8.29 shows an example of improved performance degradation of a three-cell stack with Co–Mn spinel coated interconnects [25,42].

Another potential degradation mode in multi-cell stacks with metallic interconnects is cathode–interconnect contact resistance changes. The contact between the ceramic cathode and the metallic interconnect tends to change due to thermodynamic driving forces (cathode ceramic particles tend to sinter/agglomerate to minimise their surface free energy, thus moving away the metallic interconnect) and other operating characteristics such as temperature distribution, thermal expansion mismatch as operation proceeds. These factors can lead to degradation in long-term operation. It is highly possible that during long-term operation, chemical interaction develops and electrical contact between the cathode and the interconnect evolves, ohmic resistance

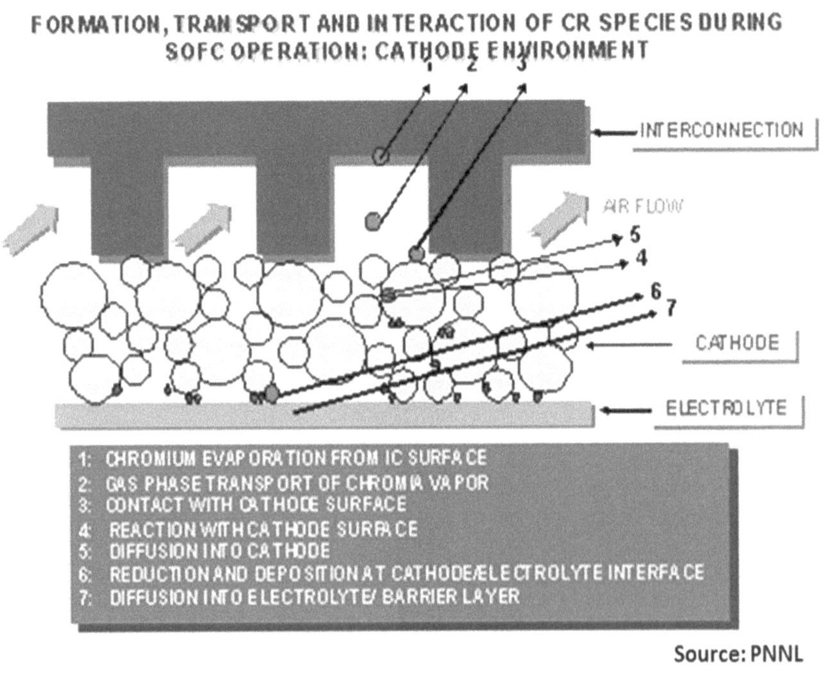

Figure 8.28 Potential steps in chromium poisoning of SOFC cathodes.

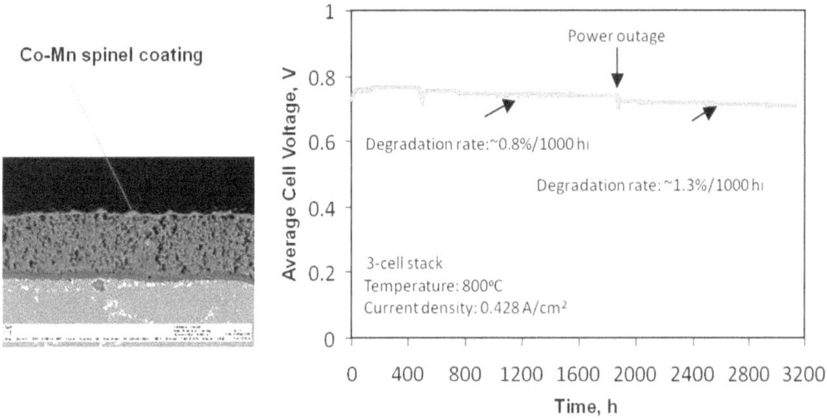

Figure 8.29 Micrograph of Co–Mn spinel coated metallic interconnect (left) and performance degradation of three-cell stack with anode-supported cells and coated interconnects (right).

increases and contact area reduces, resulting in higher ohmic losses and thus degradation (Figure 8.30) [61].

SOFC stacks have been operated for tens of thousands of hours, and durability has been demonstrated with low performance degradation rates under specified operating conditions. For example, a short planar SOFC stack (with uncoated metallic interconnects) has been in operation at 700 °C for more than 5 years with the overall voltage degradation of about 1% per 1000 h [62]. With coated metallic interconnects, a stack has been tested for more than 14,000 h with a reduced degradation rate of about 0.12% per 1000 h [62]. Figure 8.23 is an example of performance (with internal reforming) of a 96-cell stack showing 1.3% voltage degradation per 1000 h [56].

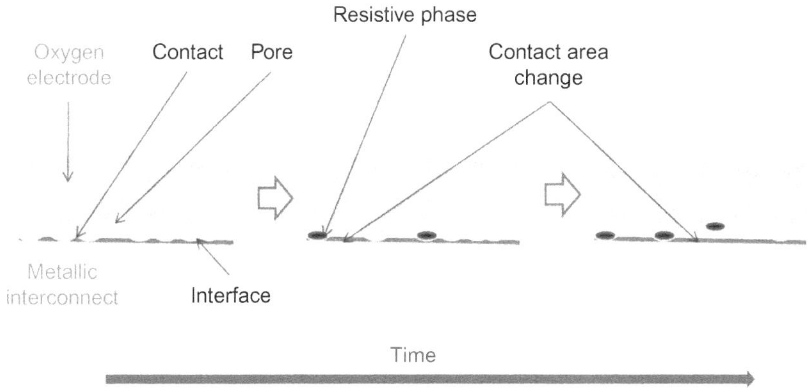

Figure 8.30 Cathode-metallic interconnect contact evolution.

8.5 Summarising remarks

The SOFC offers a number of advantages in terms of cell/stack design, fabrication and operation characteristics (such as design flexibility, multiple cell fabrication options, multi-fuel capability and operating temperature choices). The SOFC also presents several technical challenges (such as limited material selections, increased undesirable chemical interactions between components and certain operation restrictions). To date, various design options and fabrication approaches have been proposed and developed for SOFC single cells and multi-cell stacks. Cell and stack performance under realistic operating conditions has reached the level suitable for practical uses. However, several critical issues such as cell/stack reliability and reproducibility and long-term performance degradation remain to be resolved. Progress in these areas along with cell and stack cost reduction should move the technology towards widespread applications.

References

[1] N.Q. Minh, J. Am. Ceram. Soc. 76 (1993) 563.
[2] N.Q. Minh, T. Takahashi, Science and Technology of Ceramic Fuel Cells, Elsevier, Amsterdam, The Netherlands, 1995.
[3] K. Kendall, N.Q. Minh, S.C. Singhal, in: S.C. Singhal, K. Kendall (Eds.), High Temperature Solid Oxide Fuel Cells: Fundamentals, Design and Applications, Elsevier, Oxford, UK, 2003, p. 197.
[4] S.P. Simner, J.F. Bonnett, N.L. Canfield, K.D. Meinhardt, J.P. Shelton, V.L. Sprenkle, J.W. Stevenson, J. Power Sources 113 (2003) 1.
[5] S. Bebelis, N. Kotsionopoulos, A. Mai, F. Tietz, J. Appl. Electrochem. 37 (2007) 15.
[6] N.Q. Minh, in: D. Stolten, B. Emonts (Eds.), Hydrogen Science and Engineering – Materials, Processes, Systems and Technology, Wiley-VCH, 2015.
[7] N.Q. Minh, in: D. Stolten, B. Emonts (Eds.), Fuel Cells Science and Engineering – Materials, Systems, Processes and Technologies, vol. 2, Wiley-VCH, Weinheim, Germany, 2012, p. 963.
[8] S. Küln, T. Pessara, D. Rückemann, A. Weber, K. Paciejewska, A. Stoeck, S. Mnich, L. Winkler, in: Presentation at Hannover Messe, 2014. www.h2fc.com/hm14/images/tech-forum-presentations/2014-0409-1030.pdf.
[9] S.D. Vora, in: 8th Annual SECA Workshop, 2007. www.netl.doe.gov/events/conference-proceedings/2007/seca-worksop.
[10] H. Yahiro, Y. Baba, K. Eguchi, H. Arai, J. Electrochem. Soc. 135 (1988) 2077.
[11] E.D. Wachsman, K.T. Lee, Science 334 (2011) 935.
[12] E.A. Armstrong, J.W. Park, N.Q. Minh, Electrochem. Solid State Lett. 15 (2012) B75.
[13] R.T. Leah, A. Bone, A. Selcuk, D. Corcoran, M. Lankin, Z. Dehaney-Steven, M. Selby, P. Whalen, ECS Trans. 35 (1) (2011) 351.
[14] J. Will, A. Mitterdorfer, C. Kleindogel, D. Perednis, L.J. Gauckler, Solid State Ionics 131 (2000) 79.
[15] N.Q. Minh, C.R. Horne, in: F.W. Poulsen (Ed.), High-Temperature Electrochemical Behaviour of Fast Ion and Mixed Conductors, Risø National Laboratory, Roskilde, Denmark, 1993, p. 337.
[16] M.J. Day, S.L. Swartz, G.B. Arkenberg, ECS Trans. 35 (1) (2011) 385.

[17] E. Traversa, ECS Interface 18 (3) (2009) 49.
[18] M. Kuln, T.W. Napporn, Energies 3 (2010) 27.
[19] J.-W. Kim, A.V. Virkar, K.-Z. Fung, K. Mehta, S.C. Singhal, J. Electrochem. Soc. 146 (1999) 69.
[20] C.-C. Chao, C.-M. Hsu, Y. Cui, F.B. Prinz, ACS Nano 5 (2011) 5692.
[21] H. Ghezel-Ayagh, in: 12th Annual SECA Workshop, 2011. www.netl.doe.gov/events/conference-proceedings/2011/seca-workshop.
[22] Z. Zhang, D.M. Bierschenk, J.C. Cronin, S.A. Barnett, Energ. Environ. Sci. 4 (2011) 3951.
[23] M. Mogensen, hydrail.org/sites/hydrail.org/files/2_Mogensen.pdf.
[24] S.C. Singhal, MRS Bull. 25 (3) (2000) 16.
[25] N.Q. Minh, ECS Trans. 7 (1) (2007) 45.
[26] N.Q. Minh, in: S.C. Singhal (Ed.), High Temperature Materials, The Electrochemical Society, Pennington, NJ, 2002, p. 54.
[27] National Energy Technology Laboratory, Recent solid oxide fuel cell cathode studies, Report DOE/NETL-2013/1618, 2013.
[28] T.Z. Sholklapper, C.P. Jacobson, S.J. Visco, L.C. De Jonghe, Fuel Cells 5 (2008) 303.
[29] J.M. Vohs, R.J. Gorte, Adv. Mater. 21 (2009) 943.
[30] C. Ding, H. Lin, K. Sato, T. Kawada, J. Mizusaki, T. Hashida, Solid State Ionics 181 (2010) 1238.
[31] Y. Gong, D. Palacio, X. Song, R.L. Patel, X. Liang, X. Zhao, J.B. Goodenough, K. Huang, Nano Lett. 13 (2013) 4340.
[32] N.Q. Minh, ECS Trans. 57 (1) (2013) 197.
[33] GE Hybrid Power Generation Systems, Solid state energy conversion alliance (SECA) solid oxide fuel cell program, Final report, 2006.
[34] S. Park, J.M. Vohs, R.J. Gorte, Nature 404 (2000) 265.
[35] S. McIntosh, R.J. Gorte, Chem. Rev. 104 (2004) 4845.
[36] E.N. Armstrong, J.-W. Park, N.Q. Minh, ECS Trans. 45 (1) (2012) 499.
[37] H. Yokokawa, K. Yamaji, M.E. Brito, H. Kishimoto, T. Horita, J. Power Sources 196 (2011) 7070.
[38] Y.L. Liu, S. Primdahl, M. Mogensen, Solid State Ionics 161 (2003) 1.
[39] S.H. Kim, T. Ohshima, Y. Shiratori, K. Itoh, K. Sasaki, in: V. Pthenakis, A. Dillon, N. Savage (Eds.), MRS Proceedings, vol. 1041, Materials Research Society, Warrendale, PA, 2007, 1041-R03-10.
[40] A. Hagen, K. Neufeld, Y.L. Liu, J. Electrochem. Soc. 157 (2010) B1343.
[41] W. Zhou, F. Liang, Z. Shao, Z. Zhu, Sci. Rep.2 327 (2012). http://dx.doi.org/10.1038/srep00327.
[42] N.Q. Minh, S.C. Singhal, M.C. Williams, ECS Trans. 17 (1) (2009) 211.
[43] G. Agnew, R. Goettler, in: 13th Annual SECA Workshop, 2012. www.netl.doe.gov/events/conference-proceedings/2012/seca-workshop.
[44] A. Nakanishi, M. Hattori, Y. Sasaki, K. Kimura, Y. Ando, H. Miyamoto, T. Onodera, S. Kanehira, K. Takenobu, M. Nishiura, H. Oozawa, in: S.C. Singhal, J. Mizusaki (Eds.), SOFC-IX, vol. 1, The Electrochemical Society, Pennington, NJ, 2005, p. 82.
[45] J. Doyon, in: 8th Annual SECA Workshop, 2007. www.netl.doe.gov/events/conference-proceedings/2007/seca-workshop.
[46] N.Q. Minh, U.S. Patent 5,215,946 (1993).
[47] Fuel Cell Institute of Australia, www.fuelcells.org.au.
[48] Westinghouse Electric Corporation, Solid oxide fuel cell power generation system, the status of the cell technology—a topical report, Report no. DOE/ET/17089-15, U.S. Department of Energy, Washington, DC, 1984.

[49] E. Ivers-Tiffée, W. Wersing, B. Reichelt, 1990 Fuel Cell Seminar Abstracts, Courtesy Associates, Washington, DC, 1990. 137.
[50] R. Goettler, T. Ohm, in: 11th Annual SECA Workshop, 2010. www.netl.doe.gov/events/conference-proceedings/2010/seca-workshop.
[51] N.Q. Minh, C.R. Horne, U.S. Patent 5,162,167 (1992).
[52] www.fz-juelich.de/portal/EN/Research/EnergyEnvironment/fuelcells/SOFC/_node.html.
[53] N.Q. Minh, T.L. Stillwagon, U.S. Patent 5,256,499 (1993).
[54] S.C. Singhal, in: F. Grosz, P. Zegers, S.C. Singhal, O. Yamamoto (Eds.), Proceedings of the Second International Symposium on Solid Oxide Fuel Cells, Commission of European Communities, Luxembourg, 1991, p. 25.
[55] J. Piascik, D. Dalfonzo, J. Yamanis, E. Ong, U.S. Patent US 6677069 B1 (2004).
[56] H. Ghezel-Ayagh, in: 13th Annual SECA Workshop, 2012. www.netl.doe.gov/events/conference-proceedings/2012/seca-workshop.
[57] www.mhi.co.jp/en/technology/business/power/sofc/development_situation.html.
[58] N. Minh, A. Anumakonda, R. Doshi, J. Guan, S. Huss, G. Lear, K. Montgomery, E. Ong, J. Yamanis, in: H. Yokokawa, S.C. Singhal (Eds.), SOFC-VII, The Electrochemical Society, Pennington, NJ, 2001, p. 190.
[59] P. Narasimhamurthy, R. Kerr, in: 12th Annual SECA Workshop, 2011. www.netl.doe.gov/events/conference-proceedings/2011/seca-workshop.
[60] N.Q. Minh, J. Korean Ceram. Soc. 47 (2010) 1.
[61] N.Q. Minh, M.B. Mogensen, ECS Interface 22 (4) (2013) 55.
[62] L. Blum, U. Packbier, I.C. Vinke, L.G.J. de Haart, Fuel Cells 13 (2013) 646.

System designs and applications

9

N.Q. Minh
University of California, San Diego, CA, USA

9.1 Introduction

A solid oxide fuel cell (SOFC) system is a fully functional unit integrating SOFC stacks and other required components for specified power generation applications [1]. The development of an appropriate architecture for an SOFC power system involves design, analysis and optimisation to define system configurations and specify system components and operating parameters that meet the requirements of the target application. This chapter provides an overview of SOFC power system designs, discusses key design factors and operating parameters and reviews selected examples of power systems for portable, transportation and stationary applications. In addition, this chapter includes a brief discussion on system designs for SOFCs/reversible SOFCs (RSOFCs) operating in electrolysis mode (in this case, referred to as solid oxide electrolysis cells or SOECs) for hydrogen/chemical production.

9.2 Overview of SOFC power systems

An SOFC power system consists of fuel cell stack(s) and all required components for a fully functional, stand-alone unit. Figure 9.1 is an example of a simplified schematic depicting the configuration of a complete 5 kW SOFC system [2]. A photograph of a model of such a system showing the stack and other supporting components is given in Figure 9.2 [2]. In general, system components can be grouped into different subsystems depending on their roles within the system: power generation, fuel processing, fuel, oxidant and water delivery, thermal management, power conditioning and control. The components for these subsystems except the power generation are together referred to as balance of plant or BOP.

- *Power generation subsystem*: For the majority of SOFC power systems, the SOFC stack is the only component that produces electricity, thus constituting the power generation subsystem. In this case, the power generation subsystem consists of mainly SOFC stack(s). For certain designs, the power generation subsystem contains additional equipment such as gas turbine (GT) and/or steam turbine (ST) generators.
- *Fuel processing subsystem*: The fuel processing subsystem is used to precondition the fuel (other than hydrogen) before it enters the fuel cell stack. This subsystem consists of a reformer (if the system design includes external reformation) and other equipment for fuel cleanup (e.g. acid gas removal, particulate removal, etc.) if any.

High-temperature Solid Oxide Fuel Cells for the 21st Century. http://dx.doi.org/10.1016/B978-0-12-410453-2.00009-9

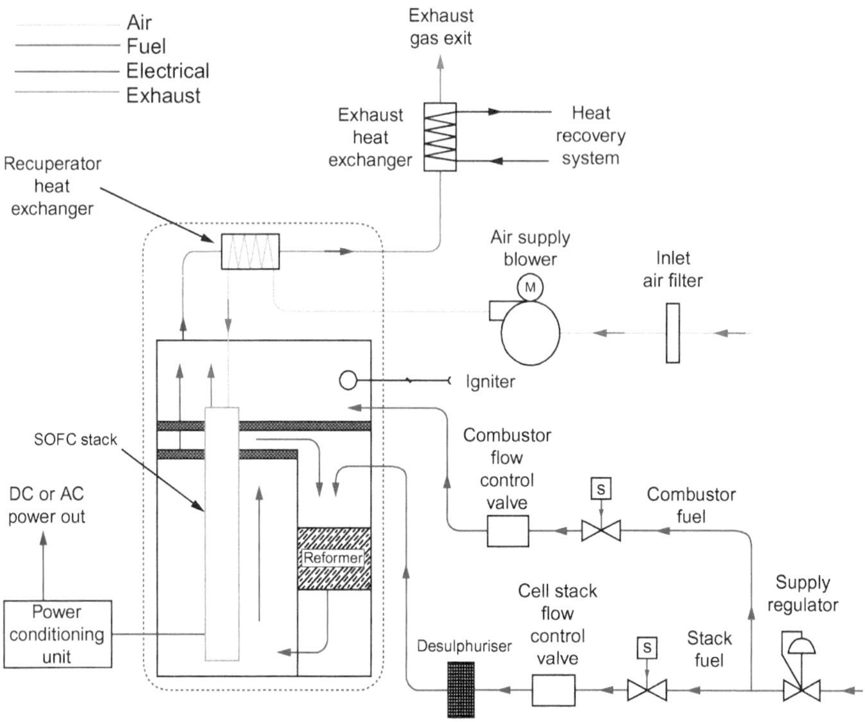

Figure 9.1 Schematic of 5 kW SOFC power system [2].

- *Fuel, oxidant and water delivery subsystem*: This subsystem mainly consists of blowers/compressors and valves/orifices to deliver required reactants to the fuel reformer (FR) and fuel and oxidant to the SOFC stack.
- *Thermal management subsystem*: The subsystem consists of a number of heat exchangers (HEXs)/recuperators (including steam generators) and combustors/burners to maintain the SOFC stack temperature at the required level and to control heat supply/removal for efficient operation of the reformer and other cleanup equipment. Insulation is also an important element in thermal management of the system to contain heat losses.
- *Power conditioning subsystem*: The power conditioning subsystem converts variable DC from the fuel cell to regulated DC or AC power appropriate for the application. Depending on the particular application, this subsystem may consist of power electronics (DC–DC converters and DC–AC inverters) and transformers.
- *Control subsystem*: This subsystem is a controller including control software that provides for system start-up, shutdown and normal operation while maintaining the system within its operating constraints when subjected to load changes or disturbances. Sensors are also an important element in the control of the system.

9.3 Type of SOFC power system

Current SOFC power systems can be divided into two general classes depending on their power generation cycles: simple cycle systems and hybrid cycle systems.

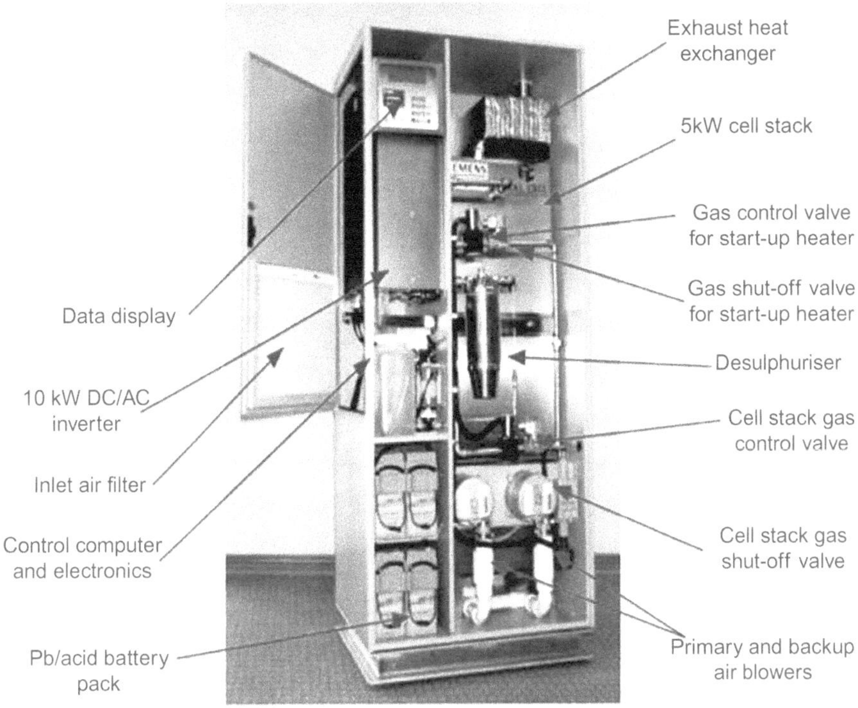

Figure 9.2 5 kW SOFC system model [2].

Simple cycle SOFC systems: The SOFC is the only power generating component for a simple cycle system. The simple cycle is the common and preferred configuration for power systems with outputs ranging from tens of watts to hundreds of kilowatts although it is also considered for megawatt-size power plants [3,4]. Simple cycle SOFC systems have been shown to achieve high system efficiencies, for example, 45–60% for 1–2 kW systems [5,6]. Figures 9.3 and 9.4 show, as examples, two simple cycle systems: a 2 kW system [7] and a combined heat and power (CHP) 100 kW system [8], respectively. In simple cycle SOFC systems, the fuel cell typically operates under atmospheric pressure. In principle, the SOFC can be pressurised, and pressurised operation can provide additional benefits [9,10]. The benefits of pressurisation include improved cell/stack performance, reduced heat losses and pressure drops and smaller equipment sizes. Although a pressurised SOFC power system has several advantages relative to an atmospheric pressure system, pressurisation is not commonly used in simple cycle systems because of the inherent cost and technical issues relating to pressurised operation, especially for systems with small power outputs. In large systems, the efficiency of simple cycles may be lower than that of hybrid cycles of similar power levels (hybrid cycles discussed below); however, the critical issues concerning integration of the fuel cell with a heat engine and pressurised operation if required can be avoided for simple cycle designs.

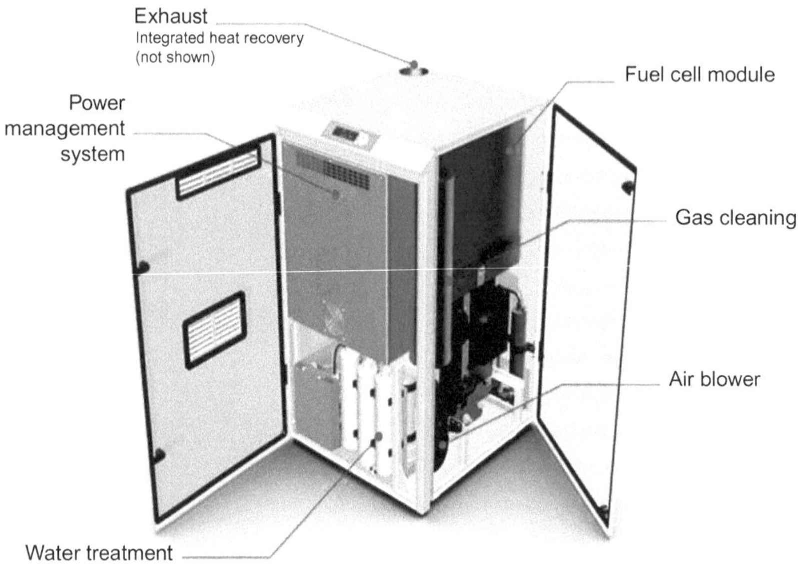

Figure 9.3 2 kW simple cycle SOFC system [7]. (CFCL entered administration in 2015)

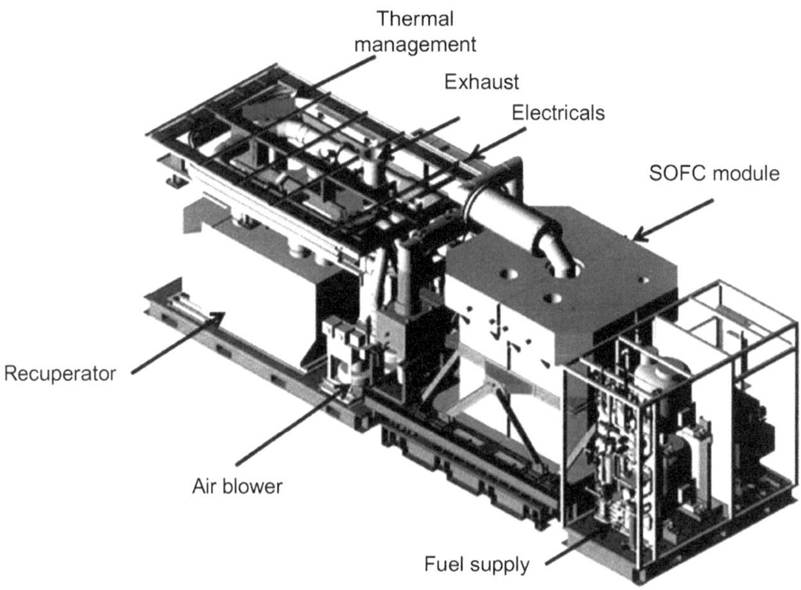

Figure 9.4 100 kW simple cycle CHP system [8].

Hybrid cycle SOFC systems: The SOFC can be combined with another power generating equipment such as a heat engine [11,12] or a battery [13,14] to form a hybrid power system. Hybridisation of the SOFC with a heat engine is the most common hybrid cycle. In a typical hybrid combination with a heat engine, the heat energy of the fuel cell exhaust is used by the heat engine to generate additional electricity. The heat engine can be GT, ST or combination of heat engines such as steam and GT combined cycle and integrated gasification combined cycle (IGCC) [15,16] with SOFC/GT hybrids the most common. Figure 9.5 shows, as an example, a 300 kW SOFC/GT system [17].

Figure 9.6 is a simplified schematic of an SOFC/GT hybrid [17]. It can be seen from Figure 9.6 that the tail gas from the SOFC is combusted and then fed to the turbine section of the GT to generate additional electricity. In a generic SOFC/GT hybrid design, the SOFC produces about 65–80% of the power and about 20–35% is from the GT. The SOFC/GT system schematic given in Figure 9.6 is based on the direct fired design. In this design, the SOFC operates under pressure. In addition to generating power, the GT also provides some BOP functions for the fuel cell such as supplying air under pressure and preheating fuel and air in a recuperator [18]. In the indirect fired design, the recuperator transfers the heat energy of the SOFC exhaust to compressed air supply for the turbine and the expanded air is supplied to the fuel cell. In this design, the SOFC operates under ambient pressure. In general, direct fired hybrid systems exhibit higher system efficiencies than indirect fired systems under equivalent design conditions [19]. Hybridisation of the SOFC with a GT significantly improves system efficiency and can be beneficial for use in large (≥hundreds of kilowatts) power plants [20]. As an example, Table 9.1 compares estimated efficiencies of three SOFC power systems: simple cycle, SOFC/GT hybrid and SOFC/GT hybrid with anode and cathode gas recycle [21]. It can be seen from the table that the hybrid cycle arrangement shows much higher efficiencies. The estimated efficiency of the SOFC/GT hybrid with gas recycle is 71.0%, about 26 points higher than that of the simple system.

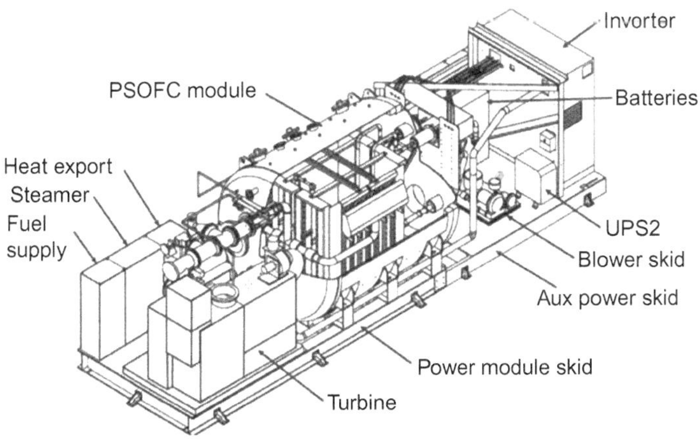

Figure 9.5 300 kW SOFC/GT hybrid system [17].

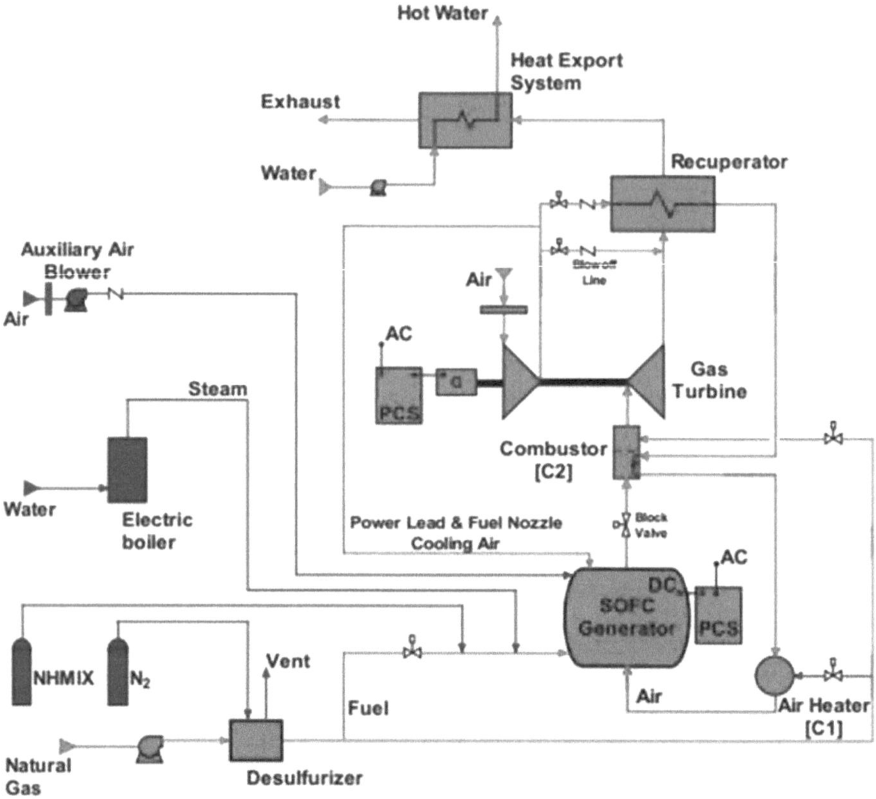

Figure 9.6 Simplified schematic of SOFC/GT hybrid system [17].

Table 9.1 Comparison of simple cycle and SOFC/GT hybrid systems

	Simple cycle	SOFC/GT hybrid	SOFC/GT hybrid with gas recycle
Efficiency (%)	44.8	61.1	71.0
SOFC power[a] (kW)	3709	3709	4099
GT power[a] (kW)	0	976	1447
Parasitic power (kW)	(389)	(100)	(288)
Net plant power (kW)	3320	4585	5258
SOFC pressure (atm)	1.3	4.6	4.6

[a]After power conversion.
Values in parentheses indicate reduction in power.

9.4 SOFC power system design

The focus in designing a fuel cell power system is to develop and optimise the system configuration to meet the specifications of its intended application. These specifications could include the following: cycle efficiency, duty cycle, fuel specification, cost (purchase and installation), reliability, maintenance, size and weight, environmental interfaces, cogeneration, acoustic noise, power quality and safety. In general, SOFC systems are designed to fulfil the key requirements for practical/commercial products, namely, performance, reliability and cost. Thus, the design aims at establishing system configurations and defining system components, including component specifications, component performance characteristics and effects of component and process variables on system performance and reliability. The detailed system design can be used to estimate/determine system costs.

An example of the system design process is schematically illustrated in Figure 9.7 [1]. The process involves the following main steps:

(i) A conceptual system design is first developed based on the system requirements and knowledge of the primary technologies and components.
(ii) The conceptual system design is analyzed, modelled and traded using assumed or known performance.

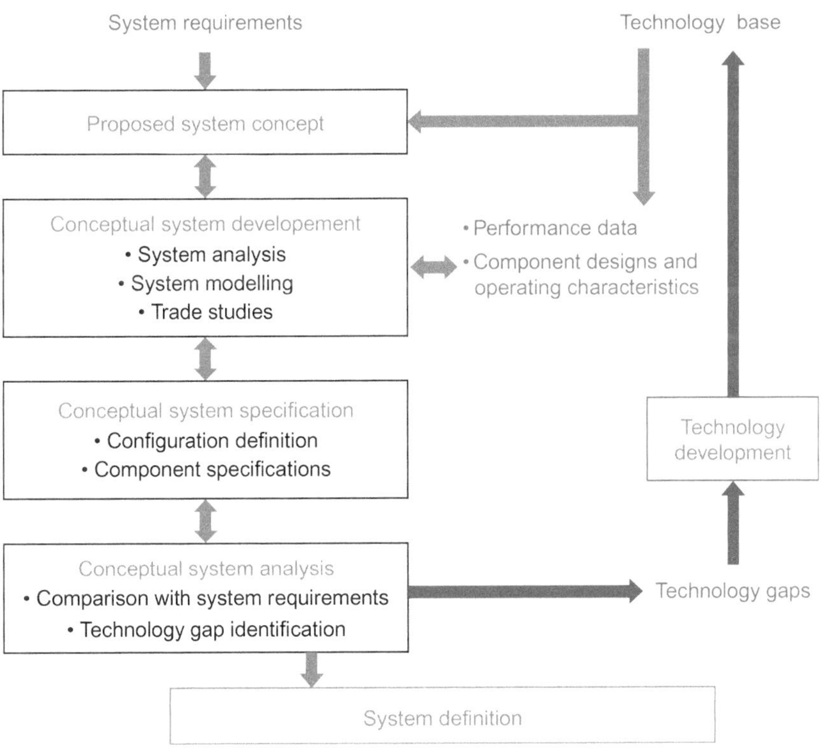

Figure 9.7 System design process.

(iii) The conceptual design is defined with specifications created for each component from system modelling and analysis and given the existing technology base. Adjustments to assumed performance may be made to refine the system model/analysis.
(iv) The conceptual system is compared to the application requirements to identify technology gaps, if any, that need to be resolved in order for the system design to be realised.
(v) The system design meeting the application requirements is then defined.

As shown in Figure 9.7, the inputs for this process include the system requirements and technology base, and the major outputs include a system design, identified technology gaps to meet the requirements and the definition of various system components.

9.4.1 Stack designs and parameters

Several stack design parameters and operating variables are critical in the design of the system (to meet the performance, reliability and cost targets of the intended application) and are discussed in this section.

Number of stack and stack arrangement: The number of stacks selected for an SOFC power system is dependent on several factors such as power level, stack design, available technology and other requirements (e.g. cost, reliability). In general, the number of stacks in a system is preferably as small as feasible since the smaller number of stacks decreases part counts (e.g. fewer cells) and reduces packaging (e.g. less insulation), resulting in lower system cost. Multiple stacks deployed in the system can be arranged in parallel or series (stages) or combination of parallel and series with respect to electrical connection and gas flow. These arrangements are a design option to optimise system configuration.

For large SOFC power systems, defining the number of stacks and stack arrangement becomes more important as it is impractical from the cost, reliability and technology standpoint to build, for example, one stack for megawatt-sized power plants. Therefore, multiple stacks have to be integrated in such systems, and determination of the optimal number of stacks and the stack size is needed. One approach to guide this determination is to examine effects of stack size (cell size and number of cells in the stack) and stack arrangement on the performance, cost and reliability of the system. An example is given here on the design of a conceptual 25 MW planar SOFC/GT hybrid system (with 20 MW from the SOFC) [22]. Cost analysis of this system shows that the system cost is a weak function of the number of stacks and the system cost is only a function of cell size and not the number of cells in the stack [22]. Thus, the optimal cell size can be determined solely by minimising the SOFC stack cost. Figure 9.8 shows stack cost as a function of cell size derived from the stack cost analysis [22]. This figure shows an optimal cell size for a minimal stack cost. Other constraints, however, may limit the cell size. Technology capability is a constraint. Reliability considerations favour large cell diameters and a smaller cell count, but the cell reliability may decline rapidly with the cell diameter, thus placing a constraint on the cell size. Therefore, the cell size is generally selected through a stack cost and reliability optimisation.

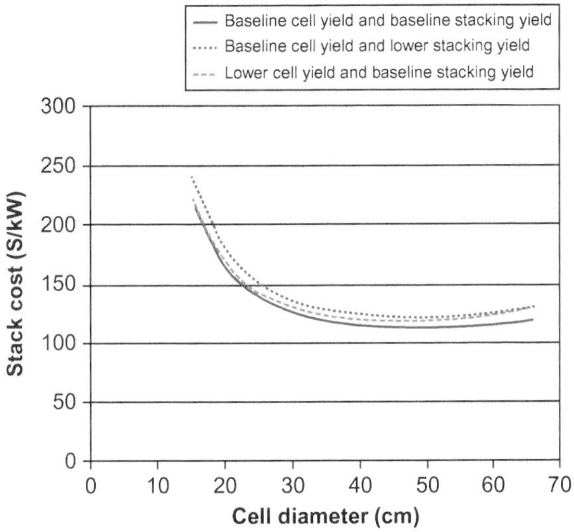

Figure 9.8 Stack cost as a function of cell diameter for a 25 MW SOFC/GT hybrid system [22].

The number of cells in a stack is determined by consideration and optimisation of the power conditioning subsystem (to minimise power conditioning losses). For the 25 MW SOFC/GT system example (with 20 MW from the SOFC), the optimal maximum stack voltage is 400 V, which translates into the optimal number of cells per stack of 400–500. Given the cell size and the number of cells per stack, the optimal stack building block for the 25 MW plant is estimated to have a nominal power rating of about 320 kW. This system thus needs 64 stack building blocks; these stacks, however, can be divided into modules to lower capital costs (multiple stacks inside a single vessel are called a module). Based on reliability considerations, the optimum number of stacks per module, and hence modules per plant, can be determined. In this 25 MW SOFC/GT system, the selected stack arrangement scheme includes eight active stack modules, each module containing eight stack building blocks. In addition, the system may contain redundant module(s) to improve system reliability.

Stack voltage and current density: Performance of a stack, reflected in its cell voltage and current density, is an important operating parameter in the design of SOFC power systems. System efficiency is obviously a strong function of cell voltage/current. The cell voltage is often traded with the SOFC power density (or current density) to achieve the most favourable system efficiency/system cost balance.

Fuel utilisation: Fuel utilisation (FU) is a stack operating parameter that requires definition and optimisation in designing SOFC power systems. An example is given in Figure 9.9 showing the efficiency of a 5 kW SOFC system as a function of FU and operating voltage for current densities of 0.3 and 0.6 A/cm^2 [23]. The uppermost line on the plot is for a current density of 0.3 A/cm^2 and operating voltage of 0.84 V. The lowest set of lines on the plot is for current densities of 0.3 and 0.6 A/cm^2 with operating voltage at 0.6 V. In this case, at an operating voltage of 0.6 V, a 10% decrease in FU results in a 3.7% drop in system efficiency. At an operating voltage of 0.84 V, a 10% decrease in FU results in a 6% decrease in system efficiency. It should be noted

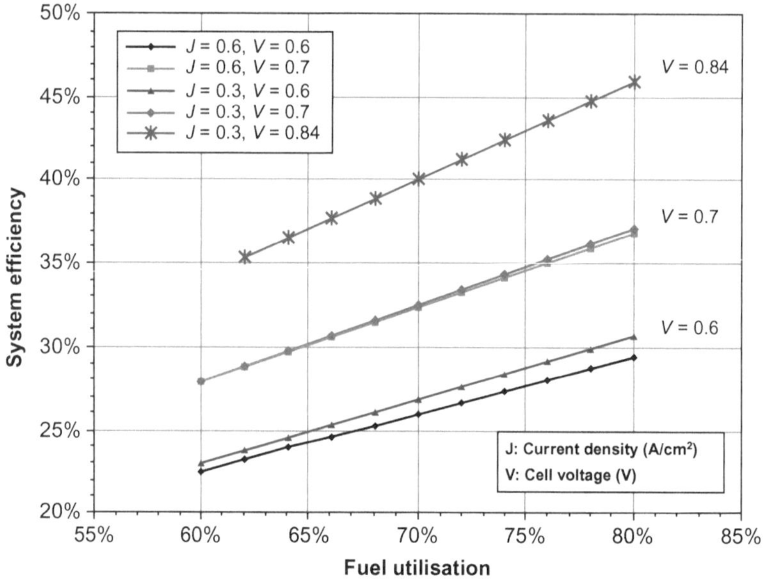

Figure 9.9 Effect of fuel utilisation on system efficiency for a 5 kW SOFC system [23].

that cell operating voltage and FU are not independent parameters, and dropping the FU will generally tend to increase the operating voltage of a cell. The increase in cell voltage will tend to offset the drop in system efficiency; however, the effect is considered to be minor from a system standpoint.

The level of FU has significant impact on other components of the system. The system components most affected by operation at low FU are the tail gas combustor, the cathode air blower and the steam generator. Operation at low FU impacts the tail gas combustor directly by increasing the temperature of the combustor and associated downstream components beyond the normal operating temperature limits of the materials used to manufacture these components. The cathode air blower is indirectly affected by operation at low FU since it will be required to provide increased flow to the system to maintain combustor temperatures below a limit. If increased cooling air flow is provided to the system, steam generator performance will also be affected. Increased cooling air flow to the system has the net effect of reducing the pinch temperature difference between the hot and cold streams supplied to the steam generator, thereby making steam generation more difficult.

Stack pressure drop: SOFC system performance at the design point depends on component efficiencies, pressure drops and heat losses. In terms of pressure drop, all major components of the system must be evaluated with regards to their contribution to the overall system pressure drop, and where feasible, design efforts must be made to lower their design pressure drop since high pressure drops result in high parasitic power consumption. In typical SOFC systems, the major contributor to the overall system pressure drop can be the pressure drop through the stack, and

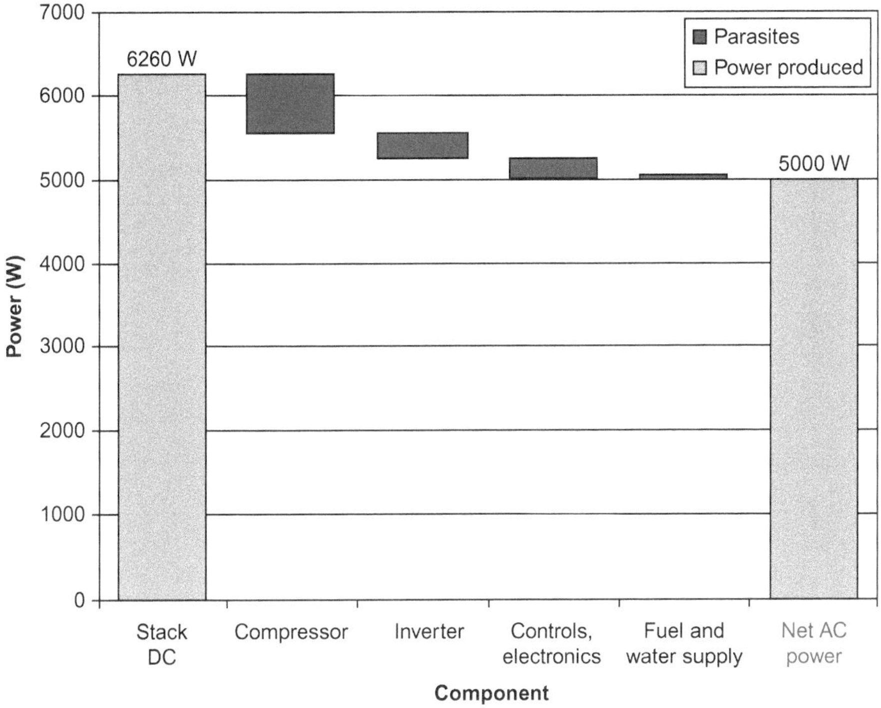

Figure 9.10 Power breakdown for a 5 kW SOFC system [23].

the major parasitic power consumption is the cathode air delivery blower/compressor (Figure 9.10) [23]. Thus, from a system standpoint, minimisation of the parasitic power consumption in the system requires minimising the pressure drop through the stack, which, in turn, lowers the power requirements for the cathode air blower. Figure 9.11 shows estimates for system pressure drop as a function of cell operating voltage and current density at a constant FU of 80% for a 5 kW system [23]. As expected, pressure drop is reduced as cell operating voltage increases due to decreasing air flow requirements.

Stack temperature gradient: One important parameter in designing SOFC power systems is stack temperature gradient or stack air temperature rise (the difference between stack air outlet and inlet temperatures). A stack air temperature rise as high as possible is desired as it lowers air cooling requirements (thus, less power required to deliver cooling air to the stack), resulting in improved system efficiency. The main limiting factor in setting the stack air temperature rise for a system is the thermal stress (as a result of temperature gradient within the cell caused by the air temperature rise) that can lead to stack structural failure. Therefore, it is desirable to develop stack designs that minimise cell temperature gradients while permitting increased air temperature rises. Typical stack air temperature rise of current SOFC systems is on the order of 100–200°C.

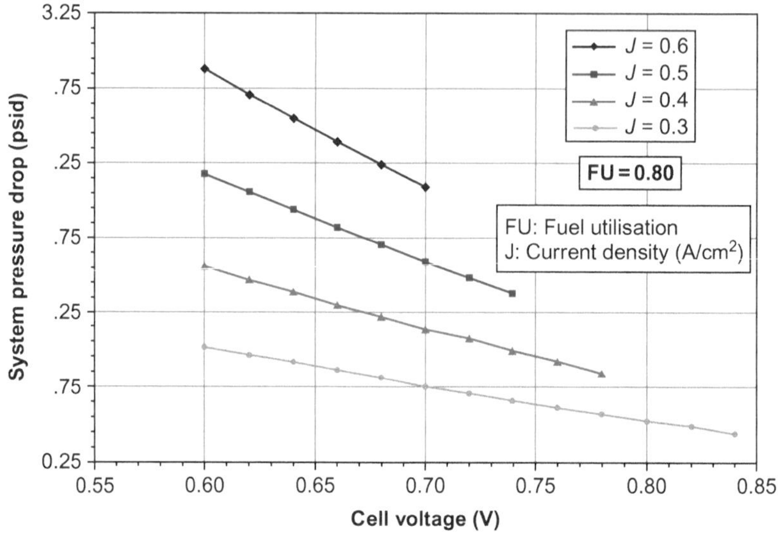

Figure 9.11 System pressure drop as a function of cell voltage and current density in a 5 kW SOFC system [23].

Figure 9.12 shows an example of the estimated cooling airflow requirements necessary to limit the stack air temperature rise to 175°C as a function of cell voltage and FU at a current density of 0.7 A/cm^2 for a 5 kW SOFC system [23]. It can be seen that, in this case, suitable operating points are those below the air blower upper limit of 60 SCFM (standard cubic foot per minute).

9.4.2 BOP component designs and selection

For BOP component design and selection, the following general guidelines are commonly used in the design of SOFC power systems to improve system performance, reliability and cost to meet the requirements of the intended application.

Use of off-the-shelf components: It is desirable to use off-the-shelf BOP components as many as possible in the design to avoid one-of-the-kind costly components.

Deployment of approaches to lower component requirements: It is advantageous to employ certain approaches in system design to lower component requirements to improve system performance and/or cost. For example, in certain system designs, only a portion of the fuel is reformed in an external reformer (in this case, referred to as a pre-reformer) and the resulting reformate plus the remaining fuel are fed to the SOFC where the fuel is internally reformed (via steam reforming) within the stack. This option is commonly employed to use the endothermic reforming reactions to reduce cooling requirements in thermal management of the SOFC. This design approach increases overall system efficiency and permits smaller reformers and air blowers/compressors, thus reduced costs for the system.

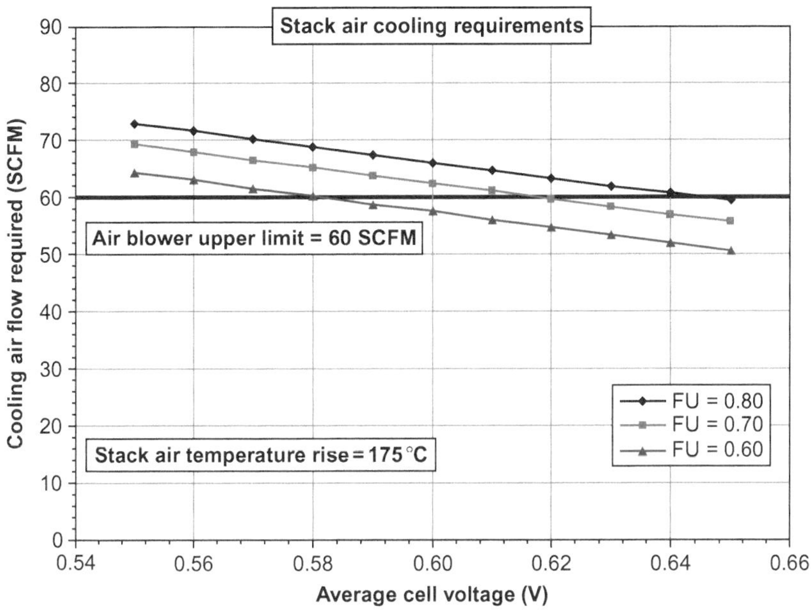

Figure 9.12 Air cooling requirements as function of cell voltage and fuel utilisation for a 5 kW SOFC power system [23].

Simplification of system design: It is preferable to reduce the number of components in the system if feasible to lower cost and increase reliability. For example, an external reformer is not needed in a system with complete (100%) internal reformation, thus reducing system costs and eliminating the need to develop procedures and conditions for integrating the operation of the reformer with the SOFC.

Component integration and packaging: It is crucial to match component design parameters to achieve stable and efficient operation of the system. For example, a critical factor in the design of an SOFC/GT hybrid system is the matching of design parameters between the two components, especially the turbine inlet temperature, the pressure ratio (P/P) and the air flow rate. In general, the desired features of the GT for hybrid systems are (i) adequate performance on low turbine inlet temperatures and low P/Ps, (ii) ability to sustain long-duration thermal cycling and slow time-response output of the fuel cell and (iii) large window of operation to allow for system turndown and avoid system shutdown [18]. It is well known that the design principles and operating characteristics of these two technologies are quite different. For example, high turbine inlet temperature and high P/P are desirable to improve the performance of stand-alone GTs but are not necessary for hybrid systems.

It is also desirable to develop component arrangements to reduce volume and minimise piping and other requirements. For example, components operating at high temperatures can be packaged together in a hot section to reduce insulation, thus costs.

9.5 Applications of SOFC power systems

SOFC power systems being developed range from watt-sized devices to multi-megawatt power plants and practical applications being considered covers all market sectors (portable/mobile, transportation and stationary) (Table 9.2). Many of the applications for SOFC power systems have progressed to hardware demonstration and prototype/pre-commercial stages while several applications, especially those with large power outputs, are at the conceptual/design stage (Figure 9.13).

9.5.1 Portable systems

SOFC systems in the 1–500 W power range have been considered and developed for mobile and portable applications. Examples of mobile/portable applications are power for consumer electronic devices, battery charging, remote power and low-level auxiliary power [24,25]. Portable SOFC systems are commonly operated on hydrocarbon fuels and particularly suitable for tactical military applications (e.g. soldier power and unmanned vehicle power) due to its potentiality for operation on logistics fuels such as JP-8 [26].

Three factors critical in the design of SOFC power systems for portable applications are stack design and stack power density, system thermal management and reforming process selection and reformer design.

- **(i)** *Stack design*: The stack must be designed and configured for minimum size and weight, optimum power and desired operating characteristics required for the portable system. For example, the microtubular and metal-supported cell design has been selected and developed for several portable SOFC systems because of its suitability for small power and rapid start-up [27–31]. Thin-film cells on silicon substrates have also been considered and developed for portable applications because of their small weight and volume [32,33].
- **(ii)** Thermal management is one of the key design aspects in portable SOFC systems. For large-scale applications, thermal management can be addressed by ancillary systems such

Table 9.2 SOFC power system applications

Market sector	Application	Power size	Status
Portable	Power for consumer electronic devices, mobile portable power	1–100 W	Demonstration, pre-commercial
	Portable power, battery chargers	200–500 W	Demonstration
Transportation	Automobile and truck auxiliary power unit (APU)	5–50 kW	Demonstration
	Aircraft APU	Up to 500 kW	Concept
Stationary	Residential/microcombined heat and power (CHP), uninterruptible power	1–10 kW	Prototype, pre-commercial
	CHP and distributed generation (DG)	100 kW–1 MW	Demonstration and concept
	Base load	100–500 MW	Concept

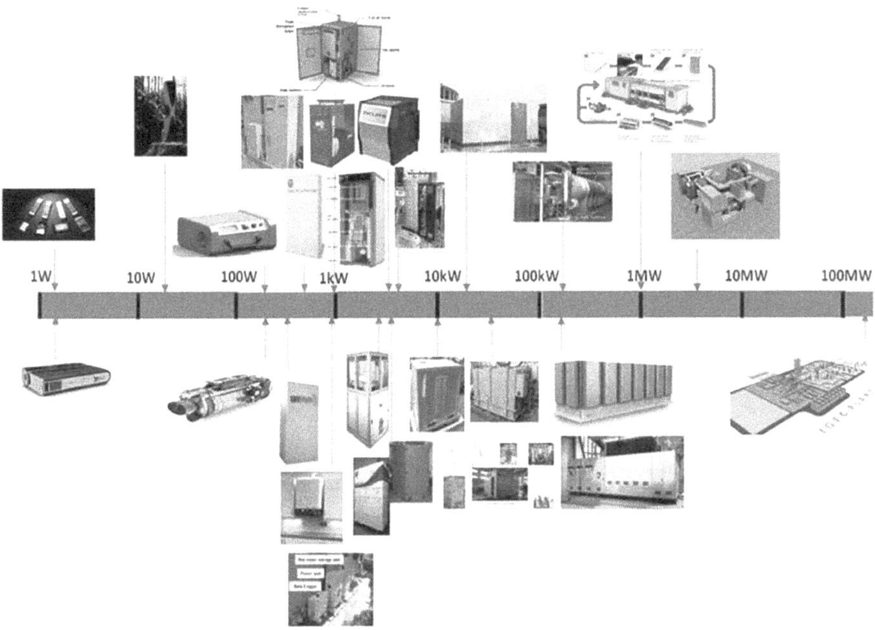

Figure 9.13 Selected SOFC power systems (hardware/prototype and concept).

as HEXs, heaters and combustors. For weight- and size-critical portable applications, such peripherals must be reduced in size or eliminated and optimised for power consumption without degrading performance. In addition, the system design must be optimised to have a small thermally integrated insulation package.

(iii) Another critical aspect in designing portable SOFC systems is selecting the appropriate reforming process and developing a compact design for the reformer [34]. The choice of the reaction process and the design of the reformer depend on the system requirements, thus the operating characteristics of the application (e.g. the type of fuel, start-up/shutdown frequency and varying power demand). The reformer most commonly used in portable SOFC systems is the catalytic partial oxidation (CPOX) [35,36]. The advantages of this type of reformer include operation without water, rapid start-up and compact design. The main disadvantage is high concentration of inert nitrogen in the reformate.

Three examples of portable SOFC system prototypes are a 2.5 W mobile power (Figure 9.14) [37,38], a 125 W generator (Figure 9.15) [39] and a 250 W remote power (Figure 9.16) [40]. The specifications of the 2.5 W mobile power include USB 2.0 interface, standard +5 V output, >2.5 W peak power, 55,000 mW-h of energy in a single butane canister, volume of 202 cm^3 (including butane canister) and weight of 235 g (including butane canister). The 125 W portable generator has an efficiency of about 25%, a power density of 50 W/kg, a specific energy of >1200 Wh/kg at 72 h and a fuel consumption of about 1.5 L/day (125 W) operating low-sulphur kerosene. The 250 W remote power has typical dimensions of length 40 × width 20 × height 36 cm^3 with a nominal mass of 10.9 kg and a fuel consumption of 112 g/h on propane. The performance of this unit is 250 W at 12–24 VDC.

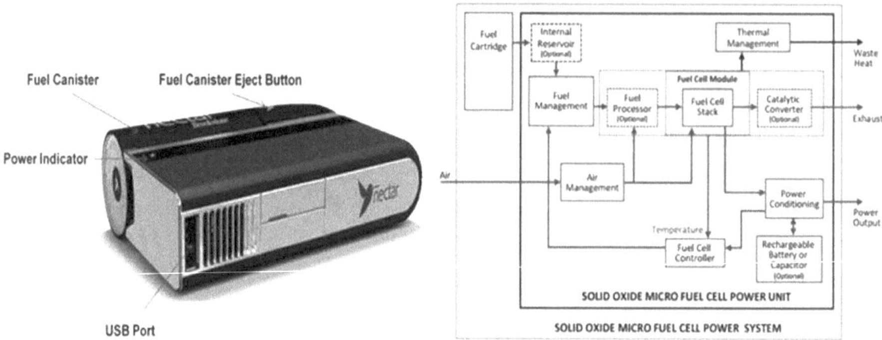

Figure 9.14 2.5 W mobile power: photograph (left) and system schematic (right) [37,38].

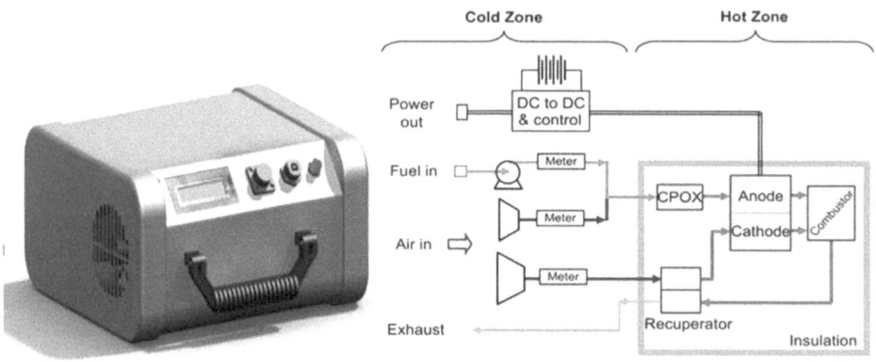

Figure 9.15 Liquid-fuelled 125 W generator [39].

Figure 9.16 250 W remote power [40].

9.5.2 Transportation systems

9.5.2.1 Automobile and truck APUs

SOFC power systems have been developed for automobile and truck APU applications [41–44]. Examples of automobile APU applications are automobile 5 kW APUs for engine-off power and 5 kW APUs in combination with lithium ion batteries for electric vehicle range extension. Examples of truck APU applications are Class 6–8 truck 3–10 kW APUs, recreational vehicle 3–5 kW APUs and long-haul Class 8 truck 10–30 kW APUs for refrigeration. SOFC-based APUs for automobiles and trucks are designed to complement the internal combustion engine, serving as an efficient generator to provide power with the engine on or off. The expected benefits of SOFC APUs include (i) power supply with high efficiency and essentially zero or very low emissions with engine on or off, (ii) operation of any electrical accessory and (iii) possible enabler for high-power consuming advancements. In addition, by-products from SOFC APUs, that is, syngas and heat, may be very valuable. For instance, the syngas can be used for enhanced combustion and aftertreatment in an internal combustion engine. The heat can be used for vehicle heating and accelerated engine and catalyst warm-up and may drive a bottoming cycle such as an expander to recover additional power.

The main components of an automobile/truck SOFC-based APU are SOFC stack; FR and other BOP parts for process air supply; thermal management; waste energy recovery; power electronics/controls and heating, ventilation and air conditioning (Figure 9.17) [41]. Since common fuels for automobile/truck SOFC APUs are gasoline and diesel, selection and development of a suitable reformer or pre-reformer is an important element in the design of SOFC APUs [45,46]. A schematic of a system design for an SOFC APU with a CPOX reformer and anode tail gas recycle is given

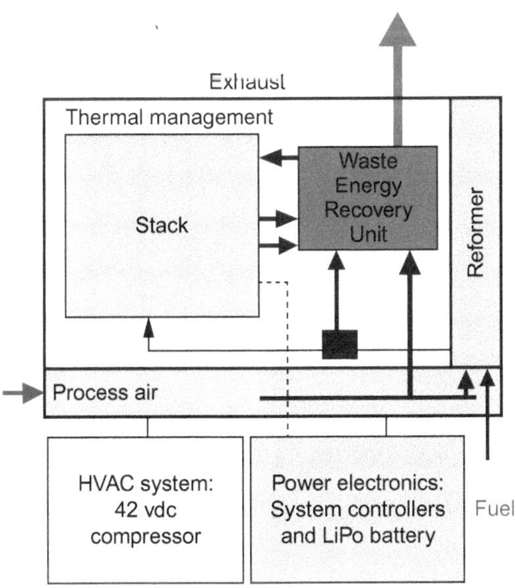

Figure 9.17 Main components of automobile SOFC APU [41].

in Figure 9.18 [47]. Figure 9.19 is a photograph of an automobile APU prototype system showing the design features and packaging [48].

Examples of APU system hardware demonstrations in automobiles [41] and trucks [47,49] are shown in Figures 9.20–9.22.

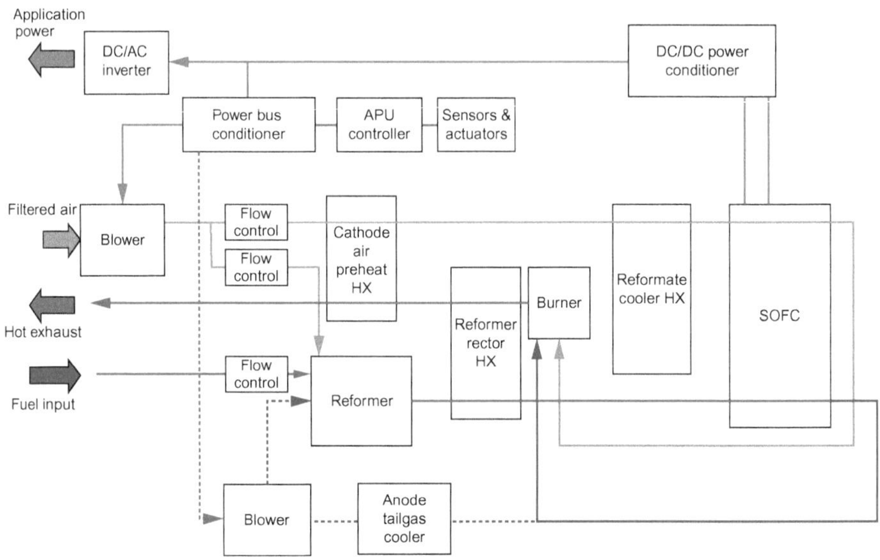

Figure 9.18 Schematic of SOFC APU with CPOX reformer and anode tail gas recycle [47].

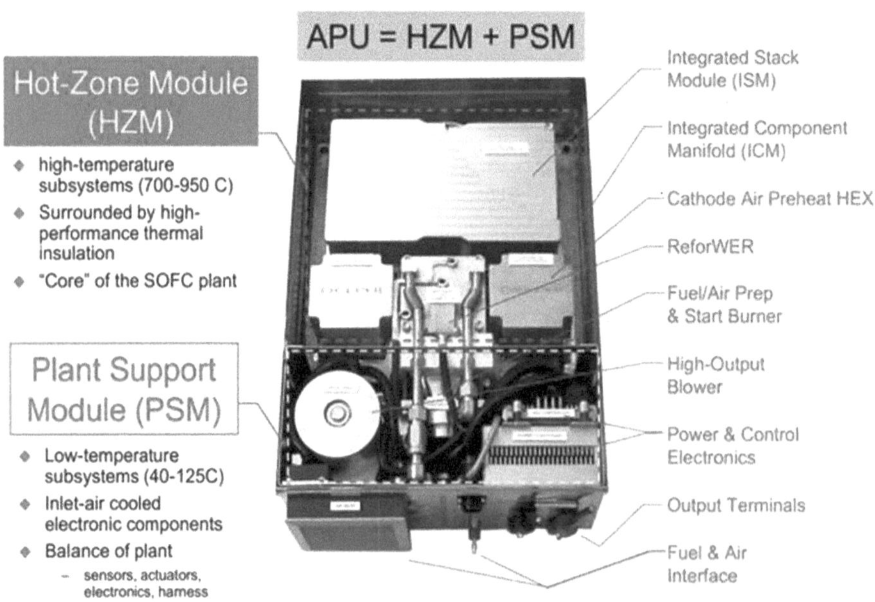

Figure 9.19 Automobile SOFC APU prototype [48].

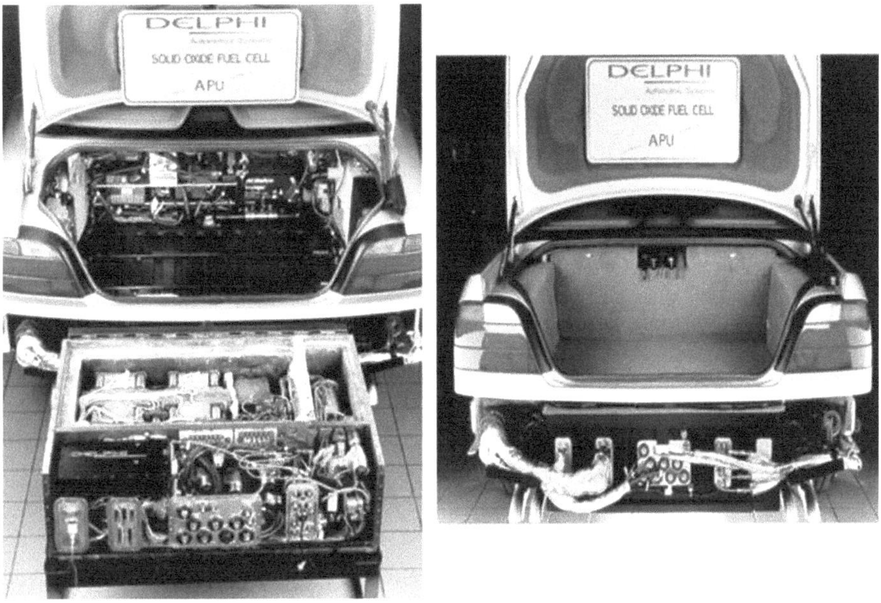

Figure 9.20 Automobile SOFC APU demonstration [41].

Figure 9.21 Truck SOFC APU demonstration [47].

Figure 9.22 SOFC APU system demonstration for Class 7/8 trucks [49].

9.5.2.2 Aircraft APUs

SOFC power systems, especially SOFC/GT hybrid systems, have been considered as a possible replacement for conventional aircraft turbine-powered APUs in future more-electric aircraft. Figure 9.23 is a simplified diagram of an aircraft architecture with a fuel cell APU [50]. The use of SOFC-based APUs has the potential of reducing fuel consumption, noise and exhaust emissions in aircraft in-flight and on-ground power generation [50–55]. Projected fuel savings with SOFC APUs when compared with GT APUs are 40% less fuel used in flight and 75% less fuel used on ground [50]. Noise and NO_x emissions with SOFC APUs are also significantly reduced [55]. The key issues regarding SOFC-based APUs relate to weight, start-up time and cost.

Figure 9.24 is an example of a 300 kW aircraft SOFC APU design [54]. In this system design, the power generation subsystem consists of twin 150 kW hybrid SOFCs and the fuel processing subsystem consists of an autothermal reforming (ATR), FR and a desulphuriser. A fuel HEX is used to cool the reformate gas stream. A portion of exhaust gases from the SOFC is recycled via the CGR and AGR blowers. The remainder of the gases are mixed and sent to the catalytic burner. A power conditioning system converts DC voltage from the fuel cell to 270 V. In the operation of this APU, the air requirements are met entirely with the available cabin air. The SOFC stack is pressurised to 3 atm and operates at about 615°C with a FU of about 80%. The projected efficiency of the system is 45% LHV under ground conditions and 64% LHV at cruise conditions. The efficiency advantage of this system over conventional APUs is 28% points, equivalent to a fuel savings of 70% over 1 day of ground

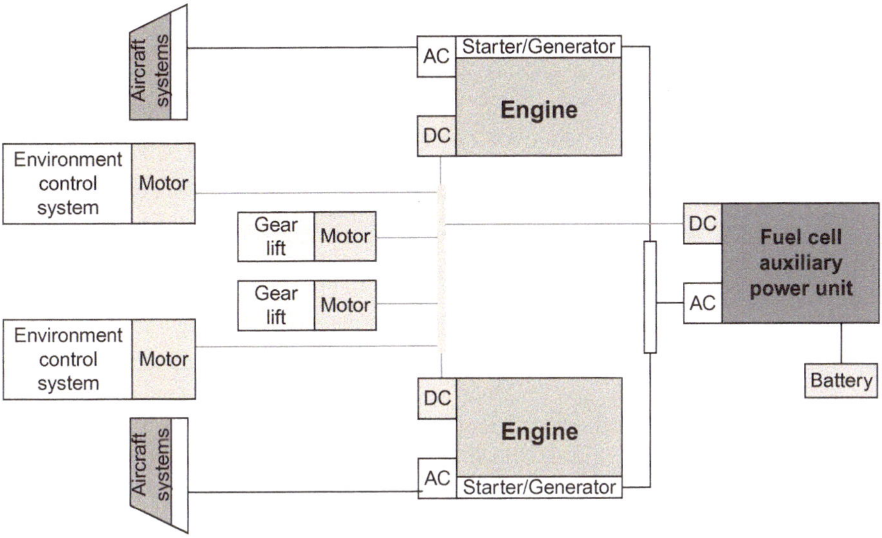

Figure 9.23 More electric aircraft architecture with fuel cell APU [50].

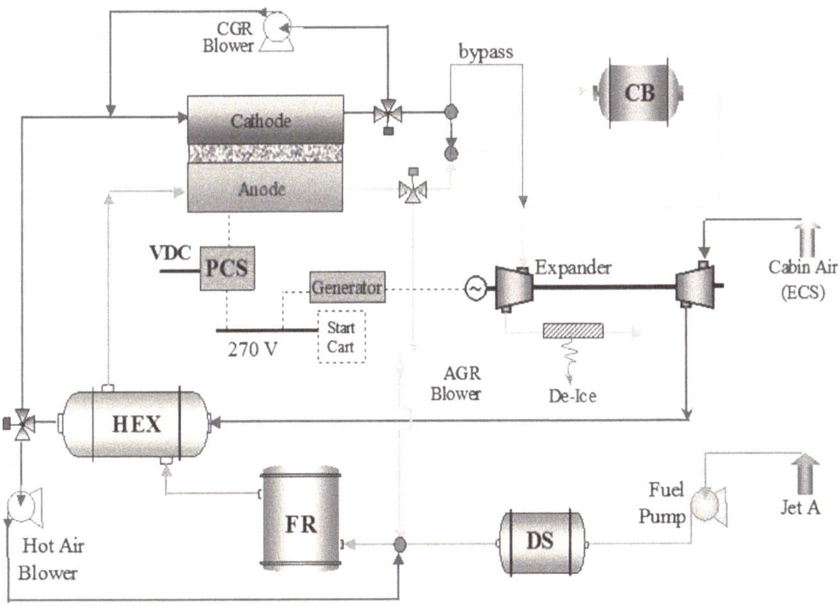

Figure 9.24 Schematic of 300 kW aircraft SOFC APU [54].

operation that translates to about 3.3% of the total fuel burn during daily aircraft operations. The SOFC APU can provide some fuel burn savings during in-flight segments when the conventional APU is not typically operated due to its poor efficiency. Thus, operating the SOFC-based APU during the climb, cruise and descent portions of the flight can provide a 1.3% saving in total aircraft fuel burn. The total fuel consumption savings of the SOFC APU from both ground and flight operations is, therefore, about 4.7% [54].

One aspect in the design of aircraft SOFC-based APUs is the CHP configuration to use the high-temperature exhaust gas to supply aircraft thermal loads, thus reducing the electric power request and, therefore, the size and weight of the APU. Another aspect is the configuration of the feed air, especially since the external air is highly affected by altitude changes. Air can be supplied by either using cabin exhaust flow or taking the air directly from outside the aircraft. The approach of feeding the SOFC with cabin air yields high system efficiency at high-altitude conditions. The lower outside air pressure at high altitudes results in higher power turbine expansion ratios, therefore, increased turbomachinery performance. The fuel cell can run at reduced power and at a more efficient operating point. Hence, the overall system efficiency increases at high altitudes, particularly when using cabin air.

SOFC-based APUs can be located in the tail end of the aeroplane, the current location of turbine-powered APUs in commercial aircraft (Figure 9.25) [50]. Another integration concept for SOFC APUs is to locate the fuel cell and the reformer in a fire compartment in the wing root since most of the electric consumers are in the centre of

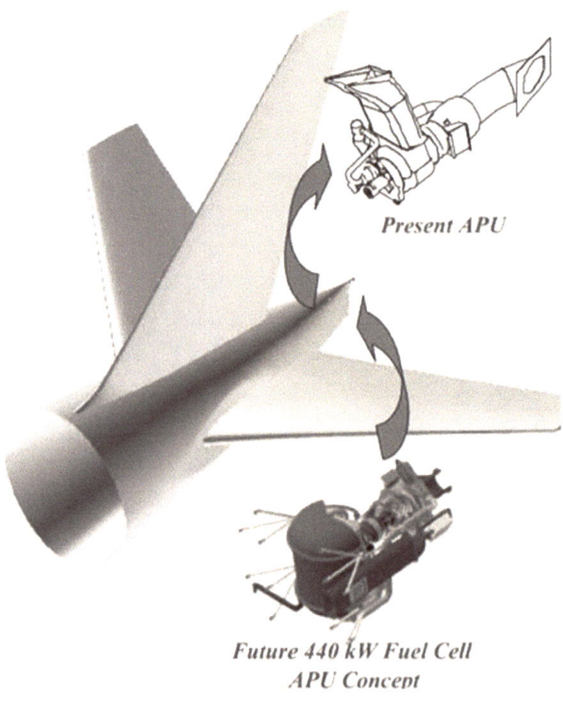

Figure 9.25 Aircraft SOFC APU concept [50].

the aircraft. An advantage of this arrangement is the option to use the hot exhaust for wing anti-ice. Air for the SOFC is taken from the recirculation and mixing plenum of the aircraft environmental control system.

9.5.3 Stationary systems

9.5.3.1 Simple cycle power systems

Simple cycle SOFC systems in the power level of 1–200 kW have been developed for residential, CHP and on-site power applications. These systems can be designed for grid connection and/or local load applications. One example is a 5 kW SOFC system with the simplified schematic shown in Figure 9.26 [23]. This system employs an ATR pre-reformer for operation on methane. A photograph of a laboratory prototype of this system along with its performance during various operation events are shown in Figure 9.27 [23].

The performance and operating conditions of this system at the peak efficiency and peak power points are listed in Table 9.3 [23].

SOFC systems can be designed to include a heat recovery component such as an adsorption chiller–heater for CHP applications [54–59,5]. One example is a stand-alone 1 kW CHP system (Figure 9.28) [60]. This system (width 62 × depth 58 × height 164 cm^3, 170 kg) is designed to supply the full heat demand of a single-family house and deliver electricity to meet the average electrical energy demand of a family. Performance characteristics of this system include electrical output of 1 kW, thermal output of 1.8–2.0 kW, electrical efficiency (fuel cell) of about 35% LHV and overall

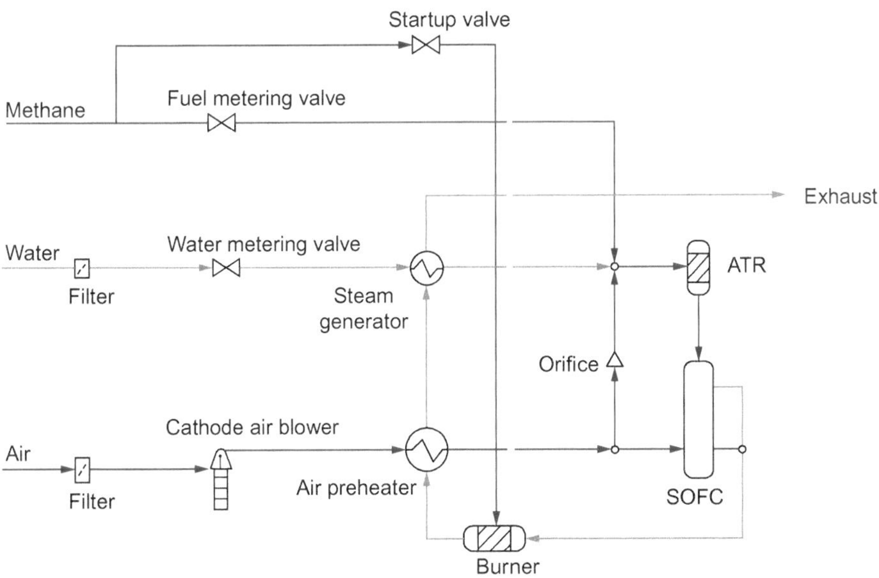

Figure 9.26 Simplified schematic of 5 kW SOFC system [23].

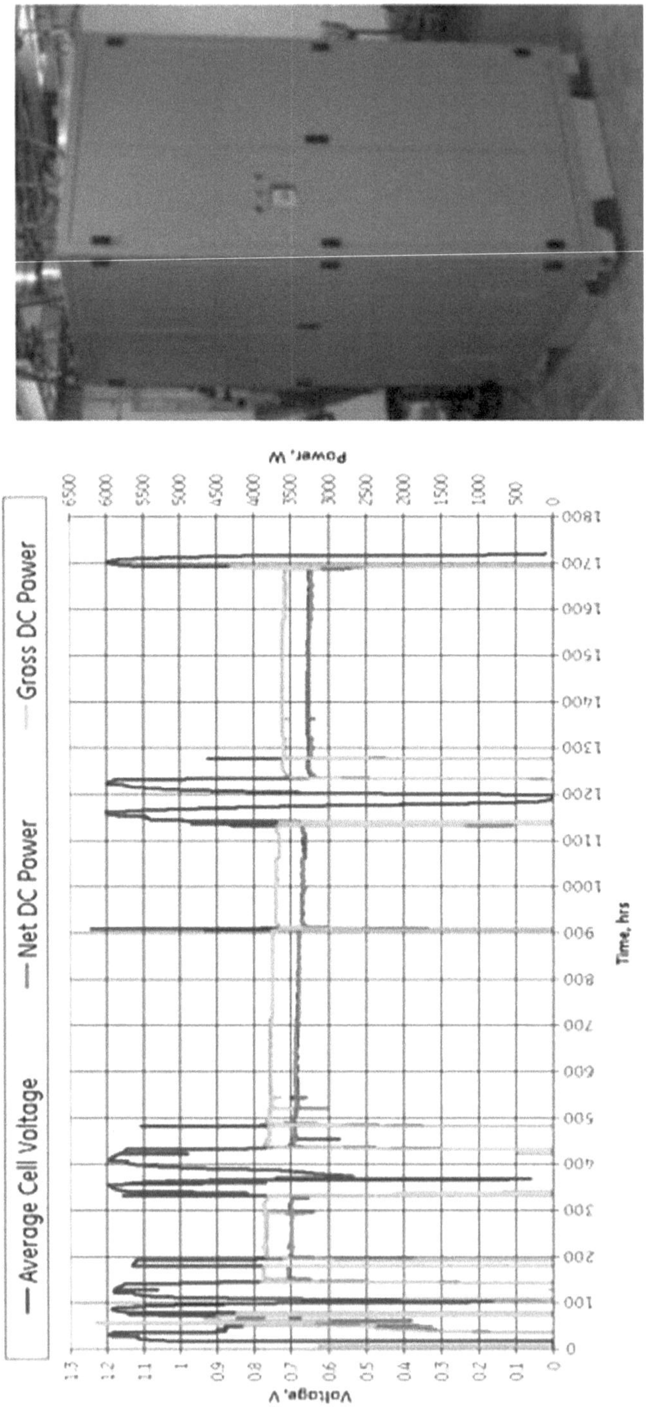

Figure 9.27 Photograph of 5 kW laboratory prototype (right) and performance in terms of average cell voltage and DC power (left) [23].

Table 9.3 **Performance and operating conditions at peak efficiency and peak power of 5 kW system**

Operating conditions	**Peak efficiency point** **Peak efficiency: 40.9%** **Gross DC power: 3.59 kW** **Net DC power: 3.26 kW**	**Peak power point** **DC efficiency: 29.0%** **Gross DC power: 6.13 kW** **Peak DC power: 5.43 kW**
Fuel utilisation (%)	78	67
Air utilisation (%)	26	18
Steam/carbon (S/C) ratio	1.2	1.3
Oxygen/carbon (O/C) ratio	0.68	0.66
Current density (mA/cm^2)	218	444
Average cell voltage (V)	0.73	0.63

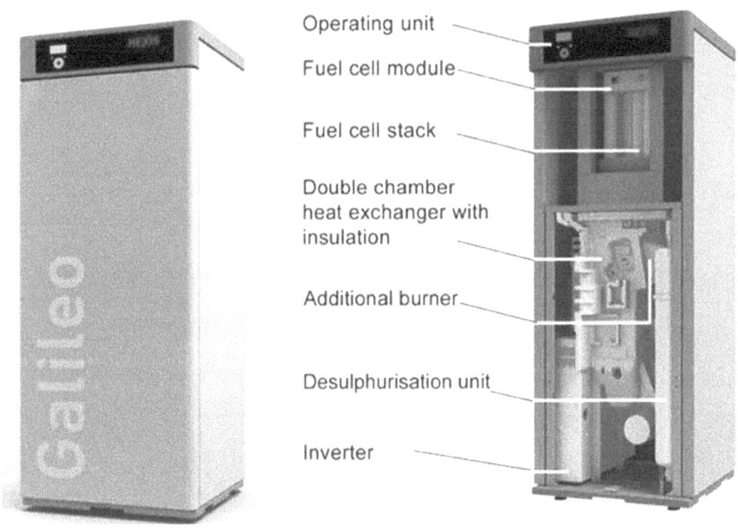

Figure 9.28 1 kW CHP SOFC system [60].

efficiency of 95–109% LHV [61]. Figure 9.29 shows performance of the system in long-term laboratory testing. Performance of the system in a single-family house test is given in Figure 9.30.

Another example of simple cycle SOFC applications is a 100 kW CHP system. The system consists of the generator module, the electrical control system and four major subsystems: fuel supply system, start-up steam supply system, thermal management

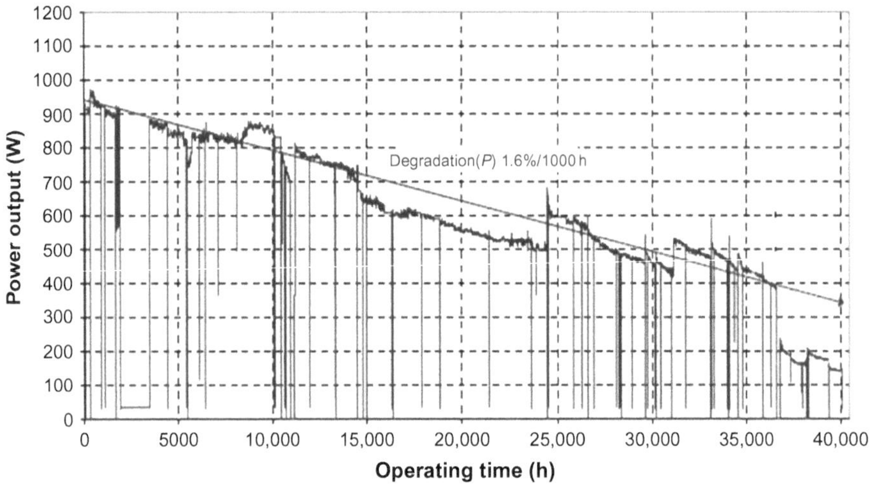

Figure 9.29 Performance of 1 kW system in long-term laboratory test [61].

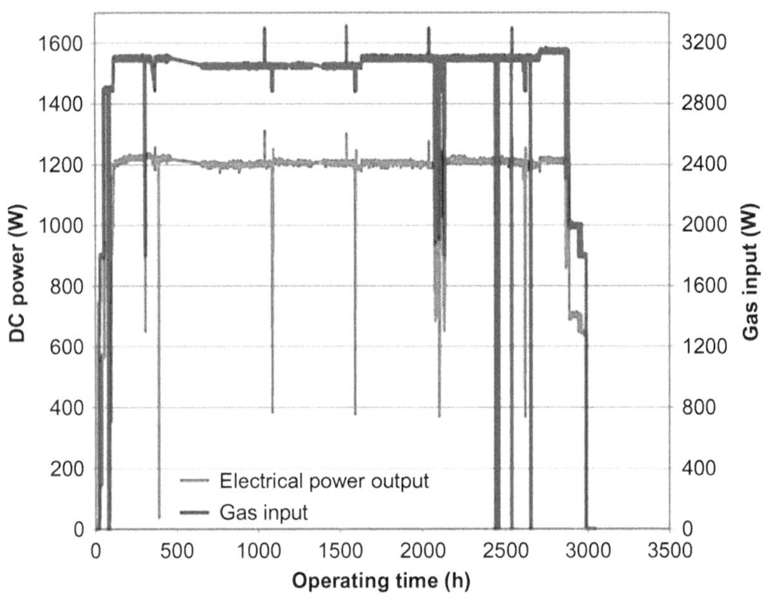

Figure 9.30 Performance of 1 kW CHP system in single-family house test [61].

system and heat export system. Figure 9.31 is a simplified schematic of the CHP system showing main components/systems [62]. (The isometric of the system is shown in Section 9.3, Figure 9.4.) The performance along with a photograph of the system is given in Figure 9.32 [63].

Two other prototype/pre-commercial SOFC systems are shown in Figures 9.33–9.34.

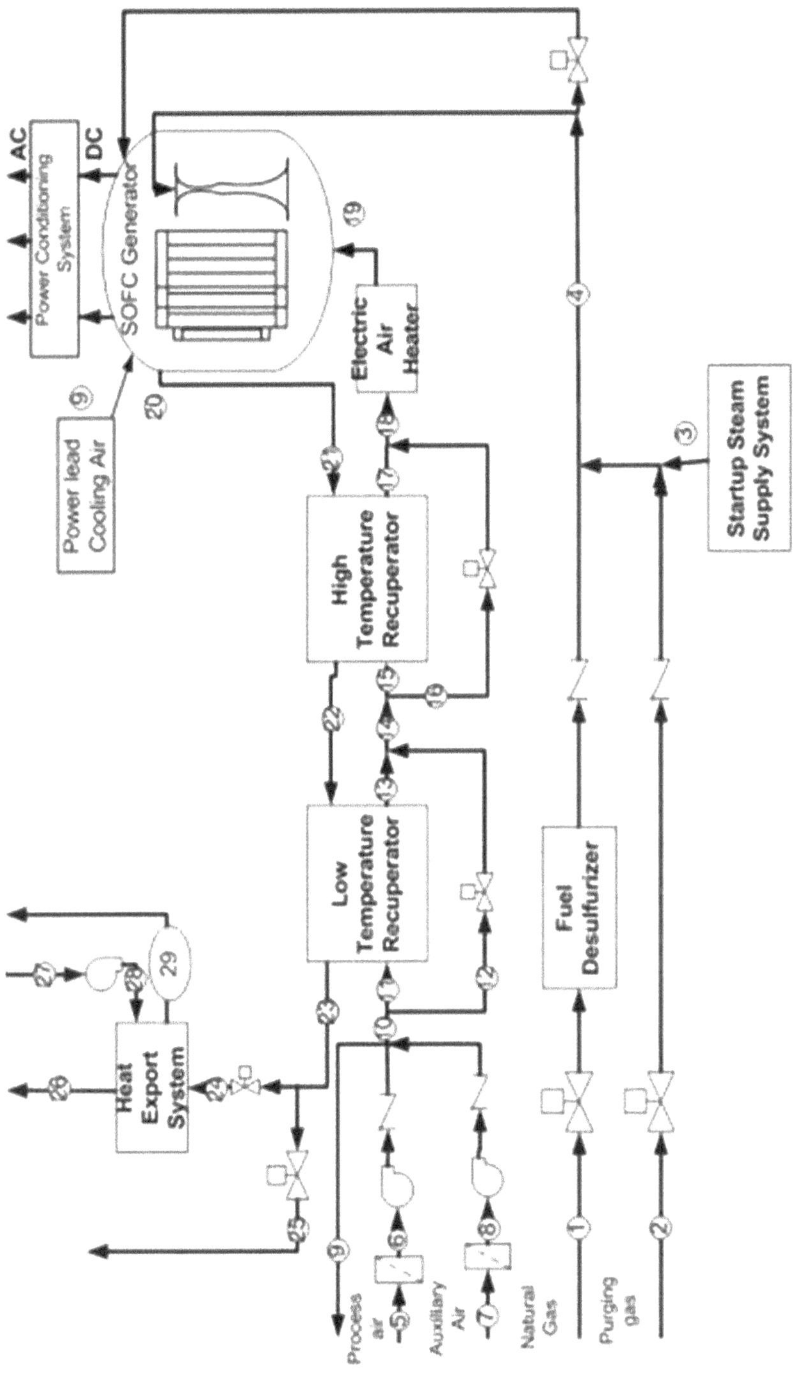

Figure 9.31 Schematic of 100 kW CHP system [62].

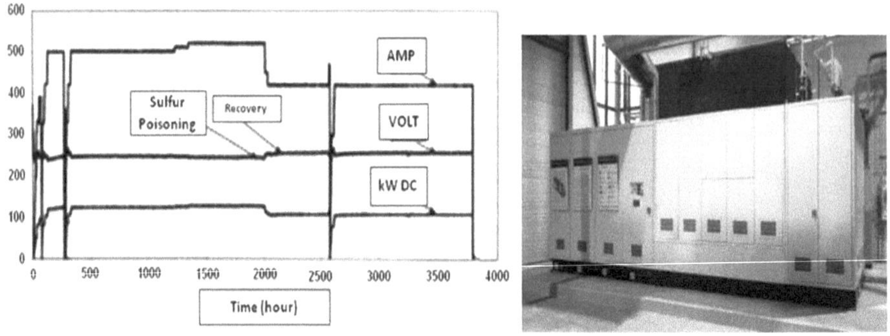

Figure 9.32 Photograph of 100 kW CHP system (right) and performance (left) [63].

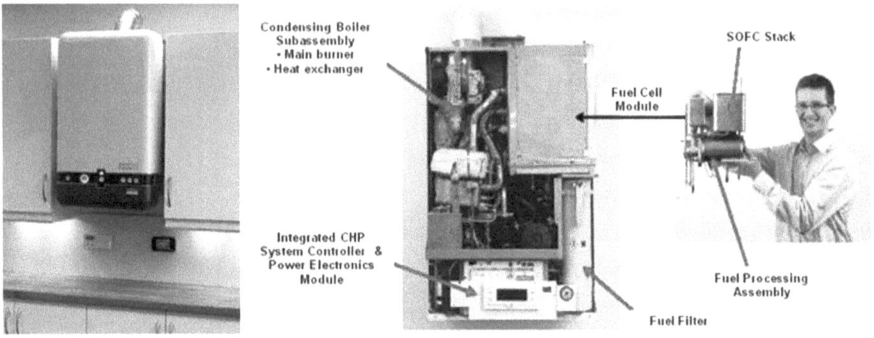

Figure 9.33 Wall-mounted 1 kW CHP system [65].

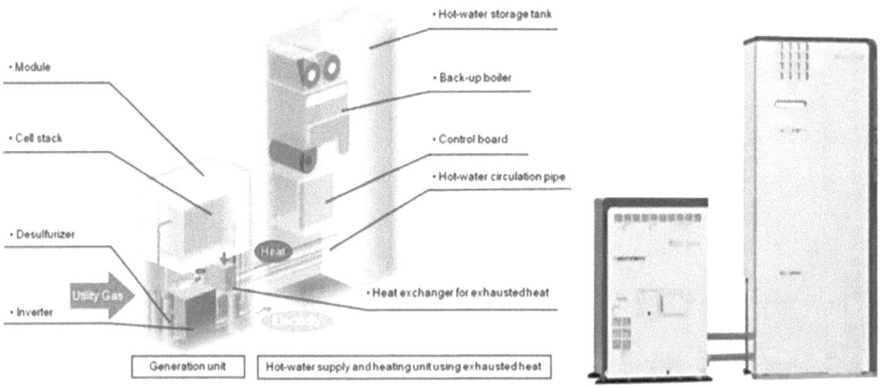

Figure 9.34 700 W CHP system [66].

9.5.3.2 SOFC/GT hybrid systems

SOFC/GT hybrid systems have been considered for DG and central power generation applications [10,22,67–76]. As indicated earlier, hybridisation of the SOFC with a GT significantly improves system efficiency (Table 9.1) and can be beneficial for use in large (≥hundreds of kilowatts) power plants (Figures 9.5 and 9.6).

Figure 9.35 is a schematic of a small (500 kW) SOFC/GT system for DG applications. This design is based on a planar SOFC, a microturbine and a steam reformer (SR) for operation with natural gas [10]. The projected efficiency of this system (based on certain assumptions and constraints) is 67% LHV (at 800°C operating temperature, 4 atm pressure, 75% FU, 0.75 V cell voltage and 100°C temperature rise for the SOFC with cathode and anode gas recycle). The net power is estimated to be 517 kW with SOFC power of 495 kW, turbine power of 78 kW and parasitic power of 56 kW. System efficiencies at part loads are shown in Figure 9.36. It can be seen that the system efficiency remains relatively flat for the net power higher than about 250 kW. The system power in this region is a strong function of two parameters: (i) the cell voltage (the higher the voltage, the higher the system efficiency) and (ii) the SOFC-specific power, that is, the ratio of the SOFC power to the cooling air flow rate (the higher the specific power, the higher the efficiency). The voltage rises at lower power loads because the SOFC stack operates at a lower current density. However, this rise is tempered by the decreasing fuel cell temperature and pressure at part load. On the other hand, the specific power tends to decrease with decreasing speed due to the recuperator temperature constraint. The combination of these effects results in a relatively flat efficiency line at higher part loads. However, the SOFC-specific power decreases at a faster pace with decreasing system power, which eventually results in lower efficiencies at low part loads.

Figure 9.37 is a drawing of a 5 MW SOFC/GT hybrid power plant, and Figure 9.38 shows the design of a 25 MW SOFC/GT hybrid system for central power generation applications [22]. The design of this 25 MW power plant uses a parallel arrangement of planar stacks with respect to both air and fuel. The air preheat is accomplished with the air recycle, while the water management with the fuel recycle to the pre-reformer. In this design, the SOFC operates at 800°C on natural gas with internal reforming and the pre-reformer is a SR. The projected system efficiency is about 65.5% LHV with a net power of 25 MW (SOFC power of 20.3 MW, GT power of 6.1 MW and parasitic power of 1.4 MW). The key operating parameter assumptions for the SOFC in this case are 0.7 V cell voltage, 70% one-pass FU, 100°C temperature rise and 5 atm pressure. The optimal number of stacks and the stack size are determined for this system using an approach to examine stack size effects on the performance, cost and reliability of the system. In this case, a cell diameter of 45.7 cm is chosen based on system cost and reliability projections [26]. From stack cost optimisation and power electronics subsystem loss studies, the optimal SOFC stack configuration for this 25 MW hybrid has 400 cells 45.7 cm in diameter. Assuming cell power density of 0.5 W/cm^2, this SOFC stack size translates into a 320 kW stack power. Given the 20 MW total SOFC stack subsystem power requirement, a minimum of 62 stacks is required to construct the 25 MW hybrid plant.

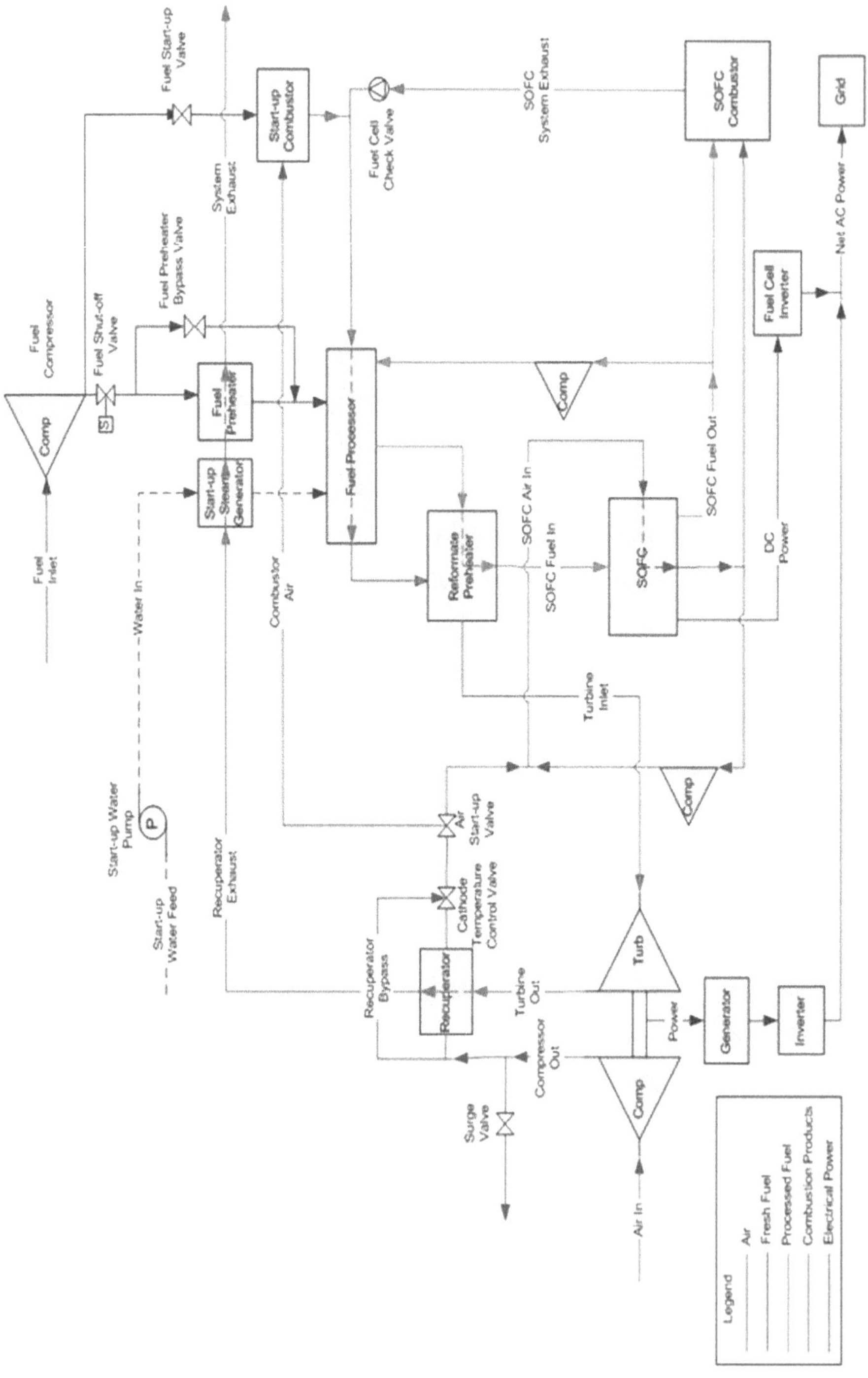

Figure 9.35 Schematic of 500 kW SOFC/GT hybrid system [10].

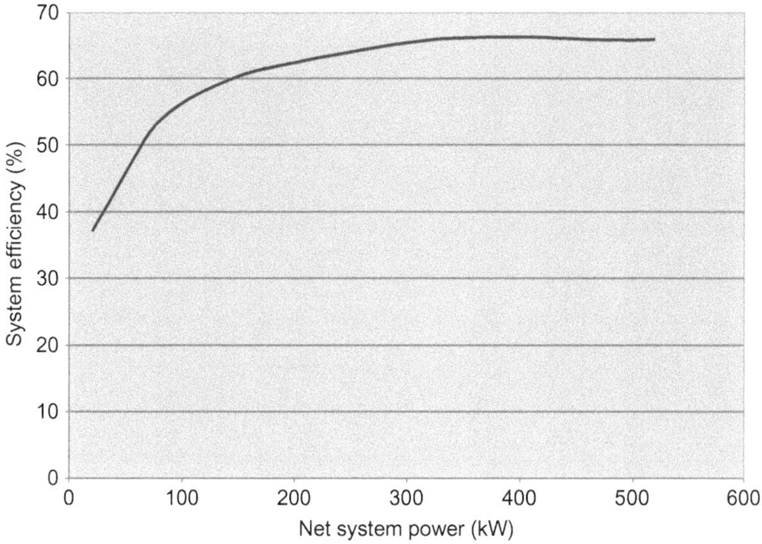

Figure 9.36 SOFC/GT hybrid system efficiency as a function of net system power [10].

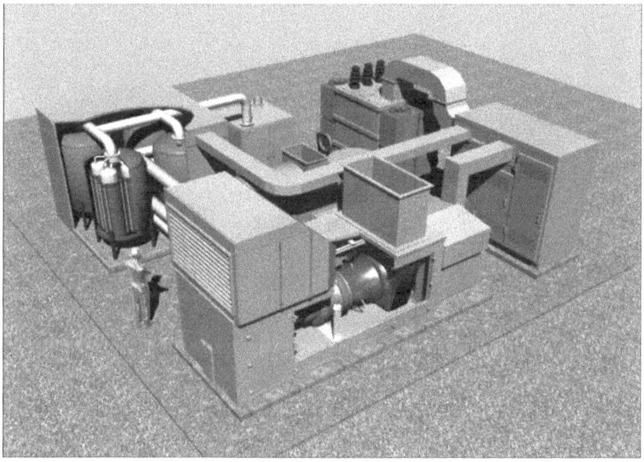

Figure 9.37 5 MW SOFC/GT system concept [22].

Operation of the SOFC/GT hybrid cycle has been demonstrated (Figure 9.39) [77], but practical exemplification of this concept is still limited. A photograph of a 200 kW SOFC/GT hybrid system is shown in Figure 9.40.

9.5.3.3 Integrated gasification fuel cell (IGFC) systems

An SOFC can be integrated with a coal gasifier to form the so-called integrated gasification fuel cell (IGFC) system for multi-megawatt base load power generation applications. The main advantages of the IGFC are its projected increased system

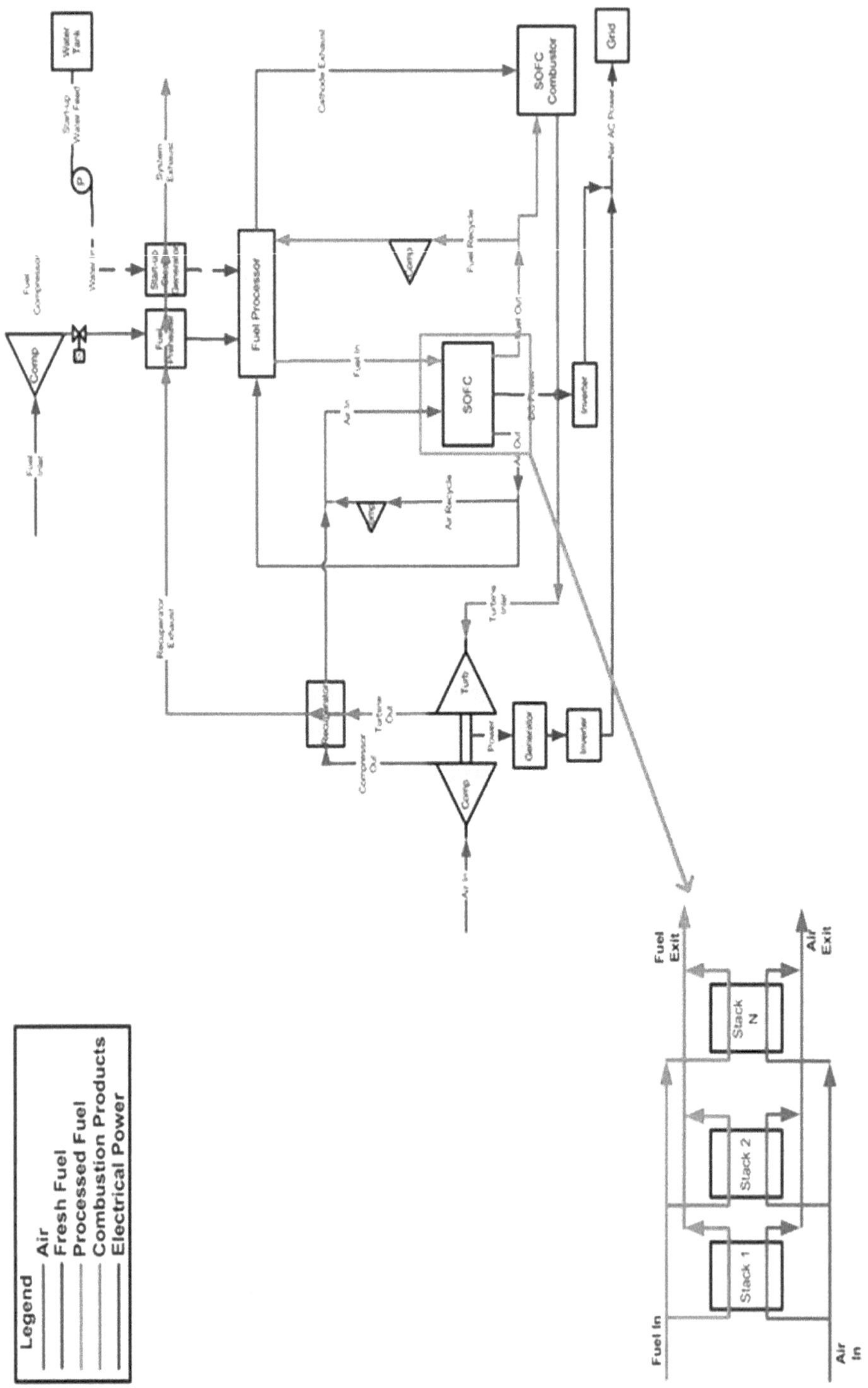

Figure 9.38 Schematic of 25 MW SOFC/GT hybrid system [22].

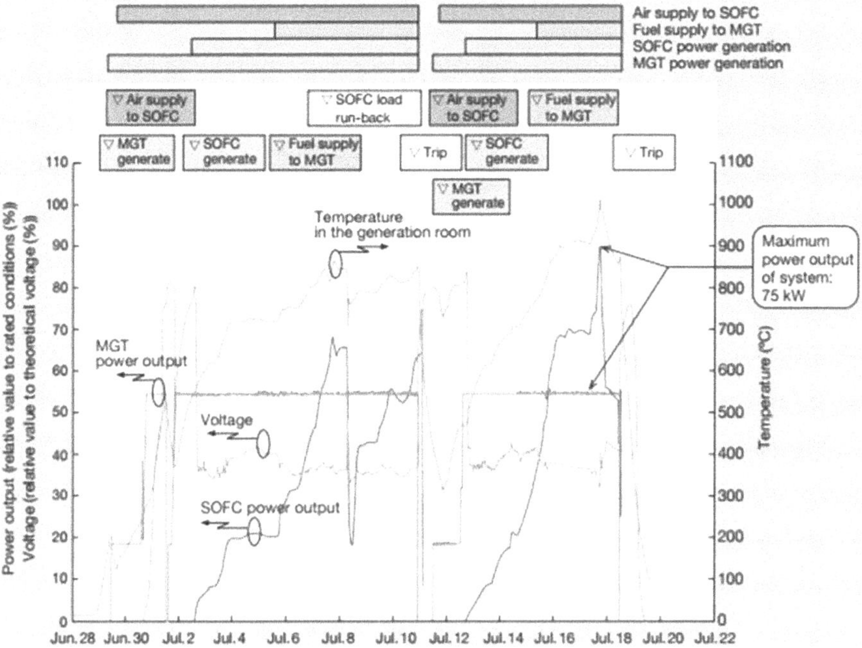

Figure 9.39 Operation of SOFC/GT hybrid system (maximum output of 75 KW) [77].

Figure 9.40 200 kW SOFC/GT hybrid system [77].

efficiencies and comparable estimated costs when compared with those of the state-of-the-art IGCC systems [78–83]. IGFC systems can operate under atmospheric pressure (with atmospheric SOFCs) or under pressure (with pressurised SOFCs or SOFC/GT hybrids). A typical IGFC consists of the following sections: gasifier, gas cleanup, air

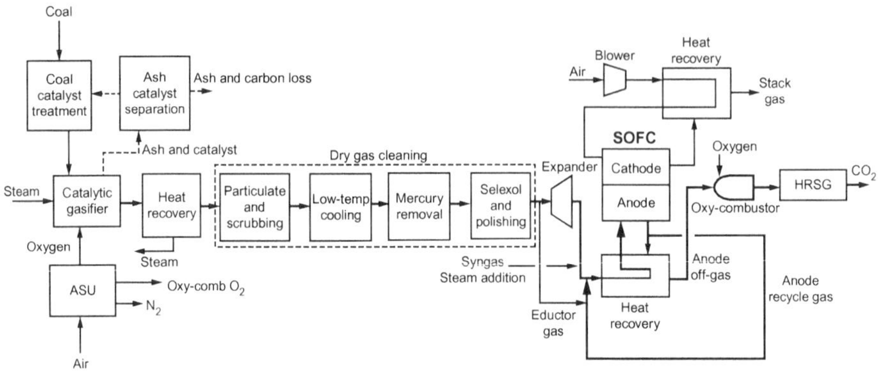

Figure 9.41 Schematic of IGFC system concept with atmospheric SOFC [84].

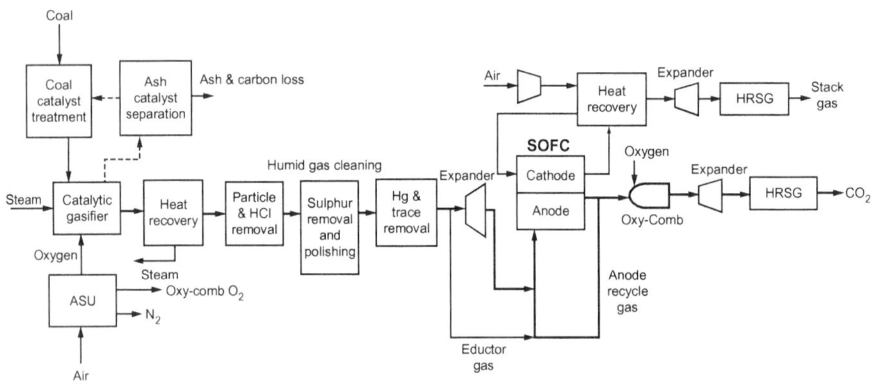

Figure 9.42 Schematic of IGFC system concept with pressurised SOFC [84].

separation unit, GT (if SOFC/GT hybrid is included in the design), ST and heat recovery steam generation (HRSG) (if steam bottoming cycle is included in the design), CO_2 separation (if included in the design) and SOFC. Figures 9.41 and 9.42 show simplified schematics of IGC system concepts with atmospheric and pressurised SOFCs, respectively [84]. An example of a conceptual presentation of an IGFC power plant is given in Figure 9.43.

A key feature in the design of an IGFC system mainly relates to air and fuel flow arrangements in the SOFC (and GT) section(s). Figure 9.44 is an example of an SOFC and GT design for a 300 MW pressurised IGFC system [78]. In this design concept, the SOFC subsystem consists of several pressure modules; each module contains several fuel cell stages (denoted as FC1, FC2, …, FCN and each stage is an SOFC stack or several stacks placed side by side). The depleted fuel along with air from the last fuel

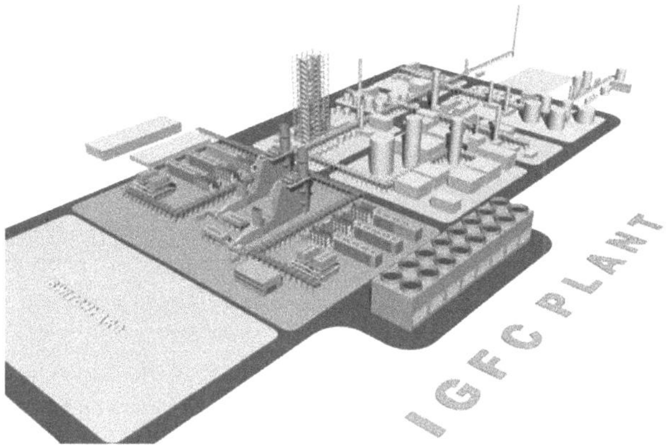

Figure 9.43 IGFC power plant.

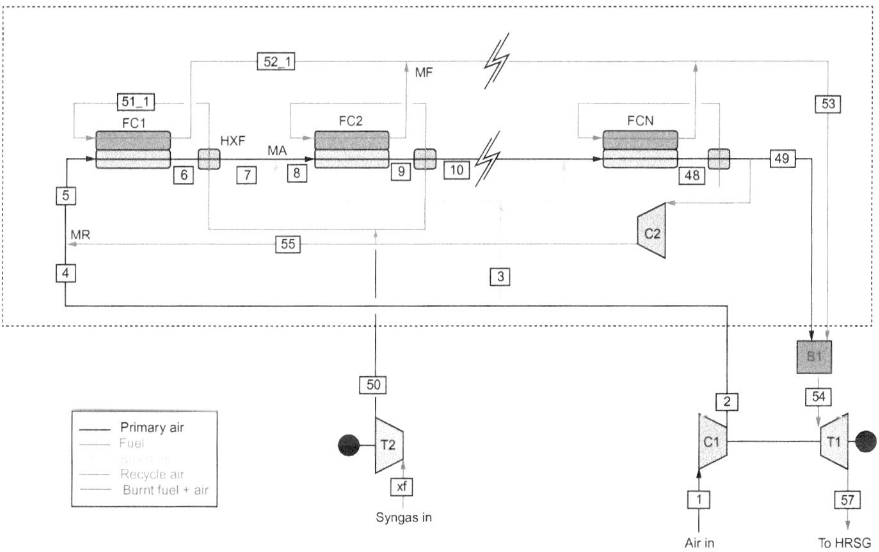

Figure 9.44 Schematic of 300 MW IGFC system design [78].

cell stage is combusted in a low-Btu combustor and fed into a GT for expansion. Gases are supplied to the modules by one or two large GTs (C1, T1, B1) and a syngas expander (T2). Several modules are served by the same GT system. Compressed air from C1 is fed to a plenum and distributed equally to the modules. Similarly, vitiated air is collected at the hot plenum and fed to the burner (B1). On the fuel side, high-pressure syngas is expanded through T2 down to the fuel cell operating pressure and distributed to the various modules and cells inside the module in a similar way. The basic idea in this design concept is that air is used for intercooling between stages. The bulk of the compressed air stream 3 is used for intercooling; only a bleed (stream 4) is used to provide air for the first stage. The temperature of the air stream at station 4 is not hot enough to be admitted directly into the stack, so a portion of the cathode exhaust is recycled, stream 55, and mixed to produce stream 5. The streams 55 and 4 mix in the mixer MR. The recycle stream is bled from the last stage cathode exhaust and a recycle blower C2 is used to overcome the pressure differential. The fuel cell stages have built-in fuel preheaters, shown as HXF. These units are integral to the stack design, and act as HEXs to raise the fuel stream temperature to the point where it can be admitted to the anode inlet. The heat source is the waste heat of the stack, which is adequate to raise the fuel temperature to the desired value. A maximum fuel cell exit temperature of 775°C is assumed in this case. The spent fuel from all the stages is collected at station 53 and sent to burner B1 to burn with the cathode exhaust 49.

Figure 9.45 shows the overall plant performance for the IGFC design concept (shown in Figure 9.44) versus fuel cell operating pressure. The HRSG inlet temperatures are also shown with maximum temperature (T_{max}) and minimum temperature (T_{min}) limit lines. At operating pressures above 10 bar (about 10 atm), the HRSG inlet temperature drops below 1000°F (537.8°C). Below this temperature, the bottoming cycle performance begins to suffer as it becomes difficult to drive a two-pressure reheat ST. At low pressures (below 6 bar), the HRSG inlet temperature exceeds the

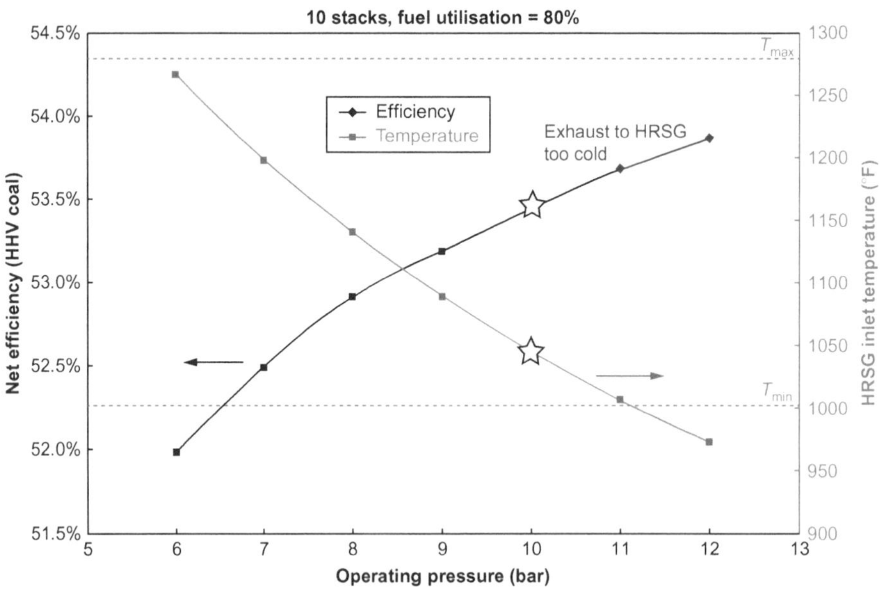

Figure 9.45 Overall performance of 300 MW IGFC system (star = selected design point) [78].

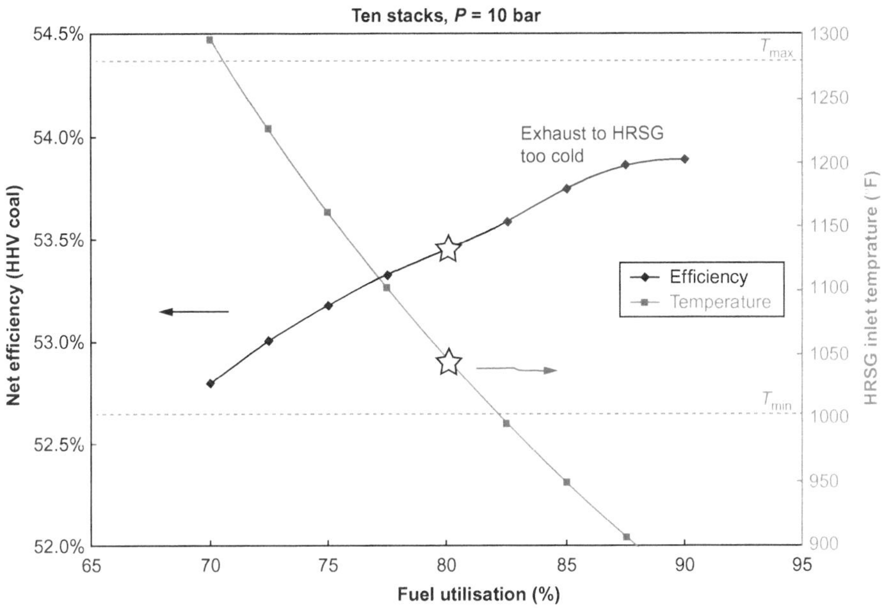

Figure 9.46 Effect of fuel utilisation on IGFC efficiency (star = selected design point) [78].

assumed maximum (1280°F or 693.3°C). The design point is shown as a star. At the design point pressure of 10 bar, the net plant efficiency is 53.4% (coal HHV basis). Figure 9.46 shows the effect of SOFC FU on plant efficiency and HRSG inlet temperature for this IGFC configuration.

9.6 Solid oxide electrolysis cell (SOEC) systems for hydrogen/chemical production

The SOFC/RSOFC functions as a SOEC when operated in electrolysis mode. (An SOFC that can operate efficiently in both fuel cell and electrolysis is referred to as a RSOFC [85].) The SOEC can be used to produce hydrogen from H_2O [86], syngas (mixture of H_2 and CO) from mixtures of H_2O and CO_2 [87] and oxygen from CO_2 [88]. The SOEC operating at high temperatures has the advantage that the electrical energy required for the electrolysis decreases as temperature increases, and the unavoidable joule heat is used in the water splitting process.

The SOEC has been considered for hydrogen production from steam [85,89] at distributed plant (e.g. 1500 kg H_2/day) and central station (e.g. 150,000 kg H_2/day) sizes [90], syngas production for industrial uses [91] and oxygen generation/recovery for space applications [92,93]. Integration of SOEC systems with nuclear [94] and renewable energy resources such as solar energy [95] has been envisioned.

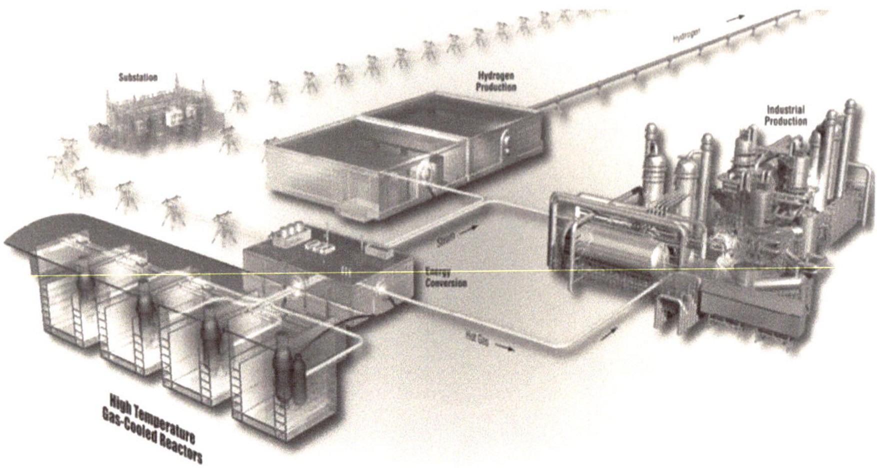

Figure 9.47 Concept for large-scale centralised nuclear hydrogen production with SOEC [96].

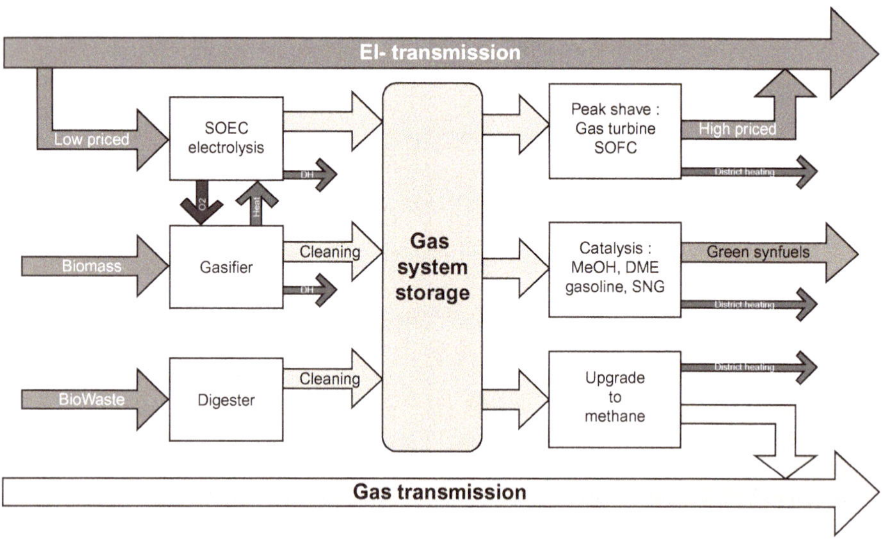

Figure 9.48 Concept combining SOEC, biomass gasifier and biomass digester [97].

Figure 9.47 shows a concept of a SOEC-based hydrogen production central system coupled with high-temperature gas-cooled nuclear reactors [96]. Figure 9.48 is an example of a concept combining SOECs with biomass-based plant [97]. In this concept, oxygen produced in the SOEC unit can be used for the gasification of biomass, and steam for the SOEC can be generated in the gasification plant. The SOEC generates hydrogen that is needed for conversion of all the carbon in the biomass to

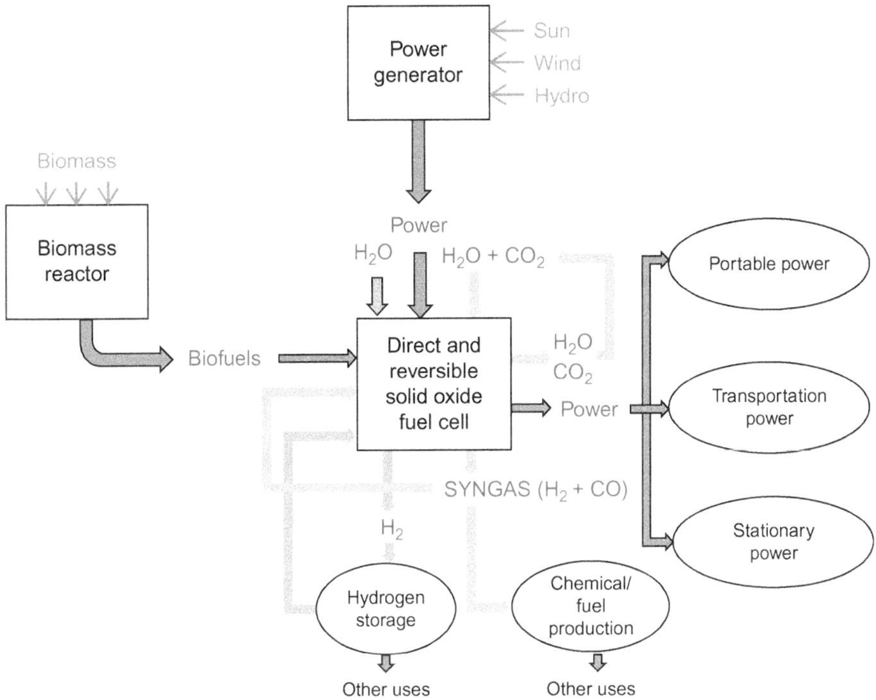

Figure 9.49 An example of a RSOFC-based sustainable energy system [85].

chemicals such as methanol, dimethyl ether, gasoline and synthetic natural gas. This concept also includes a method of upgrading digested biogas using co-electrolysis of CO_2 (present) and steam (added) in biogas. A sustainable energy system concept based on the RSOFC (i.e. SOFC+SOEC) is also possible. An example of such a system is shown schematically in Figure 9.49. In this system, the RSOFC, operating in the electrolysis mode, uses a renewable energy supply (e.g. solar, wind, hydro) to produce hydrogen (from H_2O) or syngas (H_2+CO) (from mixtures of H_2O and CO_2). The chemicals produced can be used to generate power by the same RSOFC operating in the fuel cell mode or can be stored or converted to other chemicals/fuels for subsequent uses. Similarly, the RSOFC can generate power from biomass-derived fuels and the electricity generated can then be used for a variety of power generation applications.

Design concepts have been proposed for RSOFC systems for power generation and hydrogen production. Figure 9.50 is an example of a schematic of a RSOFC system concept showing all the key components for fuel cell and electrolysis operation [90]. Fuel cell operation is discussed in previous sections, so only electrolysis operation is described here. In electrolysis mode, steam enters the system through a compressor and is heated through the two HEXs: H_2 HEX and O_2 HEX. The steam is then heated up to the operating temperature through a heater before entering the RSOFC. This

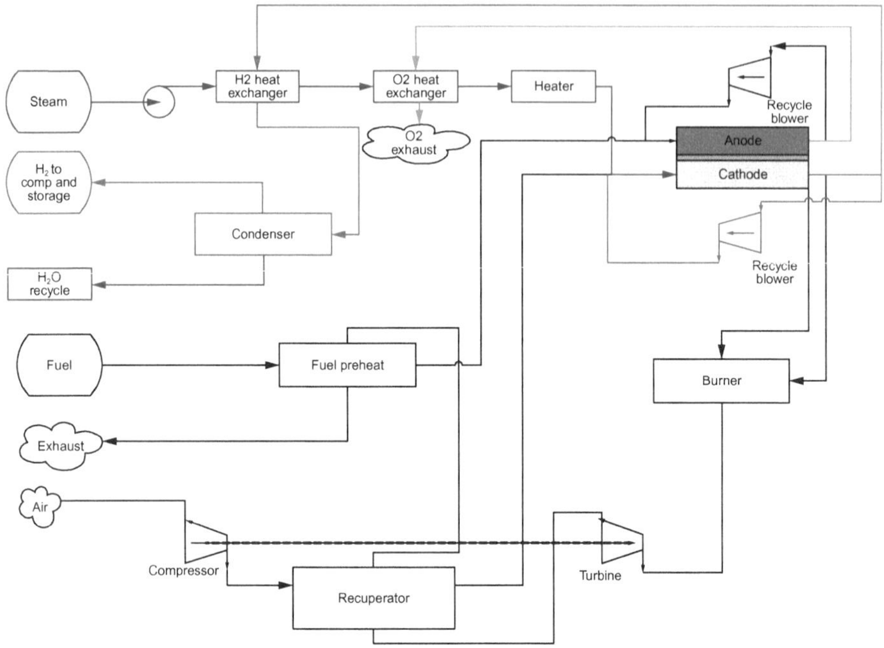

Figure 9.50 RSOFC system schematic [90].

heater also provides the power needed during start-up to get the steam up to the RSOFC operating temperature. Inside of the RSOFC stack, the steam is electrochemically converted to hydrogen and oxygen. Hydrogen is produced on the same side as the steam while oxygen is produced on the opposite side of the cell. The O_2 is exhausted through the O_2 HEX. A portion of the H_2 and H_2O stream is recycled back into the stack inlet, while the rest enters the hot side of the H_2 HEX before moving onto separation. The HEXs for both O_2 and H_2 provide recuperation of the heat energy of the exhaust gases. H_2 is separated from H_2O by condensing the steam with use of cooling water (not shown in the figure). The H_2 at this point is considered the product.

9.7 Summarising remarks

System design, analysis and optimisation are the integral elements in the definition of appropriate architecture and the specification of components and operating parameters that meet the requirements of the specified application for SOFC power systems. The key challenge is to devise an optimal system configuration that integrates operation of all components within technology constraints and, at the same time, performs most efficiently. Besides efficiency, the development of a system design should also consider the reliability and cost of the system. To date, preliminary designs of SOFC power systems of various power outputs have been developed and operation of certain systems, especially those at 200 kW power output or below, has been demonstrated. Advances in SOFC system design and operation are expected as the technology progresses.

References

[1] N.Q. Minh, in: D. Stolten, B. Emonts (Eds.), Fuel Cell Science and Engineering, vol. 2, Wiley-VCH, Weinheim, Germany, 2012, p. 963.

[2] R.A. George, in: Proceedings of 3rd Annual SECA Workshop, U.S. Department of Energy, National Energy Technology Laboratory, Morgantown, WV, 2002. www.alrc.doe.gov/publications/proceedings/02/SECA/seca02.html.

[3] N.Q. Minh, T. Takahashi, Science and Technology of Ceramic Fuel Cells, Elsevier, Amsterdam, The Netherlands, 1995.

[4] N.Q. Minh, Solid State Ionics 174 (2004) 271–277.

[5] M. Suzuki, S. Iwata, K. Higaki, S. Inoue, T. Shigehisa, I. Miyachi, H. Nakabayashi, K. Shimazu, in: S.C. Singhal, H. Yokokawa (Eds.), SOFC-XI, Part 1, ECS Transactions, vol. 25, No. 2, The Electrochemical Society, Pennington, NJ, 2009, p. 143.

[6] R. Payne, J. Love, M. Kah, in: S.C. Singhal, H. Yokokawa (Eds.), SOFC-XI, Part 1, ECS Transactions, vol. 25, No. 2, The Electrochemical Society, Pennington, NJ, 2009, p. 231.

[7] Ceramic Fuel Cells Limited, Introducing BlueGen modular generator: power + heat, www.cfcl.com.au/Assets/Files/BlueGen_Launch_information_(Web)_May-2009.pdf, 2009.

[8] C.A. Forbes, in: 2002 Combined Heat and Power Conference, 2002. www.nyserda.ny.gov/Energy-Efficiency-and-Renewable-Programs/CHP/CHP-Conferences/2002-CHP-Conference.aspx.

[9] E.R. Ray, R.A. Basel, J.F. Pierre, Pressurized solid oxide fuel cell testing, U.S. Department of Energy, Office of Scientific and Technical Information, Oak Ridge, TN, 1995. CONF-950886-11.

[10] GE Hybrid Power Generation Systems, Solid oxide fuel cell hybrid system for distributed power generation, DOE/NETL Cooperative Agreement DE-FC26-01NT40779 Phase 1 topical report, 2004.

[11] W. Winkler, in: U. Bossel (Ed.), Proceedings of First European Solid Oxide Fuel Cell Forum, vol. 2, 1994, p. 821, Lucerne, Switzerland.

[12] S. Samuelsen, Fuel cell/gas turbine hybrid systems, ASME International Gas Turbine Institute, Norcross, GA, 2004.

[13] P. Aguiar, D.J.L. Brett, N.P. Brandon, J. Power Sources 171 (2007) 186.

[14] G. Napoli, M. Ferraro, F. Sergi, G. Brunaccini, G. Dispenza, L. Andaloro, A. Yasin, V. Antonucci, ECS Trans. 42 (1) (2012) 209.

[15] F. Zabihian, A. Fung, Int. J. Eng. 3 (2009) 85.

[16] W. Winkler, H. Lorenz, in: U. Bossel (Ed.), Proceedings of Fourth European Solid Oxide Fuel Cell Forum, vol. 1, 2000, p. 413, Lucerne, Switzerland.

[17] S.E. Veyo, in: Third DOE/UN Conference and Workshop on Hybrid Power Systems, Newport Beach, CA, 2003. www.alrc.doe.gov/publications/proceedings/03/hybrid/hybrid03.html.

[18] J. Brouwer, Hybrid Gas Turbine Fuel Cell Systems, The Gas Turbine Handbook, U.S. Department of Energy, National Energy Technology Laboratory, Morgantown, WV, 2006, 127.
[19] S.K. Park, T.S. Kim, J. Power Sources 163 (2006) 490.
[20] D. Tucker, M. Shelton, A. Manivannan, Interface 18 (3) (2009) 45.
[21] N.Q. Minh, S.C. Singhal, M.C. Williams, in: M. Williams, K. Krist, N. Garland (Eds.), Fuel Cell Seminar 2008, ECS Transactions, vol. 17, No. 1, The Electrochemical Society, Pennington, NJ, 2009, p. 211.
[22] GE Hybrid Power Generation Systems, Solid oxide fuel cell hybrid system for distributed power generation, SOFC scaleup for hybrid and fuel cell systems, DOE/NETL Cooperative Agreement DE-FC26-01NT40779 final topical report, 2004.
[23] GE Hybrid Power Generation Systems, Solid state energy conversion alliance (SECA) solid oxide fuel cell program, DOE/NETL Cooperative Agreement DE-FE-26-01NT41245 final report, 2006.
[24] M. Weston, J. Matcham, Portable Power Applications of Fuel Cells, DTI Pub Urn no. 02/1493, London, UK, 2002.
[25] E. Traversa, Interface 18 (3) (2009) 49.
[26] R.J. Nowak, in: Proceedings of the 2nd SECA Annual Workshop, U.S. Department of Energy, National Energy Technology Laboratory, Morgantown, WV, 2001. www.alrc.doe.gov/publications/proceedings/01/SECA/seca01.html.
[27] K. Hayashi, O. Yamamoto, H. Minoura, Solid State Ionics 135 (2000) 343.
[28] P.K. Cheekatamarla, C.M. Finnerty, C.R. Robinson, S.M. Andrews, J.A. Brodie, Y. Lu, P. G. DeWald, J. Power Sources 193 (2009) 797.
[29] J. Rice, Solid oxide fuel cells to reduce the logistics burden, http://www.dtic.mil/ndia/2011power/Session14_12801Rice.pdf, 2011.
[30] www.ezelleron.com.
[31] www.pointsourcepower.com.
[32] J.D. Morse, A.F. Jankowski, J.P. Hayes, R.T. Graff, Proc. SPIE 3876 (1999) 223.
[33] M. Tsuchiya, B.-K. Lai, S. Ramanathan, Nat. Nanotechnol. 6 (2011) 282.
[34] M. Krumpelt, T.R. Krause, J.D. Carter, J.P. Kopasz, S. Ahmed, Catal. Today 77 (2002) 3.
[35] N. Minh, A. Anumakonda, B. Chung, R. Doshi, J. Ferrall, J. Guan, G. Lear, K. Montgomery, E. Ong, J. Yamanis, in: S.C. Singhal, M. Dokiya (Eds.), SOFC-VI, The Electrochemical Society, Pennington, NJ, 1999, p. 68.
[36] N. Minh, A. Anumakonda, R. Doshi, J. Guan, S. Huss, G. Lear, K. Montgomery, E. Ong, J. Yamanis, in: H. Yokokawa, S.C. Singhal (Eds.), SOFC-VII, The Electrochemical Society, Pennington, NJ, 2001, p. 190.
[37] www.nectarpower.com, 2014.
[38] S.B. Shaevitz, Powering the Wireless World with MEMS, www.nectarpower.com/assets/Uploads/Powering-the-Wireless-World.pdf.
[39] J.C. Poshusta, D. Bruinsma, C.W. Booten, J.L. Martin, C.M. Martin, www.fuelcellseminar.com/assets/2009/HRD42-3_1130AM_Poshusta.pdf, 2009.
[40] www.ultra-ami.com, 2014.
[41] C. De Minco, S. Mukerjee, in: Proceedings of the 2nd SECA Annual Workshop, U.S. Department of Energy, National Energy Technology Laboratory, Morgantown, WV, 2001. www.alrc.doe.gov/publications/proceedings/01/SECA/seca01.html.
[42] J.J. Botti, M.J. Grieve, J.A. MacBain, 2005 SAE World Congress, SAE International, Warrendale, PA, 2005, 2005-01-1172.
[43] T. Salameh, SOFC Auxiliary Power Units (APUs) for Vehicles, 2008 TRRF05 Fuel Cell Technology, 2008.

[44] S. Shaffer, in: Proceedings of the 5th SECA Annual Workshop, U.S. Department of Energy, National Energy Technology Laboratory, Morgantown, WV, 2004. www.alrc.doe.gov/publications/proceedings/04/seca-wrkshp/seca-wrkshp04.html.
[45] T. Aicher, B. Lenz, F. Gschnell, U. Groos, F. Federici, L. Caprile, L. Parodi, J. Power Sources 154 (2006) 503.
[46] A. Lindermeir, S. Kah, S. Kavurucu, M. Mühlner, Appl. Catal. B Environ. 70 (2007) 488–497.
[47] S. Shaffer, in: Proceedings of the 8th SECA Annual Workshop, U.S. Department of Energy, National Energy Technology Laboratory, Morgantown, WV, 2007. www.alrc.doe.gov/publications/proceedings/07/SECA_workshop/index.html.
[48] J. Zizelman, in: Proceedings of 4th Annual SECA Workshop, U.S. Department of Energy, National Energy Technology Laboratory, Morgantown, WV, 2003. www.alrc.doe.gov/publications/proceedings/03/SECA/seca03.html.
[49] www.greencarcongress.com/2010/05/cumminsapu-20100512.html.
[50] D. Daggett, in: Proceedings of 4th Annual SECA Workshop, U.S. Department of Energy, National Energy Technology Laboratory, Morgantown, WV, 2003. www.alrc.doe.gov/publications/proceedings/03/SECA/seca03.html.
[51] S. Eelman, E. Pozo del Poza, T. Krieg, in: 24th International Congress of the Aeronautical Sciences, 2004, p. 1.
[52] M. Gummalla, A. Pandy, R. Braun, T. Carriere, J. Yamanis, R. Welch, Fuel Cell Airframe Integration Study for Short-Range Aircraft – Volume I: Aircraft Propulsion and Subsystem Integration Evaluation, NASA Center for Aerospace Information, Hanover, MD, 2006. NASA/CR-2006-214457/VOL1.
[53] K. Rajashekara, J. Grieve, D. Dagget, in: Industry Applications Conference, 2006. 41st IAS Annual Meeting, Conference Record of the 2006 IEEE, 2006, p. 2185.
[54] R.J. Braun, M. Gummalla, J. Yamanis, J. Fuel Cell Sci. Technol. 6 (2009) 031015.
[55] M. Santarelli, M. Cabrera, M. Cali, J. Fuel Cell Sci. Technol. 7 (2010) 021066.
[56] R.J. Braun, S.A. Klein, D.T. Reindl, J. Power Sources 158 (2006) 1290.
[57] R.J. Braun, J. Fuel Cell Sci. Technol. 7 (2010) 031018.
[58] T. Haga, N. Komiyama, H. Nakatomi, K. Konishi, T. Suton, T. Kikuchi, in: S.C. Singhal, H. Yokokawa (Eds.), SOFC-XI, Part 1, ECS Transactions, vol. 25, No. 2, The Electrochemical Society, Pennington, NJ, 2009, p. 71.
[59] J. Love, S. Amarasinghe, D. Selvey, X. Zheng, L. Christiansen, in: S.C. Singhal, H. Yokokawa (Eds.), SOFC-XI, Part 1, ECS Transactions, vol. 25, No. 2, The Electrochemical Society, Pennington, NJ, 2009, p. 115.
[60] www.hexis.com.
[61] A. Mai, B. Iwanschitz, J.A. Schuler, R. Danzler, V. Nerlich, A. Schuler, in: T. Kawada, S. Singhal (Eds.), SOFC-XIII, ECS Transaction, vol. 57, No. 1, The Electrochemical Society, Pennington, NJ, 2013, p. 73.
[62] M. Cali, E. Fontana, V. Giaretto, G. Orsello, M. Santarelli, conference.ing.unipi.it/ichs2005/Papers/330110.pdf, 2005.
[63] R.A. George, J. Power Sources 86 (2000) 134.
[64] www.cfcl.com.au.
[65] www.cerespower.com.
[66] www.global.kyocera.com.
[67] Siemens Westinghouse Power Corporation, Pressurized solid oxide fuel cell/gas turbine power system, final report, U.S. Department of Energy, Office of Scientific and Technical Information, Oak Ridge, TN, 2000.
[68] R. Kurz, ASME J. Fuel Cell Sci. Technol. 2 (2005) 268.

[69] C.J. Steffen Jr., J.E. Freeh, L.M. Larosiliere, Solid Oxide Fuel Cell/Gas Turbine Hybrid Cycle Technology for Auxiliary Aerospace Power, NASA Center for Aerospace Information, Hanover, MD, 2005, NASA/TM-2005-213586.
[70] A. Himansu, J.E. Freeh, C.J. Steffen Jr., R.T. Tornabene, X.-Y.J. Wang, Hybrid Solid Oxide Fuel Cell/Gas Turbine System Design for High Altitude Long Endurance Aerospace Missions, NASA Center for Aerospace Information, Hanover, MD, 2005, NASA/TM-2006-214328.
[71] J. Van Osdol, E. Liese, D. Tucker, R. Gemmen, R. James, ASME J. Fuel Cell Sci. Technol. 7 (2010) 015001-1-8.
[72] F. Fathian, World Appl. Sci. J. 10 (2010) 388.
[73] J.S. Yang, J.L. Sohn, S.T. Ro, J. Power Sources 166 (2007) 155.
[74] J.S. Yang, J.L. Sohn, S.T. Ro, J. Power Sources 175 (2008) 296.
[75] N.Q. Minh, in: S.C. Singhal, J. Mizusaki (Eds.), SOFC-IX, vol. 1, The Electrochemical Society, Pennington, NJ, 2005, p. 76.
[76] C. Stiller, B. Thorud, O. Bolland, J. Eng. Gas Turbines Power 128 (2005) 551.
[77] T. Gengo, Y. Kobayashi, Y. Ando, N. Hisatome, T. Kabata, K. Kosaka, Mitsubishi Heavy Industries, Ltd. Technol. Rev. 45 (1) (2008) 33.
[78] GE Hybrid Power Generation Systems, Coal integrated gasification fuel cell system study, final report, DOE/NETL Cooperative Agreement DE-FC26-01NT40779, 2004.
[79] H. Ghezel-Ayagh, in: Proceedings of the 10th SECA Annual Workshop, U.S. Department of Energy, National Energy Technology Laboratory, Morgantown, WV, 2009. www.alrc.doe.gov/publications/proceedings/09/seca/index.html.
[80] D. Keairns, in: Proceedings of the 12th SECA Annual Workshop, U.S. Department of Energy, National Energy Technology Laboratory, Morgantown, WV, 2011. www.alrc.doe.gov/publications/proceedings/11/seca/index.html.
[81] K. Gerdes, in: Proceedings of the 13th SECA Annual Workshop, U.S. Department of Energy, National Energy Technology Laboratory, Morgantown, WV, 2012. www.alrc.doe.gov/publications/proceedings/12/seca/index.html.
[82] E. Grol, J. Wimer, System Analysis of an Integrated Gasification Fuel Cell Combined Cycle, Volume 1: Technical Assessment, U.S. Department of Energy, National Energy Technology Laboratory, Morgantown, WV, 2009, DOE/NETL-40/080609.
[83] K. Gerdes, E. Grol, D. Keairns, R. Newby, Integrated Gasification Fuel Cell Performance and Cost Assessment, U.S. Department of Energy, National Energy Technology Laboratory, Morgantown, WV, 2009, DOE/NETL-2009/1361.
[84] D. Keairns, R. Newby, E. Grol, in: Proceedings of the 10th SECA Annual Workshop, U.S. Department of Energy, National Energy Technology Laboratory, Morgantown, WV, 2009. www.alrc.doe.gov/publications/proceedings/09/seca/index.html.
[85] N.Q. Minh, M.B. Mogensen, Interface 22 (4) (2013) 55.
[86] A. Hauch, S.D. Ebbesen, S.H. Jensen, M. Mogensen, J. Mater. Chem. 18 (2008) 2331.
[87] S.H. Jensen, P.H. Larsen, M. Mogensen, Int. J. Hydrogen Energy 32 (2007) 3253.
[88] J. Guan, R. Doshi, G. Lear, K. Montgomery, E. Ong, N. Minh, J. Am. Ceram. Soc. 85 (2002) 2651.
[89] B. Yu, W.Q. Zhang, J. Chen, J.M. Xu, S.R. Wang, Sci. China Ser. B Chem. 51 (2008) 289.
[90] GE Hybrid Power Generation Systems, High performance flexible reversible solid oxide fuel cell, final technical report, 2008.
[91] Q. Fu, in: A. Indarto, J. Palgunadi (Eds.), Syngas: Production, Applications and Environment Impact, Nova Science Publishers, Hauppauge, NY, 2011, p. 209.
[92] K.R. Sridhar, R. Foertner, J. Propul. Power 16 (2000) 1105.

[93] D. Weng, S. Yates, AIAA Paper 2010-8675, American Institute of Aeronautics and Astronautics, Reston, VA, 2010.
[94] R. Rivera-Tinoco, C. Mansilla, C. Bouallou, F.F. Werkoff, Int. J. Nucl. Hydrogen Prod. Appl. 1 (2008) 249.
[95] J. Padin, T.N. Veziroglu, A. Shahin, Int. J. Hydrogen Energy 25 (2000) 295.
[96] J.E. O'Brien, X. Zhang, C.R. O'Brien, G. Tao, www.fuelcellseminar.com/media/9066/hrd34-5.pdf, 2011.
[97] M. Mogensen, S.H. Jensen, S.D. Ebbesen, A. Hauch, C. Graves, J.V.T. Høgh, X. Sun, S. Das, P.V. Hendriksen, J.U. Nielsen, A.H. Pedersen, N. Christiansen, J. B. Hansen, in: 2012 World Gas Conference, Kuala Lumpur, Malaysia, 2012.

Portable early market SOFCs

10

K. Kendall[1,2]
[1]Adelan Ltd, Weekin Works, Harborne, Birmingham, UK, [2]School of Metallurgy and Materials, University of Birmingham, UK

10.1 Introduction

Solid oxide cells, stacks and systems fall naturally into two classes:

(1) Large, high power, very efficient products which can operate on natural gas, biogas or coal gas to compete with present day turbine power stations
(2) Smaller fuel cells which have unique capabilities to produce small power where conventional combustion engines are ineffective, especially with liquid fuels.

This chapter focuses on the smaller systems, even down to 1 W and below, but certainly less than 1 kW and often smaller than 100 W. The objective is to understand the niche applications, which vary from SOFC sensors through cell-phone chargers, to hybrid power trains for unmanned air vehicles (UAVs), to auxiliary power units (APUs) for leisure vehicles and truck cabs, to combined heat and power (CHP) units for caravans and small buildings. All these applications demand rapid start-up and resistance to thermal shock, often require high power density because of weight and space restraints, need to run on real available fuels but may not be overly concerned with efficiency as a marketing criterion.

Very significant progress has been made during the past decade for developing SOFCs to satisfy these niche applications in early market areas. In particular, the recognition that improved ceramic processing, by tape casting, Micro-Electro-Mechanical-Systems (MEMS) patterning, injection moulding or extrusion, can produce thin cells, and stacks of high quality has led to manufacture of small SOFC devices, with good thermal shock resistance and rapid start-up [1,2]. The issues that remain to be solved relate to the application-related prices of these products. Sensors and phone chargers can command prices of $40,000/kW. In other markets, such as UAVs for military applications, commercial success can be attained for SOFCs costing $100,000/kW. Applications in CHP need lower prices, around $10,000/kW. The present designs still do not have power densities or start-up times fully competitive with combustion engines or polymer fuel cells (PEMFCs) which are currently used in transport applications. But potential exists in the next decades for the rapid evolution of performance increase of small SOFCs, with the prospect of outperforming both combustion engines and PEMFCs.

It is logical to start with the smallest power units because pressed discs and thimbles of yttria-stabilised zirconia (YSZ) have been around since early experiments in the 1930s [3–5] as described in Chapter 2. These turned out to be both too thick and electrically resistive for higher power applications but were readily applicable in

High-temperature Solid Oxide Fuel Cells for the 21st Century. http://dx.doi.org/10.1016/B978-0-12-410453-2.00010-5

sensors which became a massive market in the past decades of the twentieth century when catalysts were mandated for automobile exhaust clean-up. Therefore, we deal with sensors first and then go on to small portable and finally larger portable SOFC generators.

10.2 Sensor SOFCs

From the original pressed pellets of YSZ electrolyte studied by Mobius and his contemporaries in the 1950s [6–8], thimble shapes were developed by von Gunther Bauman's team at Robert Bosch company in Germany during the 1960s to enable fuel injection systems on automobiles to be effective [9,10]. This was the lambda sensor, introduced in production cars by Volvo and Saab in 1976, to give a voltage output depending on the difference in oxygen concentration in the exhaust gas at the outer electrode compared to the air reference at the inner cathode. Electrodes were typically gas permeable, porous platinum on both surfaces. The sensor was integrated in an electronic control system comprising the engine control unit (ECU) computer, and the fuel injectors which defined the amount of fuel entering each cylinder of the engine, allowing the combustion reaction to run near the stoichiometric air/fuel ratio, that is, lambda point at which there was a sudden change of output voltage, where emissions were minimised.

In 1998, Bosch introduced a planar design of sensor, shown in Figure 10.1, based on tape casting of the electrolyte to replace the pressed thimble. This was much more compact, used less material, incorporated heater elements and gave rapid response as the engine started up, plus longer lifetime rising to 100,000 miles of car operation. Failure of the sensors after such duration was usually due to catalyst degradation unless impurities such as lead or silicon existed in the fuel to degrade the platinum prematurely.

Useful reviews of zirconia sensors have been published [11–14]. Several manufacturers are active worldwide with NGK and Bosch among the leaders, making a total annual production around 100 million units. The zirconia is usually fused, then bead milled to a fine ink which can be tape cast in a suitable formulation, dried, pressed with other laminate components and fired in a furnace. The output from such a sensor in an automobile exhaust is shown in Figure 10.2, illustrating the lambda shape.

A more linear oxygen sensor was introduced by NTK in 1992 with two functions: a YSZ oxygen pump to push oxygen electrochemically into an inner chamber, which had a defined leakage rate through a small hole; plus a YSZ oxygen detector to keep the oxygen signal constant. The pump current was then a direct measure of oxygen concentration in the vehicle exhaust.

Further evolution of YSZ sensors has been taking place to allow measurement of other gases and to obtain controlled oxygen references, while improvement of speed, sensitivity, ruggedness and cost has been advancing steadily. A typical example is the

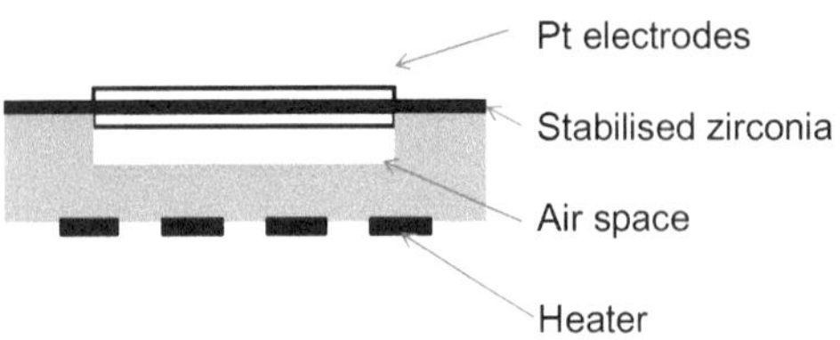

Figure 10.1 Cross-section through a planar YSZ sensor [11].

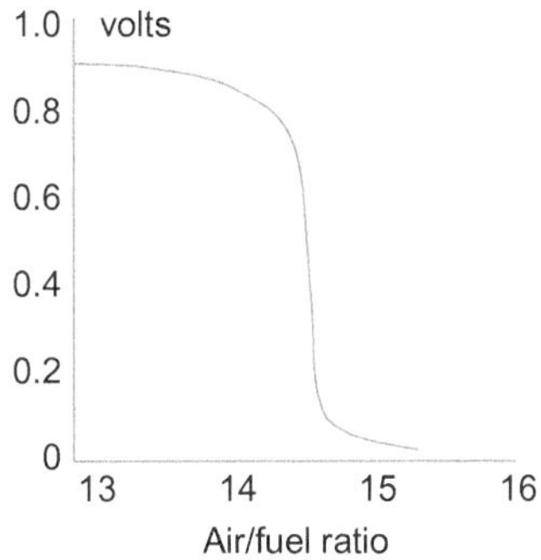

Figure 10.2 Voltage output from automotive exhaust sensor at 700 °C versus air/fuel ratio [11].

NO_x sensor based on YSZ for use above 500 °C [14]. A platinum-loaded zeolite catalyst is used to react the NO_x with oxygen in the gas stream to produce an equilibrium mixture of NO and NO_2, dependant only on temperature and background oxygen concentration. The equilibrated NO/NO_2 mixture is then measured using a YSZ sensor with metal oxide electrodes at a specific temperature. Total NOx is then determined.

Applications of YSZ oxygen sensors are widespread in many industries including aerospace, petrochemicals, oil & gas, steelmaking, glass manufacture, power, cement, food and paper & pulp [15]. Measuring the oxygen content on-line of molten metal is a typical challenging environment in which the thermal and chemical endurance of YSZ is essential. Typical YSZ sensors for dipping into molten metal (Figure 10.3) are currently quite crude and could benefit enormously from the materials and process improvements generated by SOFC developments. Other elements can then be measured by variants of the oxide ion conducting electrolyte. For example, Figure 10.3 shows a zirconia electrolyte thimble for measuring silicon content of molten stainless steel. Placing auxiliary electrodes on the outside of the YSZ sensor then allows detection of chromium, phosphorous and other contaminants.

The conclusion is that YSZ sensors have further potential to advance in a range of industries, especially since the devices are becoming smaller, more user friendly and more economic as material and processing for commercial applications are optimised.

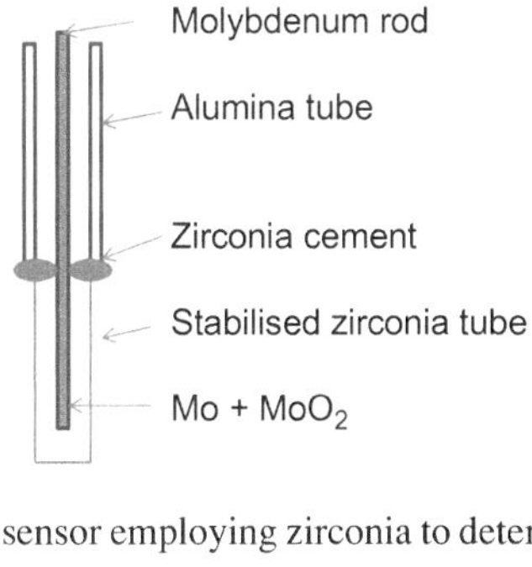

Figure 10.3 Electrochemical sensor employing zirconia to determine silicon content of molten stainless steel [15].

10.3 MEMS-based SOFCs

One of the most significant potential markets for portable fuel cells has been charging of smartphones and laptop computers where the batteries typically last for a few hours, yet the consumer needs days or weeks. Perhaps the most publicised invention in this market sector has been the methanol-powered polymer electrolyte membrane fuel cell (PEMFC) which was expected to plug into laptops around the year 2004 [16]. This type of product has been introduced in some hand-held portable devices in warehouses but has achieved its main success in soldier power [17]. Other possible fuels for portable PEMFCs have been studied including hydrogen stored in a metal hydride, hydrogen released from borohydrides and methanoic acid. All these fuels are unfamiliar and problematic. Hundred companies have been developing these devices over the past decades.

Yet the most familiar fuels available to consumers are the hydrocarbon liquids propane and butane which are carried everywhere by consumers for cigarette lighters and flame igniters. These fuels also give the highest energy density about 8 kWh/L, several times higher than the PEMFC fuels described above. Also, they are approved for transport such as aircraft in defined packages. Since these hydrocarbons need processing temperatures from 600 to 800 °C, it is logical to use SOFC technology, especially if the cells can be miniaturised to give light-weight combined with low cost. Figure 10.4 shows a butane powered phone charger from Ezelleron, based on microtubular SOFC. MEMS technology could provide the answer [18]. The key marketing

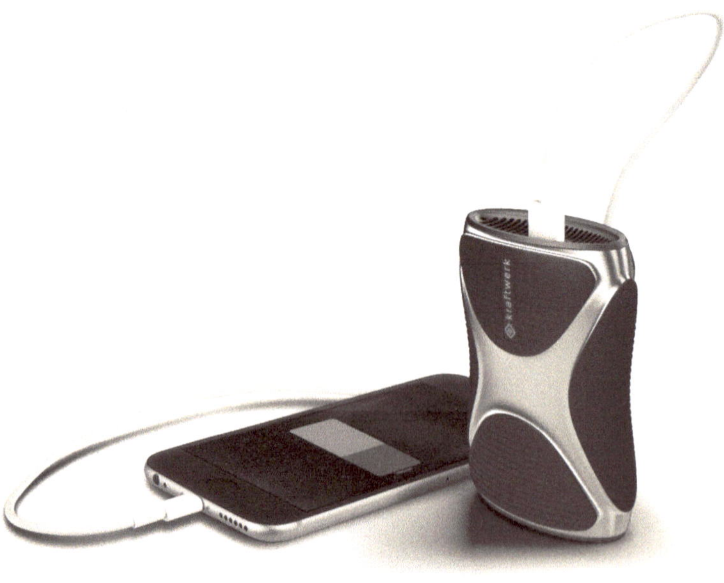

Figure 10.4 SOFC phone charger from Ezelleron.

aspects are power output, phone run-time and price. The power output criterion is significant because requirements for 4G wireless have risen significantly for smartphone users who are travelling away from grid connections and need constant internet access whereas battery capacity has only increased at 5% per year. Run-time requirement is 2 weeks and the price point is said to be near $200 for acceptance to consumers [19].

Development of MEMS processing of silicon to create portable SOFCs was carried out at Lawrence Livermore Laboratories [20] and adapted by Lilliputian, a spin-out company from MIT. The design of the etched silicon SOFC component, the so-called Generator Chip, consisted of a number of silicon slabs with the free-standing tubes joining the assembly together.

Several problems had to be overcome to achieve the objective of a hand-held device suitable for a mass-market phone charger. The main one was to limit heat loss such that the 700–800 °C operating temperature was fully contained. This required vacuum insulation together with blockage of the radiant heat pathways and preventing the escape of heat along the wires and the gas flow tubes, which were made of very thin silicon nitride to minimise heat flow. A new sealing approach was required to achieve this. Then the YSZ cells had to be integrated in the silicon/silicon nitride MEMS structure. Finally, the complex flow paths for cathode air and anode fuel/air mix had to be built in. Figure 10.5 shows a typical MEMS SOFC structure.

A key feature of the device is its integration with the battery to provide a hybrid electrical system. The battery was needed to start up the SOFC, but once the battery was fully charged, the hot system needed to shut down as indicated in the diagram of Figure 10.6 which shows the operational state loop.

The output provided by the MEMS system is shown in Figure 10.7.

Dr. Schaevitz and his colleagues from MIT went on to found Lilliputian Systems Inc. and he became CEO. The Company, headquartered in Wilmington, Massachusetts, secured exclusive licences from MIT and Lawrence Livermore National Laboratories for certain aspects of the technology and raised over $100 million in venture capital financing to commercialise its core technology, the Silicon Power Cell™. Two patents were filed in 2005–2006 [22,23].

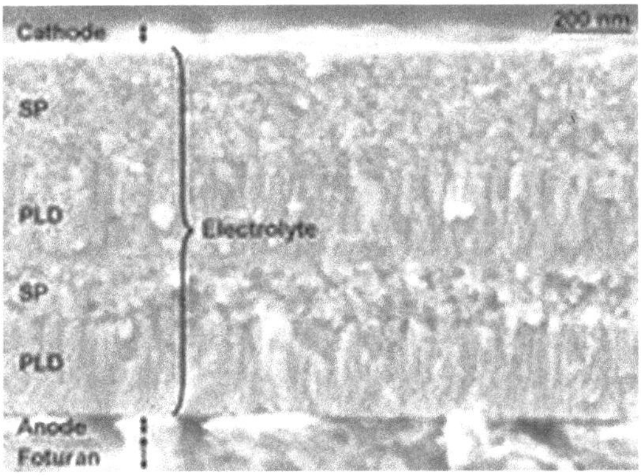

Figure 10.5 MEMS structure of SOFC [21] with permission.

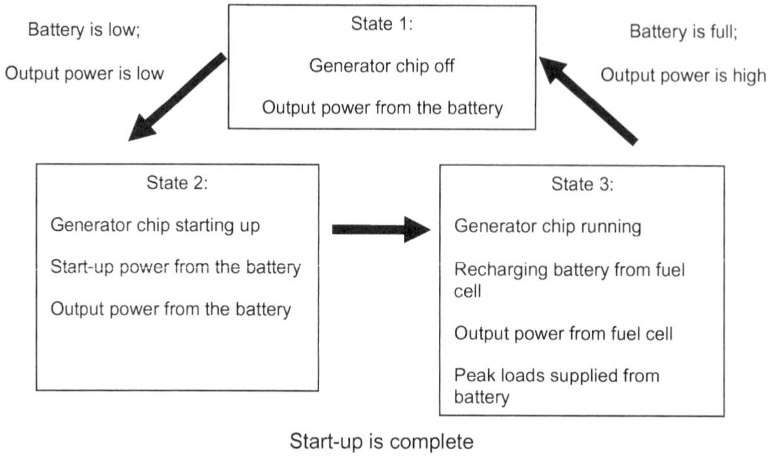

Figure 10.6 System operational state loop.

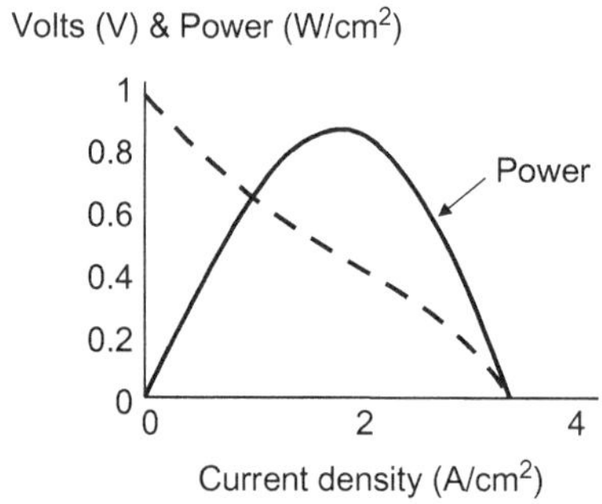

Figure 10.7 Voltage and power versus current density for an early silicon-based device from MIT.

Using this technology, Lilliputian developed the hand-held Nectar device, targeted at the $50-billion portable power market. Typically, one filling of the device with butane powered up a Smartphone 10 times and lasted for typically 2 weeks. In 2012, Lilliputian announced Brookstone as its first retail partner for the product. Brookstone was to market, distribute and sell the product through their various distribution channels such as catalogue, Brookstone.com and retail stores including airport and mall locations. Lilliputian in 2014 filed bankruptcy after 15 years of development.

A competitor in this field was Onebat from Switzerland based on patents from 2004 to 2007 [24–26]. Researchers from ETH in Zurich studied the deposition of YSZ on glass ceramic substrates such as Foturan to produce a MEMS device for portable power. Their scientific work was described in a review [21] which showed how a propane-based fuel processor could be integrated with an SOFC and battery in a portable system. Silicon was also trialled as a support material for MEMS processing.

An array of 1 μm thick cells of YSZ with porous platinum electrodes, about 100 μm in diameter, was deposited on the glass ceramic wafer which was etched with hydrofluoric acid to allow fuel and air to reach the anode and cathode, respectively. Despite the thermal expansion mismatch, the cells could be heated without cracking to around 600 °C which gave good power output.

A review of these issues has been published [27] to show how supported nano-membranes can be constructed to deliver portable SOFCs for various applications including sensors, phone chargers and larger portable power generators. Zirconia was considered for oxygen conduction in thin-film fuel cells or oxygen sensors [28], large area YSZ membranes were noted for transport studies [29], an integrated YSZ membrane for a TF-SOFC [30], mixed conducting CGO for thin-film SOFCs or chemical strain sensors [31] and proton-conducting BaY0.2Zr0.8O3 for application in fuel cells [32].

Typical processing routes are shown in Figure 10.8. The aim is to produce small regions of free-standing SOFC films by patterning a substrate, etching by various methods and then releasing the nano-membrane to provide the cell. Firstly, an easily etched substrate is selected and conventional semiconductor processing techniques are deployed. A barrier layer is needed on both sides of the substrate for pattern transfer and bulk micromachining, respectively. This backside is patterned using a lithographic technique and then selectively etched to reveal the desired feature on the front side of the substrate. After the oxide film is deposited, the final support is etched away, creating a self-supported clamped membrane. Using silicon substrates (Figure 10.8, left), some common wet etchants are potassium hydroxide (KOH), tetramethyl ammonium hydroxide (TMAH) or ethylene-diamine-pyrocatechol (EDP). These solutions etch in an anisotropic manner. In other words, there is a significantly higher etch rate along certain crystallographic directions. Much data exist on KOH and it is widely

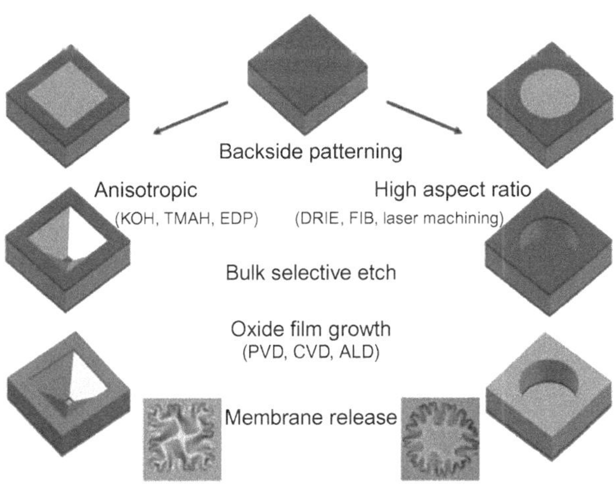

Figure 10.8 Two options for producing supported nano-membranes of YSZ [27].

used in MEMSs. Many self-supported YSZ membranes and thin-film SOFCs fabricated to date have used this approach [33–36].

10.4 Micro-tubular SOFCs

At a somewhat higher power level, greater than around 10 W, mSOFCs have been successful. Tube diameters less than 10 mm and even less than 1 mm are considered such that start-up in a few seconds is realistic. These have been made possible by improved ceramic processing methods which can increase the strength and thermal shock resistance by a significant factor, allowing portable applications to warm-up rapidly. Originally, the yttria-stabilised zirconia (YSZ) was used as the support material, but since 2000, the anode support has become dominant, with the electrolyte only 5–10 μm thick. Lower-temperature electrolytes have been tested, for example, ceria/gadolinia, but the need to do this has fallen away as the improved cathode materials, for example, LSCF operating at 700 °C have become available, replacing LSM previously used which operated above 800 °C.

Stacking the micro-tubes has been achieved in various ways, but the simplest method is to wind wires around the cathode and anode materials to provide interconnection between cells. These stacks can then be manifolded to deliver the fuel inside the tubes while providing the cathode air on the outside. Reforming of methane, propane and butane is considered, as well as the anode off-gas combustion and recuperation. Thermal insulation may be a significant issue. Small stacks can provide a few Watts for charging cell-phones, whilst larger stacks up to 1000 tubes can give several hundred Watts for auxiliary power while also utilising the heat. Typical system designs are described, and several applications analysed as in reviews which have appeared since 2003 [37–40]. Starting with the need for mSOFCs, which was seen about 1990, early studies during the 1990s are reviewed, which indicated that lower tube diameter was most significant in giving rapid start-up and higher power density compared to Westinghouse tubes which were 22 mm in diameter [41]. Various layouts of portable mSOFC systems are described. Applications for UAVs, leisure vehicles and trucks are then reviewed.

10.4.1 Need for mSOFCs

Looking back to 1990, the most successful SOFC at that time was developed by Westinghouse Electric Corporation as shown in Figure 10.9, and this was due to be commercialised during the 1990s.

Each tube was made in a multi-step process that had around 20 stages and took several days because the 50-mm YSZ electrolyte layer had to be laid down by chemical vapour deposition. Consequently, it was comparatively costly. Secondly, the diameter was about 22 mm, the length 180 cm and the wall thickness 3 mm because of the thick cathode support and anode layers, with the result that the tubes had low resistance to thermal shock and took many hours to warm-up to the operating

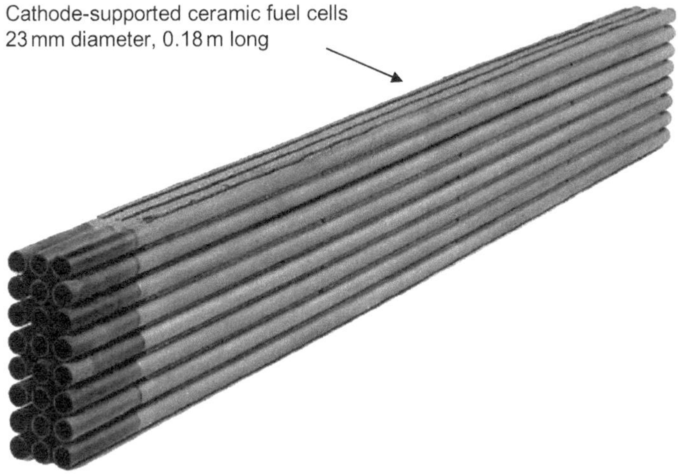

Figure 10.9 Bundle of 24 Westinghouse tubes developed by 1990 [41].

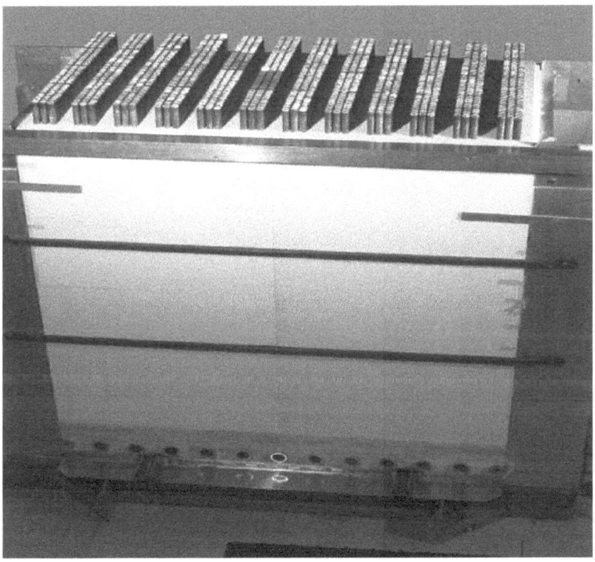

Figure 10.10 A 48 bundle stack of Westinghouse tubes giving 100 kWe.

temperature between 900 and 1000 °C. These large dimensions also meant that the volume and weight of the stack were large, giving poor system power density about 10 W/kg, much too low to power an aircraft, for example.

The 100-kW stack shown in Figure 10.10 was successfully demonstrated in several locations. But the high cost, long start time, plus poor power density proved to be insurmountable, and the production of these tubes ceased in 2005.

After Siemens took over the operation in 1998, the tubes were flattened and began to look more like the Rolls-Royce flat tube design [42] which increased power density but did not much improve the start-up time because the large flat areas still tended to crack if heated in less than 10 h [43]. These flat tubular stacks were shown to achieve cost and lifetime targets at 10 kW level for the SECA programme on coal utilisation [44] but could not warm-up quickly and were designed for 1 MW pressurised operation at 3 bar, not portable power.

It is evident from this brief history (see Chapter 8 for more details on stack design) that radical change was needed if small-scale SOFC was to flourish. The author asked Prof. Brian Steele in 1991 if such portable SOFC power was possible and he replied that below 1 kW was difficult because of heat losses [45]. This prediction proved false as shown in the next section.

10.4.2 Invention of mSOFC

The earliest reports of work on mSOFCs were in the early 1990s when the possibility of extruding thin-walled YSZ electrolyte tubes, 1–5 mm in diameter and between 100 and 200 μm wall thickness (Figure 10.11), was demonstrated while the author worked at ICI in Runcorn [46], and the ionic conductivity, leak tightness and thermal shock resistance of such electrolyte tubes were found to be good [47–49]. Micro-tubes were

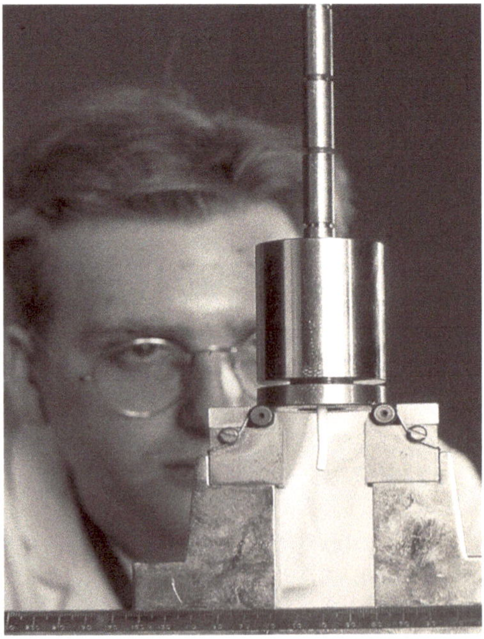

Figure 10.11 ICI lab extrusion of 8YSZ electrolyte tube using a ram extruder with a tube die on an Instron compression machine.

sent to Westinghouse in Pittsburgh who were leading SOFC development at that time, but the response was negative.

In 1994, it became clear that there are three major benefits of mSOFCs which will be discussed in the next section. The first is the improved ceramic properties which result from disagglomeration of the powder by high shear plastic mixing with a polymer matrix. The second is the increase in volumetric power density when compared with the large-diameter tubular designs discussed in Chapter 8. Power density scales with the reciprocal of tube diameter. Therefore, a 2-mm diameter mSOFC could provide 10 times more power per stack volume than a 20-mm diameter tubular cell similar to the Westinghouse design. The third and perhaps most important benefit for small-scale portable applications was the rapid warm-up time, which could be less than 10 s.

The first fuel cell testing was done by Michaela Kendall and reported in her graduate thesis in 1993, showing that with platinum electrodes, good power output could be obtained [47]. It then became clear that a simple structure of mSOFC could be constructed by dipping the electrolyte tube in a nickel/YSZ ink to form an inner anode, painting lanthanum strontium manganite (LSM) on the outside to form a cathode layer and then firing the cell and wrapping silver wires on the outside and using a nickel or stainless steel brush inside the tube as interconnectors (Figure 10.12).

About 10,000 of these cells were made from 1993 to 2003 when the switch was made to anode supported cells, and many PhD students obtained useful results on interesting problems such as fuel reactions at the anode, catalyst formulations and demonstrator system construction [50–53].

Rapid progress was made in 1994 when an EU project was started at Keele University on mSOFCs with the objective of building a 1000-tube operating system. A stack system design was produced (Figure 10.13) a patent applied for and a paper

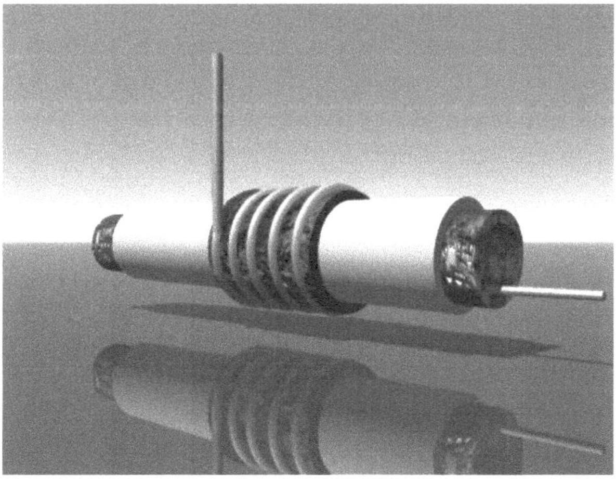

Figure 10.12 Early cell with YSZ electrolyte support, inner nickel/8YSZ anode with nickel wire collector, plus LSM cathode and silver wire. Fuel flows inside the tube and air is on the outside.

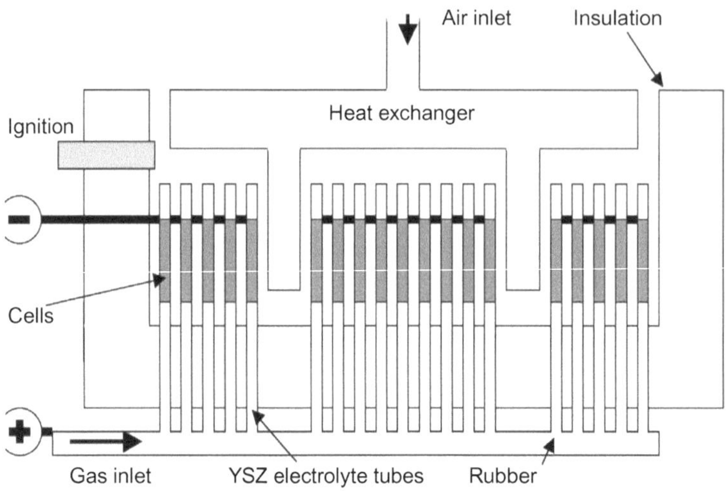

Figure 10.13 Stack design from 1994 showing the rubber sealing of the ceramic tubes, the zirconia tubes passing through the insulation, the off-gas combustion zone and recuperator to heat the incoming air.

given at the First European SOFC Forum in Luzern [54,55]. Anode/electrolyte co-extrusion was also demonstrated in that paper.

The main point of the 1994 patent was that the tubes were sufficiently shock resistant to be cold-sealed using rubber to the metal inlet manifold and then passed through the thermal insulation into the hot zone where partial oxidation (POX) of the fuel gas occurred. The anode off-gas was then combusted with cathode depleted air at the open exit ends to heat the recuperator which heat exchanged with the cold inlet air heading for the cathodes. Typically, the 2-mm diameter zirconia tube was 200 mm long, whereas the cell region only occupied a length of about 30 mm towards the outlet end of the tube. The nickel/YSZ anode, 30 mm long, was coated on the inner wall of the tube to a thickness of around 50 μm, and a nickel current collector wire was inserted and brought out from the fuel entry side of the tube. On the outside of the electrolyte tube, the LSM cathode layer, 100 μm thick, was deposited, fired, and a silver wire wound around it to obtain the cathode current.

The 40 racks of 25 tubes each are shown in Figure 10.14. The Saffil paper cushioning between each rack prevented electrical short-circuits and prevented impact damage. Alumina board insulation 50 mm thick surrounded this stack and a stainless steel recuperator was inserted on the top to inject and preheat the incoming cathode air.

Reports on testing a number of these stacks showed that the warm-up could be achieved in 20 min using the normal natural gas supply, while cycling was possible in 2 min down to 300 °C [56–58]. This device was designed to use 20 kW of natural gas for heating, while providing base-load power of around 500 W for a household. Drawbacks were the high in-plane resistance of the cells, the long current leads and the difficulty of connecting and stacking many small cells together. Only 200 W was

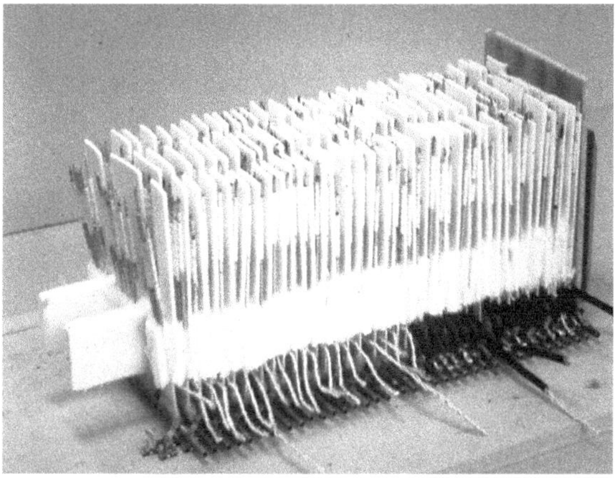

Figure 10.14 Photo of the 1000 tube stack built in 1996 showing 40 racks of 25 tubes, with fuel/air mix entering at the bottom, passing through the alumina fibre insulation, where temperature of the fuel air mix rose and POX occurred, going through the cells and combusting at the open tube ends on top.

achieved from this stack because of high connection resistances, but 500 W was clearly possible because each of the 1000 cells produced almost 1 W.

The system of Figure 10.13 was the first fuel cell device designed for a CHP system for British houses, but the market for this was not significant at that time because large barriers to grid connection were still in place, system costs were too high and lifetime was unsatisfactory. Now in 2015, there is still no significant market for this technology in the UK, and it is expected to be 2025 before costs look reasonable [59], although a number of companies such as Hexis, Ceres Power, Vaillant, Viessman, Bosch and SOFC Power are attempting to kick-start this market.

10.5 Benefit of improved ceramic processing for quality ceramics

The key problem of processing ceramic powders during the past century was that the fired products never achieved the high strengths predicted by theory [60,61]. Developments in the electronic ceramic industry in sol–gel processing and in polymer extrusion changed this by showing that high-quality ceramics in thin layers could be produced by novel means to make the multilayer structures required for the mSOFC component shown in Figure 10.15.

The realisation that ceramic nanoparticles could aggregate due to their interparticle attractions to form large clusters which acted as defects in the final product came in the late 1980s as it became clear to Lange and others that making the particles finer sometimes led to weaker ceramics [62]. A typical result is shown for titania nanoparticles in Figure 10.16 which relates to a very well dispersed nanoparticle suspension which was tested for particle size distribution immediately after bead milling, then later after drying when aggregates were seen to grow around 70 μm diameter, more than 100 times larger than the primary particles. These clusters could be filtered out and viewed in the SEM to show the clustering effect [63]. It seemed that the defects were originating from the dispersion and drying processes.

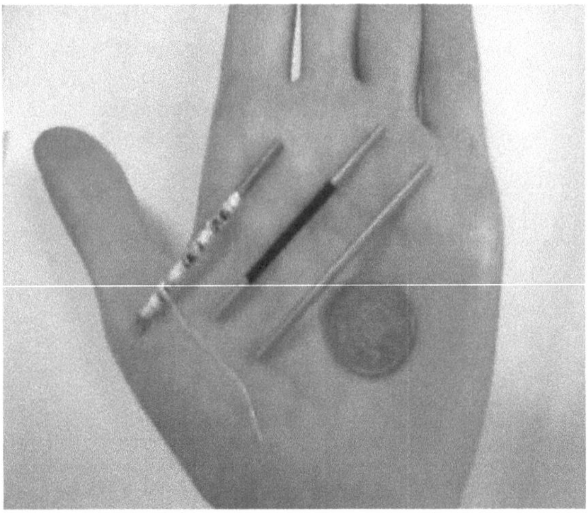

Figure 10.15 Coextruded and fired 2 mm diameter micro-tube comprising anode support and 10 μm YSZ electrolyte layer (bottom); the tube coated with cathode, for example, LSCF (middle) and the final cell with silver cathode interconnect and nickel anode wire (top).

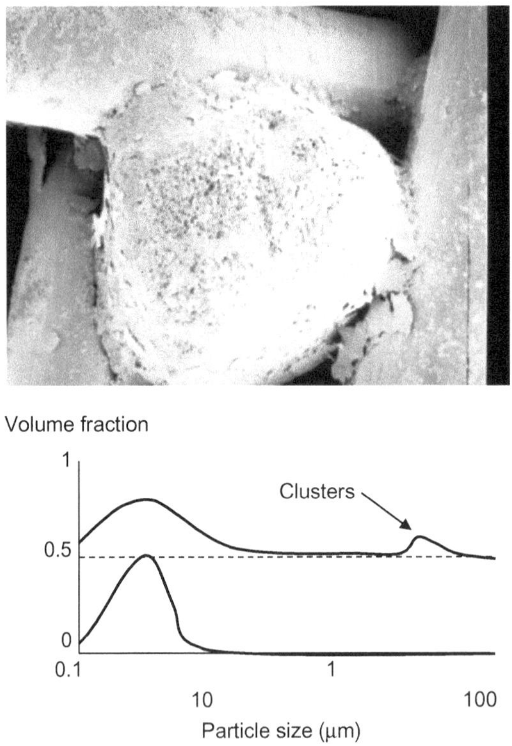

Figure 10.16 (Top) a cluster of nanoparticles filtered from a well-dispersed suspension; (bottom) Malvern Microsizer results showing original size distribution (bottom curve) and aggregation occurring after drying process (top curve).

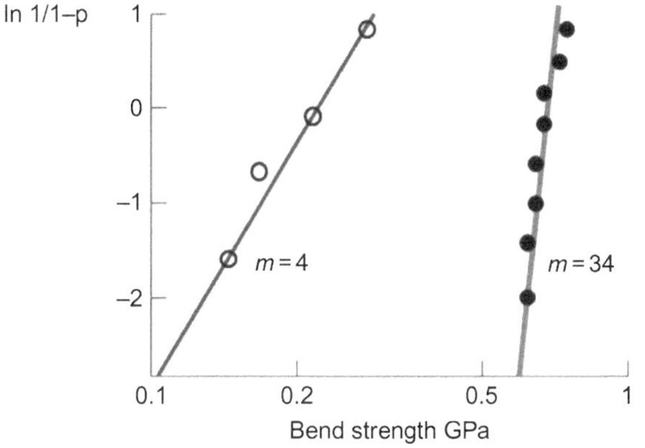

Figure 10.17 Weibull plot for bend strength of 8YSZ powder by two different processes: (left) the powder was pressed to form a test bar; (right) the powder was plastic mixed with polymer to break down the agglomerates, then extruded into test bar form.

In order to break down these clusters and prevent them recurring, polymer plastic processing was developed [64]. Powders were first selected for their weak aggregate structures, mixed with polymer solution and then sheared at high stress to destroy the weak agglomerate structure. Then, the plastic mix was extruded through a tube die to produce high-quality, thin-wall tubes. Initially, electrolyte support was used, but eventually anode support with a thin layer of YSZ electrolyte was coextruded to give wall thickness down to 100 μm. The tubes could then be fired to sinter the materials in the furnace before bend testing to measure strength and its statistical variations.

Figure 10.17 shows a Weibull plot of bend test results for two 8YSZ test batches made from identical powders. The left-hand curve shows 4 bend test results for dry YSZ powder-pressed bars with a mean strength of 230 MPa and Weibull modulus of 4. By contrast, the same powder when plastically mixed and extruded gave a bend strength of 730 MPa with a Weibull modulus of 34. This result indicates that the powder agglomerate defects have been broken down to give a much more uniform microstructure in the fired product. This was verified by polishing the ceramic surfaces and proving that the defects were now 10 times smaller.

10.6 Benefits of improved power density

One of the most significant problems of early SOFC designs was the low power density. In other words, the system, and particularly the stack, was far too heavy and voluminous for desired applications. Power density can be expressed in terms of mass as kW/kg or in terms of volume as kW/L. Some applications focus on mass. For example, an aircraft requires low weight if it is to fly successfully. The Westinghouse design of Figure 10.10 displayed a power density of about 10 W/kg which could not fly. Combustion engines are two orders of magnitude better, typically delivering 1000 W/kg. Such engines, though inefficient in thermodynamic terms, have been very successful in aircraft. PEMFCs have also been demonstrated at 1000 W/kg and have flown successfully in UAVs. The challenge has been to increase power density of SOFCs to compete with these existing technologies.

Other applications require low volume. A good example is the recreational vehicle (RV) which has to package all the auxiliary equipment like showers, cookers and power supplies into a restricted volume. Again the Westinghouse design with 22 mm diameter tubes could not attempt this application because it gave around 10 W/L, whereas PEMFC devices and combustion power units have succeeded by attaining 1000 W/L, 100 times better.

In order to understand the significance of power density in application development, it is worth considering the historic example of Stephenson's Rocket steam engine. The original heat exchanger tube design of the Rocket used a single tube 12 in. diameter exchanging heat with the boiling water. Only after George and his son Robert Stevenson, together with their colleague Henry Booth, replaced that tube with 25 2-in. diameter tubes, thus increasing the surface area by a factor 6, did the Rocket steam engine prove a success. The essential message is that SOFC tubes must be made in small diameters if they are to attain the power density to compete with combustion engines and PEMFCs.

This fundamental engineering principle applies equally to the Rocket steam engine and to the tubular fuel cell stack, as can be seen by considering a square stacking of tubes diameter D shown in Figure 10.18 with a gap D between tubes.

The stack power density is dictated by the areal power density p of the mSOFC, for example, 0.2 W/cm^2 and by the diameter of the tubes D, showing that as the diameter falls, the volumetric power density rises as

$$\text{Power density} = \frac{\pi p}{4D} \tag{10.1}$$

shown in Figure 10.19. Thus for the modest performing cell at 0.2 W/cm^2, a 0.2 cm tube diameter is required to obtain a stack power density around 1 kW/L which allows it to compete with PEMFCs and combustion engines.

The three key problems of mSOFCs are then seen by considering the reciprocal of power density (L/kW) in the diagram of Figure 10.20 which shows how power density, start-up time and number of cell connections in the stack need to be quantified. For a readily obtainable areal cell power density of 0.3 W/cm^2, the three factors converge on a 3-mm diameter cell which gives 1 kW/L, 1 min start-up and one connection

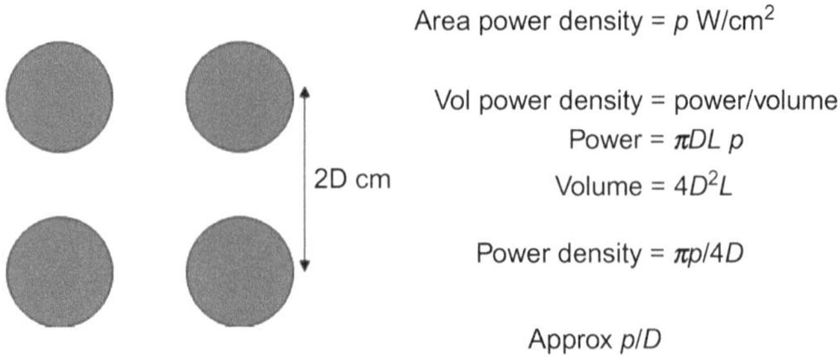

Figure 10.18 Square stacking of tubular cells with calculation of stack volumetric power density.

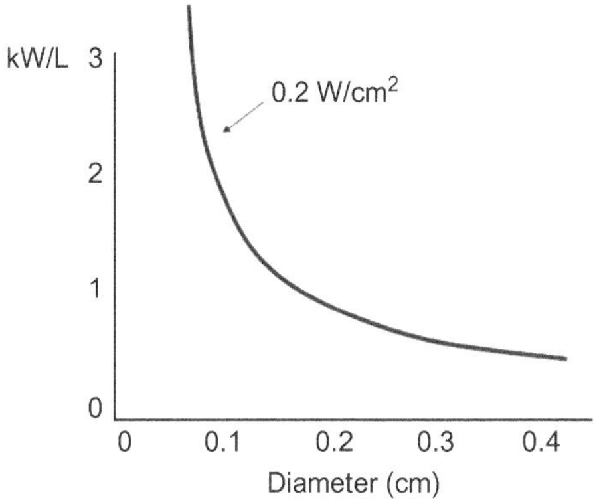

Figure 10.19 Plot of volumetric stack power density in kW/L against mSOFC diameter.

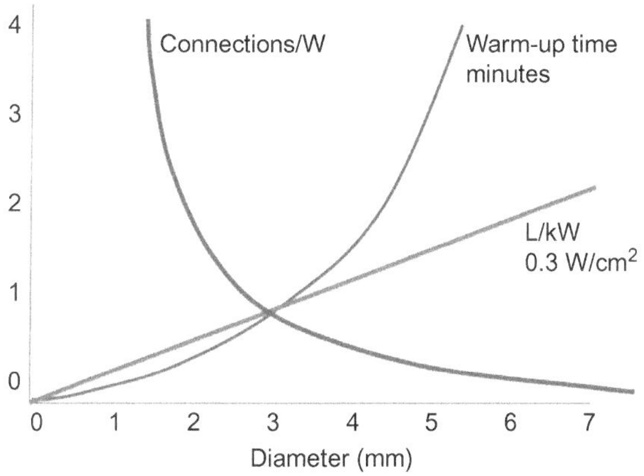

Figure 10.20 Diagram illustrating the three issues: number of connections; warm-up time and reciprocal of power density.

per Watt. It is clear from this picture that the key problem when making finer tubes is making many more electrical connections between tubes. The optimum method for connecting mSOFCs during the stack manufacturing process has not yet been invented, and this will surely play a major role in future product costs.

10.7 Rapid warm-up

It was apparent by 1994 that mSOFCs of diameter less than 2 mm could not be cracked by rapid heating/cooling, whereas planar cells made of the same materials suffered greatly from thermal shock fracture [1,2]. This was readily demonstrated by inserting planar cells in an electric toaster which heated the plates reasonably uniformly by radiant energy in a few minutes to 700 °C. YSZ plates 40 × 40 mm and 100 × 100 mm always shattered when this test was applied.

Recently, improved planar SOFC designs have produced more rapid heating performance by individually heating each of the planar cells in a stack to avoid damaging temperature gradients [65,66], but mSOFCs are still much quicker to warm up, as demonstrated in the system shown in Figure 10.21. In this hand-held, single-tube test unit, the mSOFC was a 30-mm long tube 2 mm diameter with silver anode and cathode wires attached. This was placed in the hot zone of the unit, inside the protective metal mesh. The two wires were connected to the electric motor driving the fan which started to turn when sufficient power was generated by the mSOFC. To start the device, the on switch was pushed, opening the butane valve and also activating the catalyst with a piezoelectric spark. The butane entrained some air into the catalyst mesh which began to glow red hot in 1 s as POX reaction occurred. Reformed butane then passed into the mSOFC tube, while hot air passed on the outside. Within 10 s, the mSOFC produced power causing the fan to rotate. This concept, that a ceramic tube could behave almost like an improved Bunsen burner, giving both power and heat, was one which was initiated about 2000 and continued to 2005 when the demonstrator illustrated in Figure 10.21 was shown in Luzern [67–69].

On releasing the on switch, the gas flow stopped, the catalyst cooled down in 5 s and the fan stopped. The cool-down of the mSOFC was so fast that little nickel oxidation occurred even though air was present. Thus the redox damage reported in many papers on planar SOFC did not occur [70,71]. More than 1000 cycles were achieved with this demonstrator, showing that rapid cycles were not disastrous to the nickel cermet anodes. By contrast, it is known that slow redox cycling of Hexis- or Julich cermet-based cells caused failure in one cycle. There is no doubt that a cermet anode increases in volume during its oxidation from nickel metal to nickel oxide, causing cracking as described by the fracture mechanics argument of Atkinson and his colleagues [71]. However, the precise mechanism of nickel reactions with oxygen in the presence of water vapour is complex as shown by a study of 99.995% pure nickel films evaporated onto single crystal (100) YSZ surfaces [70]. Using XPS (X-ray photoelectron spectroscopy) to analyse the nickel atoms in close contact with the YSZ electrolyte as oxidation and hydrolysis occurred, it was possible to detect NiOOH under defined electrochemical conditions. This hydroxide material could be reversibly reduced back to nickel metal under increased hydrogen concentration conditions.

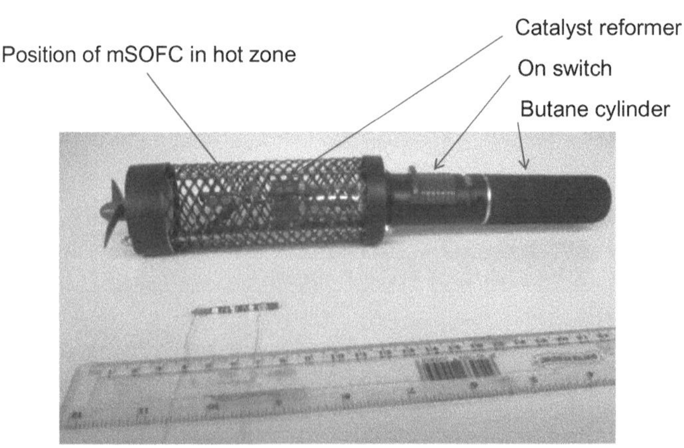

Figure 10.21 Photo of demonstrator to show rapid warm-up of the 2 mm diameter mSOFC laid next to the 30-cm ruler.

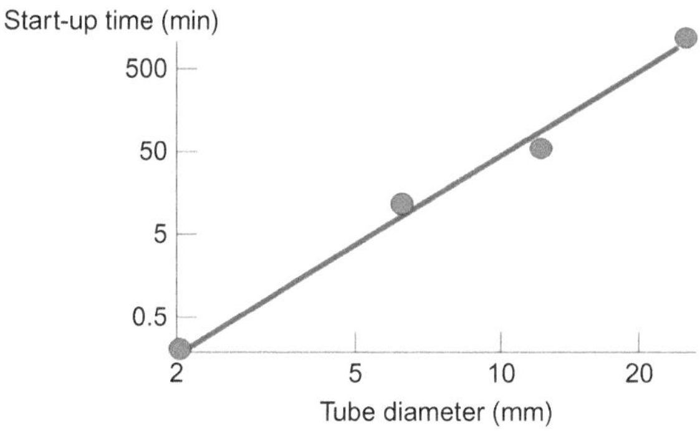

Figure 10.22 Plot of results for mSOFC start-up time versus tube diameter.

The very strong dependence of start-up time on tube diameter, shown in Figure 10.22, was first noted in 1993 [72], but the precise reasons for this phenomenon have not yet been discovered. Fracture mechanics does predict a size effect, such that smaller samples become more difficult to crack because of the lower energy contained in the stressed sample, but this argument only predicts a $D^{1.5}$ relation. It can be seen from the experimental results with a wide range of tube diameters that the diameter is much more significant than that as shown in Figure 10.22.

The result on the 2-mm diameter mSOFC demonstrator is shown as the left-hand point on the plot of Figure 10.22 which defines the very strong relationship between the tube diameter and the start-up time. Increasing the diameter from 2 to 6 mm raises the warm-up time from 10 s to 10 min, while further diameter increases to 23 mm takes the start-up to 10 h. The line drawn on the log–log plot of start-up time t versus tube diameter D is a cubic power law.

$$t = KD^3$$

Of course, there may be a large difference between the start-up time tolerated by the tubes, and that available from the system. For example, the tubes used for the 1000 cell stack of Figure 10.14 were capable of 10 s warm-up without cracking. But the system with its heavy recuperator, thick insulation and long YSZ support tubes required 10 min to rise to 800 °C with 1 kW of methane flow. Therefore, the tube diameter could have been relaxed to 6 mm for this application.

10.8 International efforts on micro SOFCs

The first international project on mSOFCs started with a connection to Prof. Wolfgang Winkler in Hamburg to develop improved micro-tubular geometries [73]. Then, a New Zealand collaboration of student exchange between Keele University and Hamilton University started where Dr. Nigel Sammes was Professor of Materials. Charles Hatchwell was the PhD candidate who arrived at Keele in 1995 to work on cathode current collection and Olivier Bellon in exchange travelled to New Zealand to carry out research on ceria-gadolinia micro-tubes [74]. This resulted in a collaborative project with Japanese, Swiss and New Zealand collaborators, including another New Zealand student Geoff Thompsett, which allowed significant work to be carried out on the micro-tubular SOFC concept as an auxiliary power generator in vehicles [75]. But the most significant outcome of this collaboration was the surge in Japanese development of 0.7 kW domestic micro-CHP devices operating on natural gas and kerosene [76], with Korea and EU following in production numbers, as shown in Figure 10.23.

This diagram illustrates the exponential rise in fuel cell installations with time since the early part of the millennium. Already more than 138,000 units are operating in Japan in 2015, almost 10% being SOFC with a prospect of a million in 2020 and 5 million in 2030. Korea is behind by an order of magnitude and the EU trails still further.

Also, the price reduction with the number of installations is shown in the learning curve on the right. The price in 2014 is still not economic but has been supported by a \$4-billion government subsidy. The problem that profit will not be made until about 2025, according to the learning curve illustrated, is a significant drawback of these present developments [59].

A collaboration between Adelan and the Austrian company ALPPS in 2002 led to a mSOFC design for a truck-cab power supply and a new company ALPPS fuel cell systems. The product was a 5-kW power supply running on diesel. Although ALPPS went into administration in 2010, one of its employees started Ezelleron in Germany in 2008 aiming at the phone charger device sector. A similar spin-off occurred when the Buffalo-based company Nanodynamics became insolvent in 2012. One of the employees, originally graduating with his PhD from Keele University where the mSOFC was invented, started up Watt Fuel Cell Corp in New York State to produce products based on mSOFCs. In 2013, they announced the launch of a propane-fuelled mSOFC power and heat supply installed in a 36-foot sailing boat. The benefits of low noise, vibration and available fuel were substantial.

Further international collaboration was developed with the 26 partners in the REAL-SOFC project led by Julich in Germany and also with the 36 partners in SOFC-NET run by ECN in the Netherlands [77]. Although these projects were mainly about the long-term life of SOFC materials, it was clear that mSOFCs had greatly

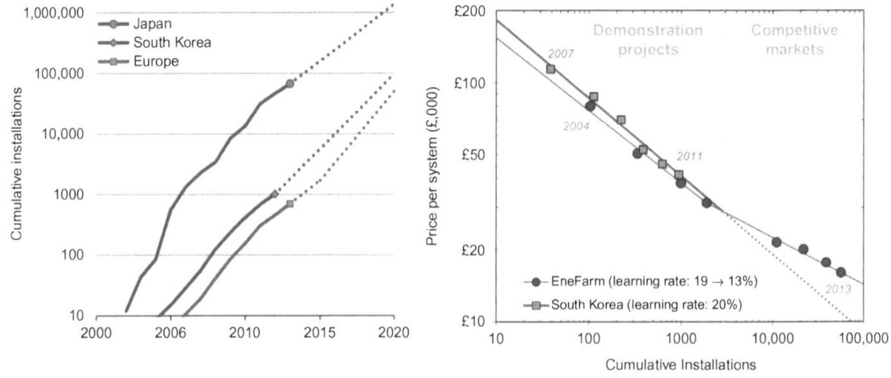

Figure 10.23 (Left) Cumulative installations of domestic fuel cell micro-CHP devices in Japan, Korea and EU; (Right) learning curve showing the price reduction with cumulative experience. Ref. [76] with permission.

superior cycling lifetime, both thermal and redox, than planar stacks. It was evident that the major issues with the Sulzer and Julich designs of planar SOFC were both the inability to change temperature rapidly and the problem of tolerating any air at the anode surfaces. Professor Sammes joined Acumentrics in Boston in 2000 to develop mSOFCs for military application. This US company supplied rugged power supplies for battlefield use where novel electrical generators were believed to be necessary. The first devices built were based on 5 mm diameter electrolyte supported SOFCs designed to provide 1 kW. In a 2-year period, this project mutated into a 5-kW unit for the SECA programme. Also Dr. Bessette took over increasing diameter further to about 15 mm. This was going against the micro-tube concept because the start-up time increased to about 1 h and the power density fell, but the project hit its cost and performance targets up to 2012 [78,79].

During the same time-frame, Adaptive Materials Inc. (AMI) started in Ann Arbor Michigan and began to extrude high-quality micro-tubular SOFCs using a polymer co-extrusion process developed by Dr. Aaron Crumm and his PhD supervisor Prof. John Halloran [80,81]. The tubes were anode supported with a 10-μm outer layer of YSZ electrolyte. Although these were batch processed and thus costly, they were of high quality and provided many research results from a collaboration with University of Birmingham since 2004 [82–87]. Tens of thousands of these 2-mm diameter tubes were produced by AMI until 2010 when they switched to 6 mm diameter because of the interconnection problems.

In 2013, AMI produced such tubes commercially at a rate of 10,000/a, having defined methods for inserting reliable metal brush contacts inside the tubes and porous silver inks plus silver wires on the outside. Simultaneously, they developed a design for processing fuel inside each tube, rather like that demonstrated in 1994 [55,88,89]. The subsequent breakthrough achieved by AMI was to demonstrate that the mSOFC stack and system so developed could drive an UAV which first flew in 2006 and has since been used by Lockheed Martin in defence applications. This application was the desirable niche portable product which excited Ultra Electronics to acquire AMI, becoming Ultra-AMI in 2011.The UAV (Figure 10.24) was the Stalker which was based on a battery electric drive converted to a hybrid SOFC battery plane. It was a 4 m wingspan vehicle weighing 10 kg at take-off carrying 1 kg of propane to fly for more than 8 h taking HD pictures. Silent above 100 m, the plane could compete with a combustion-driven craft because of its low signature and its higher reliability [90].

10.9 Demonstration projects

A UAV project based in the EU was started in 2011 [91] based on mSOFCs powered by propane. Led by HyGear in the Netherlands, the project team comprised EADS, CNR, ZUT, University of Birmingham, Catator and Adelan, a UK SME based on mSOFCs. Its objective was to build a 250-W SOFC system and to fly it in a small fixed wing aircraft weighing about 9 kg, demonstrating the extended range of an SOFC/LiPo battery hybrid compared with a pure battery craft.

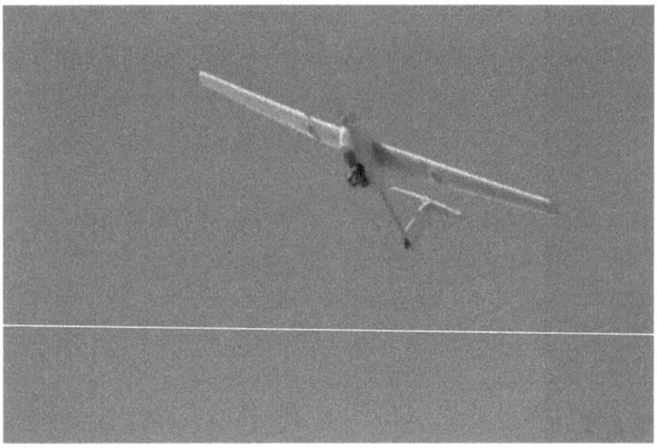

Figure 10.24 Stalker unmanned air vehicle (UAV) powered by mSOFC [90].

To test the concept of mSOFC auxiliary power in RVs, an EU project SAPIENS kicked off in November 2012 [92]. This was led by Adelan Ltd. partnered by Auto-sleepers, a UK camping van company, CUTEC in Germany, IREC in Spain, ZUT in Poland and JRC in the Netherlands. These project objectives were to design, optimise and build several mSOFC stacks and then integrate them into hybrid power systems comprising a fuel cell, battery and appliances found in RV. Propane was chosen as the fuel because of its high energy density and also because it is the preferred fuel for auxiliary systems on RVs, such as cookers, fridges and water heaters. Autogas propane is widely available at fuelling stations throughout Europe. The mSOFC was chosen because it can convert propane while also providing low noise, low emissions and heat for hot water supply. The project was primarily aimed at RVs, but other applications such as boats, trucks, ambulances and environmental testing vehicles were also considered.

In 2013, a mSOFC design for auxiliary power on trucks was defined in an EU project SAFARI [93]. The novel concept behind this project was the growth of liquid natural gas (LNG)-powered trucks in the EU. Whereas diesel fuel had been the preferred fuel, with more than 100 molecules in its formulation, many of which were poisonous to the SOFC catalysts, LNG offered a much cleaner fuel which could be readily processed by CPOX within the SOFC micro-tubes. On-board processing of diesel fuel had been a major cause of problems in the $1-billion SECA project which started in 2000 in the USA to provide APUs on trucks. Delphi eventually produced prototypes, but it was clear that the processing of the diesel fuel, which varied enormously across the world, was more difficult than the SOFC operation itself.

By contrast, LNG was mainly methane, with smaller amounts of ethane and propane, and was therefore much more manageable in a truck APU. A further reason for following this approach was the niche application: LNG trucks represent less than 0.01% of the trucks in the EU, making this a special application appropriate to a novel SOFC technology. But the prospect for increases in LNG truck manufacture is significant, with suggestions that 30% of new trucks may be LNG powered by 2050. Moreover, the EU has encouraged LNG for vehicles because it is much cleaner than diesel in its emissions, especially with sulphur levels geared to 10 ppm in 2016, plus low

Figure 10.25 Portable rapid heating planar SOFC from Almus AG [65].

particulates, less costly than other fuels, while appearing to be a secure fuel during the next century when diesel becomes less available [94].

A significant development in this truck project was the possibility of using rapid heating planar cells in a 100-W stack developed by Ulf Bossell [65] at Almus AG as shown in Figure 10.25. Each stainless steel interconnector plate in the 50 × 50 mm planar stack was electrically heated to prevent large temperature gradients causing thermal shock. The start-up time was 20 min for this design, comparable with the micro-tubular stack.

10.10 Summary

This chapter has described the development of small-scale SOFCs for several applications. Sensors represent the lowest power application, but phone chargers are now seen as a serious market near 10 W. In larger power sectors, such as CHP and APUs which require above 100 W, the need for more rapid start-up, improved ceramic processing and increased power density was evident from the Westinghouse results during the 1980s. Invention was most rapid in 1994 when the first mSOFC patent was filed and the system design formulated, followed by building a 1000 micro-tube stack in 1996. Three major benefits were observed including: improved ceramic processing methods were found to produce higher quality electrolyte and electrode materials once the problems of nanoparticle aggregation were addressed; increased power density was found to be 10 times the Westinghouse figure. The rapid start-up under 10 s for 2 mm diameter tubes was impressive. New changes leading to 2013 technology included the move to anode support coextruded with 10 μm of YSZ electrolyte, and the use of LSCF as the cathode which could operate at 700 °C. The resulting micro-tubes were assembled into stacks using wires for interconnection. System issues such as manifolding, insulation, reforming, anode off-gas combustion, recuperation and power conditioning were advanced such that portable applications became a practical proposition. In 1990, it was thought impossible that SOFCs could power phone chargers or fly in aircraft because their weight and volume were too great. Now, mSOFCs are competing with PEMFCs powered by methanol because propane and butane are accepted, cheap and widely available fuels. mSOFCs can beat PEMFCs in extended flight UAVs despite their denser ceramic components and thick thermal insulation because propane is a fuel with higher energy density than pressurised hydrogen gas stored in a heavy cylinder. Other applications in emergency power, APUs and CHP have been investigated. The future prospect for portable SOFC products looks promising in a range of product areas.

References

[1] K. Kendall, N.Q. Minh, S.C. Singhal, in: S.C. Singhal, K. Kendall (Eds.), High Temperature Solid Oxide Fuel Cells, Elsevier Science, Oxford, 2003 (Chapter 8).
[2] K. Kendall, Progress in solid oxide fuel cell materials, Int. Mater. Rev. 50 (2005) 257–264.
[3] E. Baur, R. Brunner, Uber die Eisenoxyd-Kathode in der Kohle-Luft-Kette, Z. Elektrochem. 43 (1937) 725–727.
[4] E. Baur, H. Preis, Uber Brennstoff-Ketten mit Festleitern, Z. Elektrochem. 43 (1937) 727–732.
[5] E. Baur, H. Preis, Brennstoff-Ketten mit Festleitern, Z. Elektrochem. 44 (1938) 695–698.
[6] H. Peters, H.-H. Mobius, Elektrochemische Untersuchung der Gleichgewichte CO + 0,5 O2 = CO2 und C+CO2 = 2 CO, Z. Physik. Chem. (Leipzig) 209 (1958) 298–309.
[7] H. Peters, H.-H. Mobius, Verfahren zur Gasanalyse bei erhohten Temperaturen mit Hilfe galvanischer Festelektrolytelemente, DDR-P 21673, filed 20.5.1958.
[8] J. Weissbart, R. Ruka, Oxygen gauge, Rev. Sci. Instr. 32 (1961) 593–595.
[9] B. von Gunther, An electronic fuel injection system for automobiles, Bosch Techn. J. (1967) berichte 2, Heft 3, November.
[10] S. von Hermann, Electronic fuel injection – Jetronic, Bosch Techn. J. (1) (1969) Report 3.
[11] R. Ramamoorthy, P.K. Dutta, S.A. Akbar, Oxygen sensors: materials, methods, designs and applications, J. Mater. Sci. 38 (2003) 4271–4282.
[12] K. Nishio, Fundamentals of Automotive Engine Control Sensors, Fontis Media, Netherlands, 2001.
[13] J.H. Lee, Review on zirconia air fuel ratio sensors for automotive applications, J. Mater. Sci. 38 (2003) 4247–4257.
[14] S. Akbar, P. Dutta, C. Lee, High temperature ceramic gas sensors: a review, Int. J. Appl. Ceram. Technol. 3 (2006) 302–311.
[15] M. Iwase, A. McLean, K. Katogi, K. Wakimoto, Rapid sensors for ferroalloy production and processing, in: 10th Int FerroAlloys Conf, Capetown, South Africa, 2004.
[16] G. Apanel, E. Johnson, Direct methanol fuel cells: ready to go commercial? Fuel Cells Bull. 2004 (11) (2004) 12–17.
[17] National Research Council, Energy-Efficient Technologies for the Dismounted Soldier, The National Academies Press, Washington, DC, 1997.
[18] S.B. Schaevitz, Powering the wireless world with MEMS, Proc. SPIE 8248 (2012), Proc. SPIE 8248, Micromachining and Microfabrication Process Technology XVII, 824802 (February 9, 2012); http://dx.doi.org/10.1117/12.924356.
[19] H. Wang, Y. Jiang, L. Philips, Mobile Convergence: Platform, Applications & Services, Parks Associates, Dallas, 2009.

[20] J.D. Morse, A.F. Jankowski, A novel thin film solid oxide fuel cell for microscale energy conversion, Proc. SPIE 3876 (1999) 223–226.
[21] A. Evans, A. Bieberle-Hutter, H. Galinski, J.L.M. Rupp, T. Ryll, B. Scherrer, R. Tolke, L.J. Gauckler, Micro-solid oxide fuel cells: status, challenges, and chances, Monatsch. Chem. 140 (2009) 975–983.
[22] S.B. Schaevitz, A. Franz, R. Barton, US 2005/0115889, 2005.
[23] S.B. Schaevitz, A. Franz, R. Barton, A.P. Ludwiszewski, US 2006/0263655, 2006.
[24] G. Robert, EP 1 455 409 A1, 2004.
[25] L.J. Gauckler, D. Beckel, U.P. Muecke, P. Muller, J.L.M. Rupp, WO2007/045111, 2007.
[26] L.J. Gauckler, D. Beckel, U.P. Muecke, P. Muller, J.L.M. Rupp, WO2007/045113, 2007.
[27] K. Kerman, S. Ramanathan, Complex oxide nano-membranes for energy conversion and storage: a review, J. Mater. Res. 29 (2014) 320–337, http://dx.doi.org/10.1557/jmr.2013.301.
[28] C. Ko, K. Kerman, S. Ramanathan, Ultra-thin film solid oxide fuel cells utilizing un-doped nanostructured zirconia electrolytes, J. Power Sources 213 (2012) 343.
[29] I. Garbayo, A. Tarancón, J. Santiso, F. Peiró, E. Alarcón-Lladó, A. Cavallaro, I. Gràcia, C. Cané, N. Sabaté, Electrical characterization of thermomechanically stable YSZ membranes for micro solid oxide fuel cells applications, Solid State Ionics 181 (2010) 322.
[30] S. Rey-Mermet, P. Muralt, Solid oxide fuel cell membranes supported by nickel grid anode, Solid State Ionics 179 (2008) 1497.
[31] M. Greenberg, E. Wachtel, I. Lubomirsky, J. Fleig, J. Maier, Elasticity of solids with a large concentration of point defects, Adv. Funct. Mater. 18 (2006) 48.
[32] J.H. Shim, J.S. Park, J. An, T.M. Gür, S. Kang, F.B. Prinz, Intermediate-temperature ceramic fuel cells with thin film yttriumdoped barium zirconate electrolytes, Chem. Mater. 21 (2009) 3290.
[33] C.D. Baertsch, K.F. Jensen, J.L. Hertz, H.L. Tuller, S.T. Vengallatore, S.M. Spearing, M.A. Schmidt, Fabrication and structural characterization of self-supporting electrolyte membranes for a micro solid-oxide fuel cell, J. Mater. Res. 19 (9) (2004) 2604.
[34] A.C. Johnson, B.-K. Lai, H. Xiong, S. Ramanathan, An experimental investigation into micro-fabricated solid oxide fuel cells with ultra-thin La0.6Sr0.4Co0.8Fe0.2O3 cathodes and yttria-doped zirconia electrolyte films, J. Power Sources 186 (2) (2009) 252.
[35] H. Huang, M. Nakamura, P. Su, R. Fasching, Y. Saito, F.B. Prinz, High-performance ultrathin solid oxide fuel cells for low-temperature operation, J. Electrochem. Soc. 154 (1) (2007) B20.
[36] S.J. Litzelman, J.L. Hertz, W. Jung, H.L. Tuller, Opportunities and challenges in materials development for thin film solid oxide fuel cells, Fuel Cells 8 (2008) 294.
[37] K. Kendall, Progress in microtubular solid oxide fuel cells, Int. J. Appl. Ceram. Technol. 7 (1) (2010) 1–9.
[38] K.S. Howe, G.J. Thompson, K. Kendall, Microtubular solid oxide fuel cells and stacks, J. Power Sources 196 (2011) 1677–1686.
[39] V. Lawlor, S. Griesser, G. Buchinger, A.G. Olabi, S. Cordiner, D. Meissner, Review of the micro-tubular solid oxide fuel cell Part I. Stack design issues and research activities, J. Power Sources 193 (2) (2009) 387–399.
[40] V. Lawlor, C. Hochenauer, A. Mariani, S. Griesser, S. Kuehn, K. Klein, A.G. Olabi, S. Cordiner, D. Meissner, G. Buchinger, Scrutiny oif MT-SOFC stack modelling design using CFD, Open Fuel Cell. J. 5 (2012) 1–13.
[41] S.C. Singhal, Science and technology of solid oxide fuel cells, MRS Bull. 25 (2000) 16–21.

[42] R. Goettler, T Ohrn, Update on Rolls Royce coal based SECA Program, Rolls Royce report, 27 July 2010.

[43] S.C. Singhal, Advances in solid oxide fuel cell technology, Solid State Ionics 135 (2000) 305–313.

[44] D. Driscoll, in: 13th Annual SECA Workshop, SECA Program Overview 2012, NETL, 24 July, 2012.

[45] B.C.H. Steele, 1991, personal communication.

[46] K. Kendall, in: Proceedings of the International Forum on Fine Ceramics, Japan Fine Ceramics Center, Nagoya, 1992.

[47] M. Kendall, Tubular cells: a novel SOFC design, Final year thesis. Middlesex University, 1993.

[48] K. Kendall, G. Sales, A rapid heating ceramic fuel cell, in: Ceramics in Energy Applications, Institute of Energy, London, 1994, pp. 55–63.

[49] K. Kendall, Solid oxide fuel cell ceramics through colloidal processing, Am. Ceram. Soc. Trans. 26 (1992) 125–131.

[50] R.H. Cunningham, C.M. Finnerty, R.M. Ormerod, K. Kendall, An in situ catalytic and electrochemical study of working Ni/ZrO2 Anodes in tubular SOFCs, in: U. Stimming, S.C. Singhal, H. Tagawa, W. Lehnert (Eds.), Electrochem. Soc. 1997, pp. 965–972, Pennington, NJ.

[51] K. Kendall, D.S. Williams, Catalysts for butane reforming in zirconia fuel cells, Platinum Metals Rev. 42 (1998) 164–167.

[52] K. Kendall, M. Palin, A small SOFC demonstrator for microelectronic applications, J. Power Sources 71 (1998) 268–270.

[53] K. Kendall, J. Staniforth, Biogas powering a small tubular SOFC, J. Power Sources 71 (1998) 275–277.

[54] Patent Publication No. WO 94/22178 Solid Oxide Fuel Cell Structures, by K. Kendall to Keele University, 1994.

[55] K. Kendall, M. Prica, Integrated SOFC tubular system for small-scale cogeneration, in: U. Bossel (Ed.), 1st European SOFC Forum, Luzern, Switzerland, 1994, pp. 163–170.

[56] M. Prica, T. Alston, K. Kendall, Mechanical and thermal properties of a 200 tube SOFC reactor, in: U. Stimming et al., (Ed.), Proceedings of the Fifth International Symposium on Solid Oxide Fuel Cells, Electrochemical Society Inc., Pennington, 1997, pp. 619–625.

[57] T. Alston, M. Palin, P. Windibank, M. Prica, K. Kendall, A 1000-cell SOFC reactor for domestic cogeneration, J. Power Sources 71 (1998) 271–274.

[58] K. Kendall, M. Prica, T. Alston, Micro-cogeneration from natural gas in a 1000 tube extruded zirconia SOFC, in: P. Stevens (Ed.), 3rd European SOFC Forum, European SOFC Forum, Switzerland, 1998.

[59] I. Staffell, A. Ingram, K. Kendall, Energy and carbon payback times for solid oxide fuel cells based domestic CHP, Int. J. Hydrogen Energy 37 (2012) 2509–2523.

[60] K. Kendall, N.McN. Alford, J.D. Birchall, High strength ceramics through colloidal control to remove defects, Nature 330 (1987) 51–53.

[61] K. Kendall, Revolution in Ceramic Processing, in: F. Riley (Ed.), Third European Symposium on Engineering Ceramics, Elsevier, London, 1989, pp. 97–107.

[62] F.F. Lange, Powder processing science and technology for increased reliability, J. Am. Ceram. Soc. 72 (1989) 315–324.

[63] K. Kendall, N. McN Alford, W.J. Clegg, J.D. Birchall, Flocculation clustering and the weakness of ceramics, Nature 339 (1989) 130–132.

[64] N. McN Alford, J.D. Birchall, A.J. Howard, K. Kendall, J.H. Raistrick, European Patent, Publication No. 183453, 1986.

[65] U. Bossel, Rapid start up SOFC modules, Energy Procedia 28 (2012) 48–56.
[66] U. Bossel, Universal SOFC module for rapid start-ups, in: Proceedings of 12th International Symposium on Solid Oxide Fuel Cells (SOFC XII), Montreal, Canada May 2011, ECS Trans., 35, 2011, pp. 179–186, 1.
[67] K. Kendall, Hopes for a flame-free future, Nature 404 (2000) 233–235.
[68] K. Kendall, C.M. Finnerty, J.C. Austin, T. Alston, Ceramic fuel cells to replace metal burners, J. Mater. Sci. 36 (2001) 1119–1124.
[69] K. Kendall, Formulating liquid hydrocarbons for fuel cells, in: U. Bossel (Ed.), Fuel Cell World, 2004, pp. 91–98, 28 June–2 July 2004, ISBN 3-905592-14-2, Luzern.
[70] E.V. Tsipis, V.V. Kharton, Electrode materials and reaction mechanisms in solid oxide fuel cells: a brief review, J. Solid State Electrochem. 12 (2008) 1367–1391.
[71] D. Sarantaridis, A. Atkinson, Redox cycling of Ni-based solid oxide fuel cell anodes: a review, Fuel Cells 7 (2007) 246–258.
[72] K. Kendall, M. Kendall, Micro-tubular solid oxide fuel cells (submitted to European Fuel Cell Forum 2014).
[73] W. Winkler, The influence of the mass transfer on the geometric design of SOFC stacks, in: Sixth Grove Fuel Cell Symposium Queen Elizabeth II Congress Centre in London, Journal of Power Sources 86, 2000, pp. 449–454, 1-2.
[74] C.E. Hatchwell, N.M. Sammes, I.W.M. Brown, K. Kendall, Current collectors for a novel tubular design of solid oxide fuel cell, J. Power Sources 77 (1999) 64–68.
[75] J. van Herle, R. Ihringer, K. Kendall, N.M. Sammes, G. Tompsett, K. Yamada, C. Wen, T. Kawada, M. Ihara, J. Misuzaki, Concept and technology of SOFC for electric vehicles, Solid State Ionics 132 (2000) 333–342.
[76] H.R. Ellamia, I. Staffell, P. Bujlo, B.G. Pollet, S. Pasupathia, Current staus of fuel cell based CHP systems for residential sector, J. Power Sources 293 (2015) 312–328.
[77] R. Steinberger-Wilckens, O. Bucheli, L.G.J. de Haart, A. Hagen, J. Kiviaho, J. Larsen, S. Pyke, B. Rietveld, J. Sfeir, F. Tietz, M. Zahid, Real-SOFC – a joint european effort to improve SOFC durability, ECS Trans. 25 (2) (2009) 43–56, http://dx.doi.org/10.1149/1.3205507.
[78] N. Bessette, Status of the Acumentrics SOFC Programme, in: SECA Annual Workshop, Boston MA, 2004, Asupathi, Current status of fuel cells.
[79] N. Bessette, DOE Hydrogen Programme Review, Washington, DC, 2012.
[80] AMI website www.ultra-ami.com.
[81] A.T. Crumm, J.W. Halloran, US Patent Pub. No. 6749799, Method for preparation of solid state electrochemical device, 2004.
[82] K. Kendall, C. Mallon, Sensitivity of nickel cermet anodes to reduction conditions, J. Power Sources 145 (2005) 154–160.
[83] F. Adamson, J. Billingham, A.C. King, K. Kendall, An asymptotic theory for the propagation of a surface catalysed reaction in a tube, J. Fluid Mech. 546 (2006) 363–393.
[84] K. Kendall, C.M. Dikwal, W. Bujalski, Comparative analysis of thermal and redox cycling for microtubular SOFCs, ECS Trans. 7 (2007) 1521–1525. SOFC X, The Electrochemical Society.
[85] K. Kendall, A. Dhir, Improving reliability of microtubular SOFCs for direct use on methane, ECS Trans. 7 (2007) 823–826. SOFC X, The Electrochemical Society.
[86] N. Akhtar, S.P. Decent, D. Loghin, K. Kendall, Mixed-reactant, micro-tubular, solid oxide fuel cells: an experimental study, J. Power Sources 193 (2009) 39–48.
[87] K. Kendall, T.J. Lee, Characterisation of electrical performance of anode supported microtubular SOFC with methane fuel, J. Power Sources 181 (2008) 195–198.

[88] A.T. Crumm, T. LaBreche, US Patent Pub. No. 2011/0189587A1, Interconnect Member for Fuel Cell, to Adaptive Materials Inc., 2011. Also see US 2005/0147857 on Solid Oxide fuel cell with improved current collection by A.T. Crumm, C.J. Reilly, J.R. Rice, 2005.
[89] A.T. Crumm, C.J. Reilly, T. LaBreche, US Patent Pub. No. 2005/0112452, Solid oxide fuel cell tube with internal fuel processing, 2005.
[90] www.lockheedmartin.ca/us/products/stalker-uas.html2013-12-24.
[91] SUAV project, www.suav-project.eu/Technology.html2012-11-8.
[92] SAPIENS project, www.adelan.co.uk/sapiens.
[93] SAFARI project, www.adelan.co.uk/safari.
[94] B. Kavalov, H. Petric, B. Kavalov, H. Petric, A. Georgakaki, Liquefied Natural Gas for Europe – Some Important Issues for Consideration, JRC, Netherlands, 2009. ec.europa.eu/dgs/jrc/downloads/jrc _reference_report_200907 _liquefied_natural_gas.pd.

Sources of cell and electrode polarisation losses in SOFCs

11

S.B. Adler
University of Washington, Seattle, WA, USA

11.1 Introduction

Approximately 1/3 of the reversible work potential (exergy) of a hydrocarbon-based fuel/air reaction is lost upon combustion [1]. As illustrated in Figure 11.1a, the goal of a solid oxide fuel cell (SOFC) is to recover a significant fraction of this lost work by subdividing the combustion reaction into two (or more) half-cell reactions. For example, for a SOFC running on syngas:

$$\text{Anode:}\quad H_2 + O^{2-} \Leftrightarrow H_2O + 2e^- \tag{11.1a}$$

$$CO + O^{2-} \Leftrightarrow CO_2 + 2e^- \tag{11.1b}$$

$$\text{Cathode:}\quad O_2 + 4e^- \Leftrightarrow 2O^{2-} \tag{11.2}$$

Provided these reactions (as well as all transport steps involving electrons, ions and molecules to/from the electrode) proceed near equilibrium, the exergy of the net fuel/air reaction will be converted into an electrical potential difference, $V = -\left(\mu_{e^-}^{c} - \mu_{e^-}^{a}\right)/F$, between anode and cathode, and thus can be recovered as *electrical work*.

For SOFCs, the maximum available exergy of the fuel/air reaction is usually described in terms of the *reversible cell potential*, which (in the absence of electron crossover) is equivalent to the *open circuit voltage*. The reversible potential can be derived as a function of temperature and gas-phase composition by considering the equilibrium of Equations (11.1a), (11.1b) and (11.2) (as well as the water-gas-shift reaction and all transport steps to/from the electrodes). Following the procedure outlined by Newman [2], this leads to an open circuit voltage (V_0) given by:

$$V_0 = -\frac{1}{F}\left(\mu_{e^-}^{c} - \mu_{e^-}^{a}\right)_{\text{eqb}} = -\frac{\Delta g^{\circ}_{f,H_2O}}{2F} + \frac{RT}{2F}\ln\left(\frac{\left(p_{H_2}^{a,\infty}\right)\left(p_{O2}^{c,\infty}\right)^{1/2}}{\left(p_{H_2O}^{a,\infty}\right)}\right) \tag{11.3}$$

where the gas partial pressures (assumed constant and uniform) are defined far from the active portion of the electrode; $\Delta g^{\circ}_{f,H_2O}$ is the free energy of formation of H_2O gas and R, T and F are the ideal gas constant, temperature and Faraday's constant,

High-temperature Solid Oxide Fuel Cells for the 21st Century. http://dx.doi.org/10.1016/B978-0-12-410453-2.00011-7

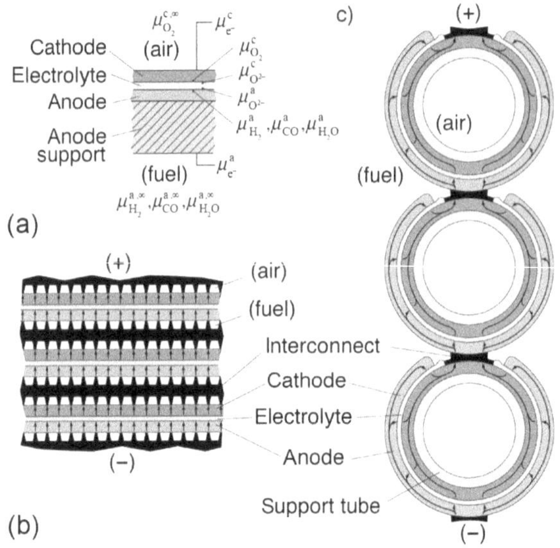

Figure 11.1 (a) Schematic of an anode-supported SOFC, indicating locations of relevant chemical and electrochemical potentials linked to cell potential losses. (b, c) Schematic of a 3-cell segment of a planar (a) and tubular (b) SOFC stack, showing current paths through the stack.

respectively. (Note that if all gas-phase reactions are at equilibrium, the open circuit voltage calculated based on Equations (11.1a) or (11.1b) will be the same.) Maximum efficiency is obtained by passing an infinitesimal current under these equilibrium conditions, yielding maximum work per coulomb equivalent equal to V_0.

Unfortunately, the idealisation described above is only partially achievable. In reality, all rate processes (both reaction and transport steps) require some level of free energy driving force to proceed at a finite rate. These driving forces consume some of the available exergy of the fuel/air reaction (*cell losses*). In addition, fuel cells operating within a *device* (e.g. a fuel cell stack) operate under conditions with lateral utilisation and temperature gradients and thus do not behave reversibly, even if single cells within that device are reversible when exposed to temperature and gas composition similar to the inlet of the stack (*stack losses*). Taken together, these losses cause the cell voltage to be less than V_0 by an amount that depends on the rate of each exergy-consuming process, which at steady state is related to the local current density (i) passed through the cell:

$$V(i) = V_0 - R_\Omega i - \eta_c(i) - \eta_a(i) - \Delta V_c{}^{(gas)}(i) - \Delta V_a{}^{(gas)}(i) \tag{11.4}$$

where R_Ω is the sum of in-series ohmic resistances, $\eta_c(i)$ and $\eta_a(i)$ are the (overall) cathode and anode overpotentials and $\Delta V_c^{(gas)}(i)$ and $\Delta V_a^{(gas)}(i)$ are additional losses associated with mass transfer and homogeneous reactions in the gas phase outside the cathode and anode. The goal of this chapter is to outline the most common sources of these losses in SOFCs, and the relationship of these losses to electrical, electrochemical and chemical rates.

11.2 Cell losses

As illustrated in Figure 11.2, when describing losses on the cell level, workers often subdivide cell losses into *ohmic losses* (associated with ion and/or electron migration within bulk phases), *gas-phase losses* (associated with transport of reactive molecules to/from the electrodes) and *overpotential losses* (associated with the electrode reactions themselves). Note that in this context, the term *overpotential* is used to describe the macroscopic behaviour of the anode or cathode as a whole (e.g. based on linearity with respect to current density, frequency response, dependence on gas conditions) as opposed to a detailed analysis of gradients occurring within the electrode itself. Further subdivision of the macroscopically observed overpotential into specific reaction and transport steps (at various levels of detail) will be discussed in subsequent sections.

11.2.1 Ohmic losses

In a planar SOFC (where lateral current distribution is often achieved via an interconnect plate), most ohmic losses are associated with ion transport in the *electrolyte*. This can be understood by considering equilibrium of Equations (11.1a), (11.1b) and (11.2) (as well as all gas-phase reaction and transport steps), but where the electrolyte is not at equilibrium. Modifying the analysis leading to Equation (11.3) accordingly leads to:

$$V = -\frac{1}{F}\left(\mu_{e^-}^{c} - \mu_{e^-}^{a}\right) = V_0 - \frac{1}{2F}\left(\mu_{O^{2-}}^{c} - \mu_{O^{2-}}^{a}\right) \tag{11.5}$$

where, as shown in Figure 11.1a, $\mu_{O^{2-}}^{c}$ and $\mu_{O^{2-}}^{a}$ are the electrochemical potential of oxygen ions in the electrolyte at the cathode and anode, respectively. For a pure electrolyte (like yttria-doped zirconia), gradients in $\mu_{O^{2-}}$ will be proportional to local current density i: $\nabla\mu_{O^{2-}} = 2Fi/\sigma_i$, where σ_i is the ionic conductivity of the electrolyte. Assuming a uniform electrolyte of thickness L_e, and uniform current density $i = I/A$, this yields:

$$V = V_0 - R_\Omega i \tag{11.6}$$

where $R_\Omega = L_e/\sigma_i$.

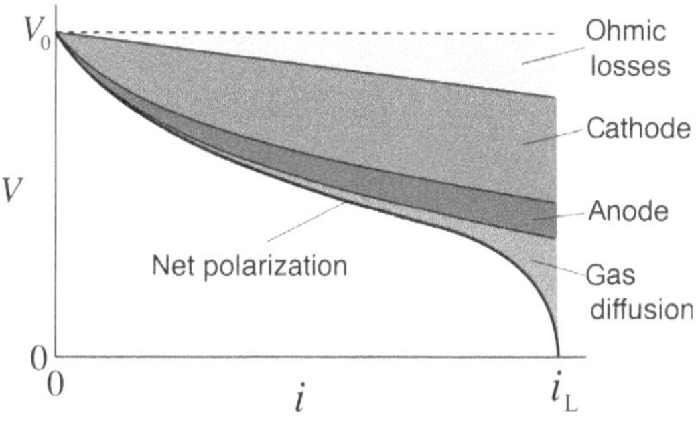

Figure 11.2 Depiction of a typical single-cell SOFC polarisation curve, showing contributions of various processes to displacement of the cell voltage (V) from the reversible cell potential (V_0) as the current density (i) is increased.

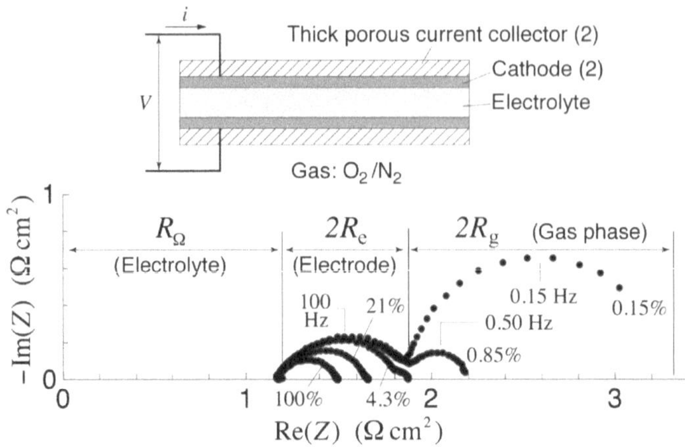

Figure 11.3 Measured impedance of a symmetrical cell composed of a dense freestanding tape-cast ceria-based electrolyte with two screen-printed and sintered single-phase porous electrodes based on divalently doped $LaCoO_3$ [3]. Both electrodes were covered by a thick (~10 μm) minimally porous silver current collector that retards O_2 diffusion. Measurements were made at 750 °C as a function of p_{O2} (at the compositions marked in the figure) using blended gases of O_2 in N_2. Division of impedance into contributions of electrolyte, electrodes and gas-phase diffusion is shown for case of 0.15% oxygen.

As illustrated in Figure 11.1, SOFCs can also exhibit ohmic losses associated with lateral distribution of electrical current over the *area* of the cell. This is a particular problem in tubular SOFCs. Unlike a planar SOFC (where current is distributed uniformly over the cell via an adjacent interconnect plate with many parallel points of contact), it is only possible to make electrical contact to a tubular cell at certain points along the perimeter. In this geometry, current must flow tangentially (or longitudinally) through a thin current collector layer, often incurring significant ohmic losses that also contribute to R_Ω (as well as influencing electrode losses).

The linear dependence of Equation (11.6) is depicted in Figure 11.2 (ohmic loss). Besides being linear, ohmic losses (which arise from a potential drop over macroscopic distances) usually involve little or no parallel capacitive current, and thus appear as a purely real (or very high frequency) impedance that is independent of gas-phase conditions. For example, Figure 11.3 shows the measured impedance of a symmetrical cathode/electrolyte/cathode cell. In the limit of very high frequency, the impedance becomes purely real, and is independent of p_{O2}, unlike the electrode and gas-phase impedances, which are spread over a range of frequencies and are p_{O2} dependent.

11.2.2 Overpotential losses

On a macroscopic level, the term *overpotential* refers to the free energy driving force needed to sustain the rate of an electrochemical reaction. For example, consider the case where the cathode (Equation 11.2) is not at equilibrium, while all anode reactions (as well as the electrolyte and all molecular transport steps to/from the cell) remain in equilibrium. A similar analysis as described previously for Equations (11.5) and (11.6) leads to:

$$V = \frac{1}{F}\left(\mu_{e^-}^{c} - \mu_{e^-}^{a}\right) = V_0 - \eta_c(i) \tag{11.7}$$

where, as illustrated in Figure 11.1a, $-F\eta_c(i) = \mu_{e^-}^{c} - \frac{1}{2}\mu_{O^{2-}}^{c} + \frac{1}{4}\mu_{O_2}^{c}$ is the overall displacement in the electron chemical potential from equilibrium with O^{2-} at the cathode and O_2 in the gas via reaction (11.2). The way overpotential is defined here, and it is a positive function of cell current density under the normal direction of operation of the fuel cell at steady state. Thus, as illustrated in Figure 11.2 (and described by Equation 11.4), the cathode and anode overpotentials will both reduce the overall cell voltage under steady polarisation.

As depicted in Figure 11.2, the anode and cathode overpotentials are generally a nonlinear function of current density. At an empirical level, a common model used to express this nonlinearity is the Butler–Volmer equation:

$$i = i_0\left(e^{\frac{\alpha_f \eta F}{RT}} - e^{\frac{-\alpha_b \eta F}{RT}}\right) \tag{11.8}$$

where i_0 is the exchange current density, and α_f and α_b are the forward and backward transfer coefficients (constants of order unity) [4]. This model generally predicts a decreasing electrode resistivity ($R_e = d\eta/di$) of each electrode with increasing current density, leading to a more rapid drop in cell voltage at small currents (often described as an "activation loss"). As described elsewhere [4,5], Equation (11.8) is only rigorously valid for charge-transfer across a two-phase interface that is limited by a single rate-determining step. However, as discussed in further detail below, many rate phenomena in SOFCs tend to mimic this behaviour, so it remains a useful empiricism for many purposes (such as stack modelling).

Equation (11.8) also predicts that the backward rate becomes insignificant in the limit high current, yielding Tafel kinetics ($\eta \sim \ln(i)$). However, unlike a low-temperature fuel cell (where the exchange current density is many orders of magnitude smaller than the current, $i \ll i_0$, and thus the cell operates in the Tafel region), the equilibrium exchange rate in a SOFC electrode is often of the same order as the current itself, meaning that the backward rate usually cannot be ignored even when the cell is operating at its maximum current density in the normal direction.

Unlike ohmic losses, electrode overpotentials are often associated with a significant electrode capacitance (C_e), due to polarisation at the electrode/electrode interface, as well as faradaic oxidation/reduction of the electrode material. As shown in Figure 11.3, this capacitance can result in a measurable imaginary impedance, which reaches a maximum when the perturbation frequency (ω) approaches the characteristic response frequency of the electrode $\omega_0 = 1/R_eC_e$. Thus if the anode and cathode have different characteristic frequencies, impedance measurements can be used to separate and quantify their resistivities.

11.2.3 Gas-phase losses

Another significant source of exergy loss in SOFCs (particularly for anode-supported planar cells) is gas-phase mass transfer. This can be understood simply by considering changes in the reversible cell potential caused by a concentration gradient in the gas phase adjacent to the cell. For example, for an anode limited by H_2 availability, cell voltage will be given by Equation (11.3), modified to include displacement in H_2 concentration from equilibrium:

$$V = V_0 + \frac{RT}{2F}\ln\left(\frac{\left(p^{a}_{H_2}\right)\left(p^{a,\infty}_{H_2O}\right)}{\left(p^{a,\infty}_{H_2}\right)\left(p^{a}_{H_2O}\right)}\right) \tag{11.9}$$

where, as shown in Figure 11.1a, $p^{a}_{H_2}$ and $p^{a}_{H_2O}$ differ from their values far from the electrode ($p^{a,\infty}_{H_2}$ and $p^{a,\infty}_{H_2O}$). If we ignore gradients in water concentration, and consider simple Fickian diffusion of H_2 through the anode support (often a combination of molecular and Knudsen diffusion), the relationship of current to $p^{a}_{H_2}$ will be given by Fick's and Farday's Laws:

$$\frac{i}{4F} = \frac{D^{eff}_{H_2}\left(p^{a,\infty}_{H_2} - p^{a}_{H_2}\right)}{L_{diff}RT} \tag{11.10}$$

where L_{diff} and $D^{eff}_{H_2}$ are the diffusion length and effective (superficial) diffusion coefficient for H_2 in the porous material covering the anode, respectively. Substitution into Equation (11.7) (neglecting gradients in H_2O) yields:

$$V = V_0 - \frac{RT}{2F}\ln\left(\frac{1}{1 - i/i_L}\right) \tag{11.11a}$$

$$i_L = \frac{4FD^{eff}_{H_2}p^{a,\infty}_{H_2}}{L_{diff}RT} \tag{11.11b}$$

where i_L is the *limiting current* at which the H_2 concentration at the electrode becomes so low that the reversible potential is driven to zero. The dependence of this gas-phase diffusion loss on current density is depicted in Figure 11.2. In the limit of low current, the gas-phase loss becomes linear in current density, with characteristic resistivity:

$$R_g = \frac{(RT)^2 L_{diff}}{8F^2 D^{eff}_{H_2} p^{a,\infty}_{H_2}} \tag{11.12}$$

As shown in Figure 11.3, this resistance can appear as a separate impedance arc that has a characteristic resistance that scales inversely with gas concentration and has a large faradaic capacitance corresponding to oxidation and reduction of the entire electrode relative to the gas phase. Note that this resistance is in *addition* to any utilisation losses (described in Section 11.4) that occur due to non-uniform or transient temperature and gas-phase concentrations within a multi-cell stack.

11.3 Ohmic and gas-phase losses within porous electrodes

From a macroscopic viewpoint, SOFC electrode reactions occur spatially somewhere near the *three-phase boundary* (TPB) among three separate phases: the gas (source/sink for molecules), the electrolyte (source/sink of oxygen ions) and the electrode material (source/sink of electrons). Thus, like many other commercial electrochemical devices (including the lead-acid battery [6]), high-performance SOFC electrodes often have a *composite microstructure* in order to maximise total available TPB area. As illustrated in Figure 11.4a, a composite SOFC electrode typically consists of an ionically conducting sub-phase, co-sintered or infiltrated with an electronically conducting sub-phase, with a microstructure designed to maximise TPB area as well as provide facile transport pathways for molecular, ionic and electronic transport to the TPB.

Although composite microstructures improve performance, they further complicate interpretation and prediction of SOFC electrode kinetics due to inherent convolution of reaction overpotential, ohmic losses and gas-phase diffusion losses within the

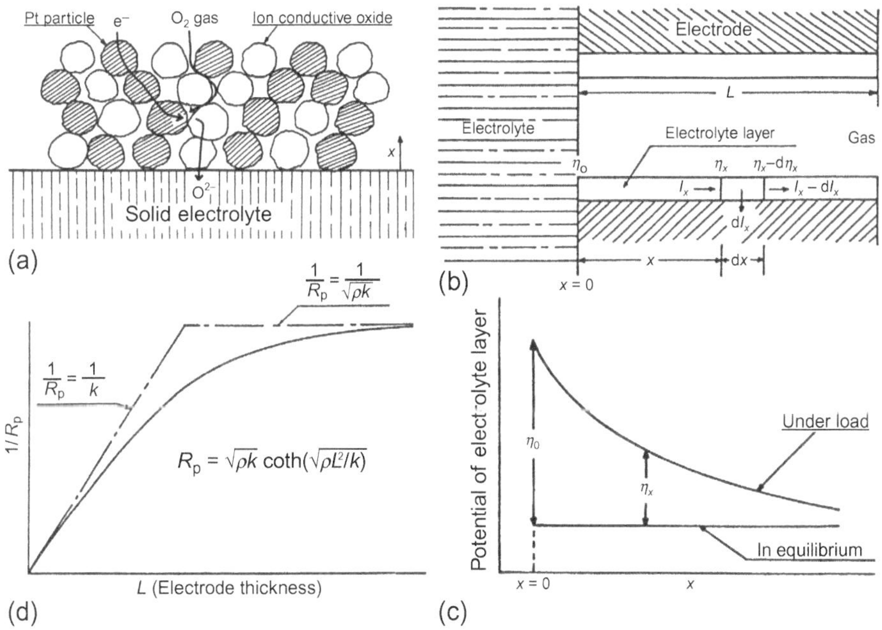

Figure 11.4 Kenjo's 1-D macrohomogeneous model for combined polarisation and ohmic losses within a porous composite electrode. (a) Illustration of the composite microstructure, showing ionic and electronic transport paths to/from the three-phase boundary (TPB). (b) Model description of the ionically and electronically conductive sub-phases and reaction at the TPB. (c) Predicted overpotential profile $\eta(x)$ within the electrode near the electrode/electrolyte interface. (d) Predicted electrode admittance as a function of the electrode thickness. Images reproduced from Ref. [7] by permission of the Electrochemical Society, Inc., and from Ref. [8] with permission from Elsevier.

electrode. One cannot linearly superpose these losses as implied by Equation (11.4). Rather, the true kinetic overpotential $\eta(x)$, electrolyte potential $\varphi_i(x)$ and gas-phase concentration vary locally with position (x) within the electrode, leading to an overall loss in cell voltage representing *co-limitation* of kinetics and transport losses.

One of the first models for a composite SOFC electrode was reported by Kenjo and co-workers [7,8], who fabricated and tested the performance of porous Pt composite electrodes containing YSZ, SDC and erbia-stabilised bismuth oxide (ESB), as well as composites of A-site-deficient $La_{0.85}MnO_3$ (LM) and YSZ. They found that the performance of these composite electrodes *increases with electrode thickness*, saturating at a thickness 10–30 μm. As shown in Figure 11.4b, the authors explained their results in terms of a macrohomogeneous model with three interpenetrating sub-phases: an electronic conductor (assumed to have infinite electrical conductivity), a gas phase (assumed to be of uniform composition) and an ionic conductor with finite ionic resistivity $\rho = 1/\sigma_i$. At the interface between these phases, oxygen reduction was assumed occur at a rate proportional to the local overpotential $\eta(x)$, resulting in a passage of current from the ionic to the electronic phase, $di_x/dx \sim \eta(x)/k$, where "k" is the interfacial resistance per unit volume. This model predicts that the composite electrode will be *co-limited* by interfacial resistance and ionic transport in the ionic sub-phase, exhibiting a utilisation region of size $\sim \sqrt{k/\rho}$, and overall resistance proportional to $\sqrt{\rho k}$ for an optimally thick electrode, in good agreement with experiment.

More sophisticated models have since been developed to predict a variety of phenomena co-contributing to overall electrode polarisation, including efficiency of the ionic and electronic transport paths [9–13], quality and quantity of internal interfacial contact between the ionic and electronic sub-phases [14,15], functional grading of the materials for structural or other purposes [16–19], molecular and Knudsen diffusion of gaseous molecular species within the pores of the electrode, as well as internal reforming and water-gas-shift reactions involving CO and hydrocarbon species [20–25]. As one might expect, these models show that interpreting the electrochemical characteristics of composite electrodes is considerably more challenging than understanding their single-phase counterparts, even if one fully understands the relationship of overpotential to current locally at the three-phase boundary.

11.4 Cell losses within a multi-cell stack

The losses described above are for a single cell exposed to uniform temperature, gas composition and current density. However, cells in a multi-cell stack exhibit modified/additional exergy losses due to gradients in temperature, gas composition and current density across each cell within a stack during operation. In general, these losses are quite complex and interrelated and difficult to characterise or describe without advanced stack modelling. However, some relevant features of these losses are described here.

One important source of irreversibility in a SOFC stack is *finite fuel utilisation*. To understand this effect, let us imagine a SOFC stack with a co-flow configuration,

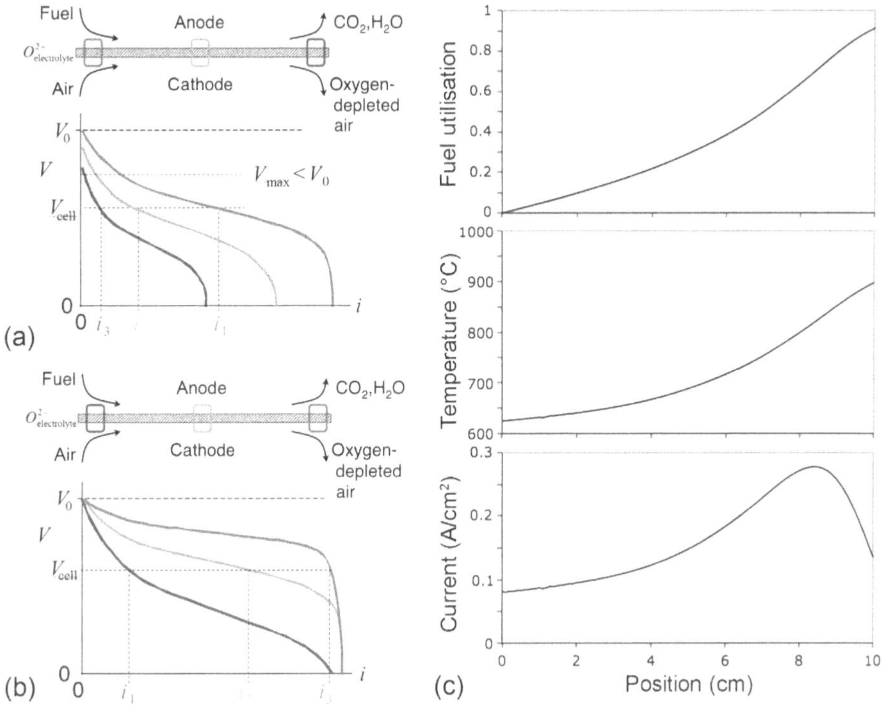

Figure 11.5 Schematic showing the effects of a 1-D concentration gradient (a) and temperature gradient (b) on the local $V-i$ characteristics of a cell, and the resulting maldistribution of current density. (c) Predicted gradients of fuel utilisation, temperature and current density within a 1-D co-flow stack operating on humidified syngas (steam to carbon ratio 2.4) with inlet temperature of 625 °C, air to fuel ratio of 36:1 and voltage per cell of 0.70 V [1].

as shown in Figure 11.5a. In this configuration, the air and fuel flow in the same direction on opposite sides of each cell, resulting in 1-D gradients in gas concentration and temperature on each side of the cell. For the moment, let us limit the discussion to the effects of concentration, assuming temperature is *uniform*.

At any point along the cell, the *local* $i-V$ characteristics will depend on the local gas concentration, as illustrated in Figure 11.5a for three points along the cell. As the fuel (and oxygen) concentration drop from left to right, the predicted electrode overpotential (at any given current density) will increase, while the limiting current and reversible cell potential will decrease (due to lower gas concentration). The net effect is that at any given cell voltage (which is common to the entire cell), the current density must decrease along the flow path. This implies that even if the cell operates reversibly towards the trailing edge, the leading edge must operate less reversibly. Also, regardless of cell losses, the maximum potential V_{max} that the cell can operate (without current reversal) is dictated by the gas concentrations in the *exit stream* (which is determined by the fuel utilisation under load), not by the inlet concentrations at open circuit.

Another source of irreversibility in a SOFC stack is associated with *temperature gradients*. As illustrated in Figure 11.5b, heat generated within the stack must accumulate in the air and fuel streams, causing a rise in temperature from left to right. Since most kinetic and transport processes are activated, this will cause the local $i-V$ characteristics to change as illustrated for three points along the flow path in Figure 11.5b. The net effect is that at any given cell voltage, the current density must increase along the flow path. As with concentration gradients, this misdistribution of current density implies a higher level of irreversibility in colder areas of the cell even if the hottest parts of the cell operate reversibly.

Prediction of the combined effect of these concentration and temperature gradients on overall average cell performance requires a detailed stack model, which accounts for (using material and energy balances) variations of temperature and gas composition within the stack, and the coupling of these changes to local cell $i-V$ characteristics as illustrated above. For example, for the co-flow situation described in Figure 11.5a and b, one can construct a relatively simple 1-D steady-state model that considers axial convection of H_2, CO, H_2O and CO_2 in the anode channel, axial convection of O_2 and N_2 in the cathode channel, faradaic transport of oxygen across the cell, equilibrium of the water-gas-shift reaction and energy conservation including export of electrical work and accumulation of thermal energy as enthalpy in the gas [1].

Assuming cell characteristics of a typical SOFC, cell length of 10 cm, inlet conditions of 625 °C, steam to carbon ratio of 2.4:1 and air to fuel ratio of 36:1 (for cooling) and 90% fuel utilisation, this model predicts the utilisation, temperature, and current density distributions shown in Figure 11.5c with a voltage per cell of 0.70 V and average current density of 0.16 A/cm^2. This cell voltage is much lower than the single-cell open circuit potential (1.01 V) under the same inlet gas conditions. Although the stack as a whole can be made to operate more efficiently by decreasing average current density, this requires much more cell area to achieve the same power and fuel utilisation. Furthermore, cell voltage cannot be increased beyond about 0.90–0.95 V without *also* reducing the fuel utilisation (causing exergy loss in post-stack combustion).

This example illustrates the essential difficulty of addressing cell performance and polarisation without also considering how that cell will be used in a stack. In general, the understanding and optimisation of cell performance requires *some* level of stack modelling, since otherwise it is difficult to translate changes in cell characteristics to changes in overall stack and system performance.

11.5 Subdivision of local overpotential into specific rate processes

For an electron-transfer reaction occurring at a two-phase metal/solution interface, the electrode overpotential has a direct interpretation as the shift in energy between reactants and products involved in the rate-determining step, leading to Butler–Volmer kinetics [2,26]. However, due to the involvement of three separate phases, the macroscopically defined "overpotential" of a SOFC reaction is generally composed of *several different driving forces* related to individual reaction and transport steps occurring locally near the TPB on the micro- or nanoscale. In order to better understand and ultimately engineer materials for improved performance or reduced degradation, it is necessary to understand the subdivision of the overpotential into these more specific driving forces.

For purposes of discussion, it is useful to categorise rate processes contributing to overpotential according to their obeyance of Faraday's Law. Here, we define an "electrochemical" process as that which involves transfer of charge through a phase or across an interface between two phases at a rate proportional to electrode current. In contrast, we define a "chemical" process as that which involves diffusion of neutral combinations of species within (or along the surface) of bulk phases, or reaction among neutral combinations of species at the gas/solid interface. Unlike an electrochemical process, which occurs in lockstep with the faradaic current, chemical processes can occur at a rate asynchronous with the faradaic current via oxidation or reduction of the electrode material under transient conditions.

11.5.1 Chemical contributions to the overpotential

The case of O_2 reduction on silver/YSZ provides a clear example of how chemical driving forces contribute to electrode overpotential in a SOFC electrode reaction. Kleitz and co-workers conducted a series of studies of silver point-contact microelectrodes, made by solidifying small (200–2000 μm) silver droplets onto polished YSZ surfaces [27]. Following *in situ* fabrication, the impedance of these silver microelectrodes were measured as a function of T (600–800 °C), p_{O2} (0.01–1.0 atm) and droplet radius. As an example, Figure 11.6a shows a Nyquist plot of the impedance under one set of conditions, which the authors resolve into two primary components, the largest (most resistive) occurring at very low frequency (0.01–0.1 Hz) and the second smaller component at medium frequencies (10–100 Hz).

Based on quantitative analysis of the impedance timescale as well as other factors, the authors conclude that the low-frequency arc corresponds to the driving force for absorption and/or diffusion of atomic oxygen inside the silver droplet. Their proposed mechanism is illustrated in Figure 11.6b. According to this model, cathodic polarisation results in electrochemical reduction of *atomic* oxygen dissolved in the silver at the silver/electrolyte interface. The resulting depletion of dissolved atomic oxygen near the interface creates a chemical potential driving force for it to diffuse towards the interface from the bulk of the silver. This ultimately leads to depletion of dissolved oxygen everywhere within the silver, including near the silver/gas interface. Depletion of oxygen in the silver relative to equilibrium with the gas then provides a driving force for dissociative absorption of gaseous O_2 into the silver.

With this mechanism, even if electrochemical reduction of atomic oxygen at the Ag/YSZ interface is equilibrated, there will still be a *Nernstian* displacement of the electrode potential from equilibrium due to a shift of the atomic oxygen concentration from equilibrium near the electrolyte interface. This displacement is essentially a *chemical* driving force because it corresponds to a *thermodynamic* potential shift that

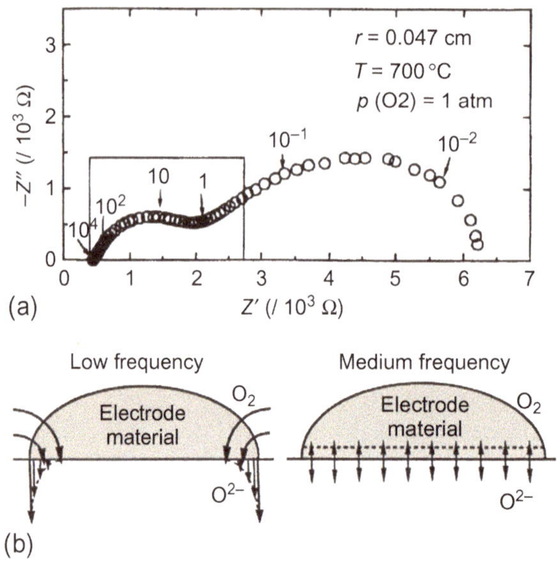

Figure 11.6 (a) Impedance of a single silver droplet solidified on YSZ, measured at 700 °C in oxygen. (b) Interpretation of the impedance in terms of bulk absorption and transport of oxygen in silver.
Figure adopted from Ref. [28] with permission from The Electrochemical Society, Inc.

drives chemical reaction and diffusion of neutral species (or neutral combinations of species within silver).

At steady state, the rate of O_2 reduction and transport within the silver reaches a level which balances the faradaic reduction of atomic oxygen at the silver/YSZ interface. However, under transient conditions, the current is balanced mostly by oxidation/reduction of atomic oxygen in the silver (while the driving force for O_2 absorption is still being established). This lack of synchronicity between chemical rates and the imposed current corresponds to a large *faradaic capacitance*, consistent with the very large observed capacitance of the low-frequency arc in Figure 11.6a (0.1–1 F/cm^2). Further evidence for this model includes both the T and p_{O2} dependence of the resistance and capacitance, which are found to be quantitatively consistent with independently measured oxygen absorption and diffusion properties of silver. Van Herle and McEvoy have subsequently confirmed these findings [29], concluding that the net rate of O_2 reduction at steady state is limited by both chemical absorption of gaseous O_2 into silver as well as the diffusion of the atomic oxygen through the silver.

In addition to this *chemical* driving force, there may also be an *electrochemical* driving force associated with reduction of dissolved atomic oxygen at the Ag/YSZ interface, and incorporation of resulting O^{2-} ions into the electrolyte. Since this ion exchange process occurs at a rate equal to the faradaic current, it is essentially electrochemical, rather than chemical, in nature. Due to double-layer formation at the Ag/YSZ interface, we would expect this faradaic current to be in parallel with a much small interfacial capacitance (10^{-4}–10^{-6} F/cm^2), possibly explaining part of the medium frequency arc shown in Figure 11.6a.

11.5.2 Mixed-conducting SOFC electrodes

In the quest for reversibility of SOFC electrode at intermediate temperatures (<700 °C), workers have often turned to electrocatalysts based on *mixed-conducting oxides* (oxides that can conduct both electrons and oxygen ions) so as to achieve more intimate contact among ions, electrons and gaseous O_2 [30]. An example is the $La_{1-x}Sr_xCo_{1-y}Fe_yO_{3-\delta}$ family of perovskites, which have been heavily studied and in some cases commercially implemented for use in intermediate-temperature SOFC cathodes [31].

Figure 11.7 illustrates some of the processes thought to govern O_2 reduction within a porous mixed-conducting SOFC cathode [32]. This illustration assumes a single-phase electrocatalyst; however, the following discussion also applies to the region near the TPB in a two-phase composite electrode. Oxygen is generally thought to dissociatively adsorb on the mixed-conducting surface (a), forming a monatomic species on the surface or becoming completely reduced and incorporated into the bulk as O^{2-}. The resulting surface intermediates and/or bulk oxygen ions then travel via ambipolar surface and/or bulk diffusion (b, c) to the electrolyte, where they are further reduced to O^{2-} and/or transferred as bulk O^{2-} into the electrolyte (d).

With the exception of the faradaic ion exchange process (d) at the electrode/electrolyte interface, which in practice is often found to be reversible for appropriately chosen materials, the O_2 reduction and transport processes shown in Figure 11.7 involve *chemical driving forces* associated with transport and reaction of molecules or neutral combinations of ions/electrons (analogous to those in silver, as described previously). For this reason, the behaviour of this class of electrode materials is closely tied to their chemical catalytic and transport properties. The following two

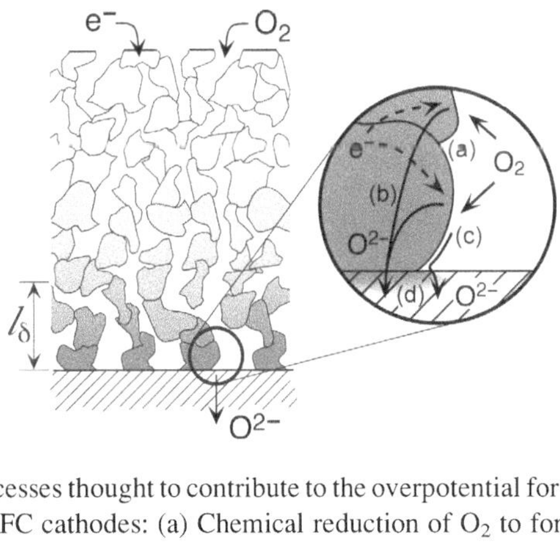

Figure 11.7 Rate processes thought to contribute to the overpotential for O_2 reduction in porous mixed-conducting SOFC cathodes: (a) Chemical reduction of O_2 to form monatomic oxygen species on the surface or within the bulk of the mixed conductor; (b) ambipolar diffusion of oxygen ions (O^{2-}) and charge-compensating electrons/holes in the bulk of the mixed conductor; (c) ambipolar diffusion of monatomic oxygen species (and mirror charges in bulk) along the surface of the mixed conductor; (d) faradaic transfer of bulk/surface oxygen ions (O^{2-}) into the electrolyte at the mixed-conductor/electrolyte interface.

sections go into further detail about the rates of these processes and how they contribute to the measured overpotential in a SOFC cathode.

More traditional SOFC electrode materials (such as $La_{1-x}Sr_xMnO_3$ (LSM) cathodes or Ni anodes) have little or no bulk oxygen ion conductivity, thus the subdivision of the overpotential into chemical and electrochemical driving forces is somewhat murkier. Nonetheless, a growing body of evidence suggests that these materials are governed by a similar set of chemical reaction and surface transport process, just on shorter length scales that are more difficult to probe directly [5,32]. Since the main focus of this chapter is conceptual, the following sections focus primarily on materials with significant bulk involvement, where this subdivision can be made more clearly, only briefly addressing some of the challenges of extending these concepts to surface-bound reactions. For more specifics about factors governing the performance and degradation of particular electrode materials (including LSM and Ni-based electrodes), see Chapters 5 and 6 (Anodes and Cathodes).

11.5.3 Chemical O_2 exchange on $La_{1-x}Sr_xCoO_{3-\delta}$

Over last 20–25 years, workers have put considerable effort into trying to better understand the driving force and rate of chemical exchange on mixed-conducting SOFC electrode materials. Of particular interest (as a model material) has been $La_{1-x}Sr_xCoO_{3-\delta}$ (LSC), which exhibits unusually high bulk vacancy concentration and O_2 exchange rates under oxidising conditions at intermediate temperatures (450–750 °C). Oxygen exchange on bulk freestanding samples of LSC has been studied using a variety of techniques, including step-relaxation measurements [33–42] and isotope tracer techniques [43–46], and, more recently, impedance studies of PLD-fabricated thin-film LSC electrodes where O_2 exchange at the surface dominates resistance, or is separable from other effects via frequency [47–62]. As shown in Figure 11.9a, these films often exhibit nearly perfect low-frequency semicircular impedances, indicating that oxygen transport within the bulk of the film is equilibrated, and thus absorption/desorption of oxygen at the gas/film interface is rate limiting (as confirmed, in some cases, using isotope tracer analysis [48,50]).

To date, most studies of O_2 reduction kinetics on LSC thin films have focused on measurement and analysis of the O_2 *exchange* rate at equilibrium with the gas. More recently, Wilson, Kreller, Adler and co-workers used nonlinear impedance spectroscopy (NLEIS) to help separate the forward and backward rate laws for O_2 exchange on dense thin films of LSC-82 and LSC-64 [62–67]. They modelled the results using various rate laws derived using a non-equilibrium thermodynamic approach (which forces rigorous consistency between the assumed rate law and the underlying thermodynamics of O_2 exchange within LSC). Regardless of Sr doping level, temperature or oxygen partial pressure, they found that LSC closely obeys the following rate law for O_2 exchange with the surface:

$$r = \Re_0\left[\delta^2/\delta_0^2 - e^{-2(A_0-1)\frac{(\delta-\delta_0)}{\delta_0}}\right]; \quad \Re_0 = k(T)p_{O2}^{gas}(\delta_0/3)^2 \tag{11.13}$$

where $\Re_0$ is the equilibrium exchange rate, $k(T)$ is a temperature-dependent rate constant (activation energy ~45 kJ/mol) that also depends on the fabrication and degradation history of the sample, δ is the oxygen nonstoichiometry in bulk of LSC near the surface, δ_0 is the value of δ in equilibrium with the ambient p_{O2}^{gas} and $A_0 = -\frac{1}{2}\partial \ln p_{O2}^{gas}/\partial \ln \delta_0$ is the thermodynamic factor for vacancy exchange (which expresses how far the oxygen vacancy concentration deviates from the dilute solution limit) [4,64]. As shown in Figure 11.8a, this rate law does an excellent job of explaining exchange data on bulk freestanding samples of LSC over a wide range of p_{O2} and Sr content [4]. This rate law is also surprisingly robust to shifts in bulk Sr content near the surface of a thin-film electrode caused by local cation stratification [62,65,66].

For a thin-film electrode at steady state (in which all transport processes are fast), the rate law in Equation (11.13) can be re-expressed in terms of the steady-state current vs. overpotential by considering Faraday's Law and the Nerstian relationship between potential and oxygen vacancy concentration at the electrolyte interface (assuming ion exchange is equilibrated):

$$i = 4F\Re_0\left(\left(\frac{1}{A_0-1}\right)W\left[(A_0-1)e^{(A_0-1)+\frac{2F\eta}{RT}}\right]\right)^2\left(1-e^{-\frac{4F\eta}{RT}}\right) \qquad (11.14)$$

where $W[x]$ is the Lambert W function. Equation (11.14) is plotted in Figure 11.8b for representative values of the thermodynamic factor A_0. Although somewhat similar in

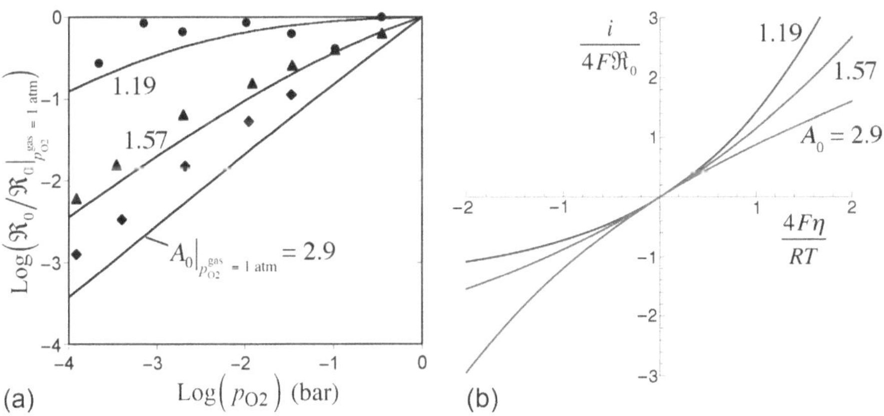

Figure 11.8 (a) Measured equilibrium O_2 exchange rate on bulk freestanding $La_{1-x}Sr_xCoO_{3-\delta}$ at 750 °C (scaled to its value at $p_{O2}=1$ atm.) vs. Sr doping (x) and p_{O2}. Circles, triangles and diamonds correspond to $x=0.2$, 0.5 and 0.7. Calculated curves are model predictions based on Equation (11.13), as a function of the thermodynamic factor A_0 at $p_{O2}=1$ atm. (which is x-dependent). (b) Predicted steady-state $i-\eta$ characteristics of a thin-film electrode based on Equation (11.14), as a function of thermodynamic factor A_0.
(a) Data are from Ref. [41], as analysed in Ref. [4].

appearance to a Butler–Volmer equation for select values of the forward and backward transfer coefficients, Equation (11.14) arises from an entirely different set of physics related to the thermodynamics and chemical kinetics of O_2 exchange, with the overpotential entering through the dependence of thermodynamic potential on δ rather than a shift in electrostatic potential of charged reactants vs. products, as assumed in deriving Equation (11.8) [2].

For a thin-film under transient conditions, oxidation and reduction of the film is expected to cause a significant faradaic capacitance. The relationship of this capacitance to oxygen nonstoichiometry properties of LSC, assuming equilibrium of oxygen transport within the film, has been derived by Kawada and co-workers [51,64]:

$$C_\delta = -\frac{8F^2L}{V_\mathrm{m}}\frac{\mathrm{d}\delta}{\mathrm{d}\mu_{O_2}} = \frac{4F^2\delta_0 L}{V_\mathrm{m}A_0RT} \tag{11.15}$$

where V_m is the molar volume of the perovskite lattice. Equation (11.15) predicts that at low vacancy concentration (when $A_0 = 1$), the capacitance scales linearly with vacancy concentration δ_0 but shows a weakening dependence as $A_0 = 1 + 4\delta_0/g_0RT$ becomes significantly greater than 1 [68]. As exemplified by the comparison in Figure 11.9b, this behaviour is in excellent agreement with the measured capacitance of thin-film LSC electrodes over a wide range of Sr doping, T and p_{O2}, provided

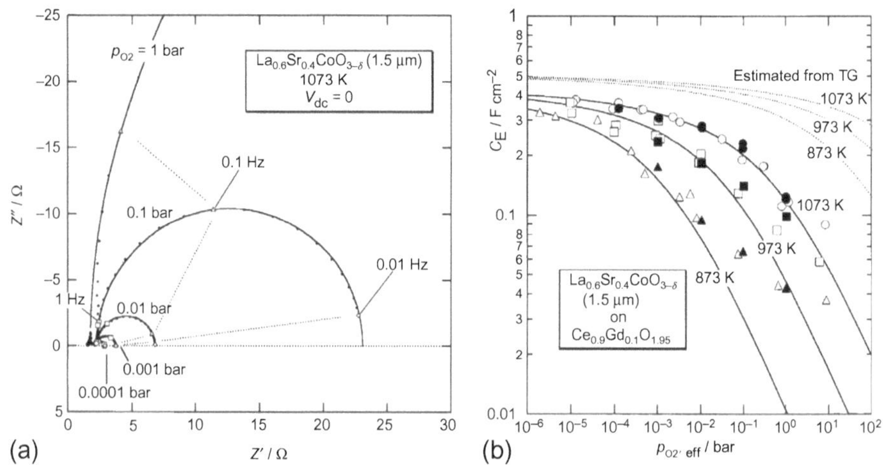

Figure 11.9 (a) Impedance of a 1.5 μm thick oriented thin film of 40% Sr-doped LSC at 800 °C as a function of p_{O2}, with fits to a RC equivalent circuit as shown. (b) Faradaic capacitance extracted from the impedance of the film in (a) as a function of temperature and oxygen fugacity f_{O2}, where f_{O2} was controlled both by changing gas-phase p_{O2} at zero bias (open symbols), as well as by changing bias potential at fixed p_{O2} (closed symbols). Lines show the predicted capacitance based on a validated model for the oxygen exchange thermodynamics of LSC, with a modified oxygen incorporation enthalpy.
Figures reproduced from Ref. [51] with permission from The Electrochemical Society, Inc.

unique deviations in the thermodynamic factor are accounted for. It is currently thought that these deviations may be caused by cation segregation/stratification within the film [62,64,65].

11.5.4 Co-limitation of O_2 exchange and transport in porous $La_{1-x}Sr_xCoO_{3-\delta}$

As discussed previously, and illustrated in Figure 11.7, O_2 reduction in a porous mixed-conducting electrode involves both O_2 reduction on the surface as well as transport of the resulting oxygen ions or surface intermediates to the electrolyte. As shown by Adler, Lane and Steele [69,70], this situation leads to a steady-state O_2 reduction rate which is *co-limited* by both kinetics and transport over a wide range of conditions.

More recently, Lu and co-workers [66,67,71] extended this model to include nonlinear reaction kinetics, as predicted by Equation (11.13), and both bulk and surface transport within/on the surface of LSC assuming a porous electrode comprised of pseudocylindrical electrode particles. This model predicts that the distribution of oxygen vacancies within the mixed conductor (as well as the form of the impedance) will depend on two dimensionless ratios:

$$\kappa = \frac{4\mathfrak{R}_0 V_m r_{cyl}}{\delta_0 D_v} \quad \text{and} \quad \nu = \frac{\Gamma_0 \theta^0_{O_s} D_{O_s} V_m}{r_{cyl} \delta_0 D_v} \tag{11.16}$$

where κ is a ratio of coefficients governing surface kinetics vs. bulk diffusion and ν is a ratio of coefficients governing surface diffusion vs. bulk diffusion [71]. Here, r_{cyl} is the pseudocylindrical radius of the particles, Γ_0 is the molar surface site concentration and $\theta^0_{O_s}$ is the apparent coverage of those sites by surface-mobile oxygen at equilibrium with the ambient $p^{gas}_{O_2}$.

Figure 11.10 shows examples of the vacancy concentration profile within and along the surface of the electrode particles predicted by this model at both high and low frequency. These results show that κ has a strong influence on the size of the active region of the electrode, where the material becomes reduced or oxidised relative to the gas phase. At low values of κ (Figure 11.10a), gradients will extend far from the interface via bulk diffusion, whereas at high κ (Figure 11.10b), gradients remain closely confined to the TPB. At low to moderate values of κ, the size of the active region is further modified by ν since, as shown in Fig 11.10c, surface diffusion allows concentration gradients to extend out along the surface at much higher values of κ than bulk diffusion can when acting alone.

Since both the faradaic capacitance and the kinetically active surface area depend on the size of this active region, one might expect κ and ν to strongly influence the characteristic resistance (R_c) and response frequency (ω_c) of the electrode as measured using impedance. Indeed, as shown in Figure 11.11, the model predicts strong opposing dependencies of R_c and ω_c on κ and ν. These dependencies allow measured values of R_c and ω_c to be interpreted in terms of kinetic and transport properties of the electrode material: $\mathfrak{R}_0$, D_v and surface diffusion coefficient.

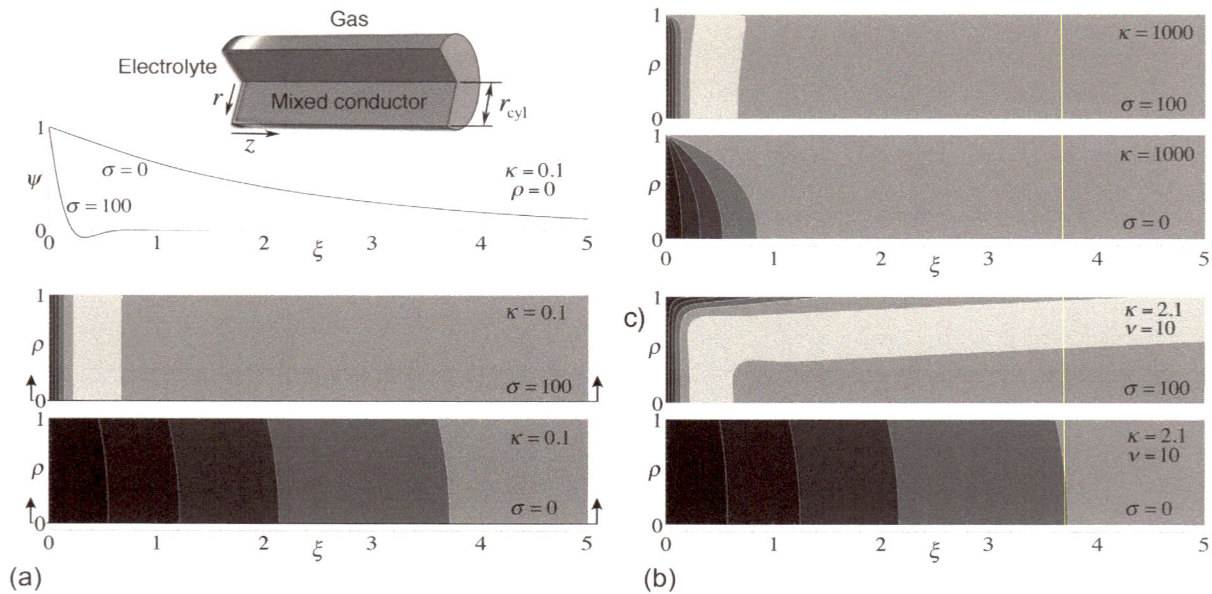

Figure 11.10 Model calculations of the vacancy concentration profile within a cylindrical mixed conductor particle near the three-phase boundary [71]. (a) Assumed electrode geometry, and calculated real part of the vacancy concentration displacement $\psi = \delta - \delta_0/\delta_0$ for $\kappa = 0.1$ (slow kinetics) and $\nu = 0$ (no surface diffusion). The line graph shows the concentration displacement at the centerline of the particle ($r = 0$) at $\sigma = 100$ (high frequency) and $\sigma = 0$ (steady state) as a function of axial position $\xi = z/r_{cyl}$. The contour plots show the 2-D profile at the same frequencies, where $\rho = r/r_{cyl}$ is radial position, and shading indicates $Re[\psi]$. (b, c) Contour plots of the vacancy concentration displacement at the same frequencies as (a), but assuming fast kinetics ($\kappa = 10^3$) or moderate κ with substantial surface diffusion ($\nu = 10$).

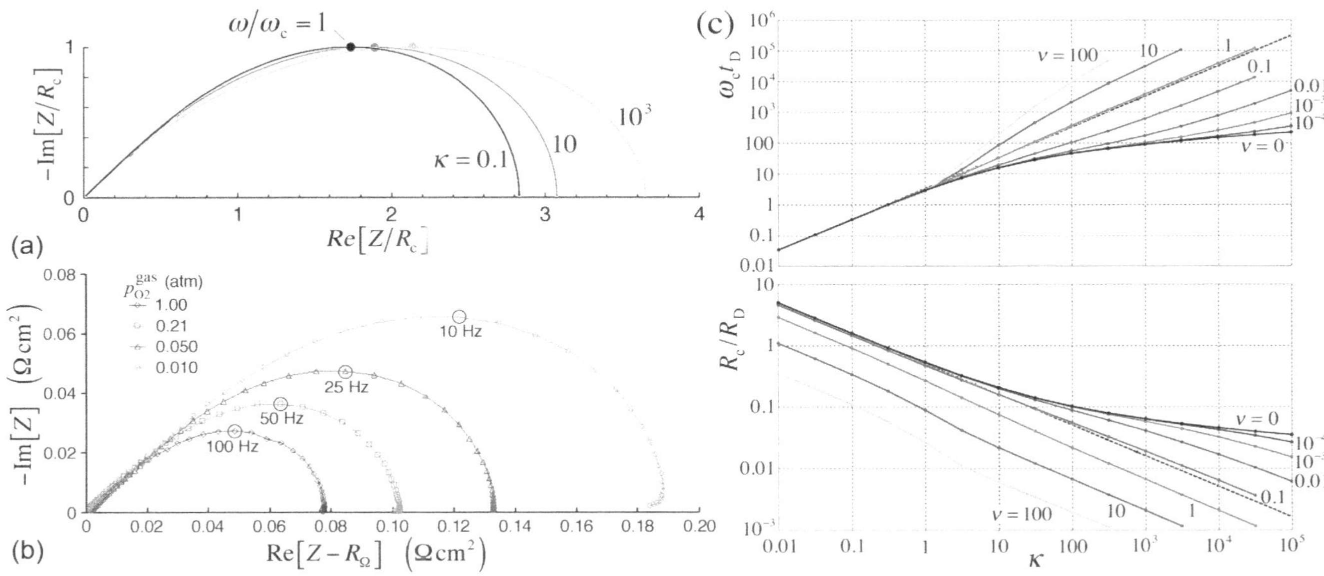

Figure 11.11 (a) Impedance predicted by the model shown in Figure 11.10 for $\nu = 0$ as a function of κ, with amplitude and frequency scaled to characteristic values (R_c and ω_c) corresponding to the point of maximum imaginary amplitude. (b) Measured impedance of a porous $La_{0.6}Sr_{0.4}CoO_{3-\delta}$ microelectrode on Sm-doped ceria at 750 °C, showing how R_c and ω_c vary with p_{O2}. (c) Correlated dependency of R_c and ω_c on κ and ν, as predicted by the model.
Data and model from Ref. [71].

Fits of this model to the measured characteristic resistance and frequency of porous LSC electrodes yields values for the apparent surface exchange rate, $\Re_0 = k(T)p_{O2}^{gas}(\delta_0/3)^2$, bulk oxygen vacancy diffusivity D_v and apparent surface diffusion coefficient. For 40% Sr-doped LSC, these fits suggest surface diffusion plays only a minor role above 650 °C. As shown in Figure 11.12a and b, the extracted values of the kinetics agree well with the predictions of Equation (11.13), while the values of D_v agree within a factor of 2 of independent measurements of bulk ion diffusion. In contrast, for 20% Sr-doped LSC, which has 1–2 orders of magnitude fewer vacancies than 40% LSC under the same conditions, fits to the model show good agreement with Equation (11.13) but suggest surface diffusion plays an increasingly important role at lower temperatures and/or higher p_{O2}, with ν spanning 0.1–10. If the extracted values of ν are interpreted in terms of a mobile adatom or interstitial species (Figure 11.11d), the product of the equilibrium constant for interstitial formation (K_I) and surface diffusion coefficient (D_I) appears to be independent of p_{O2}, with high activation energy (~240 kJ/mol). More recent results using nonlinear impedance on porous 20% and 40% doped LSC electrodes further reinforce these same conclusions [66].

The main take-home message from these studies on thin-film and porous LSC electrodes is that the overall rate of electrochemical O_2 reduction is determined locally by *chemical catalytic* and *diffusive transport* processes within/on the surface of the electrode material itself (not by an additional electrochemical process localised to the solid/solid interface). However, this does not imply that the nature of the interface and its history are not important factors in determining electrode performance. A growing body of evidence shows that the properties of mixed-conducting surfaces are extremely sensitive to subtle changes in structure and composition, many of which

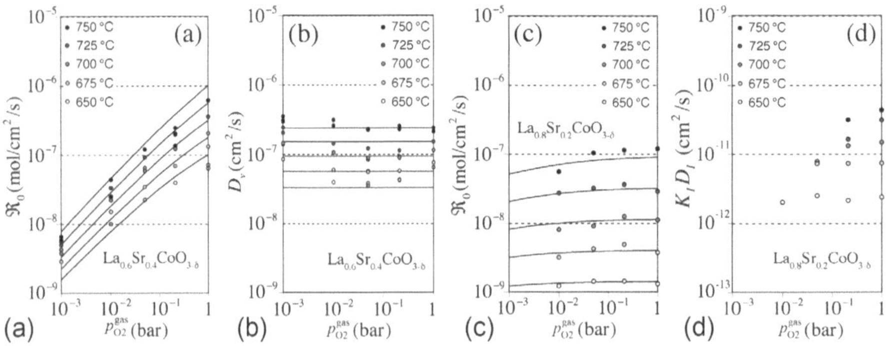

Figure 11.12 (a, b) Values of the surface exchange rate (a) and bulk oxygen vacancy diffusion coefficient (b) of 40% Sr-doped LSC, extracted from measured impedance characteristics R_c and ω_c of a porous electrode using the model results presented in Figure 11.11c. Lines through the data in (a) correspond to the predictions of Equation (11.13). Lines through the data in (b) are independently measured values of D_v from Ref. [72]. (c, d) Values of the surface exchange rate (c) and surface diffusion properties (d) of 20% Sr-doped LSC, similarly extracted from impedance data under the assumption that D_v is similar in 20% and 40% doped LSC.
Data and modelling results are from Ref. [71].

seem to driven by interactions with the gas or adjacent electrolyte [58–62,67,73,74]. It is likely that these same interactions are a major factor in the performance and degradation of porous electrodes between 100 and 10,000 h [67].

11.5.5 Difficulties in subdividing the overpotential with surface-bound reactions

The examples of Ag and LSC cathodes discussed above provide clear examples of electrode systems primarily limited by chemical catalytic and ambipolar transport processes. However, for many electrode materials of practical importance in SOFCs, subdivision of the overall reaction into individual chemical and electrochemical steps is not nearly as clear. In the extreme case of a surface-bound reaction (where no bulk transport can occur), one might expect ion transfer at the electrode/electrolyte interface to become increasingly resistive simply due to the constriction of current to a small geometric area in the vicinity of the TPB (path a–c–d in Figure 11.7). One clear example is porous Pt on YSZ, which (after correction for double-layer capacitance) appears to be limited by both chemical catalytic processes (reduction of O_2 on the Pt surface and diffusion of O_{ads} to the TPB), as well as passage of a faradaic current in a narrow region near the TPB at the electrode/electrolyte interface [32].

One of the most heavily studied and commercially relevant SOFC cathode materials, $La_{1-x}Sr_xMnO_3$ (LSM), has so far eluded attempts to break the overpotential down into specific chemical and electrochemical contributions. Although under extremely reducing conditions, LSM is known to reduce oxygen via a bulk path (similar to LSC) [32], under the relatively oxidising conditions and low overpotential of a commercial SOFC cathode, the bulk ionic conductivity of LSM is almost negligible. Indeed, calculation of the performance of LSM as a cathode based on its O_2 exchange properties (under the assumption of a bulk transport path) under-predicts its performance as a porous cathode by many orders of magnitude [69]. Thus surface transport on LSM almost certainly plays a critical role in its activity as an electrode. However, deeper interpretation of the electrochemical characteristics of LSM is hampered by a number of factors. One challenge is that under oxidising conditions, LSM exhibits significant cation nonstoichiometry and thus tends to accrete and de-accrete lattice sites at the surface upon oxidation and reduction, leading to a high level of dispersion and hysteresis in electrochemical characteristics [75]. LSM also tends to react with the YSZ electrolyte, leading to non-uniform interfacial resistance (to ion transfer) that broadly overlaps the timescale of chemical catalytic and transport processes occurring on the surface or near-surface bulk [73,76].

Another commercially relevant and well-studied electrode system is Ni on YSZ, commonly used as the anode in hydrogen- and CO-based SOFCs. As recently reviewed by Bessler and co-workers [5,77], hydrogen and CO oxidation at the Ni/YSZ interface is quite complex and appears to involve chemical reaction and transport steps occurring on both the Ni and YSZ surfaces very close to the TPB, making the contributions of these steps to the overpotential very difficult to separate from each other or from charge-transfer steps occurring at the Ni–YSZ interface. In some cases, workers have attempted to accomplish this separation by comparing measured electrochemical kinetics (mostly $i-\eta$ characteristics at moderate to high overpotential in both directions) to extremely detailed elementary kinetic and transport models of the Ni and YSZ surfaces. These models appear to support the view that H_2 and CO dissociation occurs mainly on the Ni surface, but that water desorption occurs primarily from the YSZ surface [5].

These examples illustrate some of the challenges in fully interpreting the overpotential when reaction and transport steps are confined to material surfaces near the TPB on nanometer length scales. Hopefully in the future, new experimental techniques will become available that can spatially resolve dynamics on these length scales, thereby shedding further light on the factors dominating the overpotential in common SOFC electrode reactions.

11.6 Conclusions and outlook

Twenty-five years ago, SOFC electrodes were largely described empirically, with an overpotential associated with the overall reaction occurring at/near the three-phase boundary. The work described above illustrates the extent to which we have been able to break down this overpotential into individual reaction and transport steps, in some cases quantitatively.

However, many barriers remain to our deeper understanding of SOFC electrode reactions. First, a growing body of research suggests that the key gas/solid and solid/solid interfaces involved in the reaction have compositions and structures that deviate substantially from the bulk, and which are highly sensitive to the exact composition and process history of the electrode. In order to understand electrode reactions at a deeper level, we need a new set of experimental and modelling techniques that provide much more localised information (0.1–100 nm).

Second, with so many important rates being in parallel, such as bulk and surface transport or rates on multiple surfaces, it is very difficult to directly link global aggregate $i-V$ characteristics to localised structure/composition measured *ex situ*. It is imperative for us to develop a new set of *in operando* imaging techniques that can temporospatially correlate global $i-V$ perturbations to *local* perturbations in composition and electrical state. If we had a way to *visualise* where the driving force is manifest, we would be in a much better position to identify what factors are controlling rates.

Finally, there are currently many anodes and cathode materials having initial performance far exceeding technological performance targets. The challenge over the next 25 years will be understanding degradation mechanisms and mitigating sensitivity of SOFC electrodes to environmental factors that are strongly tied to SOFC stack and system cost. Although these are challenges faced by developers for decades, further success will require those working on more fundamental aspects of SOFCs to shift their focus from initial performance under idealised conditions to what happens to these electrode materials over time in realistic environments.

References

[1] S.B. Adler, P. Holtappels, Course notes from "Introduction to Solid Oxide Fuels", in: A Short Course Presented at the 219th Meeting of the Electrochemical Society, May 1, 2011, Canada, The Electrochemical Society, Inc., Pennington, NJ, Quebec, 2011.

[2] J.S. Newman, K.E. Thomas-Alyea, Electrochemical Systems, Wiley, Hoboken, NJ, 2004 (Chapter 2).

[3] S.B. Adler, unpublished data (1997).

[4] S.B. Adler, X.Y. Chen, J.R. Wilson, J. Catal. 245 (2007) 91–109.

[5] S.B. Adler, W.G. Bessler, Elementary kinetic modeling of SOFC electrode reactions, in: H. Yokokawa (Ed.), Fuel Cell Handbook: Fundamentals Technology and Applications, John Wiley and Sons, New York, 2009.

[6] J. Newman, W. Tiedemann, AICHE J. 21 (1975) 25–41.

[7] T. Kenjo, S. Osawa, K. Fujikawa, J. Electrochem. Soc. 138 (1991) 349–355.

[8] T. Kenjo, M. Nishiya, Solid State Ionics 57 (1992) 295–302.

[9] C.W. Tanner, K.Z. Fung, A.V. Virkar, J. Electrochem. Soc. 144 (1997) 21–30.

[10] V. Dusastre, J.A. Kilner, Solid State Ionics 126 (1999) 163–174.

[11] A.V. Virkar, J. Chen, C.W. Tanner, J.W. Kim, Solid State Ionics 131 (2000) 189–198.

[12] J.D. Kim, G.D. Kim, J.W. Moon, H.W. Lee, K.T. Lee, C.E. Kim, Solid State Ionics 133 (2000) 67–77.

[13] J.-D. Kim, G.-D. Kim, J.-A. Park, Han'guk Seramik Hakhoechi 37 (2000) 227–232.

[14] M.J. Jorgensen, S. Primdahl, C. Bagger, M. Mogensen, Solid State Ionics 139 (2001) 1–11.

[15] M. Mogensen, S. Primdahl, M.J. Jorgensen, C. Bagger, J. Electroceram. 5 (2000) 141–152.

[16] F.M. Figueiredo, J.R. Frade, F.M.B. Marques, Solid State Ionics 135 (2000) 463–467.

[17] F.M. Figueiredo, F.M.B. Marques, J.R. Frade, J. Electroceram. 7 (2001) 47–56.

[18] N.T. Hart, N.P. Brandon, M.J. Day, N. Lapena-Rey, J. Power Sources 106 (2002) 42–50.

[19] C. Xia, W. Rauch, W. Wellborn, M. Liu, Electrochem. Solid-State Lett. 5 (2002) A217–A220.

[20] R.J. Kee, H. Zhu, D.G. Goodwin, Proc. Combust. Inst. 30 (2005) 2379–2404.

[21] H. Zhu, R.J. Kee, V.M. Janardhanan, O. Deutschmann, D.G. Goodwin, J. Electrochem. Soc. 152 (2005) A2427–A2440.

[22] G.K. Gupta, E.S. Hecht, H. Zhu, A.M. Dean, R.J. Kee, J. Power Sources 156 (2006) 434–447.

[23] H. Zhu, R.J. Kee, J. Electrochem. Soc. 153 (2006) A1765–A1772.

[24] H. Zhu, R.J. Kee, M.R. Pillai, S.A. Barnett, J. Power Sources 183 (2008) 143–150.

[25] S.C. DeCaluwe, H. Zhu, R.J. Kee, G.S. Jackson, J. Electrochem. Soc. 155 (2008) B538.

[26] K.J. Vetter, Electrochemical Kinetics, Theoretical and Experimental Aspects, Academic Press, Waltham, 1967.

[27] M. Kleitz, L. Dessemond, T. Kloidt, M.C. Steil, in: M. Dokiya, O. Yamamoto, H. Tagawa, S.C. Singhal (Eds.), Solid Oxide Fuel Cells IV, The Electrochemical Society, Inc., Pennington, 1995, pp. 527–536.

[28] R. Jimenez, T. Kloidt, M. Kleitz, J. Electrochem. Soc. 144 (1997) 582–585.

[29] J. van Herle, A.J. McEvoy, J. Phys. Chem. Solids 55 (1994) 339–347.

[30] B.C.H. Steele, Solid State Ionics 134 (2000) 3–20.

[31] N.P. Brandon, S. Skinner, B.C.H. Steele, Annu. Rev. Mater. Res. 33 (2003) 183–213.

[32] S.B. Adler, Chem. Rev. 104 (2004) 4791–4843.

[33] S. Sunde, K. Nisancioglu, T.M. Gur, J. Electrochem. Soc. 143 (1996) 3497–3504.

[34] J.E. ten Elshof, M.H.R. Lankhorst, H.J.M. Bouwmeester, J. Electrochem. Soc. 144 (1997) 1060–1067.
[35] M.H.R. Lankhorst, H.J.M. Bouwmeester, J. Electrochem. Soc. 144 (1997) 1261–1267.
[36] S. Diethelm, A. Closset, J. Van Herle, A.J. McEvoy, K. Nisancioglu, Solid State Ionics 135 (2000) 613–618.
[37] R. Bredesen, F. Mertins, T. Norby, Catal. Today 56 (2000) 315–324.
[38] J.A. Lane, J.A. Kilner, Solid State Ionics 136 (2000) 997–1001.
[39] S. Wang, A. Verma, Y.L. Yang, A.J. Jacobson, B. Abeles, Solid State Ionics 140 (2001) 125–133.
[40] W. Preis, E. Bucher, W. Sitte, J. Power Sources 106 (2002) 116–121.
[41] L.M. van der Haar, M.W. den Otter, M. Morskate, H.J.M. Bouwmeester, H. Verweij, J. Electrochem. Soc. 149 (2002) J41–J46.
[42] J. Yoo, A. Verma, S.Y. Wang, A.J. Jacobson, J. Electrochem. Soc. 152 (2005) A497–A505.
[43] R.J. Chater, S. Carter, J.A. Kilner, B.C.H. Steele, Solid State Ionics 53–56 (1992) 859–867.
[44] R.H.E. van Doorn, I.C. Fullarton, R.A. de Souza, J.A. Kilner, H.J.M. Bouwmeester, A.J. Burggraaf, Solid State Ionics 96 (1997) 1–7.
[45] R.A. De Souza, R.J. Chater, Solid State Ionics 176 (2005) 1915–1920.
[46] R.A. De Souza, Phys. Chem. Chem. Phys. 8 (2006) 890–897.
[47] K. Masuda, T. Kawada, A. Kaimai, K. Kawamura, Y. Nigara, J. Mizusaki, H. Yugami, H. Arashi, in: U. Stimming (Ed.), 5th International Symposium on Solid Oxide Fuel Cells, Electrochemical Society, Achen, Germany, 1997, p. 473.
[48] T. Kawada, K. Masuda, J. Suzuki, A. Kaimai, K. Kawamura, Y. Nigara, J. Mizusaki, H. Yugami, H. Arashi, N. Sakai, H. Yokokawa, Solid State Ionics 121 (1999) 271–279.
[49] Y.L. Yang, C.L. Chen, S.Y. Chen, C.W. Chu, A.J. Jacobson, J. Electrochem. Soc. 147 (2000) 4001–4007.
[50] Y. Yang, A.J. Jacobson, C.L. Chen, G.P. Luo, K.D. Ross, C.W. Chu, Appl. Phys. Lett. 79 (2001) 776–778.
[51] T. Kawada, J. Suzuki, M. Sase, A. Kaimai, K. Yashiro, Y. Nigara, J. Mizusaki, K. Kawamura, H. Yugami, J. Electrochem. Soc. 149 (2002) E252–E259.
[52] M. Sase, D. Ueno, K. Yashiro, A. Kaimai, T. Kawada, J. Mizusaki, J. Phys. Chem. Solids 66 (2005) 343–348.
[53] F.S. Baumann, J. Fleig, H.-U. Habermeier, J. Maier, Solid State Ionics 177 (2006) 1071–1081.
[54] F.S. Baumann, J. Fleig, G. Cristiani, B. Stuhlhofer, H.U. Habermeier, J. Maier, J. Electrochem. Soc. 154 (2007) B931–B941.
[55] F.S. Baumann, J. Maier, J. Fleig, Solid State Ionics 179 (2008) 1198–1204.
[56] R. Oike, K. Amezawa, T. Nakamura, Y. Tamenori, K. Yashiro, T. Kawada, "Development of in situ soft X-ray absorption spectroscopic technique under high temperature and controlled atmosphere", Solid State Ionics 262 (2014) 911–913.
[57] G.J. La O, S.J. Ahn, E. Crumlin, Y. Orikasa, M.D. Biegalski, H.M. Christen, Y. Shao-Horn, Angew. Chem. Int. Ed. 49 (2010) 5344–5347.
[58] M. Kubicek, A. Limbeck, T. Fromling, H. Hutter, J. Fleig, J. Electrochem. Soc. 158 (2011) B727–B734.
[59] E. Mutoro, E.J. Crumlin, M.D. Biegalski, H.M. Christen, Y. Shao-Horn, Energy Environ. Sci. 4 (2011) 3689–3696.
[60] E.J. Crumlin, S.J. Ahn, D. Lee, E. Mutoro, M.D. Biegalski, H.M. Christen, Y. Shao-Horn, J. Electrochem. Soc. 159 (2012) F219–F225.

[61] B.J. Ingram, J.A. Eastman, K.C. Chang, S.K. Kim, T.T. Fister, E. Perret, H. You, P.M. Baldo, P.H. Fuoss, Appl. Phys. Lett. 101 (2012) 051603.
[62] C.R. Kreller, T.J. McDonald, S.B. Adler, E.J. Crumlin, E. Mutoro, S.J. Ahn, G.J. La O, Y. Shao-Horn, M.D. Biegalski, H.M. Christen, R.J. Chater, J.A. Kilner, J. Electrochem. Soc. 160 (2013) F931–F942.
[63] J.R. Wilson, D.T. Schwartz, S.B. Adler, Electrochim. Acta 51 (2006) 1389–1402.
[64] J.R. Wilson, M. Sase, T. Kawada, S.B. Adler, Electrochem. Solid-State Lett. 10 (2007) B81–B86.
[65] T.J. McDonald, C.R. Kreller, S.B. Adler, in preparation (2015).
[66] C.R. Kreller, S.B. Adler, in preparation (2015).
[67] Y.X. Lu, C.R. Kreller, S.B. Adler, J.R. Wilson, S.A. Barnett, P.V. Voorhees, H.-Y. Chen, K. Thornton, J. Electrochem. Soc. 161 (4) (2014) F561–F568.
[68] S.B. Adler, in: T.A. Ramanarayanan, W.L. Worrell, H.L. Tuller, M. Mogensen, A.C. Khandkar (Eds.), Ionic and Mixed Conducting Ceramics III, The Electrochemcial Society, Paris, France, 1997, pp. 539–549.
[69] S.B. Adler, J.A. Lane, B.C.H. Steele, J. Electrochem. Soc. 143 (1996) 3554–3564.
[70] S.B. Adler, Solid State Ionics 111 (1998) 125–134.
[71] Y.X. Lu, C.R. Kreller, S.B. Adler, J. Electrochem. Soc. 156 (2009) B513–B525.
[72] M.H.R. Lankhorst, in: Univerdsity of Twente, Enschede, Netherlands, 1997.
[73] A. Mitterdorfer, L.J. Gauckler, Solid State Ionics 111 (1998) 185–218.
[74] J.H. Joo, R. Merkle, J. Maier, J. Power Sources 196 (2011) 7495–7499.
[75] T.T. Fister, D.D. Fong, J.A. Eastman, P.M. Baldo, M.J. Highland, P.H. Fuoss, K.R. Balasubramaniam, J.C. Meador, P.A. Salvador, Appl. Phys. Lett. 93 (2008) 151904.
[76] A. Mitterdorfer, L.J. Gauckler, High temperature electrochemistry: ceramics and metals, in: Proceedings of the Risoe International Symposium on Materials Science, 17th, Roskilde, Den., September 2–6, 1996, 1996, pp. 357–362.
[77] M. Vogler, A. Bieberle-Hütter, L.J. Gauckler, J. Warnatz, W.G. Bessler, J. Electrochem. Soc., 156 (5) (2009) B663–B672.

Testing of electrodes, cells and short stacks

12

T. Li[1], W. Wang[2]
[1]University of Electronic Science and Technology of China, Chengdu, Sichuan, China,
[2]Chinese Academy of Sciences, Ningbo, China

12.1 Introduction

This chapter describes electrochemical testing of individual electrodes and cells and gives a brief discussion of stack testing, starting from the original chapter [1] written by Mogensen and Hendriksen and then inserting many more recent results since 2003. It is extremely important in testing solid oxide fuel cells (SOFCs) to have a good understanding of the fundamental principles of electrochemistry; to this end, classical textbooks [2,3], which are based on liquid electrochemistry, are certainly helpful, but it is also extremely important to be familiar with different aspects of solid-state electrochemistry, especially those involving solid oxide electrolytes [4–12]. This chapter first describes the testing of electrodes, particularly in relation to reference electrodes. Then it goes on to cell and stack testing, especially defining the ASR (area-specific resistance) and considering large area cells. Finally, the effect of various fuels on cell testing is described.

Testing of cells and stacks is conducted for two reasons: to assess their commercial viability and for continued cell development. The information required for these two purposes is usually different, and an interpretation of the data, taking into account the effects of the test conditions, has to be done for comparison of SOFCs from different sources. Despite the fact that many cell test data have been reported [13–16], no general agreement on test procedures exists and the actual test equipment is often not described in detail. Also, the concept of ASR is not standardised when reporting the test results. Although parameters such as temperature, inlet gas composition, fuel utilisation and current density are normally given, substantial additional information is required for complete evaluation of test data and detailed analysis of cell behaviour. Four particular issues that are considered in this chapter are: definition of ASR, accuracy of temperature measurement at the cell, the effects of gas leakage through the seals and the use of reference electrodes.

The techniques and procedures for testing individual stacks should be detailed more carefully to evaluate the performance and stability. Such tests of a stack containing up to a dozen cells are both expensive and complex. There are many factors that determine the results and tests on stacks, and more care is needed to understand such widely performed tests during recent years [17–24].

High-temperature Solid Oxide Fuel Cells for the 21st Century. http://dx.doi.org/10.1016/B978-0-12-410453-2.00012-9

If an interconnect material is used for current collection from the electrodes in a single-cell test, then the test is sometimes referred to as a 'short stack', It should be noted that the contact resistance between the electrodes and the interconnect is usually significant [24,25], especially in cases where the interconnect is made of stainless steel or other chromia forming alloys, which have a tendency to form poorly conducting surface layers [17,26].

There are special problems in SOFC electrode testing. An electrode can only be characterised electrochemically if it is part of a cell with at least one reference electrode, that is, a cell with at least three electrodes in contact with the same electrolyte. However, more efforts have been made to incorporate a reference electrode into a thin electrolyte film in an electrode-supported cell; this is discussed in the next section.

A recent trend is that some fuels like natural gas, coal syngas and biogas have been paid increasing attention due to their abundant sources in nature and convenient production and storage, and so various tests have been conducted using these complex gases or simulated syngas as fuels fed to SOFCs. Among these tests, researchers focus on two issues: carbon deposition and impurities in fuel poisoning effects. To reveal these mechanisms by these sorts of tests will be highly helpful to advance the SOFC technology; for one thing, making it run at low cost; for another, laying the paths for developing new anode materials which are expected to tolerate carbon corrosion and impurities contamination.

12.2 Testing electrodes

The main problem with characterising the individual electrodes in single cells and short stacks is the insertion of the reference electrodes, which are used to judge the performance of the individual components and interfaces in the cell. Since a single-cell stack is made up of five components, an electrolyte, two electrodes and two interconnects, there are four interfaces at which reference electrodes can be inserted. Unfortunately, such reference electrodes will not work in the geometries that are normally employed.

Results of tests on cells with thin electrolyte layers using one or more reference electrodes have been reported on many occasions [27,28], but the electrodes under investigation appeared not to be polarised; the short explanation is that the reference electrodes were not working. The geometric requirements for the position of electrodes in three-electrode set-ups have been treated in detail by several researchers [29–31], and there is general agreement among these researchers. In spite of this, there still seems to be a great need in the SOFC community for basic information on how to measure electrode potentials properly; some of these details are given below.

Figures 12.1 and 12.2 illustrate the problem. An electrode-supported cell with a 'reference' electrode is often sketched as shown in Figure 12.1a. However, such a sketch is very deceiving when it is used for an assessment of the current distribution. For this purpose, the sketch should be drawn to scale, that is, the electrolyte thickness should be the relevant unit of length. When the correct length scale is used, as in Figure 12.1b, it is evident that the gap between the upper working electrode and the 'reference' electrode is huge. This means that the current distribution around

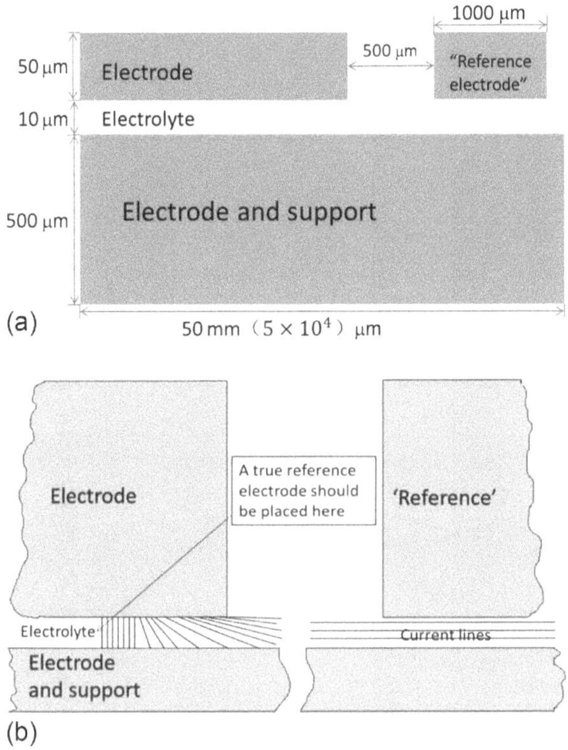

Figure 12.1 (a) A typical sketch of an electrode-supported cell. Note that such an illustration is out of scale because the gap between the top electrodes should be 50 times greater than the electrolyte thickness. (b) Expanded view of electrode corners showing the current distribution indicated by schematic current lines. The current density, apart from being approximately parallel to the electrolyte plane, is very small at the position of the 'reference' electrode, at least 50 (500/10 μm) times smaller than the current density of the cell. The correct position of the reference electrode would be inside the electrolyte of the cell and some distance away from the corner, but this is difficult in a 10 μm thick electrolyte of the 'reference' electrode becomes parallel to the electrolyte plane, that is, there is only a minute voltage difference across the electrolyte at the 'reference' electrode position.

the right-hand edge of the working electrode becomes very different from the even current distribution in the main part of the cell. Furthermore, the current in the vicinity

Figure 12.2a illustrates the potential across a cell in the open circuit voltage (OCV) condition. Note, that for a good electrolyte (ionic conduction only), there will be no potential gradient inside the electrolyte at zero current. The whole potential change across the cell is localised at the interfaces between the electrolyte and the electrodes. These regions are the so-called electrochemical double layers with thickness in the nanometer range and with high space charge concentrations as a result of the very high potential gradients. Figure 12.2b gives the potential across a cell when it is loaded, that is, a current flows through it.

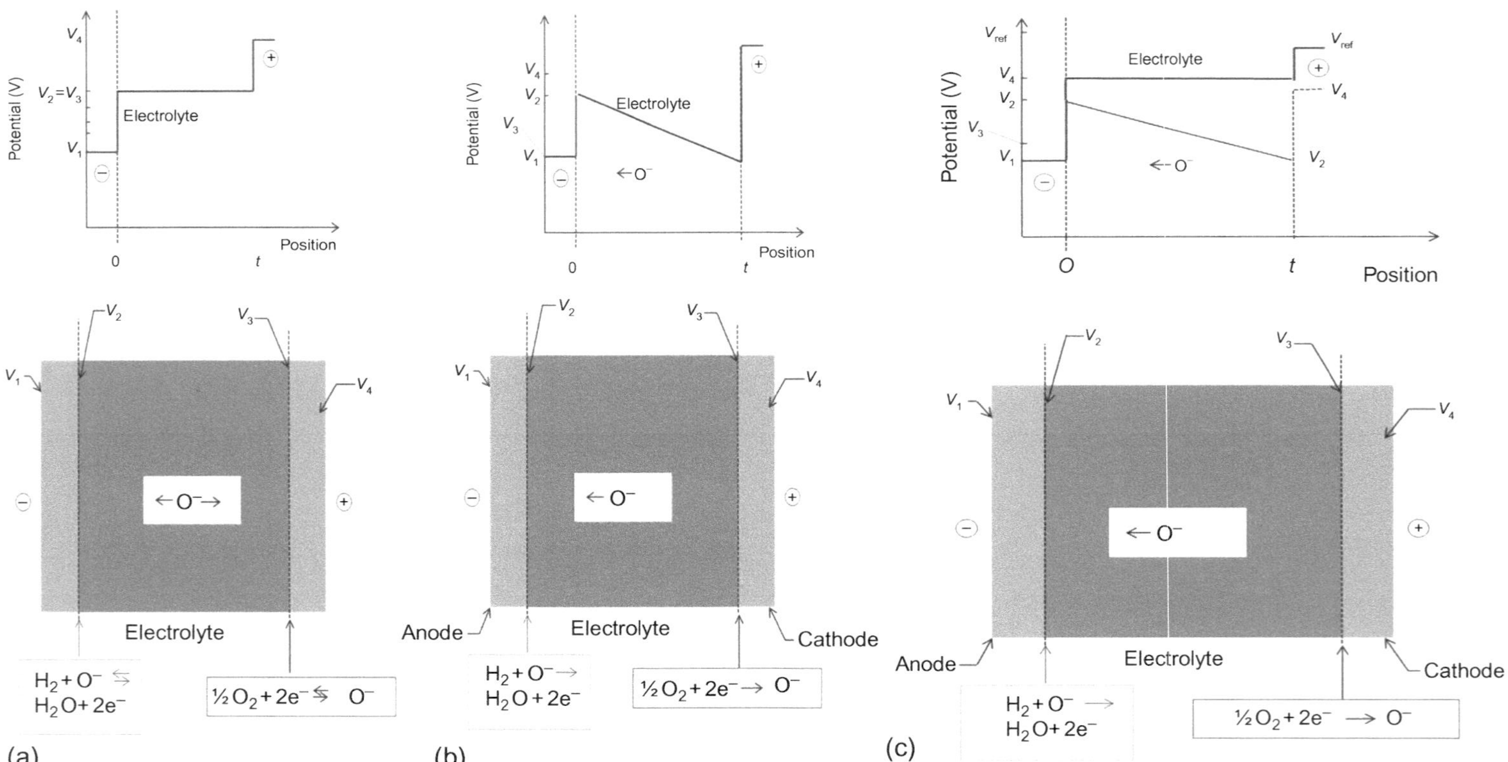

Figure 12.2 (a) Potential through the electrode-supported cell with no current and (b) through a cell with a current load. (c) The potential across the electrolyte at the 'reference' electrode position (thick line) and through the cell part with the current load (thin line). It is seen that: $(V_{\mathrm{ref}} - V_4)/i = R_{\mathrm{p}}$ anode $+ R_{\mathrm{p}}$, cathode $+ R_{\mathrm{elyt}} =$ the total polarisation of the cell apart from concentration polarisation.

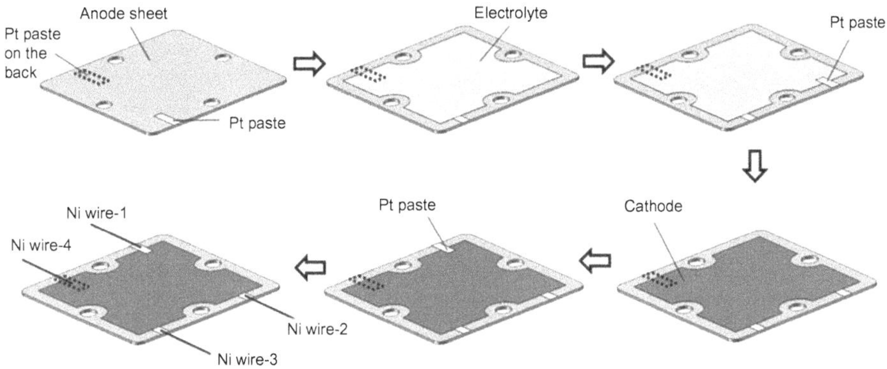

Figure 12.3 Schematic diagram of the embedment of Pt voltage probes. Embedding internal voltage probes into the TPB: the unit cell used in the experiments was a 10 × 10 cm anode-supported NiO–YSZ/YSZ/LSCF–GDC, with an active electrode area of 63 cm^2. Pt voltage probes were screen printed at the electrode/electrolyte TPB within the cell during manufacturing. The schematic process consists of six procedures of embedding Pt voltage probes into the TPB within the cell [31].

There are some attempts like embedding Pt voltage probes into the triple-phase boundary of a commercial single cell as shown in Figure 12.3. The Pt paste is placed on both sides of the electrolyte to achieve *in situ* measurement of voltage drop from anode, cathode and electrolyte. As discussed, this Pt tape can still not be regarded as a true reference electrode, but note that the Pt strap located between electrolyte and anode layer is polarised and actually the potential differences can be determined by measuring the current–voltage curves or impedance spectra.

Note that there is a potential loss across the electrolyte due to the electrolyte resistance. The potential steps at the interfaces are now smaller compared to the OCV condition due to the losses originating from the polarisation resistance of the electrode processes. Figure 12.2c illustrates the case of an anode-supported cell. The potential is given for both the position of the 'reference' electrode where no current flows across the electrolyte (i.e. the electrical potential is constant across the electrolyte), and for a position far away from the electrode edges as in Figure 12.2b.

As the potential along the anode, which is a very good electronic conductor, is the same everywhere (the anode constitutes an isopotential plane), the potential of the anode at the 'reference' electrode must be equal to the potential in the middle of the current-bearing part of the anode. Thus the two potential curves in Figure 12.2c must start at the same point. Therefore, as seen in Figure 12.2c, the potential difference between the 'reference' electrode and the upper electrode in Figure 12.1 is simply the total polarisation of the full cell. Thus it is clear that the 'reference' electrode measures only the emf of the cell with the actual gas compositions at the 'reference' and inside the support at the lateral position opposite to the reference whilst the current is flowing. Thus, no information on what happens on any of the working electrodes can be derived from such measurements. If the concentrations of the reactants and products at the lateral position of the reference electrode were the same as in the active electrode/electrolyte interfaces, then it would be possible to deduce the total concentration overpotential. This is, however, in general not the case, and this means that the voltage difference actually measured between the working and the 'reference' electrodes cannot be assigned any clear meaning. In the present context, it is also helpful to remember that a single-electrode potential cannot be measured directly: it is only possible to measure a potential difference between two electrodes.

To avoid such problems, it is necessary to test the electrodes using a suitable three-electrode set-up or a symmetrical two-electrode cell, even though both have their shortcomings. Some examples of useful set-ups for studying electrode performance are briefly presented here. The set-ups are based on zirconia pellets with an electrode arrangement suitable for three-electrode studies. Specific material choices and mounting details are given in the caption of Figure 12.3. Such pellet-like geometries, where the reference electrode can be suitably placed (in a bore as in the figure or as a ring around the pellet) are suitable for fundamental studies of electrode kinetics. The pellet-like test cell geometries depicted in Figure 12.4 suffer from two disadvantages: it is difficult to ensure that the fabrication process for the electrodes used is identical to the one used for the actual cells, and the ohmic resistance between the working and the reference electrodes is quite substantial which may result in a 'signal-to-noise' problem when very good electrodes are studied. An improved three-electrode geometry using a ring-shaped working electrode has been reported [30].

To measure a particular electrode performance in detail, a symmetrical cell with identical electrodes on each side can be used as shown in Figure 12.5. This has a

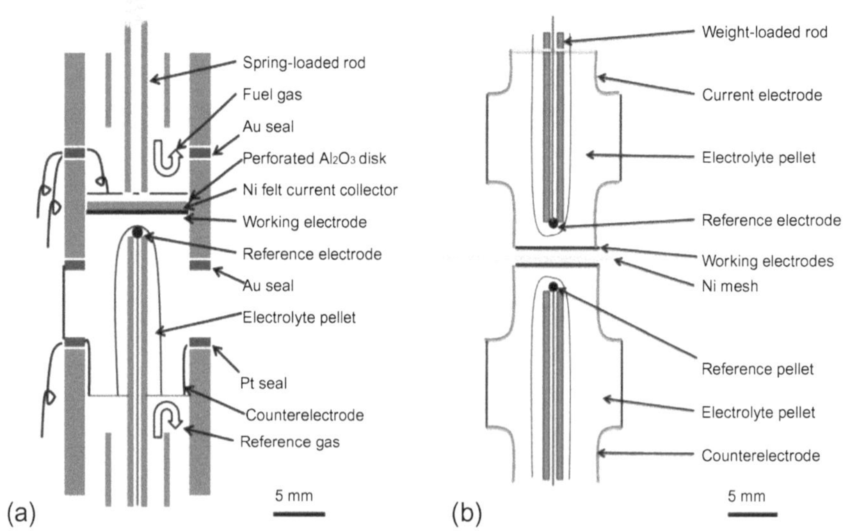

Figure 12.4 Possible three-electrode set-ups based on a cylindrical YSZ electrolyte pellet. The pellet is mounted on an alumina tube. (a) and (b) illustrate material choices and details suitable for anode studies and (c) and (d) like wire for cathode studies. Using two three-electrode pellets pressed against a common Ni mesh as depicted in (b), one may eliminate the impedance due to gas conversion. *(Continued)*

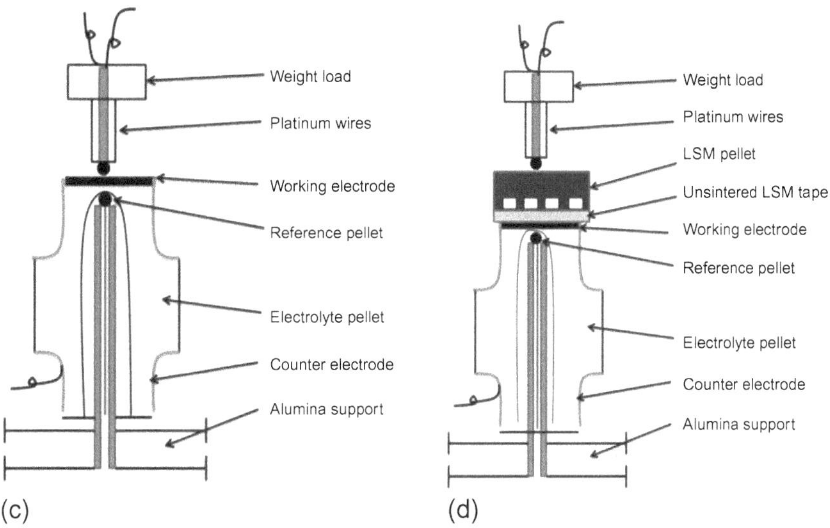

Figure 12.4 Continued. (c) In the case of a very well conducting electrode material, the current is picked up directly from the working electrode using a Pt bead. (*U*) In the case of a not-so-well conducting electrode, an improved current pick-up may be obtained by an unsintered LSM foil in contact with a dense, channelled LSM pellet [32,33].

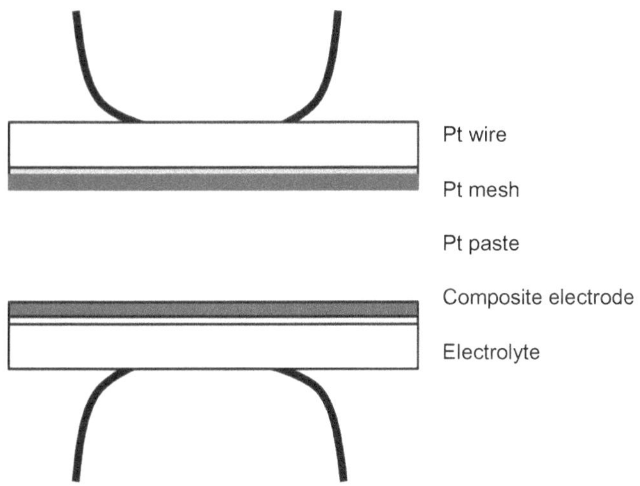

Figure 12.5 A symmetrical two-electrode cell arrangement for measurements near the OCV.

platinum mesh to make good contact with the electrodes, and two platinum wires coming out from each side of the cell, one to measure current and the other for potential determination. Such a test cell is well suited for electrode development work because there is no ambiguity about the source of electrode properties in this case. However, its use is limited to investigations close to OCV, where the electrode loss does not depend on whether it is anodically or cathodically polarised (Figure 12.5).

12.3 Testing single cells and stacks

For planar SOFCs testing, achieving good cell sealing and proper contact between electrodes and current collecting components is crucial for the right results with as little external interference as possible. In this case, an alumina test kit may be taken into consideration when it comes to assessing the commercial feasibility due to its reasonable thermal expansion coefficients, which will be first chosen if one is going to evaluate the long-term performance. However, to develop new materials with various purposes, it is more practical to test a button cell with regard to fundamental research or to test a commercial single cell as a real short stack.

Accurate evaluation of the performance of a button cell is needed in a Lab which focuses on fundamental research. There are mainly two factors, including cell sealing and contact between each component, which determine the testing results. The input and output gas flow can be simultaneously monitored to judge the air-tightness of the whole testing kit. A bubbler is sometimes applied to roughly check the output gas flow. Special glass and ceramic sealant paste is often recommended for SOFCs to metals and ceramics, and extra compaction force is usually needed to achieve a better sealing effect. A proper kind of paste plus the right force will make an OCV close to 1.2 V if the electrolyte is dense enough.

A simple kit for button cell testing can be designed and manufactured as shown in Figure 12.6. Two inner thin gas tubes with four holes are usually selected not only to input air and fuel but also to accommodate the platinum cables for voltage and current

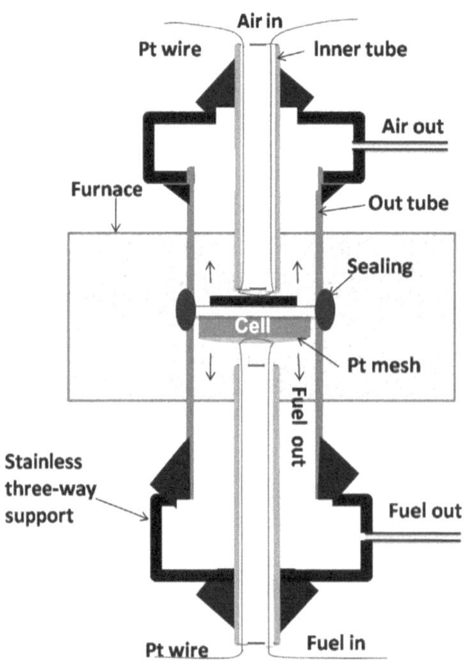

Figure 12.6 Schematic diagram for a button cell test.

collection. If the temperature distribution along the furnace has a gradient, it can be monitored using a thermocouple to detect the temperature point close to the cell position. It should be noticed that the low-level content of sealing glass (SiO_2) tends to contaminate the cell electrodes as reported in Ref. [34], so it is better to fabricate the cell with larger electrolyte areas to avoid the cell directly touching the electrode. More sophisticated design should take into account the heat exchange and recycled utilisation of the fuel.

To test cells with active areas ranging from a dozen square centimetres to hundreds of square centimetres, a full stainless design as sketched in Figure 12.7 is becoming more popular owing to its convenience in production and operation. Figure 12.8 presents testing result for a single cell using this housing, where the OCV is around 1.2 V, indicating the sealing is perfect and the cell ASR comes to about 0.6 Ω cm^2 at 750 °C. The fuel utilisation is calculated to over 90%.

For stack testing, the schematic diagram in Figure 12.9a demonstrates how to assemble a short stack. The cells are connected by ferritic stainless steel interconnect with gas channels, convex on both sides for ideal contact. Normally, the interconnect adherent to the cathode part will be pre-treated by printing Lanthanum Strontium Manganate Cathode (LSM) paste to make it stable in the oxidant atmosphere, and also

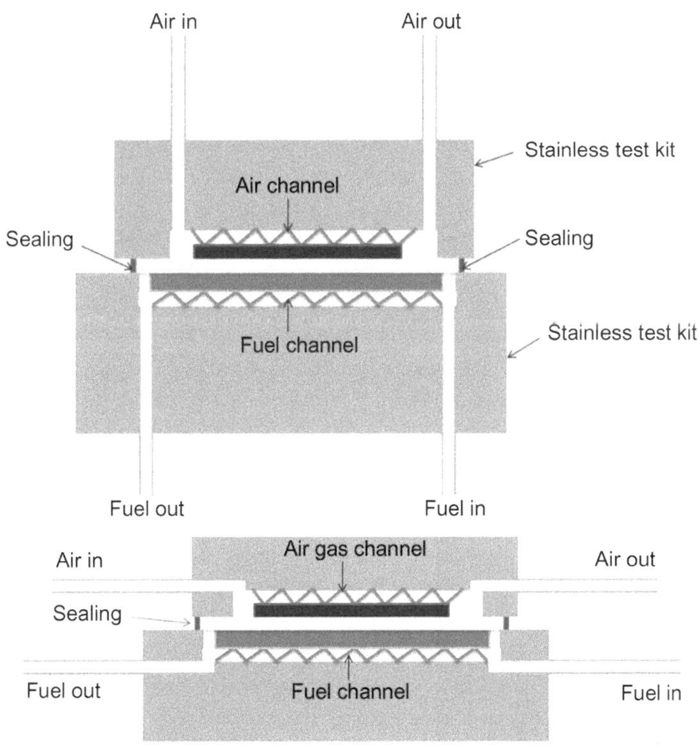

Figure 12.7 Sketch of kits for testing the cells with large active area.

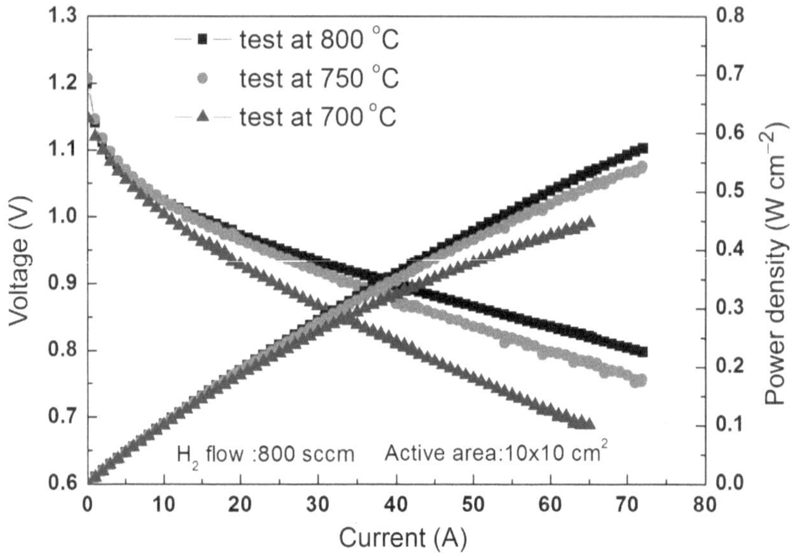

Figure 12.8 The electrochemical performance of a commercial cell with Ni/YSZ as anode, YSZ as electrolyte and LSCF as cathode tested at three different temperatures. The active area is 100 cm^2 and peak power densities come to around 0.55 W cm^{-2} at 750 °C. The corresponding fuel utilisation is calculated to over 90%.

protect it from poisoning by chromium evaporation from stainless steel. The red (light grey in the print version) and blue (dark grey in the print version) arrow shows the fuel and air flow direction. Modularisation tends to standardise the stack test. The sealing and contact between each component are also determinant factors that influence the testing effects.

In order to measure the unit cell performance, voltage probes were build in from both sides of the cell according to the schematic diagram shown in Figure 12.9b. The cell anode adheres to the interconnect via a nickel mesh, and the testing results indicate the contact between cathode and interconnect mainly accounts for the voltage loss. The protrusions on the interconnect can be designed and adjusted to its embedding depth and contact area in the cathode, which greatly affect the stack performance as shown in Figure 12.10, where the maximum power density of the unit cell increases by 56.3% with the interface contact area between the cell cathode and the metal interconnect increasing from 28.17% to 38.84%, but when the contact area is increased from 38.84% to 45.37%, the maximum power density shows no obvious changes, meaning there is 'a saturation point' for the contact area. The saturation means a bigger contact area can decrease the performance loss stemming from the ohmic contact, but it may reduce the channels for air transportation and increase the concentration polarisation. The contact depth in Figure 12.10 presents more obvious evidence for this understanding. Therefore, the interconnect and cathode current collect layer both need to be accounted for to achieve an ideal testing effect.

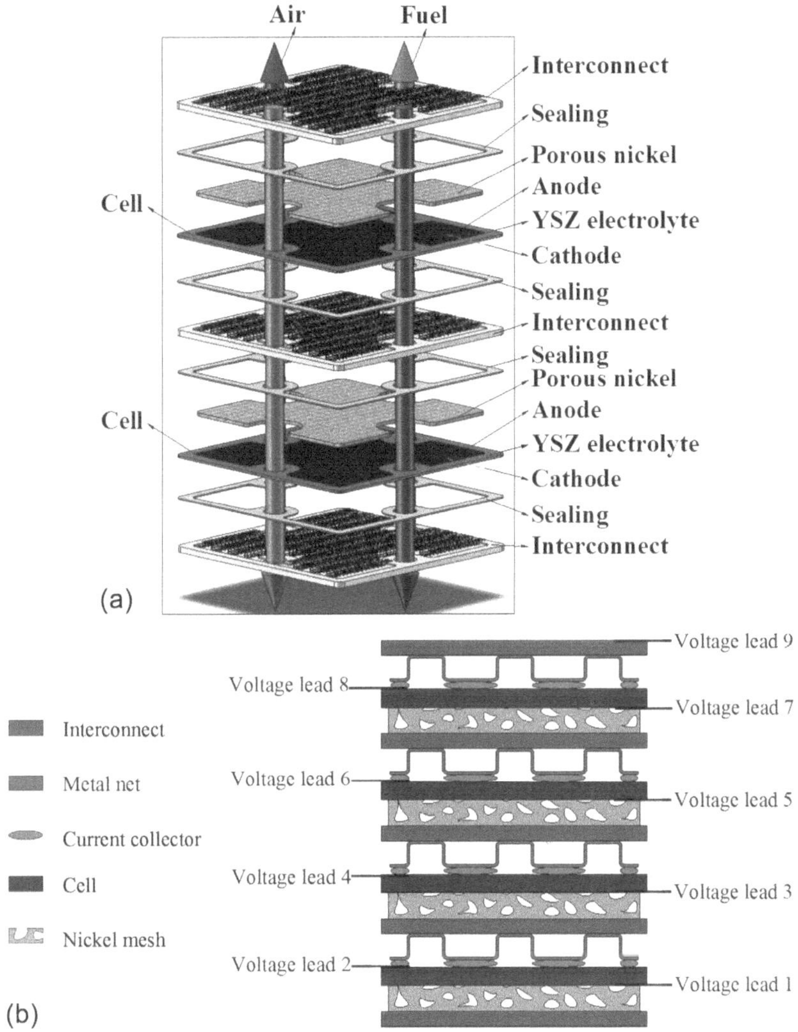

Figure 12.9 (a) Top and (b) bottom schematic diagrams of stack assembling [26,35]. This image (b) was published in Ref. [26], Copyright Elsevier (2013).

Stack ageing needs to be tested because it is an important step for SOFC commercialisation. Using the above test configuration, we can monitor and investigate the cell degradation in a stack and then evaluate the stack lifetime, where the stack ageing originates from several causes including the sealing, single-cell degradation and contact between components. By incorporating a reference electrode, the contributions from the anode, cathode and electrolyte of a unit cell to the voltage drop have been confirmed to be 76.9%, 15.4% and 7.7% after a two-unit stack runs for 200 h at a

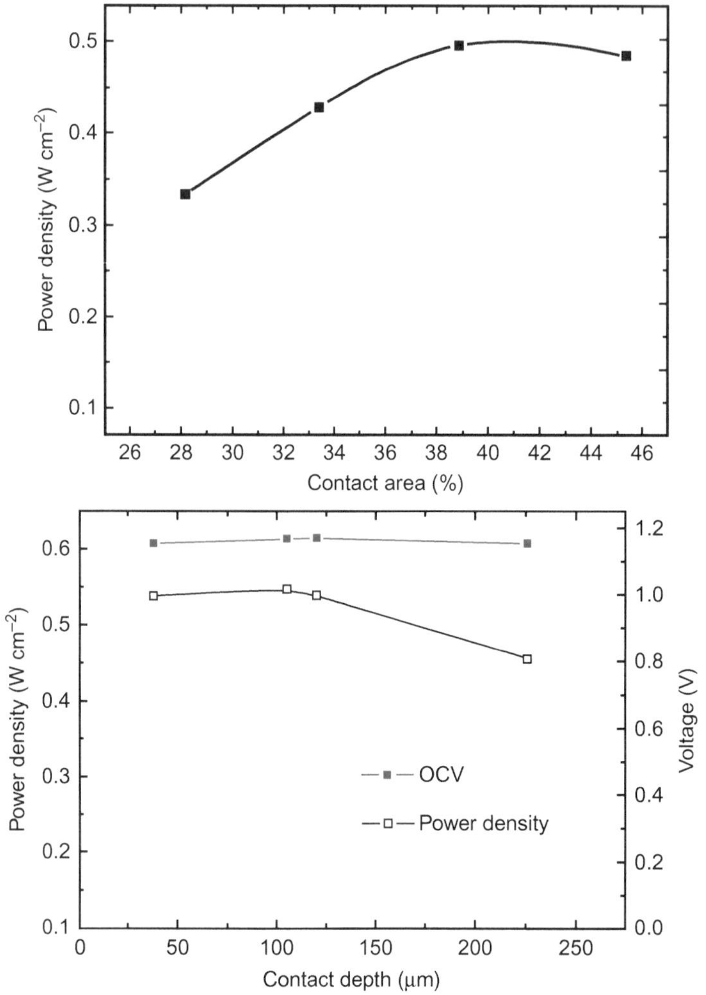

Figure 12.10 The relationship between the cell maximum power density of the tested stack and the interface contact area/depth [26].
This image (b) was published in Ref. [26], Copyright Elsevier (2013).

constant current density of 0.32 A cm^{-2}, meaning that nickel agglomeration is the main factor that determines the cell performance in a stack. Moreover, the stack shows a degradation rate of 2.27% per 100 h, equal to each cells as displayed in Figure 12.11. The first stage involved fast degradation at a rate of 3.27% per 100 h, following a slower degradation rate of 0.25% per 100 h. Cell degradation inside SOFC stacks was found to depend primarily on the contact between the cathode current-collecting layer and the interconnect. However, cell degradation inside SOFC stacks was also

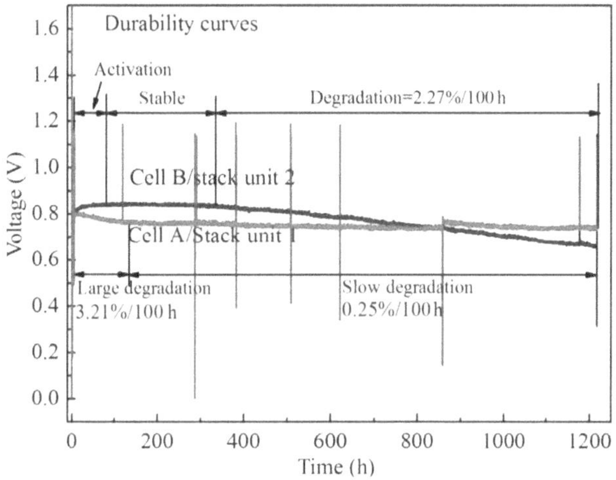

Figure 12.11 Single cell degradation in a two-unit stack, where the overall stack degradation comes to 2.27% per 100 h [36].

found to be independent of both Cr vaporisation from the stainless steel interconnect and high-temperature oxidation when the stack was sealed well.

12.4 Area-specific resistance

A fuel cell stack can be regarded as a 'black box' into which hydrogen (gas) and oxygen (air) are inputs, and electricity and exhaust gases are outputs. For such a stack, ASR is defined as:

$$\mathrm{ASR} = \frac{\mathrm{Emf} - U}{i} \tag{12.1}$$

where Emf is the electromotive force with the inlet fuel and air, and U is the cell voltage at the current density, i is the at the design point. A possible design point, for example, might be 0.6 V at 1000 °C, a fuel utilisation of 85% and an air flow of four times the stoichiometric amount (4stoichs).

The cell voltage, U, should be measured independently of the leads carrying the current, that is, separate potential probes should be used. This ASR is in most cases not very sensitive to small variations in cell voltage and fuel utilisation. By determining the ASR at a few different temperatures, an apparent activation energy, E_A, may be derived. Thus, in a voltage interval (from say 0.5 to 0.7 V) and a temperature interval (say from 650 to 1050 °C), the cell may be characterised, with fair approximation, by only two characteristic numbers, namely, ASR at one temperature and E_A.

In case the i–V curve is concave, it may be tempting to use a differential ASR (i.e. the tangent) at high current densities as this gives a nice low value. Such a number has the drawback that it does not reflect the cell performance over the full polarisation range as does the quantity defined by Equation (12.1).

Often, cell tests are conducted with very low fuel and air utilisations because these are easier to perform than the 'realistic' tests with high fuel utilisation. In case of insignificant fuel and oxygen utilisations, the relevant definition of ASR is again that of Equation (12.1), but the insignificant utilisation makes this value incomparable to ASR derived from experiments with high fuel utilisation because the concentration polarisation resistance due to the fuel and air conversion can be a considerable fraction of the total cell resistance. Thus it is, in general, necessary to specify also the fuel and oxygen utilisation together with temperature and apparent activation energy. Furthermore, using almost dry hydrogen, as is the common practice, it is not easy to conduct experiments with a real negligible fuel utilisation, since even small current densities will create enough water to change significantly the Emf of the hydrogen/water fuel gas versus air, for example, if the inlet gas contains 0.1% H_2O and the fuel utilisation is 0.1%, this changes the H_2O/H_2 ratio by a factor of 2, which, in turn, changes the Emf by 34 mV at 850 °C. Therefore, in order to be able to compare results for different fuel utilisations, the ASR value should be corrected for the effect of fuel utilisation. Before describing how this may be done, various contributions to the total ASR are examined below.

ASR may be divided into ohmic resistance, R_s, and electrode polarisation resistance, R_p. The ohmic resistance originates from the electrolyte, the electrodes materials and the current collection arrangement. This is very much dependent on geometric factors such as thickness of the cell components and the detailed geometry of the contact between current collection and electrodes, and between electrodes and electrolyte as current constrictions may be important [37]. The electrode polarisation resistance is further divided into contributions from the various rate-limiting steps. Thus, ASR can be broken down in five terms:

$$\mathrm{ASR} = R_{\mathrm{elyt}} + R_{\mathrm{connect}} + R_{\mathrm{p,elchem}} + R_{\mathrm{p,\,diff}} + R_{\mathrm{p,\,conver}} \qquad (12.2)$$

where R_{elyt} is the electrolyte resistance calculated from the measured specific conductivity and the thickness; $R_{\mathrm{connect}} = R_s - R_{\mathrm{elyt}}$ is the resistance due to non-optimised contact and current collection; $R_{\mathrm{p,elchem}}$ is the electrode polarisation originating from all the limiting chemical and electrochemical processes on the electrode surfaces, in the bulk electrode material and on the electrolyte/electrode interfaces; $R_{\mathrm{p,diff}}$ is the contribution from the gas phase diffusion and $R_{\mathrm{p,conver}}$ is the contribution due to gas conversion, i.e. fuel oxidation and oxygen reduction. This division of ASR is based on what is possible to measure and calculate reliably rather than on any physical or electrochemical basis. Some terms in Equation (12.2) can therefore be thought of as 'equivalent resistances', for example, the Emf drop due to changes in gas composition resulting from the fuel utilisation is translated to an equivalent resistance. Depending on the exact type of electrode, different types of contributions are possible as derived from more basic electrochemical point of view. For example, current constriction may be important if the electrode has coarse porous structure but of less or no importance in case of a fine structured electrode. In one type of cathode, the surface diffusion may be important, but in another the diffusion of oxide ions (and electrons) through the electrode particles may cause the main polarisation loss.

It is seen that the contribution from the concentration polarisation, $R_{\text{p, diff}} + R_{\text{p, conver}}$ is dominating. In an electrode-supported cell, the limitation of gas diffusion through the support is a cell-relevant resistance, whereas $R_{\text{p,conver}}$ results from operation demands, and it is thus of special interest to be able to correct ASR for the effect of fuel conversion.

If a significant amount of fuel is consumed in the cell (or stack) under test, a resistance derived on the basis of Emf of the inlet gas (cf. Equation 12.1) will be an overestimation of the 'true' cell resistance. The larger the fuel utilisation the larger will be the overestimation and results of cell tests performed with different fuel (and air) utilisations are thus not directly comparable. A comparison between cell test results obtained under different and non-negligible fuel utilisations must thus, to be meaningful, be based on a resistance measure, ASR_{cor} where the effects of changes in gas composition over the cell area have been taken into account. How the correction is applied depends on how the gases are fed to the cell. Here, two idealised cases are considered, namely the case where the fuel compartment may be considered a continuously stirred tank reactor (CSTR) or a plug flow reactor.

If the fuel compartment can be considered CSTR-like due to effective mixing because of a turbulent gas stream and fast gas diffusion, ASR_{cor} can be calculated from the expression

$$\text{ASR}_{\text{cor}} = \frac{\text{Emf}_{\text{avg}} - U}{i} \tag{12.3}$$

where Emf_{avg} signifies the average Emf, which in this case is the same as the Emf of the outlet gas. An example of such conditions is reported in Ref. [38].

The plug flow case is slightly more complex. Under the assumptions that the local ASR is independent of the position along the fuel and air flow channels, and the flow pattern is co-flow, ASR_{cor} may be calculated from the expression [39]

$$\text{ASR}_{\text{cor}} = \left\{ \frac{i}{X^{\text{i}}_{\text{H}_2} - X^{\text{o}}_{\text{H}_2}} \int_{X^{\text{o}}_{\text{H}_2}}^{X^{\text{i}}_{\text{H}_2}} \frac{\text{d}X_{\text{H}_2}}{\text{Emf}(X_{\text{H}_2}) - U} \right\}^{-1} \tag{12.4}$$

where

$$\text{Emf}(X_{\text{H}_2}) = E^0 - \frac{RT}{2F} \ln \left\{ \frac{X^{\text{i}}_{\text{H}_2\text{O}} + X^{\text{i}}_{\text{H}_2} - X_{\text{H}_2}}{X_{\text{H}_2} \sqrt{X^{\text{i}}_{\text{O}_2} - \frac{N_{\text{f}}}{2N_{\text{a}}} \left(X^{\text{i}}_{\text{H}_2} - X_{\text{H}_2} \right)}} \frac{1}{\sqrt{P_{\text{a}}/\text{atm}}} \right\} \tag{12.5}$$

Here X is the molar fraction, P_a is the air pressure and superscripts i, o signify inlet and outlet, respectively. N_f and N_a, are the molar flows of fuel and air. The assumption of a position-independent local ASR is an approximation, which is not always justifiable. Part of the anode polarisation resistance is dependent on fuel composition. However, often this part is small. If the cell is not isothermal, the local resistance will vary with position due to its temperature dependence. Also the actual flow pattern may be much more complex than just co-flow. Even so, if the fuel utilisation is large, $\mathrm{ASR_{cor}}$ derived from Equations (12.4) and (12.5) will always be a better characteristic of a cell than a value derived neglecting the fuel utilisation (Equation 12.1). More precise evaluation of $\mathrm{ASR_{cor}}$ requires a rigorous 3D modelling of the cell test.

For purposes of evaluating Equation (12.4), the integral may be approximated by a sum [40]:

$$\mathrm{ASR_{cor}} = \left\{ \frac{i}{N} \sum_{j=0}^{N-1} \frac{1}{\mathrm{Emf}(X_{\mathrm{H_2}}(j)) - U} \right\}^{-1} \tag{12.6}$$

where $\mathrm{Emf}(X_{\mathrm{H_2}}(j))$ is given by Equation (12.4) with

$$X_{\mathrm{H_2}}(j) = X^{\mathrm{i}}_{\mathrm{H_2}} + \frac{j + (1/2)}{N} \left(X^{\mathrm{o}}_{\mathrm{H_2}} - X^{\mathrm{i}}_{\mathrm{H_2}} \right) \tag{12.7}$$

The more terms are included in the sum, the better the approximation. A 'first-order' correction for the effects of finite fuel and air utilisations may be obtained taking only one term in the sum in which case $\mathrm{ASR_{cor}}$ should be evaluated from Equation (12.3) with

$$\mathrm{Emf_{avg}} = E^0 - \frac{RT}{2F} \ln \left\{ \frac{\bar{p}_{\mathrm{H_2O}}}{\bar{p}_{\mathrm{H_2}} \sqrt{\bar{p}_{\mathrm{O_2}}/\mathrm{atm}}} \right\} \tag{12.8}$$

where the bar indicates 'average', i.e.

$$\bar{p}_{\mathrm{H_2O}} = \frac{P^{\mathrm{i}}_{\mathrm{H_2O}} + P^{\mathrm{o}}_{\mathrm{H_2O}}}{2}, \quad \bar{p}_{\mathrm{H_2}} = \frac{P^{\mathrm{i}}_{\mathrm{H_2}} + P^{\mathrm{o}}_{\mathrm{H_2}}}{2} \quad \text{and} \quad \bar{p}_{\mathrm{O_2}} = \frac{P^{\mathrm{i}}_{\mathrm{O_2}} + P^{\mathrm{o}}_{\mathrm{O_2}}}{2} \tag{12.9}$$

If there are no significant leaks in the cell or the test equipment, then both fuel and air utilisation, and from this the compositions, may be calculated from the flow rates and the current using Faraday's law. Alternatively, the composition of the outlet fuel and air may be obtained by gas analysis. The conversion resistance, $R_{\mathrm{p,conver}}$ may be calculated using the concept of $\mathrm{Emf_{avg}}$:

$$R_{\mathrm{p,conver}} = \frac{\mathrm{Emf_{inlet}} - \mathrm{Emf_{avg}}}{i} \tag{12.10}$$

However, the anisotropic nature (temperature, gas composition, current flow) of a real cell under current flow often invalidates this simple approach. Gas conversion under electric load causes an uneven distribution of the current density with decreasing current density in the downstream direction. Fuel composition gradients over the cell originating from leaks cause different driving potentials at different points. In extreme cases, this may result in high Emf areas driving low Emf areas in electrolyser mode. Thus, internal currents may flow in the cell even at OCV. Gas leaks in the cell also affect the current density distribution under *load* and cause localised heating by combustion. For tubular cell designs with high in-plane resistance, the current density distribution may be affected: furthermore, the temperature may not be constant over the whole cell length with flowing gases adding to the inhomogeneity of the current density.

Modelling can simplify or reduce the extent of experimental task and predict likely behaviour under a broad range of test conditions. However, subsequent validation by comparison with relevant cell and stack data is always important.

An example of the magnitude of $R_{p,conver}$ under different conditions can be deduced from i - V characteristics obtained in hydrogen with different water vapour for an anode-supported cell. Also the ASR can be deduced directly from the curves (OCVU)/i curves and corrected for the fuel conversion (Equations 12.6 and 12.7). The correction is always largest for the dry gas, where $R_{p,conver}$ ca. 0.12 Ω cm^2, reflecting the gas composition dependence of the Nernst voltage.

After correcting for the effect of non-negligible fuel utilisation, the cell resistance is still significantly smaller when measured with 20% water in the feed than with 5%. This reflects a gas composition dependence of some of the loss terms in Equation (12.2). In Ref. [41], it is argued that the observed composition dependence is primarily due to the composition dependence of the diffusive losses on the anode side (diffusion over voltage), and it is shown how one may utilise characteristics obtained with different water vapour/hydrogen ratios to assess the magnitude of the diffusion loss [42].

12.5 Testing cells on alternative fuels

During recent years, SOFC researchers devoted much effort to improving the cell electrochemical performance and reducing the operational temperature. Hydrogen and natural gas are currently considered to be the main fuels for SOFCs. Hydrogen is an ideal fuel with respect to fuel cell working conditions. However, hydrogen is not present in the environment in an uncombined form and there are difficulties with production, transportation and storage [42]. Therefore, various alternative fuels such as biofuels, coal-derived syngas and some liquefied gases are being investigated and tested at typical operational conditions.

12.5.1 SOFC fueled by biofuels

Biomass can be defined as all organic materials and has abundant origins, for example, industrial waste waters with a high organic charge from the agro- and food industry, municipal waste water residue or sewage sludge, animal manure and green waste (straw and harvest residues) on farms [43]. The composition of biofuel indeed varies with its origin, but it always includes major constituents CH_4 and CO_2, by which hydrogen can be produced with the dry reforming reaction [44]. Biogas as fuel of SOFC single cell and short stack has been therefore been widely tested [45–49].

Bio-ethanol and bio-methanol have been the most commonly used biomass-derived fuels for fuel cell systems for the past decade and will continue to play an important role in the near future [50]. The reaction of CH_4 and CO_2 is thermodynamically favourable above 650 °C, but carbon formation will be a challenge if the ratio of CO_2 to CH_4 is lower than 1. The testing results in Figure 12.12 shows cell performance evolution with the CO_2/CH_4 of 32 (gas 1), 11(gas 2), 5.7(gas 3), 3.2(gas 4), 1.7(gas 5), 1.1(gas 6) and 0.6(gas 7) with a balance mixed gas of CO–H_2–H_2O, in which the cell shows no degradation when CO_2/CH_4 is above 1 from gas 1 to 6. However, sharp drop occurs once the ratio is below 1 in gas 7 even in the presence of 5% steam [51].

A SOFC can run on biogas, but like natural gas, CH_4 reforming is the determinant factor for its electrochemical performance and stability. The reforming studies can be carried out both dry, by mixing air with the methane, and wet by mixing steam. The sources for biomass are abundant and the SOFC is thus expected to be installed in some residential communities where living wastes can be converted into energy, paving a path for carbon-free energy in buildings.

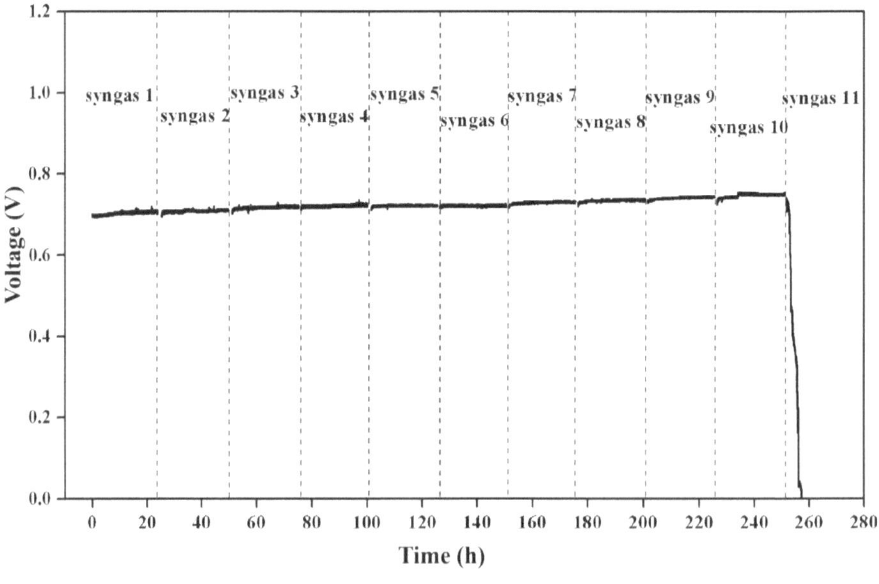

Figure 12.12 Cell voltage variation by varying the ratio of CO_2 to CH_4 from 32 to 0.6 [51].

12.5.2 SOFC fueled by ammonia

Since ammonia is a good hydrogen carrier and is carbon free, so it is a possible candidate as the fuel for SOFCs. Besides, the price of ammonia is as competitive as hydrocarbons and it can be easily liquefied; the volumetric energy density of liquefied ammonia is higher than that of liquid hydrogen, which is useful in transport and storage. It might be thus a fine substitute for hydrogen and hydrocarbons [52]. Moreover, ammonia can be completely decomposed at nickel–yttria stabilised zirconia (Ni–YSZ) anode and no nitrogen oxide was produced at the anode when oxygen-ions transferred from cathode to anode [53].

The reaction for ammonia in a typical SOFC operational condition is decomposition into nitrogen and hydrogen. This reaction is thermodynamically favourable. However, there is also possibility for ammonia reacting with oxygen to produce nitrogen and water, which is definitely not desirable as it cannot generate fuel. Ammonia as fuel makes SOFC have a lower OCV compared to that using hydrogen as fuel, so in this case the cell power densities are similar to those obtained under hydrogen since the decomposition of ammonia is both sufficient and fast in the anode chamber in the SOFC operating temperatures [53]. Using a SDC-based SOFC, the temperature can be decreased in the ranges from 650 to 550 °C, where the ammonia gas is obtained by evaporation from liquefied ammonia as fuel [54], indicating ammonia is a good potential fuel for SOFC given its convenient production and storage.

12.5.3 SOFC fueled by coal-derived gas

Coal gas can be produced at a low cost by coal gasification, so it can be a promising fuel for high-temperature fuel cells. The composition of coal-derived syngas is very complicated, containing hydrogen, carbon monoxide, nitrogen, steam, methane, ammonia, carbon dioxide, etc. In addition, there are some major impurities such as hydrogen sulphide, phosphine, hydrogen chloride, plus several trace metal impurity species like sodium, potassium and vanadium in the coal-derived gas stream [55]. Since the steam in coal syngas can reach from 10 to over 60 vol.% [56], so it is expected to suppress carbon deposition considering the contents of methane and carbon monoxide in syngas are not so high. However, various contaminants can degrade the cell to a large extent as reported [57–65], which becomes a major challenge for a fuel cell running on coal syngas. Plenty of tests are needed to study mechanism of these impurities poisoning fuel cells, offering crucial information on research and development of new anode materials.

The effects of impurities in coal syngas on the performance of Ni–YSZ anode of SOFCs have some specific features. The hydrogen sulphide (H_2S) has been paid the most attention due to its immediate and severe detrimental effect on cell performance under various operating conditions. In all cases, a small amount of H_2S ppm level causes a sharp drop in cell performance within the first few minutes of exposure, followed by a gradual but persistent deterioration in performance for hundreds of hours as shown in Figure 12.13 [66,67], where a rapid first-stage degradation occurs within a short period and three different modes follow the initial process, depending on

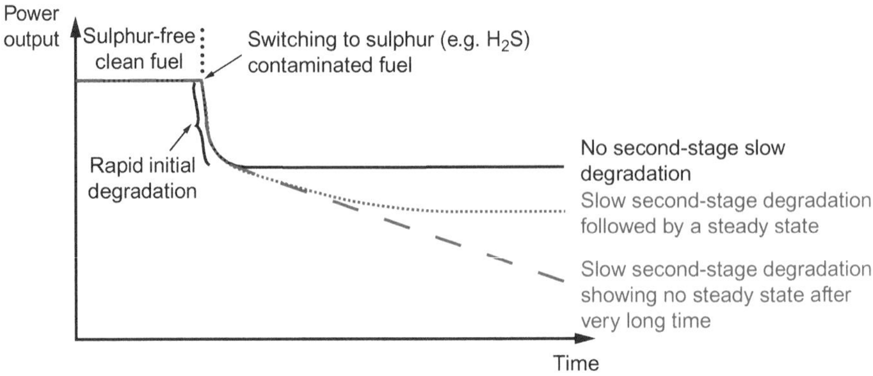

Figure 12.13 Schematic for power output versus time for SOFCs with Ni–YSZ cermet anode showing sulphur poisoning with rapid initial degradation followed by different 2nd stage behaviours: (solid line) no 2nd stage slow degradation: power output saturates right after initial rapid degradation; (dotted line) a slow 2nd stage degradation followed by a steady state; (dashed line) a slow 2nd stage degradation showing no steady state after very long time (e.g. thousands of hours) [66].

the testing conditions such a large gauge of sulphur inducing a long-term 2nd stage degradation and a higher operational temperature relieving this degradation.

All the post-mortem material analyses strongly indicate the formation of nickel sulphides such as Ni_3S_2, NiS, Ni_3S_4 and Ni_4S_3 accounts for the sulphur poisoning a SOFC [68–70]. A mechanism for the long-term permanent degradation is proposed that if local sulphur concentrations in the nickel reach approximately 33%, the resulting nickel sulphide alloy will be in the liquid state at 750 °C. Liquid phase nickel is expected to have high mobility, leading to nickel migration and disorder [71].

12.6 Summary

This chapter has considered the main types of electrochemical tests which have been applied to SOFCs and has outlined the main issues which require detailed attention for obtaining meaningful test results. One important aspect in electrode testing is to assure a correct geometry in three-electrode set-ups, very difficult in practice in case of electrode supported cells with thin electrolytes. Unfortunately, even testing the individual electrodes in sound geometry set-ups is not a perfect procedure either, because the sum of the contributions from individual cell components to the cell resistance does not add up to the actual measured total cell resistance. This is probably due to differences in the fabrication of the special cells for electrode characterisation and the practical cells.

It is recommended that cell test results be reported in a way that makes it easy to derive ASR from the *i*–*V* curves. Sufficient information should be provided so that the ASR values can be corrected for effects of finite fuel utilisation. Also, the choice of fuel composition should preferably reflect real cell operation conditions. The ASR should be derived using the Emf and a cell voltage in the range of 0.5–0.7 V and its corresponding current density. In case of a grossly non-linear *i–V* curve, a differential ASR value is of little practical use.

The design and assembly of SOFC stacks has become a major bottleneck during the past decade. The sealing, gas channel, current collector and interconnect coating are all crucial factors that determine the performance and long-term stability of stack. The contact area and depth between interconnect and cathode is detected to have obvious effects on stack output power density and durability, but the too large area will in return affect the fuel transportation and engender bad electrochemical performance. On the other hand, the thermal and gas distribution inside a stack are also important and should be taken into consideration. These can be studied by modelling and simulation, but also can be tested in situ by setting some thermocouples around the stack [72].

One of the most attractive advantages for high-temperature SOFC is flexibility to various fuels including coal syngas and biogas, which are abundant and inexpensive compared to hydrogen. So to test the cells running on these complex gases is imperative. These kinds of tests should both consider carbon deposition and a variety of contaminants poisoning effects on the catalyst. Fortunately, the performance degradation caused by carbon deposition is quite different to that induced by the major impurities like H_2S and HCl. The former tends to generate a slow and long-term performance drop, but the latter brings about a rapid degradation, followed a long time sluggish process (at low level). Moreover, sulphur contaminants are widely present in natural gas, and biogas, so these fuels should be pre-treated to achieve desulphurisation.

References

[1] M. Mogensen, P.V. Hendriksen, High Temperature SOFCs, Elsevier Science, Oxford, 2003.
[2] C.H. Hamann, A. Hamnett, W. Vielstich, Electrochemistry, second ed., Wiley, Hoboken, NJ, 2007.
[3] S.E. Group, Instrumental Methods in Electrochemistry, Ellis Horwood, Chichester, 1985.
[4] V.V. Kharton, Solid State Electrochemistry, Wiley-VCH, Weinheim, 2007.
[5] M. Kleitz, P. Fabry, E. Schouler, in: W. van Gool (Ed.), Fast Ion Transport in Solids, North-Holland, Amsterdam, 1973, pp. 439–451.
[6] P.H. Bottelberghs, in: P. Hagenmuller, W. van Gool (Eds.), Solid Electrolytes, Academic Press, New York, 1978, pp. 145–172.
[7] J.A. Kilner, B.C.H. Steele, in: O.T. Serrensen (Ed.), Non-stoichiometric Oxides, Academic Press, New York, 1981, pp. 233–269.
[8] H.L. Tuller, in: O.T. Serrensen (Ed.), Non-stoichiometric Oxides, Academic Press, 1981, pp. 271–335.
[9] H. Rickert, Electrochemistry of Solids, Springer-Verlag, Berlin, 1982.
[10] I. Riess, in: P.J. Gellings, H.J.M. Bouwmeester (Eds.), Solid State Electrochemistry, CRCPress, New York, 1997, pp. 233–268.
[11] I. Riess, J. Schoonman, CRCPress, New York, 1997, pp. 269–294.

[12] W. Weppner, CRCPress, New York, 1997, pp. 295–327.
[13] S.C. Singhal, Advances in solid oxide fuel cell technology, Solid State Ionics 135 (1–4) (2000) 305–313.
[14] J.-H. Koh, Y.-S. Yoo, J.-W. Park, H.C. Lim, Carbon deposition and cell performance of Ni-YSZ anode support SOFC with methane fuel, Solid State Ionics 149 (2002) 157–166.
[15] J.V.M.D. Gross, R.J. Gorte, A strategy for achieving high performance with SOFC ceramic anodes, Electrochem. Solid-State Lett. 10 (2007) 65–69.
[16] W.B. Guan, W.G. Wang, H.M. Li, Y.J. Xue, J.X. Wang, Y.N. Wu, J. Wang, K. Liu, SOFC research and development at NIMTE, in: 8th European Solid Oxide Fuel Cell Forum, Lucerne/Switzerland, 2008, p. B0404.
[17] W.N. Liu, X. Sun, M.A. Khaleel, Effect of creep of ferritic interconnect on long-term performance of solid oxide fuel cell stacks, Fuel Cells 10 (2010) 703–717.
[18] P. Wyss, E. Hack, P. Holtappels, Application of non-destructive testing to analyse SOFC cells and stacks, Fuel Cells 9 (2009) 907–910.
[19] L.G.J. de Haart, J. Mougin, O. Posdziech, J. Kiviaho, N.H. Menzler, Stack degradation in dependence of operation parameters; the real-SOFC sensitivity analysis, Fuel Cells 9 (2009) 794–804.
[20] M. Rautanen, O. Himanen, V. Saarinen, J. Kiviaho, Compression properties and leakage tests of mica-based seals for SOFC stacks, Fuel Cells 9 (2009) 753–759.
[21] N.H. Menzler, P. Batfalsky, L. Blum, M. Bram, S.M. Groß, V.A.C. Haanappel, J. Malzbender, V. Shemet, R.W. Steinbrech, I. Vinke, Studies of material interaction after long-term stack operation, Fuel Cells 7 (2007) 356–363.
[22] L. Blum, H.P. Buchkremer, S. Gross, A. Gubner, L.G.J. de Haart, H. Nabielek, W.J. Quadakkers, U. Reisgen, M.J. Smith, R. Steinberger-Wilckens, R.W. Steinbrech, F. Tietz, I.C. Vinke, Solid oxide fuel cell development at Forschungszentrum Juelich, Fuel Cells 7 (2007) 204–210.
[23] J. Malzbender, T. Wakui, R.W. Steinbrech, Curvature of planar solid oxide fuel cells during sealing and cooling of stacks, Fuel Cells 6 (2006) 123–129.
[24] M. Linder, T. Hocker, L. Holzer, K.A. Friedrich, B. Iwanschitz, A. Mai, J.A. Schuler, Cr2O3 scale growth rates on metallic interconnectors derived from 40,000 h solid oxide fuel cell stack operation, J. Power Sources 243 (2013) 508–518.
[25] M.J.C. Bagger, P. Vang Hendriksen, P. Halvor Larsen, M. Mogensen, B. Thorstensen (Ed.), Proceedings of the 2nd European SOFC Forum, Oberrohrdorf, Switzerland, 1996, pp. 175–184.
[26] L. Jin, W. Guan, J. Niu, X. Ma, W.G. Wang, Effect of contact area and depth between cell cathode and interconnect on stack performance for planar solid oxide fuel cells, J. Power Sources 240 (2013) 796–805.
[27] M. Godickemeier, K. Sasaki, L.J. Gauckler, in: M. Dokiya, O. Yamamoto, H. Tagawa, S. C. Singhal (Eds.), Solid Oxide Fuel Cells IV, The Electrochemical Society Proceedings, Pennington, NJ, PV95-1, 1995, p. 107.
[28] J.W. Erning, W. Schafffrath, U. Stimming, E. Syskakis, I.C. Wipperman, The Electrochemical Society Proceedings, Pennington, NJ, PV95-1, 1995, p. 492.
[29] S.B. Adler, J. Electrochem. Soc. 149 (2002) E166.
[30] M. Mogensen, P.V. Hendriksen, K. Kammer, in: J. Huijsmans (Ed.), Proceedings of the 5th European SOFC Forum, Oberrohrdorf, Switzerland, 2002, p. 893.
[31] W. Wu, W.B. Guan, G.L. Wang, F. Wang, W.G. Wang, In-situ investigation of quantitative contributions of the anode, cathode, and electrolyte to the cell performance in anode-supported planar SOFCs, Adv. Energy Mater. 4 (2014) 1400120.
[32] J. Winkler, P.V. Hendriksen, N. Bonanos, M. Mogensen, J. Electrochem. Soc. 145 (1998) 1184.

[33] M.J. Jorgensen, S. Primdahl, M. Mogensen, Electrochem. Acta 44 (1999) 4195.
[34] Y. Liu, Effects of impurities on microstructure in Ni/YSZ–YSZ half-cells for SOFC, Solid State Ionics 161 (2003) 1–10.
[35] L. Jin, W. Guan, X. Ma, H. Zhai, W.G. Wang, Quantitative contribution of resistance sources of components to stack performance for planar solid oxide fuel cells, J. Power Sources 253 (2014) 305–314.
[36] W.B. Guan, L. Jin, X. Ma, W.G. Wang, Investigation of impactors on cell degradation inside planar SOFC stacks, Fuel Cells 12 (2012) 1085–1094.
[37] J. Fleig, J. Maier, J. Electrochem. Soc. 144 (1997) 302.
[38] S. Primdahl, M. Mogensen, J. Electrochem. Soc. 145 (1998) 2431.
[39] A. Solheim, R. Tunold, R. Odegard, in: F. Gross, P. Zegers, S.C. Singhal, O. Yamamoto (Eds.), Commission of the European Communities, Luxembourg, EUR 13564EN, Solid Oxide Fuel Cells, vol. 11, 1991, p. 297.
[40] M. Mogensen, P.H. Larsen, P.V. Hendriksen, B. Kindl, C. Bagger, S. Linderoth, S.C. Singhal, M. Dokiya (Eds.), Solid Oxide Fuel Cells VI, The Electrochemical Society Proceedings, Pennington, NJ, PV99-1, 1999, p. 904.
[41] P.V. Hendriksen, S. Koch, M. Mogensen, Y.L. Liu, P.H. Larsen, in: S.C. Singhal, M. Dokiya (Eds.), Solid Oxide Fuel Cells, Electrochemical Society Proceedings, Pennington, NJ, PV2003-07, 2003, p. 1147.
[42] J.L.J. Milewski, Solid oxide fuel cell fuelled by biofuels, ECS Trans. 25 (2) (2009) 1031–1040.
[43] Y.M. Jan Van herle, O. Bucheli, Biogas as a fuel source for SOFC co-generators, J. Power Sources 127 (2004) 300–312.
[44] J.L.K. Girona, M. Petitjean, J. Fouletier, F. Lefebvre-Joud, SOFC running on biogas: identification and experimental validation of "safe" operating conditions, ECS Trans. 25 (2) (2009) 1041–1050.
[45] J. Staniforth, R.M. Ormerod, Running solid oxide fuel cells on biogas, Ionics 9 (2003) 336–342 (Springer).
[46] Y. Shiratori, T. Ijichi, T. Oshima, K. Sasaki, Internal reforming SOFC running on biogas, Int. J. Hydrogen Energy 35 (2010) 7905–7912.
[47] P. Leone, A. Lanzini, M. Santarelli, M. Calì, F. Sagnelli, A. Boulanger, A. Scaletta, P. Zitella, Methane-free biogas for direct feeding of solid oxide fuel cells, J. Power Sources 195 (2010) 239–248.
[48] D. Papurello, R. Borchiellini, P. Bareschino, V. Chiodo, S. Freni, A. Lanzini, F. Pepe, G. A. Ortigoza, M. Santarelli, Performance of a solid oxide fuel cell short-stack with biogas feeding, Appl. Energy 125 (2014) 254–263.
[49] C. Xu, J.W. Zondlo, M. Gong, F. Elizalde-Blancas, X. Liu, I.B. Celik, Tolerance tests of H_2S-laden biogas fuel on solid oxide fuel cells, J. Power Sources 195 (2010) 4583–4592.
[50] J. Xuan, M.K.H. Leung, D.Y.C. Leung, M. Ni, A review of biomass-derived fuel processors for fuel cell systems, Renew. Sustain. Energy Rev. 13 (2009) 1301–1313.
[51] T. Chen, He Miao, Tingshuai Li, Cheng Xu, Wei Guo Wang, Evaluation of carbon deposition behavior on the nickel/yttrium-stabilized zirconia anode-supported fuel cell fueled with simulated syngas, J. Power Sources 196 (2011) 2461–2468.
[52] A. Fuerte, R.X. Valenzuela, M.J. Escudero, L. Daza, Ammonia as efficient fuel for SOFC, J. Power Sources 192 (2009) 170–174.
[53] R.M. Ormerod, J. Staniforth, Clean destruction of waste ammonia with consummate production of electrical power within a solid oxide fuel cell system, Green Chem. 5 (2003) 606–609.
[54] G. Meng, C. Jiang, J. Ma, Q. Ma, X. Liu, Comparative study on the performance of a SDC-based SOFC fueled by ammonia and hydrogen, J. Power Sources 173 (2007) 189–193.

[55] J.B.P. Jayaweera, J. Perez, K.H. Lau, M. Hornbostel, A. Sanjurjo, Effect of coal contaminants on solid oxide fuel system performance and service life, Final technical report, National Energy Technology Center, 2008.
[56] F.N. Cayan, M. Zhi, S.R. Pakalapati, I. Celik, N. Wu, R. Gemmen, Effects of coal syngas impurities on anodes of solid oxide fuel cells, J. Power Sources 185 (2008) 595–602.
[57] Z. Cheng, S. Zha, M. Liu, Influence of cell voltage and current on sulfur poisoning behavior of solid oxide fuel cells, J. Power Sources 172 (2007) 688–693.
[58] K. Sasaki, K. Susuki, A. Iyoshi, M. Uchimura, N. Imamura, H. Kusaba, Y. Teraoka, H. Fuchino, K. Tsujimoto, Y. Uchida, N. Jingo, H_2S poisoning of solid oxide fuel cells, J. Electrochem. Soc. 153 (2006) A2023.
[59] J.F.B. Rasmussen, A. Hagen, The effect of H_2S on the performance of Ni–YSZ anodes in solid oxide fuel cells, J. Power Sources 191 (2009) 534–541.
[60] J. Bao, G.N. Krishnan, P. Jayaweera, J. Perez-Mariano, A. Sanjurjo, Effect of various coal contaminants on the performance of solid oxide fuel cells: part I accelerated testing, J. Power Sources 193 (2009) 607–616.
[61] J. Bao, G.N. Krishnan, P. Jayaweera, K.-H. Lau, A. Sanjurjo, Effect of various coal contaminants on the performance of solid oxide fuel cells: part II. ppm and sub-ppm level testing, J. Power Sources 193 (2009) 617–624.
[62] J. Bao, G.N. Krishnan, P. Jayaweera, A. Sanjurjo, Effect of various coal gas contaminants on the performance of solid oxide fuel cells: part III. Synergistic effects, J. Power Sources 195 (2010) 1316–1324.
[63] J.P. Trembly, R.S. Gemmen, D.J. Bayless, The effect of coal syngas containing HCl on the performance of solid oxide fuel cells: investigations into the effect of operational temperature and HCl concentration, J. Power Sources 169 (2007) 347–354.
[64] C. Xu, J.W. Zondlo, H.O. Finklea, O. Demircan, M. Gong, X. Liu, The effect of phosphine in syngas on Ni–YSZ anode-supported solid oxide fuel cells, J. Power Sources 193 (2009) 739–746.
[65] K. Haga, S. Adachi, Y. Shiratori, K. Itoh, K. Sasaki, Poisoning of SOFC anodes by various fuel impurities, Solid State Ionics 179 (2008) 1427–1431.
[66] Z. Cheng, J.-H. Wang, Y. Choi, L. Yang, M.C. Lin, M. Liu, From Ni-YSZ to sulfur-tolerant anode materials for SOFCs: electrochemical behavior, in situ characterization, modeling, and future perspectives, Energy Environ. Sci. 4 (2011) 4380.
[67] S. Zha, Z. Cheng, M. Liu, Sulfur poisoning and regeneration of Ni-based anodes in solid oxide fuel cells, J. Electrochem. Soc. 154 (2007) B201.
[68] T.S. Li, W.G. Wang, Sulfur-poisoned Ni-based solid oxide fuel cell anode characterization by varying water content, J. Power Sources 196 (2011) 2066–2069.
[69] J. Dong, Z. Cheng, S. Zha, M. Liu, Identification of nickel sulfides on Ni–YSZ cermet exposed to H_2 fuel containing H_2S using Raman spectroscopy, J. Power Sources 156 (2006) 461–465.
[70] Z. Cheng, M. Liu, Characterization of sulfur poisoning of Ni–YSZ anodes for solid oxide fuel cells using in situ Raman microspectroscopy, Solid State Ionics 178 (2007) 925–935.
[71] A. Lussier, S. Sofie, J. Dvorak, Y. Idzerda, Mechanism for SOFC anode degradation from hydrogen sulfide exposure, Int. J. Hydrogen Energy 33 (2008) 3945–3951.
[72] W.B. Guan, H.J. Zhai, L. Jin, C. Xu, W.G. Wang, Temperature measurement and distribution inside planar SOFC stacks, Fuel Cells 12 (2012) 24–31.

Cell, stack and system modelling 13

P. Pianko-Oprych[1], *Z. Jaworski*[1], *K. Kendall*[2,3]
[1]West Pomeranian University of Technology, Szczecin, Poland, [2]Adelan Ltd, Weekin Works, Harborne, Birmingham, UK, [3]School of Metallurgy and Materials, University of Birmingham, UK

13.1 Introduction

Several books have been written recently on SOFC modelling [1–3]. These mention the original modelling chapter [4] in the previous 'High-Temperature SOFCs' book (Elsevier Science, Oxford 1993), which laid out in some detail the basic methods for calculating the SOFC behaviour at molecular to macroscopic scales. The main problem over the past 10 years has been that the matching of computational results to experimental data has been rather limited. Therefore, the purpose of this chapter is to summarise the previous work, then go on to show that theory and experiment in this field can be reconciled to a certain extent.

The first part deals with the molecular diffusion theory which shows that huge concentration gradients can easily occur along the anode. The surprising result is that, even at the fuel entry point, sufficient oxygen will be diffusing from the anode exhaust to have an effect on the cell.

The second part addresses multi-scale computer modelling to demonstrate that the various processes in the cells can be integrated satisfactorily [5], especially the electrical behaviour, the flow field, the thermal/reaction and the thermal/mechanical stress effects. This part was covered quite well in the original 'High-Temperature SOFCs' book, but further results are still being obtained, especially as the cell operating temperature is reduced to 600–700 °C. The temperature gradient along the cell is still an enigmatic feature of great importance to rapid-start-up and long-life SOFCs.

Stack and system level modelling follow in the third and fourth part [6]. Clearly, this is greatly affected by the various designs discussed in Chapters 8–10. However, certain conclusions can be drawn as the stack/system is made more sophisticated by changing the fuel processing, the anode and cathode recycling, the combustor/recuperator and the heat utilisation methods. Addressing the simplest possible system in the first instance can lead to improved understanding.

Finally, we look at a novel theory which has recently been proposed to understand the oscillations often observed in an SOFC. This indicates that there are phenomena of great value which remain challenging as the complex reactions between hydrocarbons like methane and products like water are studied both experimentally and by theoretical analysis.

Firstly, we must consider the definition of parameters.

High-temperature Solid Oxide Fuel Cells for the 21st Century. http://dx.doi.org/10.1016/B978-0-12-410453-2.00013-0

Nomenclature

c	constant volume-specific heat, J/kg
c_0	constant volume-specific heat, J/kg
c_p	constant pressure-specific heat J/kg
C	concentration, kmol/m^3
d	particle diameter, m
d_p	pore diameter, m
$D_{m,j}$	diffusion coefficient of species m, m^2/s
D_{mn}	binary diffusion coefficient of species m in species n, m^2/s
D_{mK}	Knudsen diffusion coefficient of species m, m^2/s
e	electronic charge
e_i	specific internal energy, J/kg K
E	electromotive force, V
E_{Act}	activation energy of electrochemical reaction, J/kmol
$E_{act,a}$	anodic activation energy, J/kmol
$E_{act,c}$	cathodic activation energy, J/kmol
E_{aux}	auxiliary power consumption, W
E_{Nernst}	Nernst potential of cell, V
E_0	standard cell potential under standard conditions, V
F	Faraday constant, C/mol
$F_{h,j}$	thermal flux, J/m^{2}s
$F_{m,j}$	diffusive mass flux of species m, kg/m^2 s
H	enthalpy
	enthalpy flow
$h_{t,m}$	specific enthalpy of species m, J/kg
$\Delta^r H$	reaction enthalpy
h^*	specific enthalpy related to the standard state
i	current density, A/m^2
i_{elec}	electron transfer in cell electrolyte, C/m^3 s
i_{ion}	ion transfer in cell electrolyte, C/m^3 s
I	transfer current, C/m^3 s
I_{cell}	cell current, A
I_e	mean current density in electrolyte, C/m^2 s
I_0	exchange current density, C/m^2 s
k	equilibrium constant, reaction rate constant
K	thermal conductivity of gas mixture or solid materials, W/m K
K_i	permeability, kg/m^{3}s
l_e	electrolyte thickness, m
LHV	lower heating value
m	number of species, mass flow rate, kg/s
$m_{H_2,cons}$	mole flow rate of hydrogen consumed, kmol/s
$m_{H_2,in}$	mole flow rate of hydrogen fuel input to SOFC, kmol/s
M_m	molecular weight of species m, kg/kmol
M_n	molecular weight of species n, kg/kmol
n	number of electrons, –
n_{fuel}	molar fuel flow, kmol/s
n^{el}	quantity of released electrons related on the utilised fuel
	molar flow of electrons

n_F	molar quantity of the supplied fuel
	molar flow
n	molar quantity
N_A	Avogadro number, 1/kmol
N_{fuel}	mass flow of fuel, kmol/s
p	total pressure, Pa
p_o	standard pressure
p_{H_2a}	partial pressure of H_2 at the electrode–electrolyte interface, Pa
$p_{H_2O,a}$	partial pressure of H_2O at the electrode–electrolyte interface, Pa
p_{O_2a}	partial pressure of O_2 at the electrode–electrolyte interface, Pa
p_i^*	partial pressure of the ith reactant/product at the reaction sites, Pa
p_i^0	partial pressure of the ith reactant/product, Pa
P_{el}	electrical power
p_{el}	specific electrical power
$P_{out,gross}$	gross electrical power output of SOFC, W
P_{loss}	power loss
P_{rev}	reversible electrical power
q	specific heat
Q	heat
$Q_{heat\text{-}recovery}$	cogeneration heat rate, W
q_{rad}	radiative heat flux, J/m^2 s
R	electrical or ohmic resistance
R_m	universal gas constant, J/kmol K
R_e	cell reaction rate, mole/m^3 s
R_f	Air/fuel ratio
s_c	chemical energy source due to endo- or exothermic chemical processes, J/m^3 s
s_e	heat source due to electrochemical processes, J/m^3 s
s_h	energy source due to ohmic resistance and radiation, J/m^3 s
s_i	ith momentum source component, kg/m^2 s^2
s_m	mass production/consumption rate, kg/m^3 s
s_Σ	mass source, kg/m^3 s
s^*	specific entropy related to the standard state
S	total entropy
T	temperature, K
T_{surf}	MEA surface temperature, K or °C
T_∞	furnace temperature, K or °C
u_i	velocity component, m/s
U_a	air utilisation, %
U_f	fuel utilisation, %
U_i	superficial velocity, m/s
V	specific volume
V	voltage or potential
V_N	Nernst voltage
V_{cell}	cell voltage, V
$\dot{V}_i$	volumetric flow rate of the i-component, m^3/s
W_{el}	electric work, J/kg
$-W_t$	technical work
$-w_t$	specific technical work

x_i	*i*th direction, –
x_{O_2}	volume fraction of oxygen, m^3/m^3
X_{CH_4}	methane contribution to hydrogen formation, –
X_{CO}	carbon monoxide contribution to hydrogen formation, –
Y_m	mass fraction of species *m*, kg/kg

Greek letters

α_i	*i*th coefficient, –
β	symmetry coefficient,
β	ballast ratio
β_i	*i*th coefficient, –
Γ_{H_2}	fuel equivalent hydrogen content, –
γ_a	anodic pre-exponential coefficient, –
γ_c	cathodic pre-exponential coefficient, –
ε	porosity, –
η_{act}	activation polarisations, V
η_{conc}	concentration polarisation, V
$\eta_{DC/AC}$	efficiency of DC/AC conversion, –
η_{el}	electrical efficiency, %
η_{Ohm}	ohmic polarisation, V
η_{stack}	stack efficiency, %
η_{system}	system efficiency, %
$\eta_{thermal}$	thermal efficiency, %
λ	stoichiometric air to fuel ratio = 1 at stoichiometric
ξ	MEA surface emissivity, –
μ	molecular dynamic viscosity of the gas mixture, kg/m s
μ	chemical potential, V
μ^0	standard chemical potential, V
ν	fuel-related quantity
ρ	gas mixture density, kg/m^3
σ	electronic/ionic conductivity, 1/Ω m
σ_e	electrolyte anionic conductivity, 1/Ω m
σ_{elec}	electronic conductivity in anode, 1/Ω m
σ_{ion}	ion conductivity in electrolyte material, 1/Ω m
σ_{mn}	collision diameter Lennard–Jones potential model, *m*
σ_{SB}	Stefan–Boltzmann constant, $W/m^2\ K^4$
Ω_{mn}	collision dimensionless Lennard–Jones integral, –
δ	thickness, m
δ_{ij}	Kronecker delta, –
τ	tortuosity, –
τ_{ij}	stress tensor, Pa
θ	Celcius temperature, °C
φ	potential, V
φ_{elec}	electron potential, V
φ_{ion}	ion potential, V
$\Phi_{e,int\text{-}a}$	electric potential of the outermost layer of anode interconnects, V
$\Phi_{e,int\text{-}c}$	electric potential of the outermost layer of cathode interconnects, V
$\overline{\Delta h_f}$	enthalpy of formation, J/kmol

ΔH_e	enthalpy change, J/kmol
$\overline{\Delta g_f}$	Gibbs free energy, J/kmol
ΔG^0	standard Gibbs free energy change of cell reaction, J/kmol
ΔV	voltage loss
λ	air fuel ratio (=1 at stoichiometric for full oxidation)
π	pressure ratio
ζ	exergetic efficiency

Subscripts

a	anode
act	activation polarisation
c	cathode
cons	consumed
e	electrode (anode, cathode)
elec	electrolyte, electrons
f	fuel
ion	ions
i, j, k	Cartesian directions
m, n	species index

Acronyms

ADL	anode diffusion layer
APUs	auxiliary power units
ATR	autothermal reforming
CFD	computational fluid dynamics
CHP	combined heat and power system
CPOx	catalytic partial oxidation
FC	fuel cell
FEM	finite element method
HHV	fuel high heating value, J/kmol
LHV	fuel lower heating value, J/kmol
LSM	strontium-doped lanthanum manganite
MEA	membrane-electrode assembly
O/C	oxygen to carbon ratio
OCV	open circuit voltage
S/C	steam to carbon volume ratio
SOFC	solid oxide fuel cell
SR	steam reforming
YSZ	yttria-stabilised zirconia

13.2 Basic definitions

Solid oxide fuel cell (SOFC) technology represents a valuable, promising alternative to traditional energy technology, which is based on the use of fossil fuels leading to an increase in emissions of greenhouse gases and noxious pollutants. Due to many advantages of an SOFC such as high electrical efficiency, long-term durability, low emissions and low noise, it is expected that SOFC will play a major role in future as a well-suited power source for a number of stationary and portable applications ranging from combined heat and power (CHP) systems [7] to auxiliary power units (APUs) [8,9]. SOFC-based CHP systems have already reached a mature development level allowing introduction of the fuel cell technology into the market. Several reported SOFC systems reach electrical efficiencies in the range of 30–60% and overall efficiencies up to 85% [10]. However, there are different parameter sets that exert a significant impact on electrical efficiency, power and start-up behaviour of SOFC systems. Fuel utilisation U_f, air utilisation U_a, air to fuel ratio λ and cell temperature gradient are critical factors that have to be evaluated in order to reach the target stack power. SOFC cell, stack and system electrical and overall efficiencies as well as fuel and air utilisation factors used throughout this chapter are first defined.

The cell electrical efficiency, η_{el}, can be defined as a function of the cell voltage, V_c, fuel utilisation factor, U_f and inlet composition, Γ_{H_2}, according to Equation (13.1) proposed by Campanari [11]:

$$\eta_{el} = \frac{W_{el}}{\text{LHV}} = \frac{n \cdot F \cdot E}{\text{LHV}} = \frac{n \cdot F \cdot V_c + U_f + \Gamma_{H_2}}{\text{LHV}} \tag{13.1}$$

where W_{el} is the electrical energy produced per mole of fuel, LHV is known as the lower heating value and is equal to the value of the enthalpy of formation, $\overline{\Delta h_f}$, that would be produced by burning fuel and formation of water vapours [12].

The cell electrical efficiency can also be calculated according to Equation (13.2) proposed by Ameri [13]:

$$\eta_{el} = \frac{\text{Power}}{n_{\text{fuel}} \cdot LHV} = \frac{I_{\text{cell}} \cdot V_{\text{cell}}}{n_{\text{fuel}} \cdot LHV} \tag{13.2}$$

where the molar fuel flow, $n_{\text{fuel,}}$ is defined as follows:

$$n_{\text{fuel}} = \frac{n_{H_2,\text{equivalent}}}{n_{H_2,\text{in}} + n_{CO,\text{in}} + 4 \cdot n_{CH_4,\text{in}} + 7 \cdot n_{C_2H_6,\text{in}} + 10 \cdot n_{C_3H_8,\text{in}} + 13 \cdot n_{C_4H_{10},\text{in}}} \tag{13.3}$$

where $n_{H_2\ \text{equivalent}}$ denotes the equivalent hydrogen flow rate:

$$n_{H_2,\text{equivalent}} = \frac{n_{H_2,\text{cons}}}{U_f} = \frac{I}{2F \cdot U_f} \tag{13.4}$$

The fuel utilisation factor, U_f, refers to the amount of fuel oxidised electrochemically in the anode compartment of the cell stack (Equation 13.5):

$$U_f = \frac{m_{H_2,\text{cons}}}{m_{H_2,\text{in}}} = \frac{m_{H_2,\text{in}} - m_{H_2,\text{out}}}{m_{H_2,\text{in}}} \tag{13.5}$$

where $m_{H_2,cons}$ is the molar flow rate of hydrogen consumed and $m_{H_2,in}$ is the molar flow rate of hydrogen fuel input to a SOFC.

The fuel equivalent hydrogen content, Γ_{H_2}, can be calculated based on Equation (13.6) for a fuel mixture of H_2–CH_4–CO–H_2O:

$$\Gamma_{H_2} = X_{H_2} + X_{CH_4} + X_{CO} \tag{13.6}$$

where X_{CH_4} and X_{CO} are methane and carbon monoxide contributions to hydrogen formation, accounting for a complete hydrocarbon reforming and for the additional contribution of CO shift reaction.

The maximum electrical energy available is equal to the ratio of the change in Gibbs free energy, $\overline{\Delta g_f}$, to the enthalpy of formation, $\overline{\Delta h_f}$, and can be defined by Equation (13.7):

$$\eta_{max} = \eta_{therm} = \frac{\overline{\Delta g_f}}{\overline{\Delta h_f}} \cdot 100\% \tag{13.7}$$

The overall electrical efficiency, $\eta_{electrical}$, is usually described as:

$$\begin{aligned} \eta_{electrical} &= \frac{\text{electrical work output}}{\text{chemical energy input}} \\ &= \eta_{therm} \cdot \eta_{el} \cdot U_f \cdot \eta_{reformer} \cdot \eta_{system\ components} \end{aligned} \tag{13.8}$$

The air utilisation factor, U_a, is proportional to the oxygen quantity extracted from the air flow to oxidise fuel and can be defined as follows (Equation 13.9):

$$U_a = \frac{m_{O_2,cons}}{m_{O_2,in}} = \frac{m_{O_2,in} - m_{O_2,out}}{m_{O_2,in}} \tag{13.9}$$

The air to fuel ratio, R_f, is equal to the number of moles of air required to oxidise 1 mol of the given fuel accounting for air utilisation factor and can be calculated from Equation (13.10):

$$R_f = \frac{m_{air,\ inlet}}{m_{fuel,\ requested}} \tag{13.10}$$

Finally, the system performance can be defined on the net electrical efficiency, η_{net}, and the overall system efficiency, η_{system}, as proposed in Ref. [14]:

$$\eta_{net} = \frac{P_{out,\ gross} \cdot \eta_{DC/AC} - E_{aux}}{N_{fuel} \cdot \text{LHV}} \tag{13.11}$$

$$\eta_{system} = \frac{\left(P_{out,\ gross} \cdot \eta_{DC/AC} - E_{aux}\right) + Q_{heat-recovery}}{N_{fuel} \cdot \text{LHV}} \tag{13.12}$$

where $P_{out,gross}$ is the gross electrical power output of an SOFC, $\eta_{DC/AC}$ is the efficiency of DC/AC conversion, E_{aux} is the auxiliary power consumption, $Q_{heat\text{-}recovery}$ is the cogeneration heat rate and N_{fuel} is the mass flow of fuel. An extensive comparison of the efficiency of SOFC systems tested from 2000 to 2008 was reported by Staffell [15].

To achieve high efficiency of an SOFC, sensitivity analyses have to be carried out to obtain better understanding of the influence of key factors such as fuel utilisation or air to fuel ratio on the performance of the system. Assuming that high efficiency of an SOFC has to be reached, the fuel utilisation should be enhanced, which also means an increase of the electrochemical reaction rates raising the electrical power output per cell. However, at the same time, fuels (hydrogen, methane) are diluted by reaction products (water, carbon dioxide). The variation of gas concentrations and dilution effect of the hydrogen content in fuel stream along the flow path cause temperature gradients in the SOFC, which may lead to thermo-mechanical stresses and as a consequence to SOFC degradation [16]. Furthermore, carbon deposition may also cause serious damage to the power generation characteristics of SOFCs as a result of catalyst pore closure and encapsulation of the active sites. Nevertheless, from the design point of view, especially cell temperature gradient is a crucial parameter for two reasons: first – the cell materials can only withstand limited values before high stress and cracking occur and second – the power output from the fuel cell depends strongly on local temperature. Thermo-mechanical stresses can be induced within the inlet regions due to the start-up operation, when the whole system needs to be heated up to a temperature of at least 750 °C. In consequence, it may cause failure of stack components. In order to avoid local thermal gradients, Al-Masr et al. [17] suggested reducing the mass flow rate of the heating air to its minimum to prevent high stress in the stack components. Moreover, thermal stresses can be caused also by differences in material mechanical properties. Mounir et al. [18] noticed that the magnitude of the gradient stress depends on temperature difference in the anode and electrolyte as well as in electrolyte and cathode. Therefore, a key issue in the fuel cell improvement and maximising its endurance is thermal management. Heating rates or hot air flow rates have to be controlled depending on allowable maximum thermal stress or thermal gradient for specific stack design. One method to reduce the internal stresses is to reduce the operating temperature to intermediate range around 600–800 °C [19]. Vijay et al. [20] found that a new design for the SOFC planar stack, which includes a secondary air channel with flow in the counter direction to that in the cathodic air channel, was capable to reduce the steady state temperature difference across the cell to less than 2 K over a range of voltages. In addition, the flow path and temperature distribution determine also the spatial performance observed across the cell active area. Concentration or mass transport polarisation depends on the availability of reactants in the active area, which in turn is affected by the distribution of fuel and oxidant [21]. In this respect, significant local voltage deviation can be identified depending on the local fuel concentration. Thus, all these factors should be carefully considered in the design of optimal operating conditions. The local considerations may then be extended to cover the whole cell, then the full stack and finally, the overall system by multi-scale modelling.

13.3 Multi-scale modelling

Understanding the SOFC system performance at different length scales is very important. Molecular processes of diffusion and reaction must be considered first. Then each part of the cell must be modelled. The overall reactor picture comes next, followed by the system behaviour, which is the final result. Initially, we discuss molecular models, then move up to continuum calculations by numerical process simulators, coupling mass flow, heat transport and mechanical stress phenomena with electrochemical processes at the micro-scale with velocity and temperature distributions in the air and fuel channels at the macro-scale while including aspects of system components integration. This is the multi-scale modelling approach in fuel cell research as described in Ref. [4].

An impressive result shown here is Figure 13.1 which gives the one-dimensional (1D) theoretical and experimental *I–V* relations of a small-size cell (considered isothermal) for different fuel compositions at a set of temperatures [9]. The material properties were obtained by fitting the theory to the experimental data for 97% H_2+3% H_2O fuel. As shown, the simplified theory can predict variation in the *I–V* curve with fuel composition reasonably well.

13.3.1 Reaction diffusion

At the molecular level, reactions are occurring and the reactants and products are diffusing, assuming no convection, according to the well-known equations below [22],

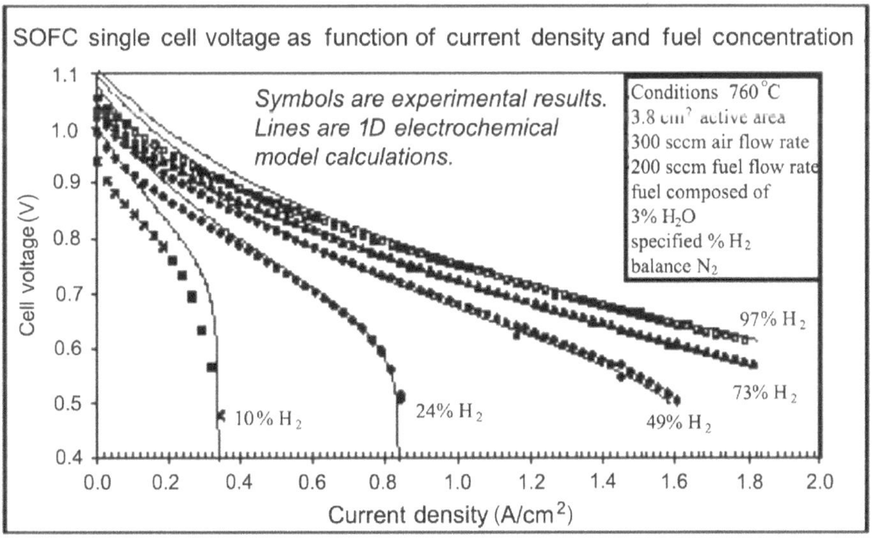

Figure 13.1 Predicted and measured cell *I–V* curves [9].

where D was the diffusion coefficient, C was oxygen concentration and $0.5C_f$ was methane concentration in the reaction:

$$CH_4 + 2O_2 \rightarrow CO_2 + 2H_2O \tag{13.13}$$

where h is the thickness of the fuel chamber, z is the thickness coordinate and a the disc radius, where r is the radial coordinate. The reaction of methane with oxygen is complex because it is catalysed by other gases such as steam produced. Also the reaction order is likely to change with position. Here we assume that the reaction is on average third order with reaction constant k.

$$D\left(\frac{\partial^2 C}{\partial r^2} + \frac{1}{r}\frac{\partial C}{\partial r} + \frac{\partial^2 C}{\partial z^2}\right) - kC^2C_f = 0 \text{ for oxygen diffusion} \tag{13.14}$$

$$D\left(\frac{\partial^2 C_f}{\partial r^2} + \frac{1}{r}\frac{\partial C_f}{\partial r} + \frac{\partial^2 C_f}{\partial z^2}\right) - kC^2C_f = 0 \text{ for methane diffusion} \tag{13.15}$$

The boundary conditions are: $C=C$ at $r=a$ and $C_f=0$ at $r=a$, where C is the atmospheric oxygen concentration and no fuel leaves the space unreacted.

These equations could not be solved analytically, but a transition layer solution could be found for comparison with experimental results in which methane (natural gas) was metered into the space at flow rate N_f.

Consider a simple radial flow shown in Figure 13.2a in which the methane and air enter the disc SOFC chamber at the centre and flow to the circumference. When no current is going through the electrolyte, the anode then shows a remarkable effect where a black disc is surrounded by a green region (Figure 13.2b) with a very sharp boundary whose diameter is small at low flow rate, but increases for high flow rates. This result reflects the reducing conditions near the fuel entry at the centre where nickel metal is the main anode component, oxidising conditions near the outer rim where nickel oxidises to its green form, with a sharp boundary where the changeover from oxygen starved to oxygen rich occurs. Figure 13.3 shows the comparison between theoretical prediction of the anode voltage versus the experimental measurements.

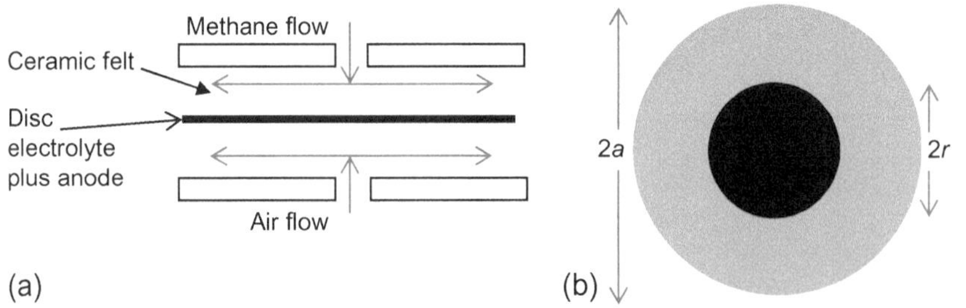

Figure 13.2 (a) Cross section through fuel cell arrangement where the spaces were filled with ceramic felt; (b) Top view of the nickel anode showing the reduced black disc surrounded by the oxidised green (grey in the print version) nickel oxide.

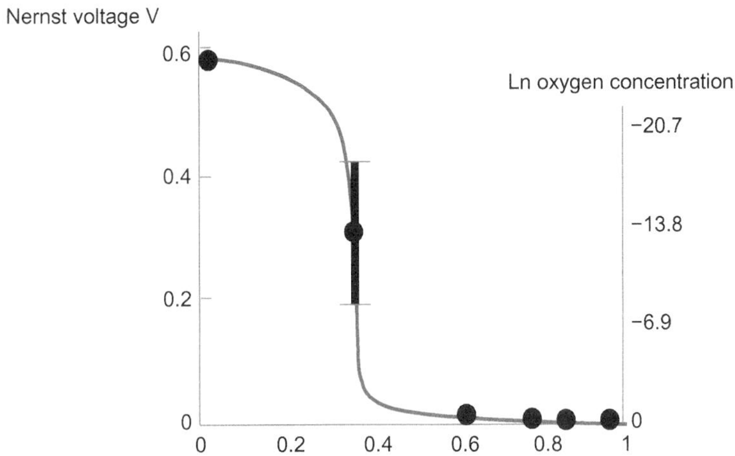

Figure 13.3 Transition layer solution of the governing equation for $\varepsilon = 1.4^{-13}$ and $\mu = 1$ compared to experimental results at 900 °C with $\mu = 1.95$ for natural gas bubbled through water at 18 °C with a 2 mm thick Zircar felt on the anode side where $\mu = N_f/\pi DCh$.

Mathematical analysis shows that there are essentially three domains in this curve: the fuel domain to the left of Figure 13.3 where there is little oxygen so that the potential is approaching 1 V, the oxygen domain to the right of Figure 13.3 where the potential is near zero and a transition zone in the middle of Figure 13.3 where the voltage falls rapidly from near 1 V to near zero as the oxygen concentration rises through 20 orders of magnitude.

This spatial voltage cliff, reminiscent of the lambda curve in an oxygen sensor used in automobile exhaust systems to find the stoichiometric condition (see Chapter 10, Figure 10.2), was measured by altering the methane flow rate and observing the potential on fine platinum electrodes positioned on the nickel anode surface. The results fitted the theory as shown in Figure 13.3. The theory showed that the oxygen concentration at the fuel inlet due to back-diffusion of air from the exit was significant, indicating that molecular diffusion of oxygen must have an influence on the anode potential gradient. A later paper [23] showed the effect of oxygen ion flow through the electrolyte.

Two simple equations emerged from this analysis:

The first was the radius of the transition zone given by

$$r = a \exp(-\pi DCh/N_f) \tag{13.16}$$

The second was the concentration of oxygen at the fuel inlet caused by the back-diffusion given by

$$C = 15.134\frac{\pi D^2 h}{N_f a^2 k} \exp\left(\frac{2\pi DhC}{N_f}\right) \tag{13.17}$$

During these experiments, oscillations were seen in the voltage signal emerging from the electrodes, suggesting that there were several metastable states along the anode. Kendall attempted to study this phenomenon during the 1990s and observed oscillations with methane, but only succeeded in 2010 when he hired Jonathan Sands [24] as a PhD student to investigate the effect with Prof. Needham and Dr. Uddin in Mathematics at Birmingham. These oscillations are dealt with in Section 13.5.

13.3.2 Multi-scale modelling by computer

Multi-scale modelling has been applied generally to chemical engineering. More specifically, it can be applied to SOFCs as presented by Zakrzewska, Jaworski [25] and Bessler [26]. This multi-scale approach covers the two main types of modelling methods: (i) CFD methods and (ii) process simulator tools.

Cordiner et al. [27], as one of the first, used this approach to speed up the design phase both for component development and SOFC system integration. The overall model included two modules, which allow to represent the biomass gasifier and SOFC behaviour. The gasifier was simulated by a 0D model assuming that the reactor residence time was high enough to let the pyrolysis products convert into gas products, while the SOFC based on a 3D CFD model, which included gas evolution into gas channels and porous electrodes. The internal reforming was modelled by the introduction of the homogeneous phase methane reforming and water–gas–shift (WGS) reactions. The output parameters for the gasifier were the amount of air and gas composition. The CFD results for the gasifier module were validated against experimental data for the case of adiabatic gasifier ($dH_{int}=0$) [27]. A calorific value of the output gas at varying operating temperature and the effect of input water content on the module output in terms of dry basis volume fractions at 800 °C equal to 0.2 were compared. It was found that the effect of input water content on the methane fraction was negligible, while for hydrogen and carbon dioxide fractions increased with increasing water content [27]. An error between experimental and numerical results was equal to 10.7% at 750 °C, 8.7% at 800 °C and it dropped to 6.1% at 900 °C [27]. The SOFC model delivered a quantitative data for the following output parameters: current field in the electrolyte as well as cathode and anode outlet compositions, temperature and residual calorific value. To incorporate the gasifier with the whole system, an integration of heat flux provided by the SOFC anode off-gas combustion was performed. The ability to change the LHV at the SOFC outlet allowed a high utilisation of 57% to be obtained with quite high efficiency of 34.3% (27).

An extension of the previous studies was developed by Bessler [26] and is summarised in Table 13.1. A comprehensive approach assumes five levels of modelling of SOFC: diffusion through the surface, laminar flow through porous media at the electrode level, transport of oxidant and fuel at the fuel cell level, electrical circuits of the cell and thermal balancing, integration of the entire system with balance-of-plant components.

Table 13.1 **Multi-scale modelling approach of chemical and transport processes in a SOFC [26]**

Level	Length and time scales	Geometry	Chemistry	Transport	Modelling approach
Surface	10^{-8} m 10^{-6} s	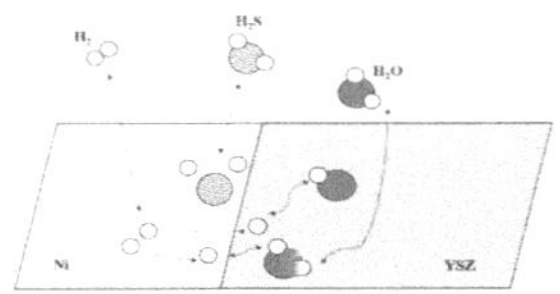	Surface chemistry	Surface diffusion	Mean field elementary kinetics with CANTERA
Electrode	10^{-4} m 10^{-2} s	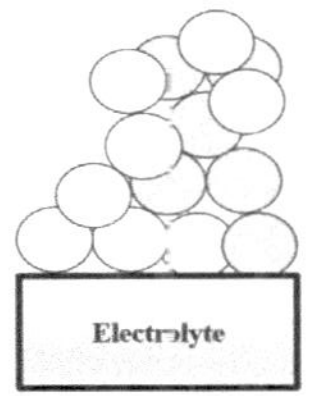	Thermal and redox cycling	Porous multiphase mass and charge transport	Continuum transport and percolation theory with DENIS
Fuel cell	10^{-2} m 10^{0} s	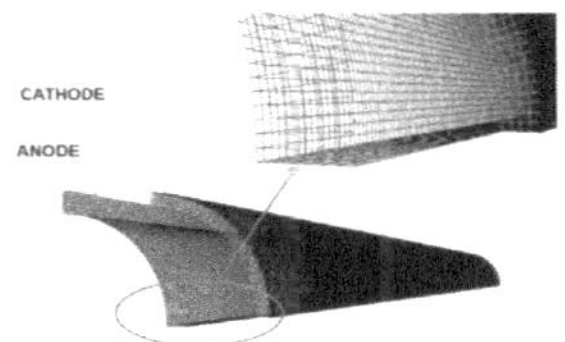	Gas-phase chemistry	Laminar flow Heat convection	Heat, mass and charge transport with DENIS
Stack	10^{-1} m 10^{2} s			Laminar flow Heat conduction radiation	CFD codes: ANSYS-Fluent, COMSOL
System	10^{0} m 10^{4} s	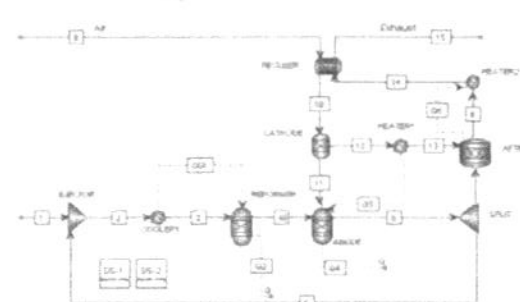		Mass, energy fluxes	Process simulators: Aspen Plus™, HYSYS, Matlab-SIMULINK

13.3.3 Cell level modelling

Numerical modelling is a highly useful tool for understanding the multi-physical nature of the phenomena that occur in SOFCs at different scales of length and time. At the cell level, two major different modelling approaches can be distinguished: micro-models and macro-models. The majority of published papers [28–30] based on micro-models played a critical role in understanding the phenomena occurring inside the fuel cell at the electrode microstructure level and had a very important role in the development of the SOFC technology. Many of these studies were performed to estimate the key parameters that influence electrode properties and to determine electrode performance [31–33], while macro-models deal with the macroscopic or overall operational behaviours of SOFCs [34].

13.3.3.1 Micro-modelling of electrodes

The micro-modelling provides a valuable tool for predicting the electrochemical performance of SOFC electrodes. The approach can also assist in optimising the cell architecture. Both experimental [35,36] and modelling studies [28] recognised that the micro-structural parameters of electrodes have significant effect into the transport and reaction processes within the electrodes. Thus, Kreller et al. [28] distinguished three main groups of micro-structural representations that are used in micro-kinetic modelling of porous SOFC electrodes as shown in Figure 13.4. In the first macro-homogeneous approach, all characteristics of a microstructure are represented in terms of volume-average macro-homogeneous parameters such as porosity or transport tortuosity [29]. These models can predict well when the active region of an electrode is large in comparison to individual micro-structural features, such as particle dimensions [30,37].

In the second approach, microstructure is represented by pseudo-particles, which have a size and arrangement chosen to match a sub-set of average micro-structural properties such as porosity and surface area. A pseudo-particle model is able to capture the effect of 3D gradients near the three-phase boundary, when the size of the

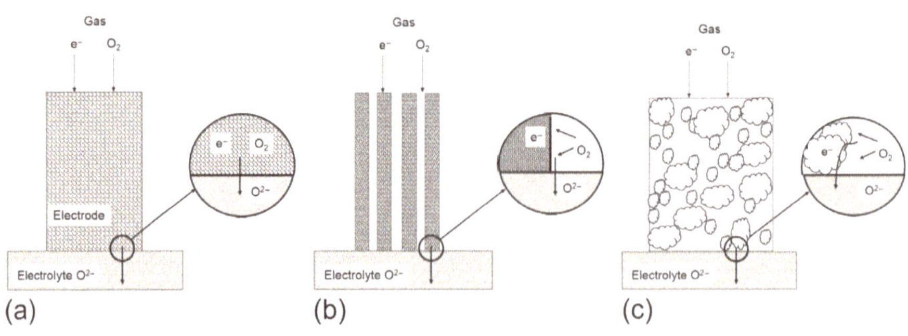

Figure 13.4 Various approximations of a microstructure used in micro-kinetic modelling of porous SOFC electrodes: (a) macro-homogeneous approach, (b) pseudo particles, (c) 3D geometry based on images of microstructures [28].

utilisation region is competitive with the microstructure [38]. However, this approach does not take into account many other details of the microstructure, such as particle size dispersion, interfacial contact angle and interfacial contact area.

In the third approach, an experimentally characterised sample of a 3D microstructure based on FIB-SEM reconstruction is employed as a model for the microstructure as a whole. These models accurately represent all the important aspects of a microstructure for its kinetics, but simulation times are very long [28].

It was also found that parameters such as volume fraction, phase distribution, average aggregate size, morphology and percolation of each phase can significantly affect the properties of composite electrodes [28]. Therefore, knowledge of parameters such as particle diameter, porosity and thickness has become very important/valuable. Determination of these parameters is possible by recent progress in experimental techniques such as scanning electron microscopy (SEM) images, FIB-SEM or X-ray micro-tomography [37–40] that led to direct three-dimensional reconstructions of electrode micro-structures. For example, Tabei et al. [41] reconstructed a 3D microstructure of Ni/YSZ electrode using Monte Carlo methodology. The 3D reconstruction was based on analysis of 2D SEM micrographs of real composite materials. FEM simulations were performed to estimate elastic and thermal properties of the reconstructed composite materials. A reconstructed porous Ni-YSZ composite showed mechanical behaviour in between full isotropic and full orthotropic with average Young's moduli of 17 GPa at room temperature and 10.2 GPA at 1000 °C, which were in good agreement with experimentally reported values [41]. Izzo et al. [37] developed reconstructions of electrodes using X-ray tomography and provided quantitative information about the complexity of electrodes.

The X-ray holotomography and FIB/SEM (focused ion beam with imaging by SEM) techniques were used also by Delette et al. [42] to reconstruct three-dimensional SOFC/SOEC electrode microstructures. In this investigation, elastic coefficients and thermal expansion coefficient (CTE) were obtained based on a homogenization technique for different stochastic volume elements extracted from the whole reconstruction, including the support and the functional layer of a Ni/YSZ supported cell as well as LSCF electrode. FEM (finite element method) simulations were done with fixed strain (KUBC, kinematic uniform boundary conditions – homogeneous deformation) applied to the six faces of the Representative Volume Element, RVE. The RVE was an elementary domain that contains sufficient numbers of solid phase particles and pores. Periodic boundary conditions (KPBC) were also used to estimate Young's modulus and Poisson coefficient values. It was found that KUBC leads to higher values of the modulus than KPBC. The Young's modulus obtained by KUBC approach was equal to 34 GPA and was very close to the experimental value of 37 GPA. Results obtained on the thick porous support Ni/YSZ were in good agreement with the experimental values also for $\varepsilon = 46\%$ [42]. The porosity and effective Young's modulus values were close to the values measured; thus it confirms that the reconstruction sizes were large enough to catch the main features of the electrode microstructure. The numerical model predicted also the effective thermal expansion of the porous electrodes. However, it was noticed that it depended only on the composition of the solid phase. A very good agreement with the experimental values of CTE measured or

Ni/YSZ cermet by Mori et al. [43] was also found. In addition, the results obtained by FEM gave the deviation lower than 0.25% from the values of CTE obtained from analytical expressions developed by Turner [44] and Kerner [45].

The effect of porosity and particle size on micro-structural properties was also investigated by Bertei et al. [46]. In the study, transport and reaction effective properties of SOFC electrodes were estimated for microstructures numerically generated through a Monte Carlo random walk method. The predicted results of tortuosity factor, permeability or effective conductivity were compared with independent experimental data and a satisfactory agreement for both gas and solid phase properties were obtained. The tortuosity factor in gas phase was determined by applying the mean-square displacement method using 20,000 tracers. A comparison of the tortuosity factor as a function of porosity between the simulated results calculated by Bertei et al. [46] as well as by Zalc et al. [47] and the experimental data obtained by Currie [48] is shown in Figure 13.5. It can be concluded that the tortuosity factor increases monotonically as the porosity decreases, in particular more strongly for porosity below 0.2. Figure 13.5 shows a good agreement between the simulated results obtained by Bertei et al. [49] and Zalc et al. [47]. In addition, comparison with experimental data of Currie [48] was also satisfactory, even though limited to a small range of porosity close to the rigid packing limit.

In the micro-modelling studies, the transport and electrochemical processes are usually evaluated by using simple empirical lumped-parameter kinetic models such as the Butler–Volmer equation or highly detailed micro-kinetic models that are based on finite-element analysis considering in detail microstructure, material properties, elementary kinetics and transport phenomena [28]. By analysing individual processes in more detail, for example, adsorption/desorption at various surfaces, chemical and electrochemical reactions, flow processes through complex electrode channels

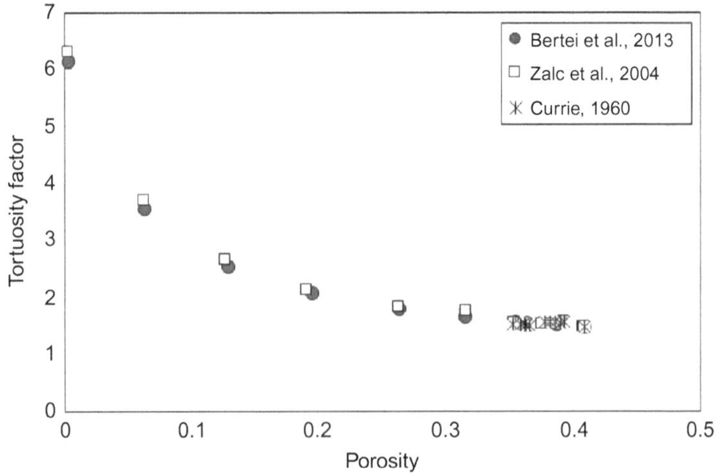

Figure 13.5 Results showing electrode predictions versus experimental results for tortuosity factor as a function of porosity in monosized random packing of spheres [46–48].

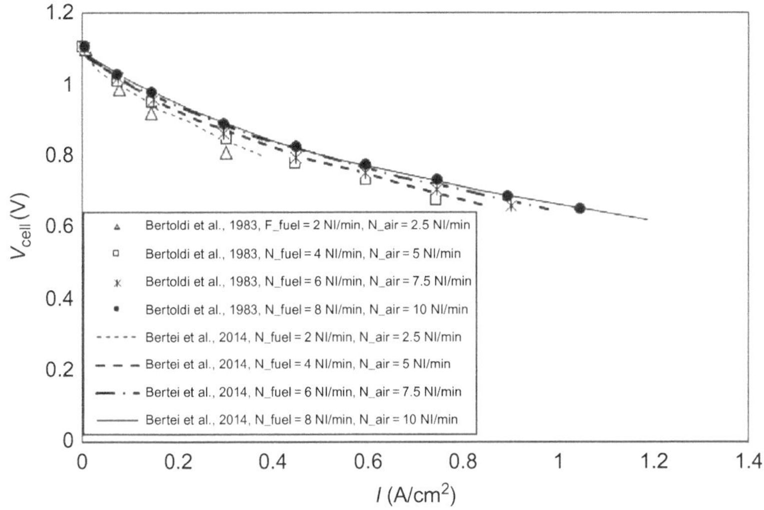

Figure 13.6 Comparison with experimental data for polarisation curves of the 4 cell SOFC stack at different flow rates of fuel and air [49].

depending on particle packing of the powder processed electrodes [38,39,50–63], predictive models have been produced and there is some agreement with experimental evidence as shown in Figure 13.6.

The results obtained by Bertei et al. [49] and presented in Figure 13.6 are an example of integration a detailed micro-structural modelling of the porous layers for estimation of the effective properties with the cell level electrochemical model, which allow to predict the polarisation behaviour in planar SOFCs. A good agreement between simulated [49] and experimental [64] polarisation curves was obtained in a 4-cell short SOFC stack with counter flow configuration. Its voltage decreased monotonically with increasing current density as a consequence of the higher activation, ohmic and concentration overpotentials. It can also be noticed that the agreement was almost perfect for fuel flow rate of 8 Nl/min, while worsened for smaller fuel flow rates. In addition, the comparison of predicted effective properties with experimental data or those evaluated on SEM images revealed discrepancies in the range of 30% as a maximum. However, we should keep in mind that it was a reasonable accuracy taking into account the complexity of the system. Thus, it should be underlined that the main advantage of the developed approach was exemption from using empirical correlations and adjusted parameters, which was usually a weak point of existing cell level models.

13.3.3.2 Macro-modelling: Cell level

Modern macro-modelling at the cell level is performed by finite volume discretization methods using various commercially available computational fluid dynamics (CFD) codes such as ANSYS-Fluent, COMSOL, CFD-RC, STAR-CD, PHOENICS and

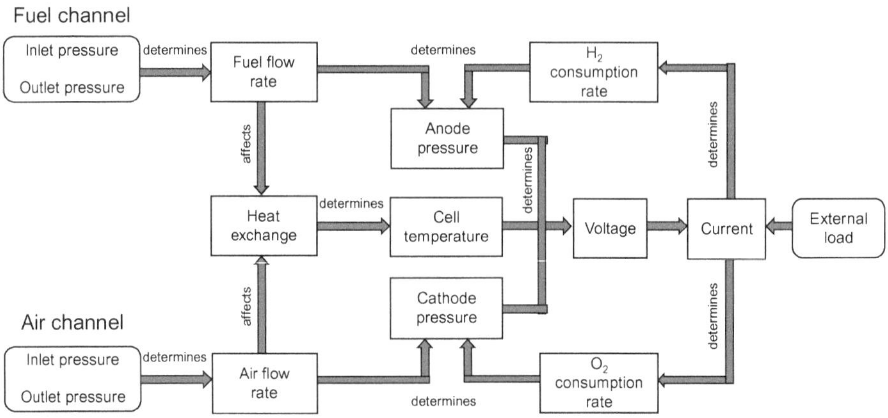

Figure 13.7 Scheme of logic correlations of SOFC process variables [3].

FLOW3D [50–53] or OpenFOAM software [53–55]. CFD offers a great potential for modelling the electrochemical and gas-phase reactions in quite short computing times. A huge increase in computational power during last few years has contributed to reduction of computing time allowing estimation of fuel cell performance at different oxygen utilisation rates, fuel to oxygen ratios or different fuel compositions. Nevertheless, still the main challenge for SOFC modelling at the macro-modelling level is an efficient coupling of the fluid flow, conduction, convection, radiative heat transfer and electrochemical reactions inside fuel cell system. The main modelling output comprises partial pressures, local temperatures, ionic and electronic conduction. Thus, understanding of these interactions is a critical issue for optimising fuel cell/stack designs and their further development. In general, all SOFC processes may be summarised schematically in a block diagram presented in Figure 13.7.

The input parameters at the cell level include cell dimensions, cell components, physical properties of reactants and flow compositions, inlet and outlet temperatures of fuel, air as well as current density, while at the stack level, macro-models require fuel/air flow rates and fuel cell configuration in the stack. Inlet and outlet pressures determine fuel and air flow rates in the anode and cathode channels. The flow rates significantly affect the heat exchange process inside the stack and cell temperatures. Reactants pass through a porous electrode to reach reaction sites. Chemical and electrochemical processes are controlled by concentration gradients between triple-phase boundaries and gases flowing in electrode channels. On the other hand, an external load determines the current that an SOFC produces and the rate of reactant consumption in electrochemical reactions [2]. In fuel cells, transport of mass, momentum, energy and charge take place simultaneously through gas channels and porous electrodes and therefore these phenomena have to be concurrently modelled. Electrical power and voltage are obtained as output parameters at the cell level, while at the stack level, component mole fractions and temperature distributions are obtained.

The need for integrating models of fuel and air flows with heat transfer and species transport along with electrochemical and reforming kinetics, when hydrocarbons are used as fuel, was reflected in several numerical studies [65–67]. Andersson et al. [65] is one of the first to describe the mechanism behind the transport processes within SOFCs in terms of momentum, mass, heat and charge transport. In addition, approaches to model electrochemical as well as internal reforming reactions were discussed in detail. An extensive review of studies on mathematical modelling of SOFCs was also presented by Hajimolana et al. [56] with respect to tubular and planar configurations divided into five subsystems and considered polarisation losses, mass, energy, momentum conservations, diffusion through porous media, electrochemical phenomena in the electrode/electrolyte region and shift/reforming reactions inside SOFCs. Lin et al. [68] developed a new comprehensive mechanistic model that coupled interdependent processes of species transport, heat transport, chemical reaction, electrochemical reaction as well as ionic and electronic conduction. The model allows to predict non-uniform velocity distributions across channels in an SOFC. The authors [68] proposed a non-uniformity index to quantitatively evaluate fuel velocity distribution within channels in the planar SOFC because of its influence on cell performance. Zhang et al. [69] applied 3D CFD model for anode-supported planar SOFC analysis to three different flow arrangements: co-, counter- and cross-flow involving multi-channels and compared the versions in terms of overall cell performance. The simulations were performed for various transport parameters and resulted in a good agreement with experimental data. It was found that the temperature for the three flow arrangements was unevenly distributed and significant temperature gradients existed along the fuel cell length, which may decrease the overall cell performance as well as the working life time [69].

Global reactions that take place in a hydrogen-fed SOFC consist of the oxidation of hydrogen in the anode and the reduction of oxygen in the cathode, releasing a water molecule as a product in the anode:

$$\text{Anode side}: \; H_2 + O^{2-} \rightarrow H_2O + 2e^- \tag{13.18}$$

$$\text{Cathode side}: \; \frac{1}{2}O_2 + 2e^- \rightarrow O^{2-} \tag{13.19}$$

$$\text{Overall reaction}: \; H_2 + \frac{1}{2}O_2 \rightarrow H_2O \tag{13.20}$$

The described reactions are global. However, the reaction kinetics are temperature dependent and more details regarding reaction mechanisms can be found in papers [70,71].

As already mentioned, accurate predictions of SOFC behaviour require calculating various coupled phenomena concurrently, according to the following mathematical model.

The mass transport of the chemical species m in the system is represented by (Equation 13.21):

$$\frac{\partial \varepsilon \rho Y_m}{\partial t} + \frac{\partial\left(\rho u_j Y_m + F_{m,j}\right)}{\partial x_j} = s_m \tag{13.21}$$

where Y_m is the mass fraction, s_m is the mass production/consumption rate due to chemical and/or electrochemical reactions and $F_{m,j}$ is the diffusive mass flux of species m. The mass flux is modelled using Fick's law:

$$F_{m,j} = -\rho D_{m,j} \frac{\partial Y_m}{\partial x_j} \tag{13.22}$$

where $D_{m,j}$ is the diffusion coefficient of species m, modelled often as:

$$\frac{1}{D_m} = \frac{\tau}{\varepsilon}\left(\frac{1}{D_{mn}} + \frac{1}{D_{m\mathrm{K}}}\right) \tag{13.23}$$

where D_{mn} is the binary diffusion coefficient of species m in species n, $D_{m\mathrm{K}}$ is the Knudsen diffusion coefficient of species m and τ is the tortuosity. Thus, the two diffusion coefficients can be determined by the following Equations (13.24) and (13.25) [61]:

$$D_{mn} = \frac{3}{16}\sqrt{\frac{2(R\cdot T)^3}{\pi}\left(\frac{1}{M_m} + \frac{1}{M_n}\right)} \cdot \frac{1}{N_\mathrm{A}\cdot p\cdot \sigma_{mn}^2 \cdot \Omega_{mn}} \tag{13.24}$$

$$D_{m\mathrm{K}} = \frac{d_\mathrm{p}}{3}\sqrt{\frac{8\cdot R\cdot T}{\pi \cdot M_m}} \tag{13.25}$$

where, d_p is the pore diameter, p is the total pressure, σ_{mn} is the collision diameter Lennard-Jones potential model, Ω_{mn} is– the collision dimensionless Lennard-Jones integral, T is the temperature, R is the universal gas constant, N_A is Avogadro's number and M_m and M_n are the molecular weights of species m and n, respectively.

The velocity and pressure fields in the computational domain are governed by the mass conservation (Equation 13.26) and the momentum balance of the ith component (Equation 13.27):

$$\frac{\partial \varepsilon\rho}{\partial t} + \frac{\partial \rho u_j}{\partial x_j} = s_\Sigma \tag{13.26}$$

$$\frac{\partial \rho u_i}{\partial t} + \frac{\partial\left(\rho u_j u_i - \tau_{ij}\right)}{\partial x_j} = -\frac{\partial p}{\partial x_i} + s_i \tag{13.27}$$

using the following notation: ε – the porosity, ρ – the gas mixture density, u_i – the velocity component in the x_i direction x_i, p – the pressure, s_Σ – the mass source, s_i – the ith momentum source component, τ_{ij} – the stress tensor component. The mass source, s_Σ, is the total mass consumption/production of all chemical species in a unit volume, which is non-zero in the catalyst layers due to electrochemical reactions. The porosity is usually assigned a value of unity in the gas channels, a value between

0 and 1 in the electrodes and a value of approximately 0 in the electrolyte. The stress tensor components for a Newtonian fluid were given by Equation (13.28):

$$\tau_{ij} = \mu\left(\frac{\partial u_i}{\partial x_j} + \frac{\partial u_j}{\partial x_i}\right) - \frac{2}{3}\rho\frac{\partial u_k}{\partial x_k}\delta_{ij} \tag{13.28}$$

where μ denotes molecular dynamic viscosity of the gas mixture and δ_{ij}, the Kronecker delta. In the porous parts of the electrodes–electrolyte structure, the gas permeation is defined by Darcy's law (Equation 13.29):

$$\frac{\partial p}{\partial x_i} = -K_i U_i \tag{13.29}$$

where U_i is the superficial velocity of direction x_i, K_i is the permeability defined as (Equation 13.30):

$$K_i = \alpha_i|u| + \beta_i \tag{13.30}$$

where α_i and β_i are the coefficients assumed to be the same in all directions and defined by Equations (13.31) and (13.32) given by Bird et al. [61]:

$$\alpha = \frac{1.75 \cdot \rho \cdot (1-\varepsilon)}{\varepsilon^3 \cdot d} \tag{13.31}$$

$$\beta = \frac{150 \cdot \mu \cdot (1-\varepsilon)^2}{\varepsilon^3 \cdot d^2} \tag{13.32}$$

where d is the particle diameter.

Finally, conservation of energy is accounted for in simulations by Equation (13.33):

$$\frac{\partial(\varepsilon\rho e)}{\partial t} + \frac{\partial(\rho u_j e + F_{\mathrm{h},j})}{\partial x_j} = \tau_{ij}\frac{\partial u_i}{\partial x_j} + s_\mathrm{h} + s_\mathrm{e} + s_\mathrm{c} \tag{13.33}$$

where $F_{\mathrm{h},j}$ is thermal flux, s_h is the energy source due to ohmic resistance and radiation, s_c is the chemical energy source due to endo- or exothermic chemical processes, s_e is the heat source due to electrochemical processes and e is the specific internal energy.

The thermal flux, $F_{\mathrm{h},j}$, comprises Fourier conduction and heat transported by diffusive mass fluxes described as follows:

$$F_{\mathrm{h},j} = -k\frac{\partial T}{\partial x_j} + \sum_m h_{\mathrm{t},m} F_{m,j} \tag{13.34}$$

where k is the thermal conductivity of a gas mixture or of solid materials and $h_{\mathrm{t},m}$ is the specific enthalpy of species m.

The energy source term, s_h, may be used to take into account heat generated by resistances to the flow of ions and/or electrons in the solid parts of a cell, which is called Joule heating or ohmic heat loss and can be determined from Equation (13.35):

$$s_h = \frac{i^2}{\sigma} \tag{13.35}$$

where i is the local electronic/ionic current density and σ is the electronic/ionic conductivity. Zhang et al. [72] found that the ohmic heat loss is about 2.3–4.1% and 8.0% of the total heat released in a planar electrolyte-supported cell and a tubular cathode-supported cell, respectively.

The specific internal energy can be described by Equation (13.36):

$$e = c \cdot T - c_0 \cdot T_0 \tag{13.36}$$

where: c, c_0 – constant volume specific heat, J/kg.

Modelling radiative heat transfer in SOFC is a complex process and the radiative transport within electrodes and electrolyte as well as surface to surface heat transfer have to be accounted for by accurate calculations. However, thermal properties such as absorption coefficient, refractive index, scattering coefficient, emissivity, reflectivity are difficult to determine. Detailed discussions of heat transfer by radiation in SOFCs are given in Refs. [62,63,73]. Some of the authors assumed that radiative heat transfer in SOFC electrodes is minimal and has a negligible effect on the temperature field [63,74,75]. Other authors used a simple radiation model, in which the surfaces of the MEA radiate to a black-body enclosure at the nominal furnace temperature and the radiative heat flux, q_{rad}, from the MEA surfaces can be evaluated as follows (Equation 13.37):

$$q_{rad} = \xi \sigma_{SB} \left(T_{surf}^4 - T_{\infty}^4 \right) \tag{13.37}$$

where: σ_{SB} – the Stefan–Boltzmann constant, T_{surf} – the MEA surface temperature, T_{∞} – the furnace temperature, ξ – the MEA surface emissivity.

The heat released by electrode reactions can be computed as:

$$s_e = -R_e(\Delta H_e + 2 \cdot F \cdot V_{cell}) \tag{13.38}$$

where: R_e – the cell reaction rate based on the rate of change of H_2 molar concentration, ΔH_e – the reaction enthalpy change, F – the Faraday's number, V_{cell} – the operating cell voltage. The electrochemistry and the electric field interact solely at the electrolyte interface. This interface is treated as an impermeable wall by most of the CFD codes such as ANSYS – Fuel Cell Tools module. The effect of electrochemistry is accounted for by a "jump" condition. In this jump condition, all occurring electrochemical effects have to be included to couple electrochemical behaviour to the potential field. These effects cover the voltage jump due to the Nernst potential,

E_{Nernst}, the voltage reduction due to activation of both the anode and cathode, η_{act}, ohmic electrolyte losses, η_{conc}, and ohmic losses in solid conducting regions such as current collectors, η_{Ohm}.

$$V_{cell} = E_{Nernst} - \eta_{conc} - \eta_{act} - \eta_{Ohm} \tag{13.39}$$

For a H_2-fed SOFC, combining the two reversible potentials gives the Nernst potential of the cell, E_{Nernst}, at the open-circuit conditions as follows:

$$E_{Nernst} = E_0 + \frac{R \cdot T}{2 \cdot F} \ln\left(\frac{p_{H_2a}}{p_{H_2O,a}}\right) + \frac{R \cdot T}{4 \cdot F} \ln\left(p_{O_2c}\right) \tag{13.40}$$

where p_{H_2a}, $p_{H_2O,a}$, p_{O_2a} – the partial pressure of the species H_2, H_2O, O_2 at the electrode–electrolyte interface, E_0 – the standard cell potential under standard conditions with the standard Gibbs free energy change of the cell reaction, ΔG^0, defined by Equations (13.41) and (13.42), respectively:

$$E_0 = \frac{-\Delta G^0}{2 \cdot F} \tag{13.41}$$

$$\Delta G^0 = \mu^0_{H_2O} - \mu^0_{H_2} - \frac{1}{2}\mu^0_{O_2} \tag{13.42}$$

where: μ – the chemical potential, given by $\mu = \mu^0 + R \cdot T \cdot \ln p$ for ideal gas.

Concentration polarisation, η_{conc}, is due to mass transport resistance in the electrodes and can be calculated as follows [76]:

$$\eta_{conc} = \eta_{conc,a} + \eta_{conc,c} = \frac{R \cdot T}{n \cdot F} \ln\left(\frac{p^*_{H_2} \cdot p^0_{H_2O}}{p^*_{H_2O} \cdot p^0_{H_2}}\right) + \frac{R \cdot T}{n \cdot F} \ln\left(\frac{\sqrt{p^*_{O_2}}}{\sqrt{p^0_{O_2}}}\right) \tag{13.43}$$

where: p^*_i denotes the partial pressure of the ith reactant and product at reaction sites, while p^0_i indicates the partial pressure of the feed values of reactants and products.

Activation polarisations (activation losses), η_{act}, are the results of the kinetics involved in electrochemical reactions. The polarisations become an important loss, when the current is low, because at a low current density reactants have to overcome an energy barrier named activation energy to drive electrochemical reactions at electrodes–electrolyte interfaces and this barrier leads to the polarisation. The Butler–Volmer equation is used to describe the relation between the activation polarisation overpotential, η_{act}, and the transfer current, I:

$$I = I_0\left\{\exp\left(\beta\frac{n \cdot F \cdot \eta_{act}}{R \cdot F}\right) - \exp\left[-(1-\beta)\frac{n \cdot F \cdot \eta_{act}}{R \cdot F}\right]\right\} \tag{13.44}$$

where β – the symmetry coefficient, n – the number of participating electrons involved in electrochemical processes, F – the Faraday's constant, I_0 – the exchange current density, the calculation of which for both electrodes uses the following experimental correlations [77]:

$$I_{o,a} = \gamma_a \cdot \left(\frac{p_{H_2}}{p_{ref}}\right) \cdot \left(\frac{p_{H_2O}}{p_{ref}}\right)^{-0.5} \cdot \exp\left(\frac{-E_{act,a}}{R \cdot T}\right) \tag{13.45}$$

$$I_{o,c} = \gamma_c \cdot \left(\frac{p_{O_2}}{p_{ref}}\right)^{0.25} \cdot \exp\left(\frac{-E_{act,c}}{R \cdot T}\right) \tag{13.46}$$

where: γ_a and γ_c are the anodic and cathodic pre-exponential coefficients, $E_{act,a}$ and $E_{act,c}$ – are the anodic and cathodic activation energies.

The ohmic polarisation is mainly caused by the resistance to the transport of oxygen anions through the electrolyte, η_{Ohm} and can be estimated as:

$$\eta_{Ohm} = \frac{\delta \cdot l_e}{\sigma_e} \cdot I_e \tag{13.47}$$

where: l_e – the electrolyte thickness, σ_e – the electrolyte anionic conductivity, I_e – the mean current density in the electrolyte estimated as follows:

$$I_e = \frac{I_a + I_c}{2} \tag{13.48}$$

To transport oxide ions from the electrolyte to the anode (Equation 13.19) and to block electron flow from the anode to the cathode are the main functions of a SOFC electrolyte. The flow of electronic charges through external circuit balances the flow of the ionic charge through the electrolyte and thus the electrical power is produced. This transport can be described by considering the ionic and electronic charge transport Equations (13.49) and (13.50):

$$\nabla \cdot i_{ion} = Q_{ion} \tag{13.49}$$

$$\nabla \cdot i_{elec} = Q_{elec} \tag{13.50}$$

and the Ohm's law for ion and electron transfer in an electrolyte (Equation (13.51) and (13.52)):

$$i_{ion} = -\sigma_{ion} \nabla \varphi_{ion} \tag{13.51}$$

$$i_{elec} = -\sigma_{elec} \nabla \varphi_{elec} \tag{13.52}$$

where: φ – the ionic/electron potential, σ_{ion} – the ion conductivity in the electrolyte material, σ_{elec} – the electronic conductivity in the anode, $\sigma_{elec,a}$, or in the cathode, $\sigma_{elec,c}$. The interfaces/boundaries can be defined as electric potential ($V = V_0$), electric isolation ($n \cdot i = 0$) or ground conditions ($V = 0$). The effects of electrochemical reactions are either defined as source terms in governing equations or as interface conditions at the electrolyte/electrode interface as was discussed by Bessler et al. [78].

More information about popular and frequently used fuel cell module and governing equations can be found in the manual for the CFD code [79] or in Refs. [34,68,80].

The mathematical model presented in this section allows to calculate the 3D distribution of gas-phase species, current density, potential as well as pressure and flow velocities for SOFCs. Each of the previous equations including the reaction rates need to be solved at each iteration. Due to the complexity of the model, which takes the form of a partial differential equation system, it is computationally stiff and therefore efficient and specialised numerical techniques are needed to solve the equations. In addition, various sub-models are coupled through appropriate source terms, which means that transport properties influence species distributions and consequently influence charge-transfer kinetics, which again influences concentration gradients. More details related to the solution algorithm and iteration procedure can be found in [80].

Typical CFD results compared with experiments for a microtubular SOFC operating at 750 °C and for a planar SOFC operating at 800 °C are shown in Figures 13.8 and 13.9, respectively.

The CFD results presented in Figure 13.8 show a good agreement with the experimental values obtained by Meadowcroft et al. [8] in the operation voltage range of the mSOFC. However, a significant difference in the depletion region at high current density can be noticed. The MEA approach used in that case overestimated hydrodynamic conditions in the anode channel and as a consequence led to overly optimistic results of the anode mass transport.

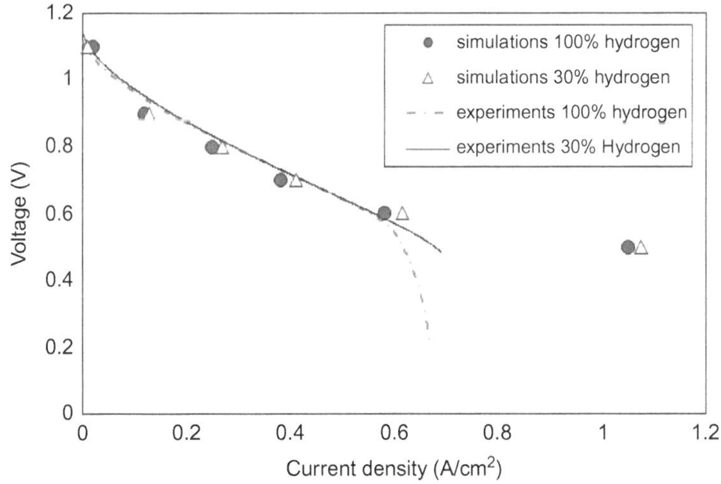

Figure 13.8 Comparison of the predicted [60] *V–I* curve from the CFD modelling with the experimental data of [8] for the single mSOFC 3D model at 750 °C with 30% and 100% hydrogen.

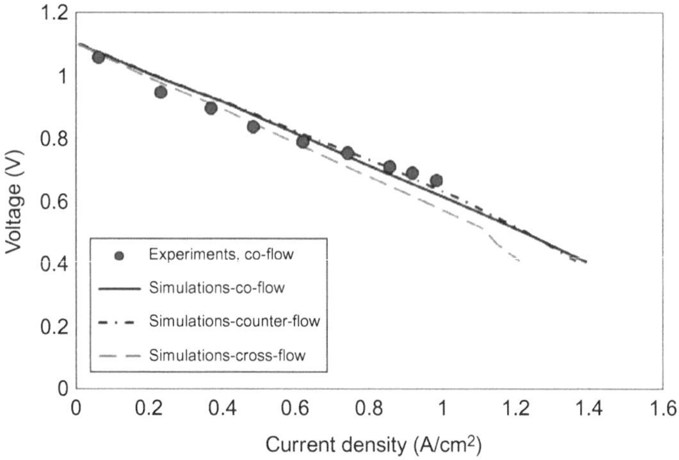

Figure 13.9 Comparison of the predicted *V–I* curve from the CFD model with experimental data at the same operating conditions for the single planar SOFC 3D model at 800 °C of co- and counter-flow [69].

Figure 13.9 presents the *V–I* curves predicted from the 3D CFD model compared with the experimental data showing a good agreement was achieved. It can also be noticed that the co-flow and counter-flow arrangements gave a better performance than that of the cross-flow.

Other, more general modelling approaches can be found also in the literature [81–83]. They classified models into three main categories: grey, black and white-box models [81]. The grey-box models based on a combination of *a priori* knowledge concerning the process and mathematical relations, which describe the behaviour of a SOFC system. Black-box methods predicted fuel cell performance without prior knowledge of many physical, chemical and electrochemical parameters. Black-box models are usually based on artificial intelligence such as Artificial Neutral Networks or neuro-fuzzy systems [82,83]. White box models were of two types: physical models and equivalent circuit models based on electrochemical impedance spectroscopy (EIS) technique [84].

More details regarding steady state macro level models including the multiphysics nature of SOFCs can be found in references [37–40,69,70,80,85].

Many studies carried out at the macro-modelling level have focused on performance of fuel cells during operation taking into account the influence of fuel/air utilisation. However, we cannot forget about a risk of carbon deposition using hydrocarbon reformates or fuel cell mechanical instability due to stress applied to ceramic components. Carbon formation and deposition on the anode surface were discussed by Schluckner et al. [34], who simulated a planar industrial-sized SOFC fed with diesel reformat. A very high degree of consensus of simulated and experimentally obtained data was reached for the assumed range of temperature and fuel compositions.

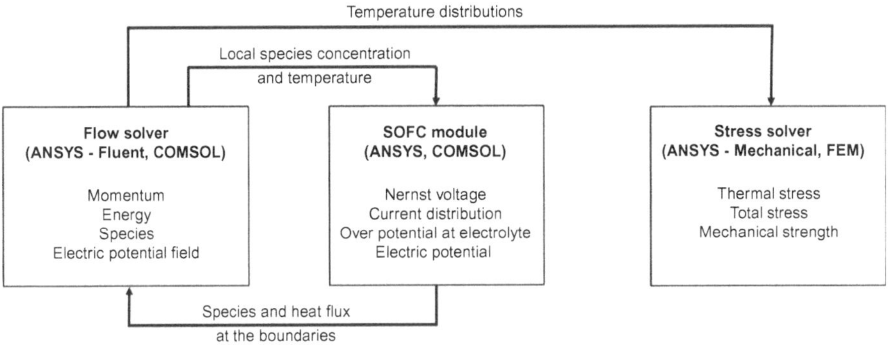

Figure 13.10 Coupling of a flow solver, a SOFC module and a stress solver [86].

Degradation of SOFC component materials, which results from high fuel cell operating temperature is another limitation to their lifetime and that topic has been investigated quite often by different investigators. An extended overview about the heating up stage over time of a planar SOFC and its thermo-mechanical reliability was given by Peksen et al. [53]. A transient-coupled 3D CFD and computational solid mechanics models were used to examine, by simulating the heating up process, the critical locations susceptible to thermo-mechanically induced stress over time. To understand the thermo-mechanical behaviour of fuel cell, the temperature distribution was an important indicator. Therefore, the accuracy of the predicted temperature distributions was validated with the measured ones using N-type thermocouples. The CFD model included coupled conduction, convection and radiation and thermal field data obtained from CFD were implemented in the FEM analyses over time. An example of a similar coupling between CFD and FEM is shown in Figure 13.10 [86].

Comparison of the simulation and measurement results of the SOFC thermal behaviour is presented in Figure 13.11. The slightly lower CFD predictions were attributed to the radiation process, which probably was more intensive in reality. However, Peksen et al. [53] noticed also that the radiation model predicted the results more accurate at higher temperatures.

Mechanical reliability and durability of SOFCs were studied by Nakajo et al. [87] as well. A thermo-electrochemical model including degradation was coupled with a finite-element thermo-mechanical model. Temperature profiles generated by the thermo-electrochemical model were implemented in the gPROMS tool. The analysis was focused on the planar anode-supported SOFC, and the Weibull analysis provided assessment of the probability of failure of each MEA layer. The anode probability of failure was predominantly governed by the temperature profile pattern, whereas that of the LSM–YSZ cathode depended upon the local CTE mismatch with the other cell layers. In the case of a glass-ceramic sealant the cathode and anode probability of failure decreased more than one order of magnitude during a short annealing period of 10 h, when 100% reformed methane was fed. However, the anode probability of failure increased in counter-flow, when the methane conversion fraction of 25% was used [87].

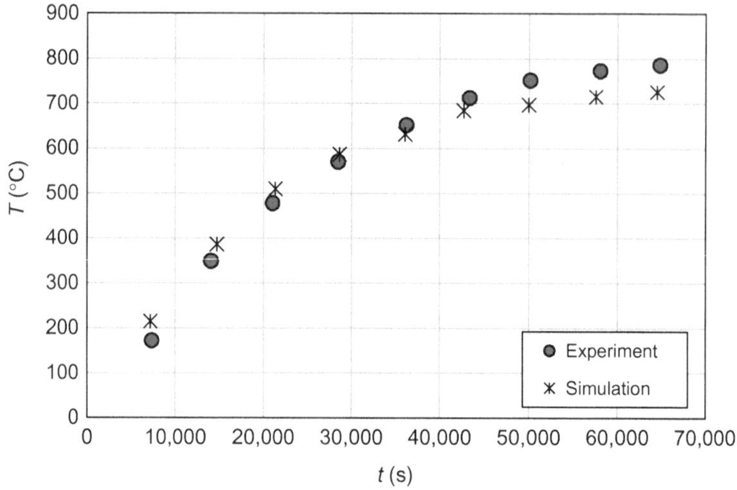

Figure 13.11 Comparison of temperature distributions (°C) of a planar SOFC over time (s) with the CFD results obtained by Peksen et al. [53].

The effects of temperature on the probability of survival of cell components were discussed also by Mounir et al. [18]. The temperature profiles can be divided into two zones and in the first zone the probability of survival of cell components was constant, while in the second zone, the probability of survival decreased quite rapidly. In the second zone, the survival probability of fuel cell components decreased due to the increasing thermal gradients. It was also noticed that the probability of survival of the electrolyte was lower than of the other fuel cell components. The authors [18] believed that this was related to the extreme tensile stresses in the electrolyte due to its low values of CTE. In addition, the SOFCs were submitted to the moderate thermal stress at the inlet fuel flow and it underwent a high thermal stress at the outlet.

Chiang et al. [88] used Star-CD to investigate the effects of different operating voltage, temperature gradients and different thermal expansion coefficients between components on thermal stresses. Calculations were performed using the Star-CD software with the es-sofc module and included combustion of hydrogen with oxygen, gas–water–shift reaction, capability for steam reformation of methane and addition of nitrogen leak into the system. To calculate the thermal stress distributions in fuel cell, the finite element programme MARC were used. Simulation results showed that thermal stresses were much higher at a lower operating voltage. The maximum principal stresses at a voltage of 0.5 V was equal to 11.47 MPa, while for voltage of 0.7 V, it was equal to 6.90 MPa. Moreover, the maximum principal stress increased from the value of 37.6–153.4 MPa as the maximum temperature gradients increased from 50 to 200 °C. The average strength of an anode-supported positive–electrolyte–negative (PEN), depending on its composition, was around 71–187 MPa [88]. It was also found that principal stresses were lower for the inlet gas temperature of 790 °C. Hence, it would be beneficial to preheat inlet gases to a certain extent to minimise temperature gradients and subsequently reduce thermal stresses of cell components. Chiang et al. [88] reported that temperature difference should be controlled within 50 °C on the PEN surface for the sake of structure integrity.

Selimovic et al. [89] focused on determination of the stress caused by different thermal expansion coefficients at high temperature gradients occurring in the SOFC layers during steady-state and transient operations, taking into account heat-up, start-up and shut-down conditions. A 2D model based on the finite volume method was developed in FORTRAN for simulation of a planar electrolyte-supported SOFC with internal reforming. The steady state model allowed to calculate the temperature and current density distribution, species concentration and channel flows. The solid model within Structural Mechanics Module in the commercial finite element tool FEMLAB was used to estimate thermal stress in the three layered structure of SOFC. A significant temperature gradient in solid cell material of 12 °C/mm was noticed, which corresponds to the maximum principal stress of 17 MPa in the electrolyte layer located close to the interface with the anode and close to the fuel inlet. The location of the highest stress was explained by the steep temperature drop caused by endothermic reforming and also by larger difference in CTE between the anode and electrolyte material [89]. In addition, transient SOFC simulations for heat-up, start-up and shut-down operations were considered. It was found that heating up the cell with the temperature difference of 100 K took around 2.5 h and the stress occurring during this process was beyond the limit, while assuming the same conditions for shut-down operation took 4 h and the maximum principle stress was similar to that occurring at the steady state of 17 MPa [89].

Last but not least, one of the major challenges faced by the macro level modelling is the ability to simulate the operation of SOFC in dynamic conditions. The model has to be able to predict the behaviour of the operation of SOFC such as start-up, load change or change of the inlet temperature of the streams. Al-Masri et al. [17] developed and tested a 3D CFD model that was able to simulate the transient heat-up process of a full scale planar type APU SOFC short stack. The temperature field was computed over time and areas with the maximum temperature values were recognised. The measured temperature at the cathode outlet was compared with the CFD results and a good agreement was noticed as shown in Figure 13.12 with the maximum temperature difference of 6% [17].

To sum up, only few research studies [17,88–91] examined transient behaviour or thermal stresses in SOFCs, which may have resulted from the fact that precise data of the mechanical properties of the MEA materials are rare. However, a challenge with SOFC is in obtaining mechanically stable structure, which is resistant to failure when subjected to moderate stress. Therefore, for the future, continuation of this direction of numerical research seems necessary in order to improve fuel cell performance and its reliability.

13.3.4 Stack level modelling

For efficient stack performance, uniform inlet gas flow and proper flow channel design are one of the critical issues in achieving more uniform temperature

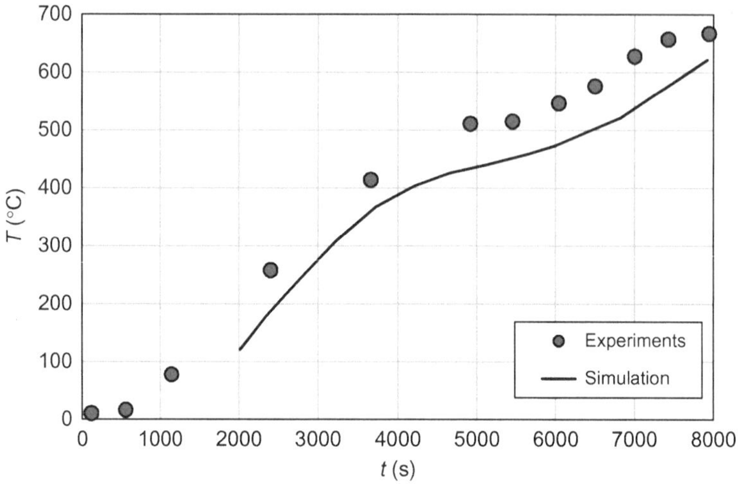

Figure 13.12 Cathode outlet temperature vs. time for planar SOFC with a footprint of 154 × 132.5 mm [17].

distribution. High-temperature gradients in the SOFC stack may lead to fuel cell cracks and stack failure. Thus, the CFD and FEMs tools have been applied by many researchers [62,63,74,75,87,89,92–95] and [96–100] to perform simulations of heat transfer and thermal stress in a SOFC stack and validate the stack design.

13.3.4.1 Heat transfer

Accurate modelling of heat transfer offers three benefits. Firstly, it provides knowledge regarding the temperature fields, enables definition of potential areas and sources of heat loss as well as allows to develop an effective way of waste heat utilisation. The second advantage results from knowledge of local temperature values, which strongly influence chemical kinetic rates and transport properties. Finally, knowledge of temperature gradients is needed in estimation of thermal stresses, which is helpful in predicting the risk of fuel cell damage and failure [101].

Therefore, a SOFC stack thermal model should include the following components: heat release and absorption stemming from reforming, shifting or electrochemical reactions, heat transfer by convection between the solid and the gas phases and radiation heat transfer between the solid part and the gas [102]. Several papers [101–105] considered in details thermal models for SOFC stacks.

One of the first numerical investigations on heat transfer in the SOFC stack for both planar and tubular configurations was carried out by VanderSteen and Pharoah [101]. It demonstrated that radiation has to be considered as part of an accurate SOFC stack model. The thermal transport model included conduction, convection and radiation in the participating media based on a commercial CFD code. For the tubular SOFC, it was shown that including radiation in the thermal model results in a lower average wall temperature, which means that radiation decreased temperature gradients in the cell. This was a significant result, because it clearly demonstrated that ignoring radiation heat transfer in a thermal model of a fuel cell can result in the calculated temperature being over 30° too high when operating at 900 °C. Similar results were obtained for the planar SOFC stack, where the CFD results revealed that temperature gradients were greatly affected by radiation at high operating temperatures.

Thus, it is clear that a key issue in fuel cell improvement and maximising SOFC endurance is its thermal management. Temperature of SOFCs can be controlled by varying the air flow rate supplied to the cathode channels [106]. Average fuel cell temperature will be quite effectively controlled by this method; however, temperature distributions across the stack still depend very strongly on its design. Important role played by the stack design in thermal distributions was presented by Vijay et al. [20], who proposed a new design for a planar SOFC stack including a secondary air channel with flow in the opposite direction to that in the cathodic air channel in order to improve the stack cooling. Vijay et al. [20] proved that the new design was capable to reduce the steady state temperature difference across the cell to less than 2 K over a range of voltages. It was also shown that the secondary air stream led to lowering of the temperature non-uniformity and allowed to reduce thermal stress in the stack.

A significant change in stack temperature due to inclusion of radiation effects was confirmed also by Hartvigsen et al. [107], but no detail of the radiation model used in the analysis was presented. Moreover, Yakabe et al. [93] found that the implementation of radiation heat exchange within the flow channels resulted in a flatter distribution of the temperature profile along the fuel cell. However, the model neglected absorption, emission and scattering in the media and it took into account only radiation effects for surface-to-surface transfer.

The effect of radiation heat transfer from and to the fuel cell stack surface were investigated numerically also by Tanaka et al. [105]. Numerical simulations of a planar SOFC stack for co-flow at various ambient temperatures between 800 and 1000 °C were performed in the steady-state conditions. The effect of heat convection with the gas flow and heat transfer between the solid and gas were taken into account by including them in the heat generation term of the heat conduction equation. It was found that the influence of radiation heat transfer was very large near the surfaces of the cell stack and cell temperature at the top, bottom and side surfaces was mainly determined by the ambient temperature [105]. The central part of the cell stack was almost free from the influence of radiation heat transfer. An interesting dependence of the single cell voltage on the ambient temperature was noticed during the investigations. Since the single cell voltage obtained by the cell was 0.751 V, when the ambient temperature was equal to 1123 K and 0.768 V at temperature of 1223 K, the average single cell voltages of the cells in the central region of the stack almost coincided with the corresponding single cell voltage obtained by the single fuel cell simulation. Authors explained it by the fact that the thermal conductivity of the cell materials was very small and the effect of the radiation heat transfer from the top and bottom surfaces have not reached the central region of the cell stack. However, as a result of low ambient temperature, the fuel cells near the top and bottom surfaces were cooled by the radiation heat transfer and the averaged single cell voltage of these cells was reduced [105].

The importance of thermal radiation in the overall heat transfer within the electrode and electrolyte layers of SOFCs was assessed also by Daun et al. [75]. The authors [75] compared the temperature field obtained by means of a conduction model with those obtained using two coupled conduction and radiation models. The air and fuel bulk inlet temperatures were set equal to 950 K and also a uniform current density of 4000 A/m^2 was assumed. Two different radiation solvers were applied to estimate the radiation source term in the electrolyte: the Schuster–Schwarzchild two-flux method and the collision-based Monte Carlo method. Temperature profiles obtained by the method were slightly higher than the Schuster–Schwarzchild solution since the former accommodates radiation heat transferred in the z – direction while the latter does not. Nevertheless, the differences were much smaller than the expected accuracy of the simulation. This also means that radiation, at least within the electrolyte, can be excluded from the heat transfer model without significant error [75]. A similar conclusion was also presented by Damm and Fedorov [62,63,74] who found that neglecting radiation may result in errors less than 1 K in the temperature field of an anode-supported SOFC. In contrast to these numerical analysis, Murthy and Fedorov [108] concluded that ignoring radiation effects caused errors larger than 100 K within an electrolyte-supported SOFC. These results were also confirmed by Daun et al. [75], who examined influence of radiation heat transfer on temperature fields of cathode- and electrolyte-supported SOFCs. The CFD results showed that radiation have a larger effect on temperature fields in the electrolyte-supported cells in comparison to the electrode supported SOFCs. This finding suggests that radiation probably became more important as the conduction resistance between the electrodes increases.

Simulation results presented by Kattke et al. [109] showed that radiation heat transfer was the dominant mechanism of SOFC stack cooling for tubular stacks. Radiation accounted for 62–93% of total heat rejection from the external tube surface. The dominance of radiation led to a strong relationship between the power output of a tube and its view factor from the tube to the relatively cold cylinder wall surrounding the bundle.

Despite some inconsistencies in the simulation results, enormous amount of information can be extracted from the CFD analysis of heat transfer in SOFC stacks and such analysis is crucial for improvements in stack design considering also how difficult is to measure local temperature within complex SOFC configurations.

13.3.4.2 Thermo-mechanical models

Typical high operating temperature of SOFC, which is usually above 650 °C, lead to stresses caused by differences in mechanical and thermal properties of the SOFC materials. Therefore, a significant effort has been made by many researchers to investigate several materials composing anodes, cathodes, electrolytes, interconnectors and sealants, which are joined together to form an assembly. Thermo-mechanical stress–strain formulation of the planar and tubular SOFCs was highlighted by Nakajo et al., [87, 95], Selimovic et al., [89], Peksen [92, 94], Fischer and Seume [96,97], Hasanov et al. [98], Li et al. [99] and Lawlor et al. [100], respectively. Two main problems were investigated. Mechanical stability of the membrane electrolyte assembly (MEA) was the first tested parameter, because if the MEA fails, then the whole stack will definitely break down. The second investigated parameters were sealants, which are needed to prevent gas leakage.

Wei et al. [86] examined thermo-mechanical stress distributions in a new design of a hexagonal stack consisting of planar anode-supported SOFCs. The material of the fuel cell support was Grancrete. Simulations were performed based on coupling the commercial CFD code ANSYS FLUENT with Fuel Cell Tools module and ANSYS Mechanical. The flow solver provided species and temperature distributions, while the stress solver based on the temperature distribution results calculated the induced stress including the von Mises stress in metal parts and the maximum principal stress in ceramic material. The results revealed that the use of Grancrete cell support effectively reduced the maximum principal stress of the SOFCs and the maximum von Mises stress was lower that the yield strength of steel, which confirms that the SOFC stack should not be damaged [86].

Nakajo et al. [90] presented an extended study on modelling that combines thermo-electrochemical models accounting for cell degradation and a contact thermo-mechanical model that considers rate-independent plasticity, creep of the components materials and shrinkage of the nickel based anode during thermal cycling of SOFC stacks. The modelling approach was described in detail [110]. A percolation model predicted reduction of the triple-phase boundary length due to the growth of nickel particles in the anode. A semi-empirical relation was assumed for the evolution of the nickel particle radius that depended upon temperature, steam and hydrogen partial pressure due to the mechanical constraints imposed by the YSZ network. XRD stress measurements in the electrolyte of anode/electrolyte bilayers from the literature were used to validate calculated residual stresses in cell layers as a function of temperature. The author [110] observed that a detrimental evolution of a temperature profile can induce unexpected failure if control strategy of electrical load demand is not adapted for degradation in the long term. The direct influence of creep deformation on the integrity of the cell was significant with glass-ceramic sealants, resulting in computed factors of 0.02 up to 1.3 for the anode probability of failure, and of 4 to 136 for that of the cathode. However, it was rather limited for compressive sealing gaskets. The residual stress in the cell and creep relaxation in the metallic interconnect increase in the long-term the deflection of a SRU (standard repeating unit) in a stack. The tensile stress in the high temperature zones was relieved, whereas that in the cold ones increased [110].

Extended studies on sealant modelling in SOFCs were performed by Peksen and Lin et al. [92,111]. To determine the thermo-mechanically induced stress distribution within the glass-ceramic sealants, the Von Mises and the maximal principal stress distributions were analysed. The results revealed that the sealant underwent high mechanical elastic strain at the mid cell region, whereas the thermal strain showed a uniform distribution across each component at the same location. Due to the fact that components undergo plastic strain, the amount of deformation increased within the steel and as a result led to higher tensile stress within the sealant at the same location and the maximum principal stress was noticed there. This was the reason that the sealants get susceptible to failure [92].

The effects of applied load on the thermal stress distribution in a stack with two different types of sealants: glass-ceramic and mica were also characterised by 3D multi-cell model by Lin et al. [111]. Thermal stresses in a 3-cell planar SOFC stack at 800 °C and at various stages of operation (loads of 0.06, 0.1, 0.6 and 6 MPA) were studied using a commercial FEA code ABAQUS. The stress distribution in the mica sealants had critical values of 26 and 30 MPa at the two after-assembly and shut down stages, which were lower that its tensile strength of 37 MPa. The critical stress at the steady-operation stage was smaller, close to 14 MPa. Moreover, the highest value in the glass-ceramic sealants for the after assembly, steady-operation and shut down stages was roughly 20 MPa, which occurred at the inner edges and corners. The authors [111] found that replacement of the glass-ceramic sealant by a mica sealant in joining the connecting metallic PEN (electrode–electrolyte–negative electrode plate) supporting frame with interconnect would cause a dramatic stress change in the PEN.

The results of the thermo-mechanical analysis strongly depend on the thermo-physical processes considered in the model and the material properties, which are temperature dependent. Thus, to receive accurate predictions, temperature dependence of the material properties has to be taken into account. Thermal properties required for the CFD analyses were summarised by Peksen [94].

There have been significant efforts towards the numerical investigations of SOFC degradation mechanisms, in particular the thermal stress as a major mechanical degradation aspect limiting the industrial development of SOFCs. It should be underlined that all these reported studies represent important progress towards thermal stress analysis of SOFC structures; however, due to lack of reliable experimental calibrations, evaluation of these theoretical speculations is rather difficult.

13.4 System level modelling

This section gives a brief outline of the fuel cell technology introducing basic features of system level models with different fuel processing and operating conditions as well as recirculation techniques of the anode and cathode streams. A description of a simple SOFC system is presented first, followed by more detailed overviews of five main types of integrated component models capable of accurately predicting the SOFC stack performance and remaining components of the Balance of Plant (BoP). Finally, a short comparison of SOFC systems with different fuel processing methods and simulation effort is given.

The simplest possible scheme of a SOFC system is shown in Figure 13.13. Natural gas or LPG enters the plant and after removing the sulphur in the desulphuriser is directed into the catalytic partial oxidation (CPO_x) reactor for fuel processing. Then hydrogen is delivered to a fuel cell stack, where the fuel is distributed through manifolds to the anodes. Air is preheated and delivered into the cathode compartments of individual cells of the stack. After electrochemical oxidation of hydrogen and reduction of oxygen, the direct current (DC) power produced in the process is inverted to alternating current (AC) in the DC/AC converter. The unreacted anode off gas such as

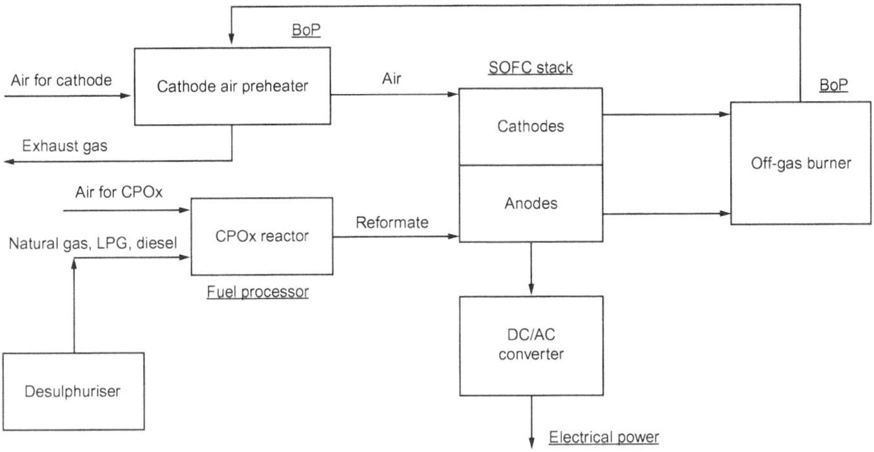

Figure 13.13 Scheme of the simplest SOFC system.

hydrogen and CO that exit the SOFC stack are directed to the off-gas burner. The products of the afterburning process are then directed to a cathode air preheater to recuperate the reaction heat. The start-up system is omitted for clarity. This pared down scheme was described in several patents and papers [112,113] starting in 1994.

The important benefits of this design are:

- SOFC stack and afterburner release heat, which recuperates the cold air inlet,
- the CPO_x works exothermically to provide the polymer manifold cool cell inlet with the 700 °C operating temperature,
- no water is needed for fuel processing, but there is an efficiency penalty,
- cost-effective parts with low weight and volume,
- rapid start-up capability (10 s with 2 mm diameter microtubes).

Improvements to efficiency can be provided by moving to steam reforming (SR), and by incorporating anode/cathode gas recycle, so this will be considered after the CPO_x.

13.4.1 Catalytic partial oxidation

In the first approach, catalytic partial oxidation (CPO_x) the exothermic method for processing, the hydrocarbon fuel into hydrogen and CO is applied as postulated in 1994 [112–114]. The partial oxidation reaction can take place on the nickel anode, which is a very effective catalyst, but it is usually better to insert a catalyst upstream of the electrodes to prevent thermal shocking of the cells and to preheat the inlet fuel gas to around 700 °C. The CPO_x system can be fed using hydrocarbons such as methane, propane, butane, LPG, biogas or diesel. However, fuels more complex than butane tend to kill the cells quite quickly. Diesel in particular has never been very successfully processed by CPO_x and that is why the use of SOFCs in cars, trucks, trains and ships has generally failed. Temperature control requires a good fixing of the air/fuel stoichiometric ratio, λ, defined by Equation (13.52) [3]:

$$\lambda = \frac{x_{O_2}\dot{V}_{air}}{2\dot{V}_{CH_4} + 3.5\dot{V}_{C_2H_6} + 5\dot{V}_{C_3H_8} + 6.5\dot{V}_{C_4H_{10}}} \tag{13.52}$$

where: $\dot{V}_i$ – volumetric flow rate of the i-component, x_{O_2} – volume fraction of oxygen.

The air to fuel λ ratio should be in a narrow operation range of 0.1–0.2. When lambda is set too high, $\lambda > 0.3$, gives a high CPO_x outlet temperature of around 950 °C, which is too high for most purposes, especially when the cells are designed to run at 700 °C as in [112]. To obtain correct gas compositions, a sensitive λ control is necessary for the CPO_x reformer reactions (13.53):

$$C_xH_y + \frac{x}{2}O_2 \leftrightarrow xCO + \frac{y}{2}H_2 \quad \Delta_r H^o < 0, \text{ exothermal} \tag{13.53}$$

The main advantages of the CPO_x system are reduced number of components and low investment costs. In particular, the CPO_x reaction can take place on a ruthenium coated alumina or other suitable support over a distance of 5 mm which leads to a very compact design [114]. The main disadvantage is limited electrical efficiency of 35% in appropriate power ranges. Therefore, CPO_x reformers are likely to be used in SOFCs for remote power applications where such low level of efficiency is acceptable due to other advantages such as low weight.

An example of a portable SOFC system with CPO_x producing electric power of 150 W from propane for military applications was evaluated numerically by Cui et al. [115]. The stack assembled 18 microtubular SOFCs in a serial connection. A CPO_x reformer was integrated with the remaining components of the BoP such as heat exchangers and after burner. CPO_x reactions were simulated by a RGibbs block available in the commercial process simulator tool such as Aspen One.

The heat losses of CPO_x, stack and after burner were calculated via three Design Specs function. The modelling results showed that decreasing the environmental temperatures could decrease stack and CPO_x temperatures and slightly drop of the stack power due to temperature variations.

Detailed characteristics of all major system components of a SOFC based micro-CHP system for natural gas with CPO_x as reforming technology were also presented by Campanari [11] in the framework of the European project FC-DISTRICT. Heat and mass balance for the overall system was generated using Aspen simulation. Simulations were carried out at a nominal stack power of 1.5 kW_{el} with turndown ratio of 3:1. The air/fuel ratio in the CPO_x reforming process was around 0.31. Different natural gas types such as methane, propane and synthetic biogas were used as a fuel. The electrical efficiency for stack power of 1.5 kW_{el} was calculated as approximately 32%, while the thermal efficiency for combined hot water storage and floor heating application at 1.5 kW_{el} was around 55%. The net system efficiency was in the range of 83–87% for 1.5 kW_{el} stack power for examined fuel types. However, as with most of the modelling on fuel cells, there was no factual data, only predictions which proved to be false later [116].

13.4.2 Steam reforming

The second fuel-processing technique widely used for hydrocarbons is steam reforming (SR) or wet reforming presented in Figure 13.14. Water, deionised and demineralised, is fed to an evaporator and then into the desulphurised fuel stream, which is preheated by exhaust gas before entering the SR catalytic reactor. The reaction of hydrocarbon and steam is endothermic, and so the product H_2 and CO requires reheating to enter the SOFC stack.

The main advantage of the method is that the electrical efficiency can be increased over CPO_x to approximately 45–60%. This improved efficiency has been observed for example in the BlueGen SOFC system made by CFCL and now operating in several locations with stated electrical efficiency of 60% [117,118].

The ideal SR process for hydrocarbon fuels can be described by reactions (13.54) and (13.55):

$$C_xH_y + xH_2O_{(g)} \leftrightarrow xCO + \left(\frac{y}{2} + x\right)H_2 \tag{13.54}$$

$$CO + H_2O_{(g)} \leftrightarrow CO_2 + H_2 \tag{13.55}$$

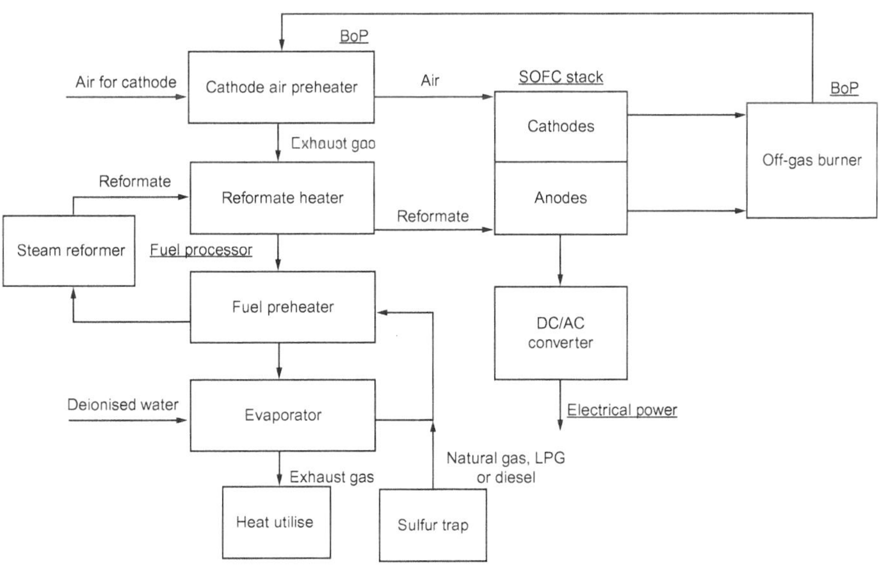

Figure 13.14 Schematic diagram of the SOFC system based on steam reforming.

Typical SR reactions for methane can be written as following Equations (13.56) and (13.57):

$$CH_4 + H_2O \leftrightarrow CO + 3H_2 \tag{13.56}$$

$$CO + H_2O \leftrightarrow CO_2 + H_2 \tag{13.57}$$

Steam to carbon volume ratio can be described by Equation (13.58):

$$S/C = \frac{\dot{V}_{H_2O}}{\dot{V}_{CH_4} + 2\dot{V}_{C_2H_6} + 3\dot{V}_{C_3H_8} + \cdots} \tag{13.58}$$

An internal reforming in a SOFC stack is often used to decrease system complexity through a reduction of air demand. However, the main problem of internal reforming is the mismatch between the heat requirement for SR and the heat available from the fuel cell section. The internal reforming operation could lead to local sub-cooling around the entrance area of reformer part, which can result in mechanical failure due to thermally induced stresses. An additional problem related to internal reforming is carbon deposition on the anode site, which occurs due to cracking reactions. The risk for carbon formation can be reduced by increasing the steam to carbon ratio (S/C) of the fuel feed stream. Steam in an amount greater than the stoichiometric requirement of reaction (4.5) has to be injected into the fuel prior to high temperature heating and admittance into a reformer. However, the minimum amount of steam needed for carbon free operation is difficult to estimate. In addition, carbon formation could result in the deactivation of anode material, which leads to the loss of fuel cell performance. The quantity of carbon deposited on the anode is affected by the operating temperature and the methane to steam ratio [119,120].

Dynamic simulations of behaviour of a SOFC stack with SR reactor were carried out by Komatsu et al. [121]. The SOFC power generation module consisted of an indirect internal reforming developed by Siemens Power Co. Ltd. and a cell stack with tubular SOFCs. Pure methane was used as a fuel, which was reformed in the reformer with steam, and then water gas shift reaction followed. The fuel and steam flow rates were manipulated to satisfy the set fuel utilisation factor of 0.85 and the S/C ratio of 2.5. The obtained results indicated efficient system operation, while the S/C values varied with the load changes and need to be observed.

The SOFC system performance was estimated by Thanomjit et al. [122] with respect to key operating parameters such as SR temperature and steam to ethanol ratio. Ethanol was converted into hydrogen using different reforming processes such as SR, partial oxidation and authothermal reforming. The SR method was identified to be the most effective one. However, it was also noticed that the ethanol reformer and preheaters required high energy supply. In order to improve the thermal system efficiency a pinch analysis was performed using Aspen Plus simulator. As a reformer unit, the equilibrium reactor module RGibbs was used in simulations. A SOFC unit was divided into two sections: the first one marked as anode was represented by the Requil reactor with defined electrochemical reaction, while the cathode was represented by the separator module Sep available in the process simulator tool. The ethanol reformer and SOFC stack were operated at temperatures of 700 °C and 800 °C, respectively. It was found that increases in the reforming temperature and steam to ethanol ratio improved the hydrogen yield. However, when the reformer was operated at the temperature higher than 800 °C, the hydrogen efficiency decreased since a reverse water gas shift reaction became stronger [122].

13.4.3 Anode off-gas recycle

Endothermic anode off-gas recycling can also provide water for SR, giving a high efficiency above 50% obtained by recirculation of anode off-gases either by hot recycle as the fuel is injected, or by cold recycle which requires an extra heat exchanger. Recirculation of 30–40% of the anode off-gas stream is sufficient for SR with appropriate oxygen to carbon ratio defined in Equation (13.59) [123]:

$$O/C = \frac{2\dot{V}_{CO_2} + \dot{V}_{CO} + \dot{V}_{H_2O}}{\dot{V}_{CH_4} + \dot{V}_{CO_2} + \dot{V}_{CO}} \tag{13.59}$$

Estimation of the O/C ratio is a tricky task especially at high temperatures. Also, high temperature components for recycle of hot gases are disadvantages of the anode off-gas recirculation system. Dynamic system behaviour and long-term stability of the power unit still remain significant challenges.

An anode off-gas recycle SOFC system for *n*-dodecane as the surrogate for diesel fuel was evaluated by Lee et al. [124], based on analytical, parametric study. The whole reforming unit consisted of a direct internal reforming component inside the SOFCs, an external reformer, four heat exchangers, a carbon dioxide adsorbent, a flue gas condenser and an anode off gas recycling unit. The model was validated by the comparison of the current density values with the experimental data of Sasaki et al. [125] and a good agreement was obtained. The system performance was evaluated as a function of steam to carbon ratio, fuel cell temperature, anode off-gas recycle ratio and CO_2 adsorption percentage. The authors found that the system efficiency started at around 70% and then monotonically decreased to the average of 50% at the peak power density before dropping down to zero at the limiting current density point. In addition, a parametric study for a combined fuel reforming showed that the anode off gas recycle ratio increased with increasing steam to carbon ration and additionally eliminated the need for external water consumption. Moreover, the main finding was that the optimal system performance was when the SOFC operated around 900 °C with steam to carbon ration higher than 3, which allowed to maximum CO_2 capture and a minimum anode off gas recirculation Lee et al. [124].

A cold blower [126] or ejector system [127] can also be used to inject the anode off gas into the incoming fuel stream. Alternatively, autothermal reforming (ATR) [128,129] can be used, especially popular when processing dirty fuels like kerosene and diesel. The mechanism of autothermal reforming can be represented by the following chemical reaction scheme (13.60):

$$C_xH_yO_p + \text{Steam} + \text{air} \rightarrow H_2 + \text{carbon oxides} + N_2 \tag{13.60}$$

Authothermal reforming is a hybrid technology that combines the catalytic aspect of SR technique with the oxidation feature of partial oxidation technology. When using oxidation catalysts, a portion of hydrocarbon is oxidised with a controlled addition of oxygen. The oxygen is controlled to limit the amount of heat release and/or to avoid sintering of the catalyst. In the method the heat of partial oxidation provides the needed thermal energy to drive the endothermic reforming reactions, transforming the gaseous fuel to hydrogen and carbon monoxide.

Autothermal reforming system was well characterised by Ersoz et al. [130], where it was used to reform desulfurised JP-8 fuel to provide H_2 and CO as fuel to planar and tubular SOFC stacks. It was found that increasing S/C ratio and ATR operating temperature helped suppress coke formation, but at the same time increased fuel dilution. This caused decay of Nernst voltage and efficiency. The maximum and minimum total efficiency of the system was 70.2% and 61.7% at the 700 and 800 °C ATR operating temperatures at S/C ratio of 0.4 and 1.0, respectively. A theoretical comparison was made by Ersoz et al. [130], but without experimental calibration.

A comparison of SOFC systems with different fuel processing methods is presented in Table 13.2 [130–136].

The described numerous steps involved in fuel preparation or surface reaction to electrochemically oxidise a fuel make the SOFC system complex and difficult to identify its optimum configuration and evaluate its operating efficiency. To gain insight into fundamental physics of fuel cell performance at the system level was possible

Table 13.2 Comparison of SOFC systems with different fuel processing methods [130–136]

	Catalytic partial oxidation (CPO_x)	Steam reforming (SR)	Anode off-gas recirculation with steam reforming
Electrical efficiency	30–35%	45–60%	50–60%
Overall efficiency	85–95%	85–95%	85–95%
Power to heat ratio	0.5–0.7	1.0–2.2	1.3–2.3
Heat requirement	Exothermic	Endothermic	Endothermic
Air/fuel ratio	0.27–0.31	–	–
Sensitive to gas and water quality	No	Yes	Partly
Start-up	Fast	External igniter need	External igniter need
Transient conditions	Easy to control	Relatively stable	Relatively stable
System costs	Low	High	High
Number of components	Minimal	Large	Large
Dedicated applications	μCHP system 0.5–5 kW	Residential 1.5 kW	Large system 50 kW

with the help of process simulators such as HYSYS or Aspen Plus™, which allow to predict the physical, thermodynamic and other properties of various materials used in SOFC system. Both software packages are quite similar and based on blocks corresponding to unit operations and chemical reactions, through which most of the operations can be simulated. A complete process flow sheet is reconstructed by interconnecting blocks using material, work and heat streams, also including gas turbine air blowers [137–141].

The system level models reveal significant dependences between fuel conversion in the reformer, fuel/air utilisations, temperature parameters and the system efficiency and SOFC lifetime. Nevertheless, the optimal SOFC design can be reached at different fuel cell voltage and fuel utilisation values depending on specific system configurations and whether thermal energy is recuperated. Therefore, system level modelling could be particularly useful for cogeneration system design and maximising power efficiency.

13.5 Oscillations in SOFCs running on methane

In experiments carried out on reaction diffusion, oscillating fuel cell currents were noted [22]. These have now been reproduced experimentally using methane in a microtubular SOFC with low humidification around 2% [24]. Figure 13.15 shows that as the methane flow was stepwise reduced with time, meanwhile increasing nitrogen to keep the overall flow rate constant, the current at 0.7 V gradually dropped until, at about 0.2 A, large oscillations were observed, repeatable many times.

There have been several previous reports on oscillating SOFCs. Kellogg et al. [142] used a single chamber SOFC with standard nickel cermet anode, in other words mixing the air and fuel together over both anode and cathode, and found waves of nickel oxidation and reduction. This had nothing to do with methane

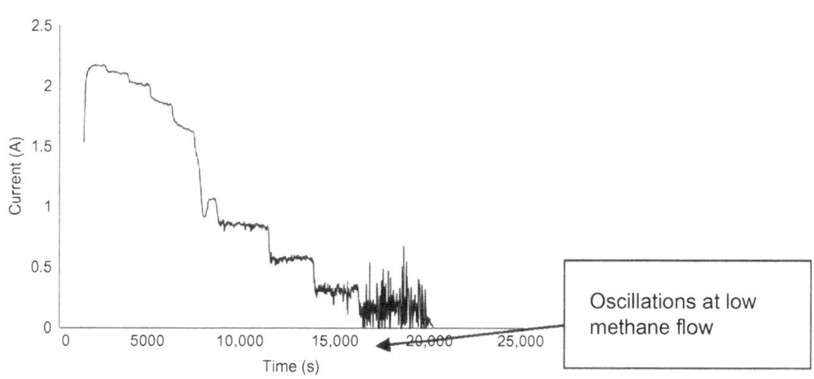

Figure 13.15 Oscillations in current observed as methane flow rate was reduced in steps.

because hydrogen was used as the fuel. Marina et al. [143] studied the effect of various contaminants in the fuel stream of an SOFC. The contaminants, H_2Se and H_2S, were added to synthetic coal gas (H_2, CO, CO_2, H_2O) and tested separately. They found that for high enough current densities and parts per million of H_2Se, voltage oscillations settled in after a considerable number of hours of degradation from a steady state. No oscillations were found when the cell was operated potentiostatically, or with sulphur in the fuel stream under the same operating conditions. The oscillations favoured higher concentrations of H_2S and could also be induced in a fuel stream containing purely hydrogen and steam along with the contaminant. This shows that cyclic behaviour involving carbon is not required for the oscillations under these conditions. The mechanism they proposed was a periodic adsorption/desorption of Se on the anode catalyst layer. Voltage oscillations were found by Noponen et al. [144] in an SOFC with pre-reformed natural gas containing sulphur. An online gas analyser was used to record the species, and it was found that minute fluctuations in methane and hydrogen occurred regardless of whether the cell was exhibiting voltages oscillations or not. Thus they believe that the primary cause was to do with inconsistent injection of sulphur into the reformate gas prior to entering the fuel cell, which would in turn would adsorb/desorb on the nickel surface causing variations in overpotential. Huang et al. [145] found current oscillations in an SOFC using LSCF-GDC, or Ni added LSCF-GDC for the anode, using methane as the primary fuel. They found that the presence of Ni actually reduced the likelihood of oscillations, and when oscillations in current were detected, the amplitudes were much smaller. They attributed the current oscillations to periodic reduction/oxidation cycles of the anode material.

The present experimental arrangement is shown in Figure 13.16. Methane was fed to the nickel YSZ cermet tube with an LSCF cathode in air at 700 °C. The effects of methane dilution in this set-up had previously been described by Kendall et al. [146].

The theoretical explanation of this behaviour was analysed in terms of reaction diffusion theory based on the non-linearity of the methane performance [147]. The model is based on fundamental chemical kinetics and Fickian mass transfer through a porous medium, which provides a suitably accurate description of the processes occurring at the anode. The cell is assumed to be an internal reforming SOFC, where

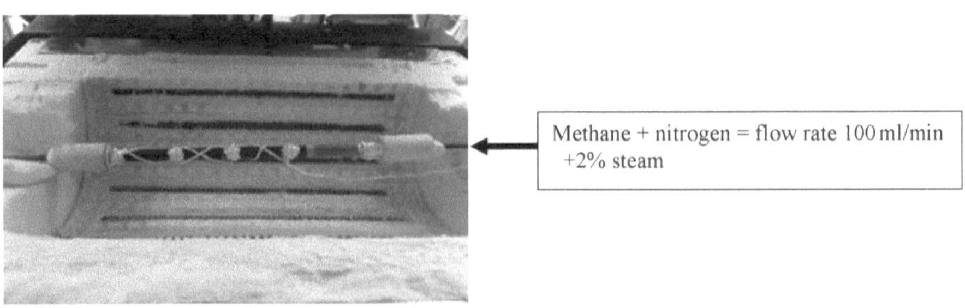

Figure 13.16 Microtubular SOFC inside furnace for operation on methane at 700 °C.

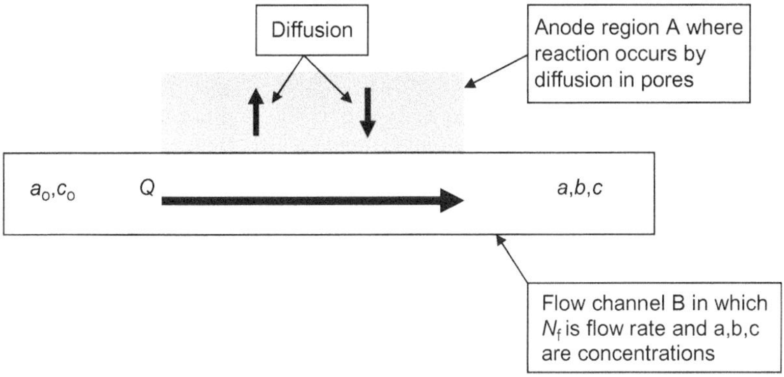

Figure 13.17 Diagram showing the two regions A and B. Diffusion occurs in A while flow occurs in B.

methane is steam reformed at the nickel catalyst near the surface of the anode. The hydrogen produced by the reforming reaction is then electrochemically oxidised, resulting in the main contribution to the current drawn from the cell. Carbon monoxide will also be produced by the SR of methane, which can then proceed to either participate in the water gas shift reforming reaction, or be directly consumed by electrochemical oxidisation at the nickel anode. The geometry of the model is shown in Figure 13.17.

For internal reforming the reactions are:

$$CH_4 + H_2O \rightarrow CO + 3H_2 \quad : k_1[CH_4][H_2O] \tag{13.61}$$

$$CO + H_2O \rightarrow CO_2 + H_2 \quad : k_2[CO][H_2O] \tag{13.62}$$

$$2H_2 + O_2 \rightarrow 2H_2O \quad : k_3[H_2]^2[O_2] \tag{13.63}$$

$$2CO + O_2 \rightarrow 2CO_2 \quad : k_4[CO]^2[O_2] \tag{13.64}$$

where k is the reaction rate constant for each step. The CO reaction is assumed to be very fast, leaving only Equations (13.61) and (13.63) as dominant processes. Now defining the concentrations:

$$[CH_4] = a \tag{13.65}$$

$$[H_2] = b \tag{13.66}$$

$$[H_2O] = c \tag{13.67}$$

$$[O_2] = x(\text{viewed as constant}) \tag{13.68}$$

The inlet flow contains only CH_4 and H_2O, at inlet concentrations a_o and c_o, respectively. We restrict attention to the situation when the flow rate N_f is sufficiently rapid, relative to the Fickian transfer rates between zone B and zone A, so that zone B provides a steady transfer pool of CH_4 and H_2O, at the inlet concentration, for Fickian transfer into zone A. This is a reasonable assumption as at higher velocities the depletion zone of a reactant stream flowing over a catalytic surface becomes very small in comparison with the width of the channel.

Then we get three coupled, nonlinear ordinary differential equations, namely:

$$\frac{da}{dt}=\frac{D_aA}{Vh}(a_o-a)-k_1ac \tag{13.69}$$

$$\frac{db}{dt}=\frac{-D_bAb}{Vh}+4k_1ac-2k_3b^2x \tag{13.70}$$

$$\frac{dc}{dt}=\frac{D_cA}{Vh}(c_o-c)+2k_3b_2x-2k_1ac \tag{13.71}$$

The parameters D_a, D_b and D_c are, respectively, the effective diffusion coefficients for the Fickian transfer of methane, hydrogen and water (as steam) to the reaction site, and h is the scale thickness of the anode. The coefficient A/V is the surface area of transfer to volume ratio, whilst a_o and c_o represent the inlet concentrations of species a and c respectively. The first term of each rate equation represents the transfer of the particular species into the reaction zone A, whilst the remaining terms represent the production/consumption of the species as given by the reduced reaction scheme:

$$CH_4+2H_2O\rightarrow CO_2+4H_2 \quad : k_1[CH_4][H_2O] \tag{13.72}$$

$$2H_2+O_2\rightarrow 2H_2O \quad : k_3[H_2]^2[O_2] \tag{13.73}$$

By investigating these equations both analytically and numerically, a dynamical system was defined in which all the equilibrium points could be identified with associated bifurcation structure in which oscillations could occur [147]. One particular bifurcation diagram is shown in Figure 13.18.

Starting with an intermediate concentration of methane, which corresponds with the far right of Figure 13.18, the only stable state of the fuel cell system is a steady power output. This power output corresponds with the branch, e^-, and can be linked to the current via Faraday's law. As the concentration decreases (moving left along the e-branch), the e-branch becomes unstable and the system exhibits oscillations. Further decreasing the methane concentration, we move left along the P^- branch, which indicates the oscillations increasing in amplitude, whilst the average current output decreases. When the methane concentration gets below a critical point, in this case represented by a_0^{\infinity}, the oscillation disappears and the fuel cell system produces little, to no current. This trend is shown in the experimental results from Figure 13.19.

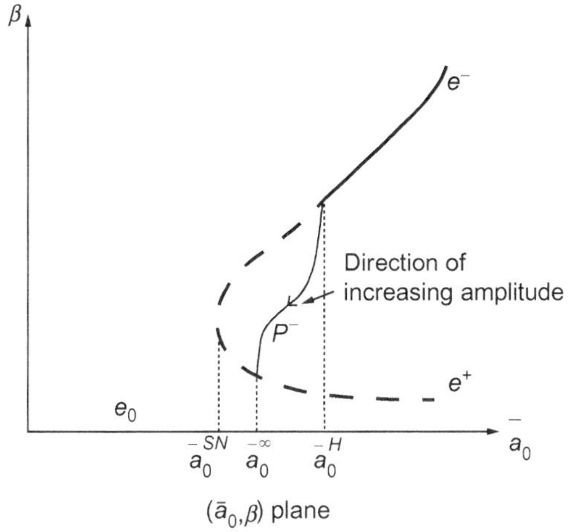

Figure 13.18 *Bifurcation diagram at one point in the unfolding plane showing both a Saddle-node and Hopf bifurcation.*

- a_0 is the concentration of methane fed to the anode in the fuel stream, Beta is the concentration of hydrogen which we assume to mainly contribute towards the electrical current output, P^- is the oscillation, e^- is the stable steady state producing a current output.
- e^+ is a saddlepoint (from dynamical systems theory) and doesn't physically correspond with an operating point. It simply contributes towards the dynamic behaviour of the fuel cell in response to a change in operating conditions, e_0 is a stable steady state that produces little, to no power output, Solid lines represent stable states, whilst dashed lines indicate unstable states.

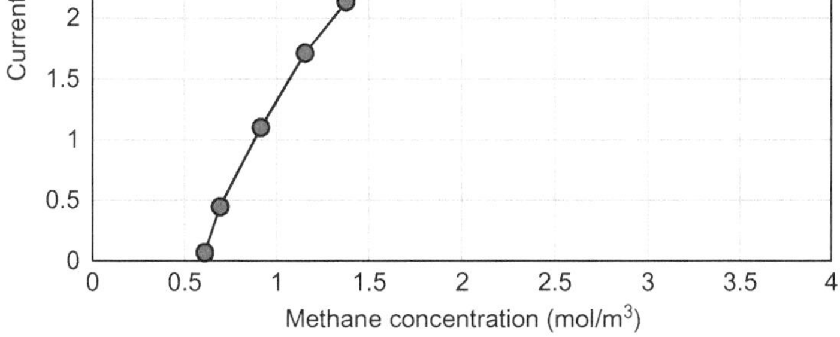

Figure 13.19 Steady state current output plotted against methane concentration showing non-linearity at 0.7 V and 700 °C showing similarity to Figure 13.18.

13.6 Summary and future prospect

Modelling of SOFCs has been progressing significantly since the original book chapter was published in 2003 [4]. The key areas of concern in this chapter are the molecular modelling aspects, the computer predictions of cell, stack and system performance and the explanation of new observations such as the oscillations observed in methane-fuelled SOFCs.

Reaction diffusion equations have been used successfully to describe the transition zone seen in anodes where the oxygen concentration can range over 20 orders of magnitude. This is akin to a flame front in combustion studied by electrochemical sensors.

Cell modelling has progressed as the complicated three-phase porous electrodes have been characterised and measured by various means. A full theoretical answer to optimisation of the anode and cathode regions has yet to be produced.

Stack modelling depends greatly on the geometry selected, whether tubular, planar or flat tube. Several commercial packages are now available to design such stacks.

System models are becoming more satisfactory. BoP components in the SOFC system model are represented by user defined units such as reformer, burner, heat exchanger, cooler, pump and compressor. The proper units can be defined using Fortran programming code or by those available in the commercial software packages such as ASPEN Plus or HYSYS and relations between them. In this chapter, the full modelling of the simplest possible SOFC system has led to good insights about theory and experimental convergence. More complex SR and anode off-gas recycle can then be addressed.

Finally, new phenomena such as oscillating SOFCs have been seen and attempts have been made to describe them mathematically. It is hoped that theory can advance further to understand these new effects satisfactorily.

Acknowledgements

The research programme leading to these results received funding from the European Union's Seventh Framework Programme (FP7/2007-2013) for the Fuel Cells and Hydrogen Joint Technology Initiative under grant agreement no [325323]. Acknowledgements are due to the partners of SAFARI project. The work was also financed from the Polish research funds awarded for the project No. 3043/7.PR/2014/2 of international cooperation within SAFARI in years 2014–2016.

References

[1] K.R. Bove, S. Ubertini, Modeling Solid Oxide Fuel Cells: Methods, Procedures and Techniques, Springer, Cleveland, OH 44135, USA, ISBN 978-1-4020-6995-6, 2008.

[2] J. Milewski, K. Świrski, M. Santarelli, P. Leone, Advanced Methods of Solid Oxide Fuel Cell Modeling, first ed., Springer-Verlag, London, ISBN 0857292617, 2011.

[3] B. Huang, Y. Qi, A.K.M. Murshed, Dynamic Modelling and Predictive Control in Solid Oxide Fuel Cells: First Principle and Data-Based Approaches, John Wiley & Sons, Chichester, UK, ISBN 978-1-118-50103-0, 2013.

[4] M.A. Khaleel, J.R. Selman, in: S.C. Singhal, K. Kendall (Eds.), Cell, Stack and System Modeling in High Temperature SOFCs, Elsevier Science, Oxford, 2003.
[5] S. Srinivasan, Fuel Cells: From Fundamentals to Applications, Springer, USA, 2006.
[6] M. Bavarian, M. Soroush, I.G. Kevrekidis, J.B. Benziger, Mathematical modeling, steady-state and dynamic behavior and control of fuel cells: a review, Ind. Eng. Chem. Res. 49 (17) (2010) 7922–7950.
[7] D. Schimanke, O. Posdziech, B.E. Mai, S. Kluge, Th. Strohbach, Ch. Wunderlich, Demonstration of a highly efficient SOFC system with combined partial oxidation and steam reforming, ECS Trans. 35 (1) (2011) 231–242.
[8] A.D. Meadowcroft, S. Howroyd, K. Kendall, M. Kendall, Testing micro-tubular. SOFCs in Unmanned Air Vehicles (UAVs), ECS Trans. 57 (1) (2013) 451–457.
[9] H.L. Tuller, J. Schoonman, I. Riess, Oxygen Ion and Mixed Conductors and Their Technological Applications, Kluwer Academic Publishers, Dordrecht, 2000.
[10] V. Liso, A. Ch. Olesen, M.P. Nielsen, S.K. Kaer, Performance comparison between partial oxidation and methane steam for SOFC micro-CHP systems. Energy 36 (2011) 4216–4226, http://dx.doi.org/10.1016/j.energy.2011.04.022.
[11] S. Campanari, Thermodynamic model and parametric analysis of a tubular SOFC module, J. Power Sources 92 (2001) 26–34.
[12] J. Ugartemendia, J.X. Ostolaza, I. Zubia, Operating point optimization of a hydrogen fueled hybrid Solid Oxide Fuel Cell-Steam Turbine (SOFC-ST) plant, Energies 6 (2013) 5046–5068.
[13] M. Ameri, R. Mohammadi, Simulation of an atmospheric SOFC and gas turbine hybrid system using Aspen Plus software, Int. J. Energy Res. 37 (2013) 412–425.
[14] I. Frenzel, A. Loukou, D. Trimis, F. Schroeter, L. Mir, R. Marin, B. Egilegor, J. Manzanedo, G. Raju, M. de Bruijne, R. Wesseling, S. Fermandes, J.M. Ch. Pereira, G. Vourliotakis, M. Founti, O. Posdziech, Development of an SOFC based micro-CHP system in the framework of the European project FC-DISTRICT, Energy Procedia 28 (2012) 170–181.
[15] I. Staffell, A review of small stationary fuel cell performance. http://www.academia.edu/1073987/A_review_of_small_stationary_fuel_cell_performance, 2009.
[16] W.G. Bessler, S. Gewies, C. Willich, G. Schiller, K.A. Friedrich, Spatial distribution of electrochemical performance in a segmented SOFC: a combined modeling and experimental study, Fuel Cells 10 (2010) 411–418.
[17] A. Al-Masri, M. Peksen, L. Blum, D. Stolten, A 3D CFD model for predicting the temperature distribution in a full scale APU SOFC short stack under transient operating conditions. Appl. Energy 135 (2014) 539–547, http://dx.doi.org/10.1016/j.apenergy.2014.08.052.
[18] H. Mounir, M. Belaiche, A. El Marjani, A. El Gharad, Thermal stress and probability of survival investigation in a multi-bundle integrated-planar solid oxide fuel cells IP-SOFC (integrated-planar solid oxide fuel cell). Energy 66 (2014) 378–386, http://dx.doi.org/10.1016/j.energy.2014.01.017.
[19] P. Chinda, S. Chanchaona, P. Brault, W. Wechsatol, A planar anode-supported Solid Oxide Fuel Cell model with internal reforming of natural gas, Eur. Phys. J. Appl. Phys. 54 (2011) 23405.
[20] P. Vijay, S. Hosseini, M.O. Tade, A novel concept for improved thermal management of the planar SOFC, Chem. Eng. Res. Des. 91 (2013) 560–572.
[21] W. Bi, D. Chen, Z. Lin, A key geometric parameter for the flow uniformity in planar solid oxide fuel cell stacks, Int. J. Hydrogen Energy 34 (2009) 3873–3884.
[22] R.C. Copcutt, A.C. King, K. Kendall, Proc. R. Soc. Lond. A452 (1996) 2639–2653.

[23] A.C. King, J. Billingham, K. Kendall, R.C. Copcutt, Analysis of a model for a loaded, planar, solid oxide fuel cell, SIAM J. Appl. Math. 60 (2) (2000) 574–601.
[24] J.D. Sands, Current oscillations arising from nonlinear chemical dynamics in solid oxide fuel cells, PhD thesis, University of Birmingham, 2015.
[25] B. Zakrzewska, Z. Jaworski, Towards multiscale modelling in product engineering, Comput. Chem. Eng. 35 (2011) 434–445.
[26] W.G. Bessler, Multi-scale modeling of solid oxide fuel cells: from patterned anodes to a power plant system, ECS Trans. 35 (2011) 859–869.
[27] S. Cordiner, M. Feola, V. Mulone, F. Romanell, Analysis of a SOFC energy generation system fuelled with biomass reformate, Energy: production, distribution and conservation, Milan, 2006, pp. 103–119.
[28] C.R. Kreller, M.E. Drake, S.B. Adler, H.Y. Chen, H.C. Yu, K. Thorton, J.R. Wilson, S.A. Barnett, Modeling SOFC cathodes based on 3-D representations of electrode microstructure, ECS Trans. 35 (1) (2011) 815–822.
[29] V. Yurkiv, A. Latz, W.G. Bessler, Modeling and simulation the influence of solid carbon formation on SOFC performance and degradation, ECS Trans. 57 (1) (2013) 2637–2647.
[30] J.R. Wilson, W. Kobsiriphat, R. Mendoza, H.Y. Chen, T. Hines, J.M. Hiller, D.J. Miller, K. Thornton, P.W. Voorhees, H. Mizusaki, Three dimensional reconstruction of Solid Oxide Fuel Cell electrodes using focused ion beam – scanning electron microscopy, Solid Oxide Fuel Cells 10 (SOFC-X), The Electrochemical Society, Nara, Japan, 2007, pp. 1879–1887.
[31] W.G. Bessler, J. Warnatz, D.G. Goodwin, The influence of equilibrium potential on hydrogen oxidation kinetics of SOFC anodes, Solid State Ionics 177 (2007) 3371–3383.
[32] D. Cui, Q. Liu, F. Chen, Modeling of anode-supported SOFCs with samaria doped-ceria electrolytes operating at 500–600 °C, J. Power Sources 195 (2010) 4160–4167.
[33] F. Bidrawn, R. Kungas, J.M. Vohs, R.J. Gorte, Modeling impedance response of SOFC cathodes prepared by infiltration, J. Electrochem. Soc. 158 (5) (2011) B514–B525.
[34] C. Schluckner, V. Subotic, V. Lawlor, C. Hochenauer, Three-dimensional numerical and experimental investigation of an industrial-sized SOFC fuelled by Diesel reformat – part I: creation of a base model for further carbon deposition modeling, Int. J. Hydrogen Energy 39 (2014) 19102–19118.
[35] T. Tsai, S.A. Barnett, Effect of LSM-YSZ cathode on thin-electrolyte solid oxide fuel cell performance, Solid State Ionics 93 (1997) 207–217.
[36] V. Haanappel, J. Mertens, D. Rutenbeck, C. Tropartz, W. Herzhof, D. Sebold, Optimisation of processing and microstructural parameters of LSM cathodes to improve the electrochemical performance of anode-supported SOFCs, J. Power Sources 141 (2) (2005) 216–226.
[37] J.J.R. Izzo, A.S. Joshi, K.N. Grew, W.K.S. Chiu, A. Tkachuk, S.H. Wang, W. Yun, Nondestructive reconstruction and analysis of SOFC anodes using X-ray computed tomography at sub-50 nm resolution, J. Electrochem. Soc. 155 (5) (2008) B504–B508.
[38] H. Iwai, N. Shikazono, T. Matsui, H. Teshima, M. Kishimoto, R. Kishida, D. Hayashi, K. Matsuzaki, D. Kanno, M. Saito, H. Muroyama, K. Eguchi, N. Kasagi, H. Yoshida, Quantification of SOFC anode microstructure based on dual beam FIB-SEM technique, J. Power Sources 195 (2010) 955–961.
[39] P.R. Shearing, J. Golbert, R.J. Chater, N.P. Brandon, 3D reconstruction of SOFC anodes using a focused ion beam lift-out technique, Chem. Eng. Sci. 64 (2009) 3928–3933.
[40] G. Reiss, H.L. Frandsen, W. Brandstatter, A. Weber, Numerical evaluation of microstructural parameters of porous supports in metal-supported solid oxide fuel cells, J. Power Sources 273 (2015) 1006–1015.

[41] S.A. Tabei, A. Sheidaei, M. Baniassadi, F. Pourboghrat, H. Garmestani, Microstructure reconstruction and homogenization of porous Ni-YSZ composites for temperature dependent properties, J. Power Sources 235 (2013) 74–80.
[42] G. Delette, J. Laurencin, F. Usseglio-Viratta, J. Villanova, P. Bleuet, E. Lay-Grindler, T. Le Bihan, Thermo-elastic properties of SOFC/SOEC electrode materials determined from three-dimensional microstructural reconstructions, Int. J. Hydrogen Energy 38 (2013) 12379–12391.
[43] M. Mori, T. Yamamoto, H. Itoh, H. Inaba, H. Tagawa, Thermal expansion of nickel-zirconia anodes in solid oxide fuel cells during fabrication and operation, J. Electrochem. Soc. 45 (4) (1998) 1374–1381.
[44] P.S. Turner, Thermal-expansion stresses in reinforced plastics, J. Res. Natl. Bur. Stand. 37 (1946) 239–250.
[45] E.H. Kerner, The elastic and thermos-elastic properties of composite media, Proc. Phys. Soc. 69 (1956) 808–813.
[46] A. Bertei, B. Nucci, C. Nicolella, Microstructural modeling for prediction of transport properties and electrochemical performance in SOFC composite electrodes, Chem. Eng. Sci. 101 (2013) 175–190.
[47] J.M. Zalc, S.C. Reyes, E. Iglesia, The effects of diffusion mechanism and void structure on transport rates and tortuosity factors in complex porous structures, Chem. Eng. Sci. 59 (2004) 2947–2960.
[48] J.A. Currie, Gaseous diffusion in porous media. Part 2. Dry granular materials, Br. J. Appl. Phys. 11 (1960) 318–324.
[49] A. Bertei, J. MErtens, C. Nicolella, Electrochemical simulation of planar solid oxide fuel cells with detailed microstructural modeling, Electrochim. Acta 146 (2014) 151–163.
[50] Z. Qu, P.V. Aravin, S.Z. Boksteen, N.J.J. Dekker, A.H.H. Janssen, N. Woudstra, A.H. M. Verkooijen, Three dimensional computational fluid dynamics modeling of anode-supported planar SOFC, Int. J. Hydrogen Energy 30 (2011) 10209–10220.
[51] P. Fan, G. Li, Y. Zeng, X. Zhang, Numerical study on thermal stresses of a planar solid oxide fuel cell, Int. J. Therm. Sci. 77 (2014) 1–10.
[52] T.X. Ho, P. Kosinski, A.C. Hoffmann, A. Vik, Modeling of transport, chemical and electrochemical phenomena in a cathode-supported SOFC, Chem. Eng. Sci. 64 (2009) 3000–3009.
[53] M. Peksen, 3D transient multiphysics modelling of a complete high temperature fuel cell system using coupled CFD and FEM, Int. J. Hydrogen Energy 39 (2014) 5137–5147.
[54] V. Novaresio, M. Garcia-Camprubi, S. Izquierdo, P. Asinari, N. Fueyo, An open-source library for the numerical modeling of mass-transfer in solid oxide fuel cells, Comput. Phys. Commun. 183 (2012) 125–146.
[55] M. Garcia-Camprubi, S. Izquierdo, N. Fueyo, Challenges in the electrochemical modelling of solid oxide fuel and electrolyser cell, Renew. Sust. Energ. Rev. 33 (2014) 701–718.
[56] S.A. Hajimolana, M.A. Hussain, W.M. Ashri Wan Daud, M. Soroush, A. Shamiri, Mathematical modeling of solid oxide fuel cells: a review, Renew. Sust. Energ. Rev. 15 (2011) 1893–1917.
[57] A. Pramuanjaroenkij, S. Kakac, X.Y. Zhou, Mathematical analysis of planar solid oxide fuel cells, Int. J. Hydrogen Energy 33 (2008) 2547–2565.
[58] T.X. Ho, A three-dimensional model for transient performance of a solid oxide fuel cell, Int. J. Hydrogen Energy 39 (2014) 6680–6688.
[59] B. Lin, Y. Shi, M. Ni, N. Cai, Numerical investigation on impacts on fuel velocity distribution nonuniformity among solid oxide fuel cell unit channels, Int. J. Hydrogen Energy 40 (2015) 3035–3047.

[60] P. Pianko-Oprych, E. Kasilova, Z. Jaworski, Assessing the effect of electrochemically-driven uniformities of heat flux in a microtubular fuel cell on mSOFC temperature distribution, in: 11th European SOFC and SOE Forum 2014, 1-4th July 2014, Lucerne, Switzerland, A1322, 2014, pp. 1–10.
[61] R.B. Bird, W.E. Stewart, E.N. Lightfoot, Transport Phenomena, Wiley, New York, 2002.
[62] D.L. Damm, A.G. Fedorov, Local thermal non-equilibrium effects in porous electrodes of the hydrogen fuelled SOFC, J. Power Sources 159 (2006) 1153–1157.
[63] D.L. Damm, A.G. Federov, Spectral radiative heat transfer analysis of the planar SOFC, J. Fuel Cell, Sci. Technol. 2 (2005) 258–262.
[64] M. Bertoldi, T. Zandonella, V.A.C. Haanappel, J. Mertens, J. Remmel, L.G.J. de Haart, Demonstration of a 4-Cell SOFC Stack Under Different Experimental Conditions, Elsevier, Amsterdam, 1983.
[65] M. Andersson, J. Yuan, B. Sunden, Review on modeling development for multiscale chemical reactions coupled transport phenomena in solid oxide fuel cells, Appl. Energy 87 (2010) 1461–1476.
[66] F. Arpino, N. Massarotti, Numerical simulation of mass and energy transport phenomena in solid oxide fuel cells, Energy 34 (2009) 2033–2041.
[67] D.P. Xenos, P. Hofmann, K.D. Panopoulos, E. Kakaras, Detailed transient thermal simulation of a planar solid oxide fuel cell (SOFC) using gPROMS™, Energy 81 (2015) 84–102.
[68] B. Lin, Y. Shi, M. Ni, N. Cai, Numerical investigation on impacts on fuel velocity distribution nonuniformity among solid oxide fuel cell unit channels, Int. J. Hydrogen Energy 40 (2015) 3035–3047.
[69] Z. Zhang, D. Yue, G. Yang, J. Chen, Y. Zheng, H. Miao, W. Wang, J. Yuan, N. Huang, Three dimensional CFD modeling of transport phenomena in multichannel anode-supported planar SOFCs, Int. J. Heat Mass Transf. 84 (2015) 942–954.
[70] V.M. Janardhanan, O. Deutschmann, CFD analysis of a solid oxide fuel cell with internal reforming: coupled interactions of transport, heterogeneous catalysis and electrochemical processes, J. Power Sources 162 (2006) 1192–1202.
[71] M. Andersson, J. Yuan, B. Sunden, SOFC modeling considering hydrogen and carbon monoxide as electrochemical reactants, J. Power Sources 232 (2013) 42–54.
[72] X. Zhang, G. Li, J. Li, Z. Feng, Numerical study on electric characteristics of solid oxide fuel cells, Energy Convers. Manag. 48 (3) (2007) 977–989.
[73] J.D.J. VanderSteen, J.G. Pharoah, Modeling radiation heat transfer with participating media in solid oxide fuel cells, J. Fuel Cell Sci. Technol. 3 (2006) 62–67.
[74] D.L. Damm, A.G. Fedorov, Radiation heat transfer in SOFC materials and components, J. Power Sources 143 (2005) 158–165.
[75] K.J. Daun, S.B. Beale, F. Liu, G.J. Smallwood, Radiation heat transfer in planar SOFC electrolytes, J. Power Sources 157 (2006) 302–310.
[76] S. Grosso, L. Repetto, P. Costamagna, IP-SOFC model, in: R. Bove, S. Ubertini (Eds.), Modeling Solid Oxide Fuel Cells, Methods, Procedures and Techniques, Fuel Cells and Hydrogen Energy, Springer, Cleveland, OH 44135, USA, ISBN 978-1-4020-6995-6, 2008 (Chapter 6).
[77] P. Costamagna, K. Honegger, Modeling of solid oxide heat exchanger integrated stacks and simulation at high fuel utilization, J. Electrochem. Soc. 145 (11) (1998) 3995–4007.
[78] W.G. Bessler, S. Gewies, M. Vogler, A new framework for detailed electro-chemical modeling of solid oxide fuel cells, Electrochim. Acta 53 (2007) 1782–1800.
[79] ANSYS Fuel Cell Tools module, 2015.
[80] H. Zhu, R.J. Kee, A general mathematical model for analyzing the performance of fuel cell membrane electrode assemblies, J. Power Sources 117 (2003) 61–74.

[81] K. Wang, D. Hissel, M.C. Pera, N. Steiner, D. Marra, M. Sorrentino, C. Pianese, M. Monteverde, P. Cardone, J. Saarinen, A review on solid oxide fuel cell models, Int. J. Hydrogen Energy 36 (2011) 7212–7228.
[82] J. Arriagada, P. Olausson, A. Selimovic, Artificial neural network simulator for SOFC performance prediction, J. Power Sources 112 (2002) 54–60.
[83] J. Milewski, K. Swirski, Modelling the SOFC behaviours by artificial neural network, Int. J. Hydrogen Energy 34 (13) (2009) 5546–5553.
[84] J. Nielsen, J. Hjelm, Impedance of SOFC electrodes: a review and a comprehensive case study on the impedance of LSM:YSZ cathodes, Electrochim. Acta 115 (2014) 31–45.
[85] V.M. Janardhanan, O. Deutschmann, Numercial study of mass and heat transport in solid oxide fuel cells running on humidified methane, Chem. Eng. Sci. 62 (2007) 5473–5486.
[86] S.S. Wei, T.H. Wang, J.D. Wu, Numerical modeling of interconnect flow channel design and thermal stress analysis of a planar anode-supported solid oxide fuel cell stack, Energy 69 (2014) 553–561.
[87] A. Nakajo, F. Mueller, J. Brouwer, J. Van herle, D. Favrat, Mechanical reliability and durability of SOFC stacks. Part I: modelling of the effects of operating conditions and design alternatives on the reliability, Int. J. Hydrogen Energy 37 (2012) 9249–9268.
[88] L.K. Chiang, H.Ch. Liu, Y.H. Shiu, Ch.H. Lee, R.Y. Lee, Thermo-electrochemical and thermal stress analysis for an anode-supported SOFC cell, Renew. Energy 33 (2008) 2580–2588.
[89] A. Selimovic, M. Kemm, T. Torisson, M. Assadi, Steady state and transient thermal stress analysis in planar solid oxide fuel cells, J. Power Sources 145 (2005) 463–469.
[90] A. Nakajo, Ch. Stiller, G. Harkegard, O. Bolland, Modeling of thermal stresses and probability of survival of tubular SOFC, J. Power Sources 158 (2006) 287–294.
[91] M.F. Serincan, U. Pasaogullari, N.M. Sammes, Thermal stresses in an operating micro-tubular solid oxide fuel cell, J. Power Sources 195 (2010) 4905–4914.
[92] M. Peksen, A coupled 3D thermofluid–thermomechanical analysis of a planar type production scale SOFC stack, Int. J. Hydrogen Energy 36 (2011) 11914–11928.
[93] H. Yakabe, T. Ogiwara, I. Yasuda, M. Hishunuma, 3-D model calculation for planar SOFC, J. Power Sources 102 (2001) 144–154.
[94] M. Peksen, 3D multiscale-multiphysics SOFC modelling status at the Institute of Electrochemical Process Engineering, FZ Jülich, Prog. Energy Combust. Sci. 48 (2015) 1–20.
[95] A. Nakajo, F. Mueller, J. Brouwer, J. Van herle, D. Favrat, Progressive activation of degradation processes in solid oxide fuel cells stacks: part I: lifetime extension by optimisation of the operating conditions, J. Power Sources 216 (2012) 449–463.
[96] K. Fischer, J.R. Seume, Thermo-mechanical stress in tubular solid oxide fuel cells: part I – transient operating behaviour and the relevance of material creep, IET Renew. Power Gener. 6 (3) (2012) 177–193.
[97] K. Fischer, J.R. Seume, Thermo-mechanical stress in tubular solid oxide fuel cells: part II – operating strategy for reduced probability of fracture failure, IET Renew. Power Gener. 6 (3) (2012) 194–205.
[98] R. Hasanov, A. Smirnova, A. Gulgazli, M. Kazimov, A. Volkov, V. Quliyeva, O. Vasylyev, V. Sadykov, Modeling design and analysis of multi-layer solid oxide fuel cells, Int. J. Hydrogen Energy 36 (2011) 1671–1682.
[99] J. Li, W. Kong, Z. Lin, Theoretical studies on the electrochemical and mechanical properties and microstructure optimization of micro-tubular solid oxide fuel cells, J. Power Sources 232 (2013) 106–122.
[100] V. Lawlor, S. Griesser, G. Buchinger, A.G. Olabi, S. Cordiner, D. Meissner, Review of the micro-tubular solid oxide fuel cell: part I. Stack design issues and research activities, J. Power Sources 193 (2009) 387–399.

[101] J.D.J. VanderSteen, J.G. Pharoah, The effect of radiation heat transfer in solid oxide fuel cell modelling, in: Combustion Institute, Canadian Section, Spring Technical Meeting, Queen's University, 9-12 May, 2004, 2004.
[102] J. Wei, Models development and system simulation of solid oxide fuel cell/gas, UMI Microform 3232512, 2006, pp. 4–28 (Chapter 4).
[103] T. Ota, M. Koyama, Ch.J. Wen, K. Yamada, H. Takahashi, Object-based modeling of SOFC system:dynamic behavior of micro tubular SOFC, J. Power Sources 118 (2003) 430–439.
[104] J. Qu, A. Fedorov, C. Haynes, An integrated approach to modelling and mitigating SOFC failure, Georgia Tech, Atlanta, GA 30332–0405, Final technical report, 1 July 2006, DOE: DE-FC26-02NT41571.
[105] T. Tanaka, Y. Inui, A. Urata, T. Kanno, Three dimensional analysis of planar solid oxide fuel cell stack considering radiation, Energy Convers. Manag. 48 (2007) 1491–1498.
[106] M. Fardadi, F. Mueller, F. Jabbari, Feedback control of solid oxide fuel cell spatial temperature variation, J. Power Sources 195 (13) (2010) 4222–4233.
[107] J. Hartvigsen, S. Elangovan, A. Khandkar, Modeling, design and performance of solid oxide fuel cells, Proc. of Zirconia V, 105 (2002).
[108] S. Murthy, A.G. Fedorov, Radiation heat transfer analysis of the monolith type solid oxide fuel cell, J. Power Sources 124 (2003) 453–458.
[109] K.J. Kattke, R.J. Braun, A.M. Colclasure, G. Goldin, High-fidelity stack and system modeling for tubular solid oxide fuel cell system design and thermal management, J. Power Sources 196 (2011) 3790–3802.
[110] A. Nakajo, P. Tanasini, S. Diethelm, J. Vanherle, D. Favrat, Electrochemical model of solid oxide fuel cell for simulation at the stack scale II: implementation of degradation processes, J. Electrochem. Soc. 158 (9) (2011) B1102–B1118.
[111] Ch.K. Lin, L.H. Huang, L.K. Chiang, Y.P. Chyou, Thermal stress analysis of planar solid oxide fuel cell stacks: effects of sealing design, J. Power Sources 192 (2009) 515–524.
[112] K Kendall, EU Pat App WO 94/22178. Solid oxide fuel cell structures, published 29 Sept 1994; K. Kendall, M. Kendall, EU Pat AppWO 2005/112158 A2, Portable fuel cell device, published 24 Nov 2005.
[113] K. Kendall, M. Prica, Integrated SOFC tubular system for small-scale cogeneration, in: U. Bossel (Ed.), 1st European SOFC Forum, Luzern, 1994, pp. 163–170.
[114] K. Kendall, D.S. Williams, Catalysts for butane reforming in zirconia fuel cells, Platinum Met. Rev. 42 (1998) 164–167.
[115] D. Cui, Y. Du, K. Reifsnider, F. Chen, Environmental effects on a thermally self-sustained SOFC hot zone, ECS Trans. 35 (1) (2011) 297–312.
[116] K. Kendall, UK Patent application number 1509444.4, Stackable fuel cell, 2015.
[117] O. Posdziech, B.E. Mai, C. Wunderlich, S. Voss, Status and market opportunities of solid oxide fuel cell based cogeneration systems, in: International Gas Union Research Conference, 2011, pp. 1–17.
[118] R. Payne, J. Love, M. Kah, CFCL's BlueGen product, ECS Trans. 35 (2011) 81–85.
[119] C.M. Finnerty, N.J. Coe, R.H. Cunningham, R.M. Ormerod, Carbon formation on and deactivation of nickel-based/zirconia anodes in solid oxide fuel cells running on methane, Catal. Today 46 (1998) 137–145.
[120] C.M. Finnerty, R.M. Ormerod, Internal reforming over nickel/zirconia anodes in SOFCS oparating on methane: influence of anode formulation, pre-treatment and operating conditions, J. Power Sources 86 (2000) 390–394.
[121] Y. Komatsu, S. Kimijima, J.S. Szmyd, Numerical analysis on the dynamic behavior of a solid oxide fuel cell with a multivariable control strategy, ECS Trans. 35 (1) (2011) 1045–1054.

[122] C.H. Thanomjit, Y. Patcharavorachot, A. Arpornwichanop, Design and thermal analysis of a solid oxide fuel cell system integrated with ethanol steam reforming, in: Proceeding of the 22nd European Symposium on Computer Aided Process Engineering, 17-20th June 2012, London, 2012.
[123] J. Kupecki, J. Jewulski, K. Badyga, Selection of a fuel processing method for SOFC-based micro-CHP system, Rynek Energii 6 (97) (2011) 157–162.
[124] T.S. Lee, J.N. Chung, Y.Ch. Chen, Design and optimization of a combined fuel reforming and solid oxide fuel cell system with anode off-gas recycling, Energy Convers. Manag. 52 (2011) 3214–3226.
[125] K. Sasaki, K. Watanabe, K. Shiosake, K. Susuki, Y. Teraoka, Multi-fuel capability of solid oxide fuel cells, J. Electroceram. 13 (2004) 669–675.
[126] J. Kupecki, J. Milewski, J. Jewulski, Investigation of SOFC material properties for plant-level modeling, Cent. Eur. J. Chem. 11 (5) (2013) 664–671.
[127] W. Zhang, E. Croiset, P.L. Douglas, M.W. Fowler, E. Entchev, Simulation of a tubular solid oxide fuel cell stack using AspenPlusTM unit operation models, Energy Convers. Manag. 46 (2005) 181–196.
[128] G. Bae, J. Bae, P. Kim-Lohsoontorn, J. Jeong, Performance of SOFC coupled with nC 4 H 10 autothermal reformer: carbon deposition and development of anode structure, Int. J. Hydrogen Energy 35 (2010) 12346–12358.
[129] T. Tanim, D.J. Bayless, J.P. Trembly, Modeling a 5 kWe planar solid oxide fuel cell based system operating on JP-8 fuel and a comparison with tubular cell based system for auxiliary and mobile power applications, J. Power Sources 245 (2014) 986–997.
[130] A. Ersoz, A. Sarioglan, S. Ozdogan, Investigation of an integrated autothermal reforming and SOFC micro cogeneration system for power generation, in: WHEC, 13–16 June 2006, Lyon, France, 2006.
[131] D.R. McIlveen-Wright, M. Moglie, S. Rezvani, Y. Huang, A techno-economic analysis of biomass gasifiers integrated with high and intermediate temperature solid oxide fuel cells, Int. J. Energy Res. 35 (2011) 1037–1047.
[132] F. Zabihian, A.S. Fung, Performance analysis of hybrid solid oxide fuel cell and gas turbine cycle: application of alternative fuels, Energy Convers. Manag. 76 (2013) 571–580.
[133] F. Zabihian, A.S. Fung, Thermodynamic sensitivity analysis of hybrid system based on solid oxide fuel cell, Sustainable Energy Technol. Assess. 6 (2014) 51–59.
[134] T. Anderson, P. Vijay, M.O. Tade, An adaptable steady state Aspen Hysys model for the methane fuelled solid oxide fuel cell, Chem. Eng. Res. Des. 92 (2014) 295–307.
[135] J. Kupecki, J. Jewulsk, Multi-level mathematical modelling of Solid Oxide Fuel Cells, Clean Energy for Better Environment, 2012, (Chapter 4), p. 53.
[136] J. Milewski, A. Miller, Mathematical model of SOFC for power plant simulations, in: Proceedings of ASME Turbo EXPO 2004, 14–17 June, Vienna, Austria, GT2004-53787, 2004.
[137] R. Schefflan, Teach yourself the basics of Aspen Plus™, John Wiley & Sons, Inc., New Jersey, ISBN 978-0-470-56795-1, 2011.
[138] Ch. Bo, Ch. Yuan, X. Zhao, C.B. Wu, M.Q. Li, Parametric analysis of solid oxide fuel cell, Clean Techn. Environ. Policy 11 (2009) 391–399.
[139] E. Riensche, E. Achenbach, D. Froning, M.R. Haines, W.K. Heidug, A. Lokurlu, S. von Andrian, Clean combined-cycle SOFC power plant – cell modelling and process analysis, J. Power Sources 86 (2000) 404–410.
[140] T. Suther, A.S. Fung, M. Koksal, F. Zabihian, Effects of operating and design parameters on the performance of a solid oxide fuel cell–gas turbine system, Int. J. Energy Res. 35 (7) (2011) 616–632.

[141] Aspen Plus™ Users Guide, 10.2, 2000. Aspen Tech Ltd., Cambridge MA, USA.
[142] I.D. Kellogg, U.O. Koylu, V. Petrovsky, F. Dogan, Effectiveness of anode in a solid oxide fuel cell with hydrogen/oxygen mixed gases, Int. J. Hydrogen Energy 34 (12) (2009) 5138–5143.
[143] O.A. Marina, L.R. Pederson, C.A. Coyle, E.C. Thomsen, D.J. Edwards, Polarization-induced interfacial reactions between nickel and selenium in Ni/Zirconia SOFC anodes and comparison with sulfur poisoning, J. Electrochem. Soc. 158 (2011) B36.
[144] M. Noponen, M. Halinen, J. Kiviaho, J. Saarinen, Feasibility of autothermally reformed natural gas on anode supported Solid Oxide Fuel Cells, J. Fuel Cell Sci. Technol. 3 (4) (2006) 438.
[145] T.J. Huang, M.C. Huang, W.J. Chen, C.L. Chou, Oscillation of electrical current during direct methane oxidation over Ni-added LSCF–GDC anode of solid oxide fuel cells, Chem. Eng. J. 153 (2009) 164–169.
[146] K. Kendall, C.M. Finnerty, G. Saunders, J.T. Chung, Effects of dilution on methane entering an SOFC anode, J. Power Sources 106 (2002) 323–327.
[147] J.D. Sands, D.J. Needham, J. Uddin, A fundamental model exhibiting non-linear oscillatory dynamics in SOFCs, Proc. R. Soc. A, 470, 2164 (April 2014).

Fuels and fuel processing in SOFC applications

14

S. Sui[1], G.H. Xiu[2]
[1]Shanghai Jiao Tong University, Shanghai, China, [2]23-602 Room, 89 Mudan Rd., Shanghai, China

14.1 Introduction

Power generation and the related environmental impacts have become important issues across the world [1–4]. Today, electrical power is provided mainly by conventional power generation technologies that rely on fossil fuel combustion, which generates soot and sulphur compounds, in addition to other noxious emissions. The combustion of fossil fuels is widely understood to contribute to both global warming and local air pollution. Advanced clean energy systems must be developed urgently, allowing us to make the shift from a fossil fuel-based economy to a new paradigm in a progressive manner [5–7].

The most attractive feature of solid oxide fuel cells (SOFCs) is their fuel flexibility at very high efficiency, even cells running directly on hydrocarbons, including liquid logistics fuels. SOFC stacks can operate on pure CO and tolerate relative high level of impurities, thus without the need for a complex and expensive fuel purification that is necessary for low-temperature fuel cells [8,9]. Low-temperature proton exchange membrane fuel cells (PEMFCs) are poisoned by even a small quantity of carbon monoxide (~ppm level) in the fuel, therefore placing significant demands, and hence cost, on a complex external fuel processor (EFP). The fuel processing comparison of SOFCs and PEMFCs is shown schematically in Figure 14.1. By contrast, within the SOFC stack, the hydrocarbon fuel is catalytically converted (internally reformed) [10], generally to synthesis gas (syngas) together with some carbon dioxide, and then the carbon monoxide and hydrogen are electrochemically oxidised to carbon dioxide and water at the anode, with production of electrical power and high grade heat.

Internal reforming of the fuel can either be achieved indirectly using a separate catalyst chamber within the SOFC stack, called indirect internal reforming, or directly using a catalyst layer on the anode, called direct internal reforming [10,11]. Direct and indirect internal reforming are shown schematically in Figure 14.2. Internal reforming of the fuel within the SOFC stack is preferred for both thermodynamic equilibrium and heat utilisation efficiency. It also makes the system compact and reduces the complexity of the system, hence lowing cost [1,3,10,12–14].

The high operating temperature of a SOFC also leads to production of highly valuable heat as a by-product in addition to the electrical power [7,11]. This high-quality heat can be used in various ways, for example, in combined heat and power (CHP)

High-temperature Solid Oxide Fuel Cells for the 21st Century. http://dx.doi.org/10.1016/B978-0-12-410453-2.00014-2

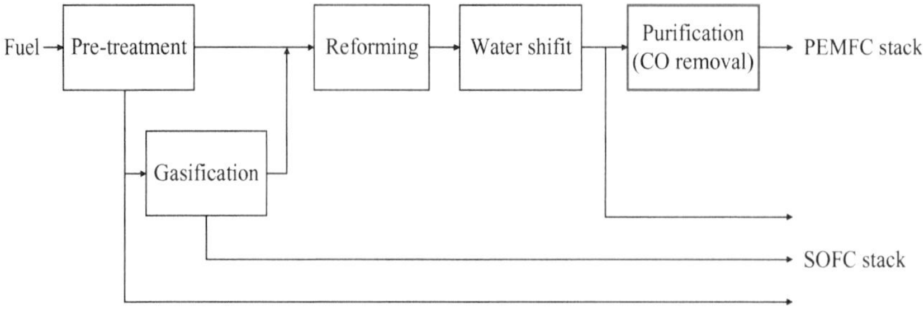

Figure 14.1 Schematic of fuel processing for a PEM/SOFC fuel cell systems.

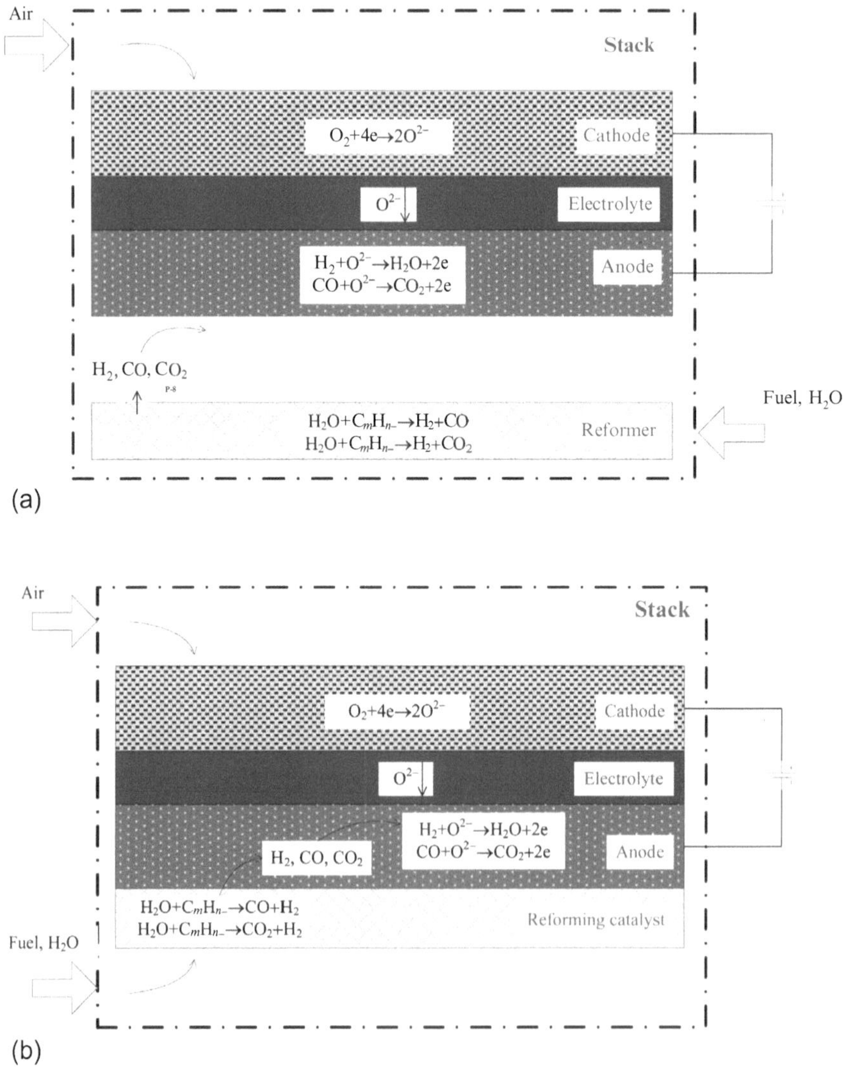

Figure 14.2 Schematic of internal reforming in an SOFC stack showing both (a) indirect and (b) direct reforming approaches.

systems, or to drive a gas turbine to generate more electricity, as well as feeding the endothermic fuel reforming process. This significantly increases the overall energy efficiency of a SOFC system compared to lower-temperature fuel cells. Figure 14.3 provides the heat flows in an advanced SOFC system, showing the heat utilisation by the internal reformer, gas turbine, boiler and heat exchanger.

Figure 14.4 shows schematically the fuel processing scheme for the SOFC CHP system running on natural gas [9]. The advantage of such CHP units for both residential and commercial use is the reliability in the fuel supply, which is becoming increasingly important, especially in premium power applications like computing.

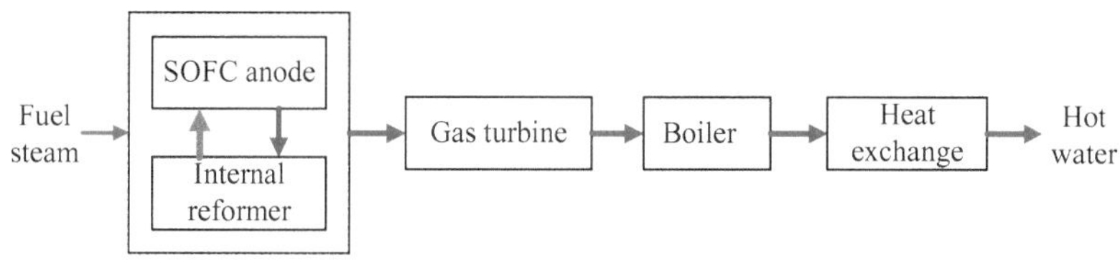

Figure 14.3 Heat flows in an integrated internally reforming SOFC system (blue line (grey in the print version) – fuel feeds; green line (light grey in the print version) – syngas; red line (dark grey in the print version) – heat flow).

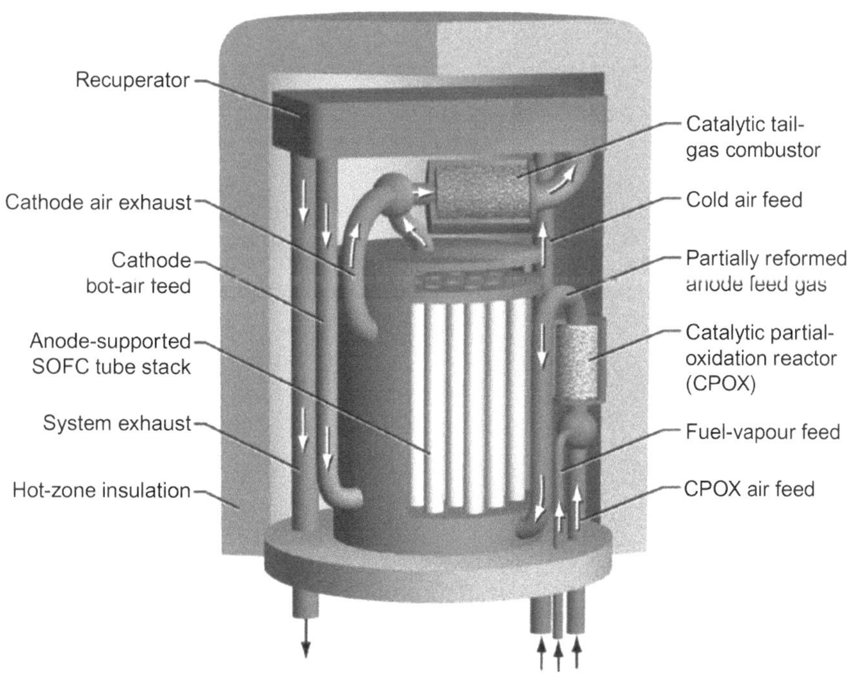

Figure 14.4 Schematic of the fuel processing scheme for the tube-stack SOFC CHP system running on natural gas [9].

Another key advantage of SOFCs over other types of fuel cells is the flexibility in the choice of fuel, which derives from the elevated operating temperature and the tolerance to carbon monoxide [15–18]. SOFCs also show greater tolerance to fuel impurities which poison PEMFCs, and to variations in the fuel composition [19]. This opens up opportunities in SOFC systems for the possible use of renewable fuel sources, such as biogas, vegetable matter and landfill gas, which are precluded for other types of fuel cells. The flexibility in the fuel choice, and in particular the ability to operate SOFCs directly on practical hydrocarbon fuels, in addition to the modular nature of fuel cells, makes SOFCs ideally suited for small-scale, stand-alone and remote applications, for example, in islands and farms. Fuels such as coal or biogas are ideally suited for SOFC applications in larger combined cogeneration (heat and power) and multi-MW gas turbine hybrid systems [7,17,20].

Thus fuel processing has been and continues to be extensively studied, and the development of a successful, cost-effective fuel processing strategy is vital for economic operation of any SOFC system.

This chapter starts by discussing possible fuel kinds for SOFCs. Then the main fuel processing methods and their principles are discussed in detail, where some technologies are still developing. The issues and problems in fuel processing, such as carbon deposition and fuel impurities, are discussed and removal solutions are described. Several typical reforming examples are provided to understand integration of fuel processing into SOFC systems at various scales. Finally, the chapter concludes with a brief consideration of present technology and future prospects.

14.2 Range of fuels

Figure 14.5 shows a range of practical fuels in SOFC application.

SOFCs are ideally suited to operation on syngas. In this chapter, syngas is defined broadly to include at least H_2 and CO and may also include CH_4, H_2O, CO_2 and N_2.

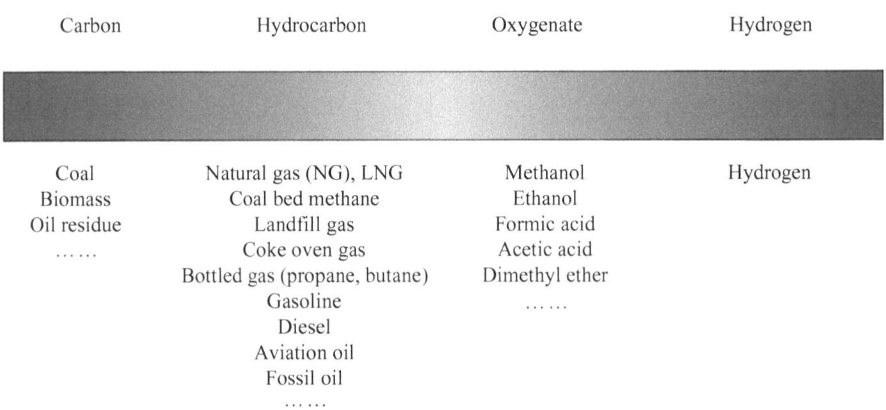

Figure 14.5 Range of potential practical fuels for SOFCs.

Many technologies have been developed to utilise different fuels to produce hydrogen or syngas, such as reforming of natural gas or liquefied petroleum gas (LPG), reforming of liquid fuels such as diesel and kerosene, gasification of coal or biomass or hydrogen produced via water splitting/electrolysis, methanol cracking or ammonia dissociation [6,7,11,21].

Syngas can be produced by alternative processes from a variety of feed stocks, but its pretreatment is needed to remove impurities that are poisons for the pre-reforming or reforming catalyst, particularly transition metal-based catalysts.

14.2.1 Methane

Typical methane-rich gases include natural gas, liquefied natural gas (LNG), coal-bed methane, coke oven gas and landfill gas.

Among the fuels, the most common fuel, especially for stationary applications, is natural gas, which is low cost, clean, abundant and readily available, with a supply infrastructure already in existence in many places. Natural gas varies in composition, being predominantly methane, but also containing significant quantities of higher hydrocarbons. Natural gas can be internally reformed by co-fed steam or oxygen within the SOFC stack at temperatures as low as 600 °C, which means that even operating at lower-temperature SOFCs can be run on natural gas without the need for a complex external fuel reformer [6,8].

The presence of higher hydrocarbons and sulphur-containing compounds in natural gas is significant in terms of deleterious carbon deposition and catalyst poisoning (hydrogen sulphide), respectively [19,22–28].

For high stack and system efficiencies, pre-reforming should only convert the higher hydrocarbons in the fuel, leaving all or most of the methane unconverted. An excess of steam or oxygen is generally required to prevent carbon deposition and coking, which results in deactivation and poor cell durability. Generally, sulphur compounds that are added to the natural gas as odorants are removed upstream of the SOFC stack to prevent poisoning.

14.2.2 Higher hydrocarbons

Higher hydrocarbons include following typical fuels such as propane, butane, gasoline, diesel, LPG and aviation oil.

LPG is a mixture of light hydrocarbons that are gases at normal temperatures and pressures but liquefy at moderate pressures or reduced temperatures. LPG occurs naturally in crude oil and natural gas production fields and is also produced in the oil-refining process. The main component gases of LPG are propane (C_3H_8) and butane (C_4H_{10}).

In remote, small-scale, stand-alone applications where there is no existing natural gas supply, LNG or bottled gas (consisting of propane/butane) offers significant practical advantages over natural gas. Such bottled gas can be internally reformed within the SOFC stack just like natural gas, and a small tubular SOFC device powered by butane had been demonstrated [15,18]. However, the problems of carbon deposition and coking are considerably more severe with higher hydrocarbons than with methane [27,28]. Thus SOFCs operating on bottled gas are more susceptible to deactivation and poor durability due to carbon deposition, and thus the challenge to avoid carbon deposition is greater.

The SOFC power systems represent a viable alternative to conventional power generation methods in remote areas with no natural gas supply, where diesel is generally used, which is both inefficient and highly polluting. It is also likely that bottled gas will be used as an emergency stand-by fuel source for larger-scale SOFCs operating on natural gas, in the event of an interruption to the natural gas supply. However, internally reforming such diesel or kerosene fuels within an SOFC system represents a major challenge in terms of avoiding coking on any of the active components of the cell, though autothermal reformers have been developed based on platinum group metal catalysts, which can efficiently convert diesel into a hydrogen-rich fuel [12,28]. Such catalysts have been successful in Japan where kerosene is a standard fuel for many domestic houses.

14.2.3 Oxygenate fuels

Typical oxygenate fuels include methanol, ethanol and dimethyl ether (DME).

DME has also been tested as a fuel for SOFCs [29]. DME is an attractive fuel because, though it is a gas at room temperature, it is easily liquefied under practical conditions (it is liquid above 3 bar) making it easier to store and handle than hydrogen.

In addition to being dictated by the specific application, the choice of fuel is also partly governed by the operating temperature of the cell. For intermediate temperature SOFCs operating at temperatures as low as 500 °C, methanol is considered the most likely fuel, since its operating temperatures are below those at which natural gas and higher hydrocarbons can be effectively reformed to synthesis gas, whereas methanol can be efficiently reformed below 300 °C [27,30]. Intermediate temperature SOFCs operating directly on methanol offer some potential for transportation and small portable applications. At such low operating temperatures, it is generally the cathode which limits the performance of the SOFC rather than issues of fuel processing if methanol is used as the fuel, though the development of composite cathodes is leading to cathode materials with improved performance at lower temperatures [31,32].

The technology of converting ethanol to hydrogen is the same as that for converting it to a fuel containing methane for application to internal reforming SOFCs. As with conversion of natural gas, there are three routes for converting ethanol to the desired fuel, that is, hydrogen or a methane-rich mixture. Ethanol is easily amenable to steam reforming (SR) and has a lower thermodynamic propensity for carbon formation than natural gas and other hydrocarbon fuels due to the presence of oxygen [6], but the C—C bond tends to promote graphite formation.

14.2.4 Solid fuels

Coal, oil residue and biomass stand for the typical solid fuels.

SOFCs can be operated on the output from coal gasification systems [7,9,11,13, 17,20,33]. The sulphur content of these gases poisons the anode, and can also poison any internal reforming catalyst, causing loss of performance and eventual cell deactivation. Thus the sulphur has to be at least partially removed from the gas prior to entering the SOFC stack [7].

Recently, the possibility of using waste biogas generated from vegetable matter and landfill gas in SOFC systems has been demonstrated [34–36]. Biomass gasification can be efficiently integrated with SOFCs to properly deploy the energy content of this renewable source and increasing the ratio of electric to thermal converted energy [37–40]. As with utilising the output from coal gasification systems, the high sulphur content of these gases is a problem in terms of poisoning and poor cell durability, and again the sulphur has to be at least partially removed from the gas prior to entering the SOFC stack [26,33,41]. Although the efficiencies of SOFCs operating on biogas, landfill gas or the output from coal gasification systems are lower than that on natural gas, these provide the potential of significantly cleaner and more efficient power generation compared to combustion engine generators, as well as offer a valuable use for generally poor-quality biogas that is currently wasted [34,35,40,42,43].

Figure 14.6 shows a top level block flow diagram for the large-scale coal-based SOFC power plant [17]. The power plant consists of three major sections or islands. The gasification island includes a coal gasifier and the supporting subsystems/components. An air separation unit is used to provide oxygen for an O_2-blown gasifier. After removal of particulates, sour syngas from the gasifier is cleaned and processed in the gas cleanup and CO_2 sequestration island. The sweet syngas is then utilised for power generation in the SOFC power island. Supplemental power is also generated by expanding the high-pressure syngas in a fuel gas expander upstream of the atmospheric pressure fuel cell stack.

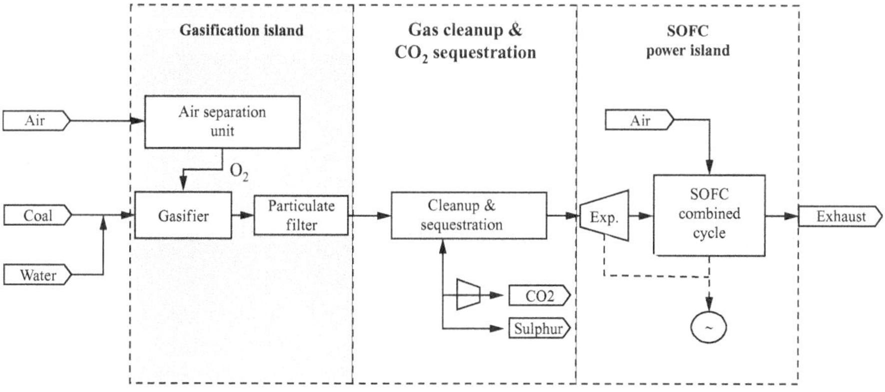

Figure 14.6 Coal-based integrated gasification fuel cell power plant block flow diagram: the SOFC/turbine hybrid concept is anticipated to provide high electrical efficiency while sequestering CO_2 for low emissions [17].

14.2.5 Hydrogen

Hydrogen production has been extensively researched globally. Hydrogen can be produced via water splitting/electrolysis, methanol cracking or ammonia dissociation.

Thermochemical water splitting converts water into hydrogen and oxygen using a series of thermally driven chemical reactions. A concept to use a nuclear reactor was presented as the energy source for thermochemical processes in order to produce hydrogen [44,45]. A combined SOFC–nuclear reactor power generator has been proposed [46]. Hydrogen works extremely well in the SOFC, but ΔG drops as temperature rises, giving lower electrical efficiency than methane which retains a high ΔG.

14.3 Fuel reforming principle

SOFCs convert the chemical energy of a fuel directly into electrical energy through electrochemical reactions that are driven by the chemical potential difference between the anode and cathode of the cell. Oxygen ions migrate through the electrolyte to the anode where they are consumed by fuel oxidation [7,18]. Among all kinds of fuel, the small molecular, gaseous hydrogen and carbon monoxide are more active than the others. Therefore, fuel reforming processes are combined with SOFCs through either internal or external reforming.

SOFCs consume syngas produced from hydrocarbon at the anode by oxide ions electrochemically pumped through the solid electrolyte and are described as the following Equations (14.1) and (14.2):

$$H_2 + O^{2-} \rightarrow H_2O + 2e^- \tag{14.1}$$

$$CO + O^{2-} \rightarrow CO_2 + 2e^- \tag{14.2}$$

Although SOFCs can be coupled to a separate fuel processor, external reforming negates the major efficiency advantage and cost–benefits of SOFC systems over low-temperature fuel cells. Internal reforming of hydrocarbon fuels in SOFC systems significantly increases system efficiency by recuperating waste heat from the stack into the fuel supply, whilst at the same time substantially reduces the complexity and cost of the system by elimination of the external reformer, associated heating arrangements and by reduction in the stack cooling air requirements and related equipment. Thus internally reforming SOFCs offer significantly higher system efficiencies and reduced complexity compared to low-temperature fuel cells, as well as offer flexibility in the choice of fuel [7,8,12,13,30]. Figure 14.7 shows possible reaction pathways in an internally reforming SOFC running on natural gas and steam.

Typically, there are seven main catalytic or non-catalytic processing routes for converting methane or higher hydrocarbons to syngas, a H_2-rich gas, or a CO- and H_2-rich gas mixture:

(1) Steam reforming
(2) Dry reforming

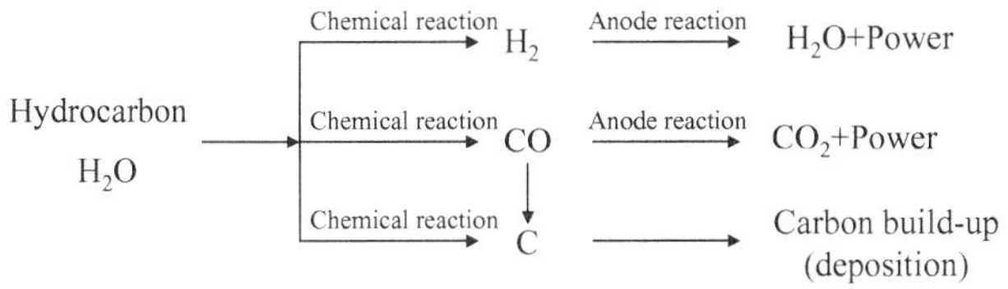

Figure 14.7 Possible reaction pathways in a directly reforming SOFC.

(3) Partial oxidation (POX)
(4) Autothermal reforming (ATR)
(5) Tri-reforming
(6) Plasma reforming with/without catalyst
(7) Direct electrocatalytic oxidation (DEO)

14.3.1 Steam reforming

Hydrocarbon SR is described in Equations (14.3) and (14.4), in which SR of methane is well known as SMR (Steam Methane Reforming). If higher hydrocarbons are used, they are firstly converted to methane, which is conventionally known as pre-reforming. SR is a strongly endothermic reaction [6,47,48].

$$CH_4 + H_2O \rightarrow CO + 3H_2 (\Delta H_{298} = 206.1\,\text{kJ/mol}) \quad (14.3)$$

$$C_nH_m + nH_2O \rightarrow nCO + (n + m/2)H_2 (\Delta H_{298} > 0) \quad (14.4)$$

The forward reaction of reaction (14.3) and the thermodynamic equilibrium of reaction (14.4) are favoured by high temperature and low pressure.

The water-gas-shift (WGS) reaction (Equation 14.5) can occur, whereby some of the CO is converted to CO_2, with production of one mole of hydrogen for every mole of CO converted:

$$CO + H_2O \rightarrow CO_2 + H_2 (\Delta H_{298} = -41.15\,\text{kJ/mol}) \quad (14.5)$$

whereas the thermodynamics of the forward reaction of Equation (14.5) is favoured by low temperature and unaffected by pressure.

Undesirable reactions during SR are as the following [6]:

$$CO + H_2 \rightarrow C + H_2O (\Delta H_{298} = -131.29\,\text{kJ/mol}) \quad (14.6)$$

$$2CO \rightarrow C + CO_2 (\Delta H_{298} = -172.42\,\text{kJ/mol}) \quad (14.7)$$

$$CH_4 \rightarrow C + 2H_2 (\Delta H_{298} = 74.87\,\text{kJ/mol}) \quad (14.8)$$

$$C_nH_m \rightarrow nC + (m/2)H_2 \quad (14.9)$$

For internal reforming in SOFCs, the above undesirable reactions cause carbon deposition resulting from hydrocarbon pyrolysis (Equations 14.8 and 14.9), especially on the nickel cermet anode [7,12,49–53], as well as on other active components within the system.

The thermodynamic minimum steam to carbon (S/C) ratio for all hydrocarbons is 1.0. However, practical minimum S/C ratios required to avoid the undesirable carbon-forming reactions generally tend to be upward of 1.7 for natural gas and higher for the higher hydrocarbons, for example, 2.2 for naphtha. In industrial practice, excess steam is used with S/C ratios ranging from 2.5 for the lighter hydrocarbons to 7 for the higher hydrocarbons. Common catalysts for SR are group VIII metals, in which Ni is the most widely used. Rh, Ru and other noble metal catalysts are more active but due to their cost are not used. Nickel SR catalysts have proved industrial performance over many decades with regard to their ease of manufacture, activation, stabilisation and chemical reactivity [6].

The hydrocarbon fuel reacts with steam over a supported nickel catalyst. The overall process consumes heat and can therefore be combined with thermal management of the fuel cell stack. The product components vary depending on process conditions. When carried out at relatively low temperatures, the product gas mixture contains hydrogen, carbon dioxide, carbon monoxide and methane, a suitable fuel mixture for SOFCs. At high temperatures, the product gas is hydrogen-rich and contains little or no methane and is suitable for all fuel cells after appropriate gas cleaning, for example, WGS reactor (Equation 14.5). Lots of integration technologies have been developed to get relatively pure hydrogen, for examples, sorption-enhanced reaction process for hydrogen production [54,55] and membrane reactor process [56].

As water is formed as the product of the electrochemical oxidation of hydrogen (Equation 14.1) at the anode, this water can be recirculated and introduced into the hydrocarbon fuel feed, rather than continuously adding water to the system. This is shown schematically in Figure 14.8. In an SOFC where the exit gas from the anode is recirculated, the additional steam from anode off-gas contributes to the total steam and the steam/hydrocarbon ratio is raised (Figure 14.9).

14.3.2 Dry reforming

Carbon dioxide, formed by WGS reaction (Equation 14.5) and electrochemical oxidation of carbon monoxide (Equation 14.2), present in the exit gas of the anode can be recirculated in the fuel supply at the cell inlet, as shown in Figure 14.8. It is well

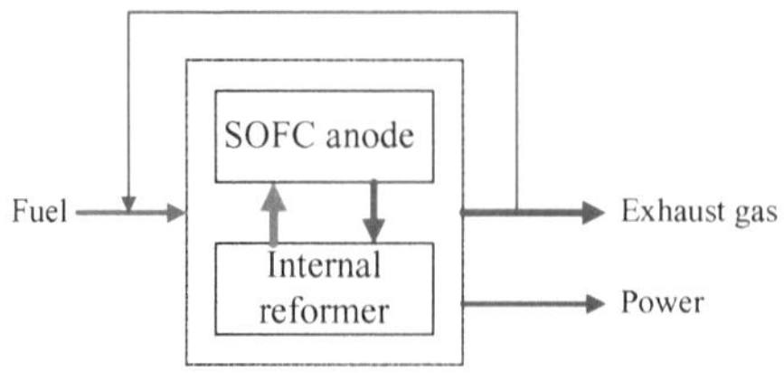

Figure 14.8 Schematic of an internally reforming SOFC with recirculation of the anode exit gas.

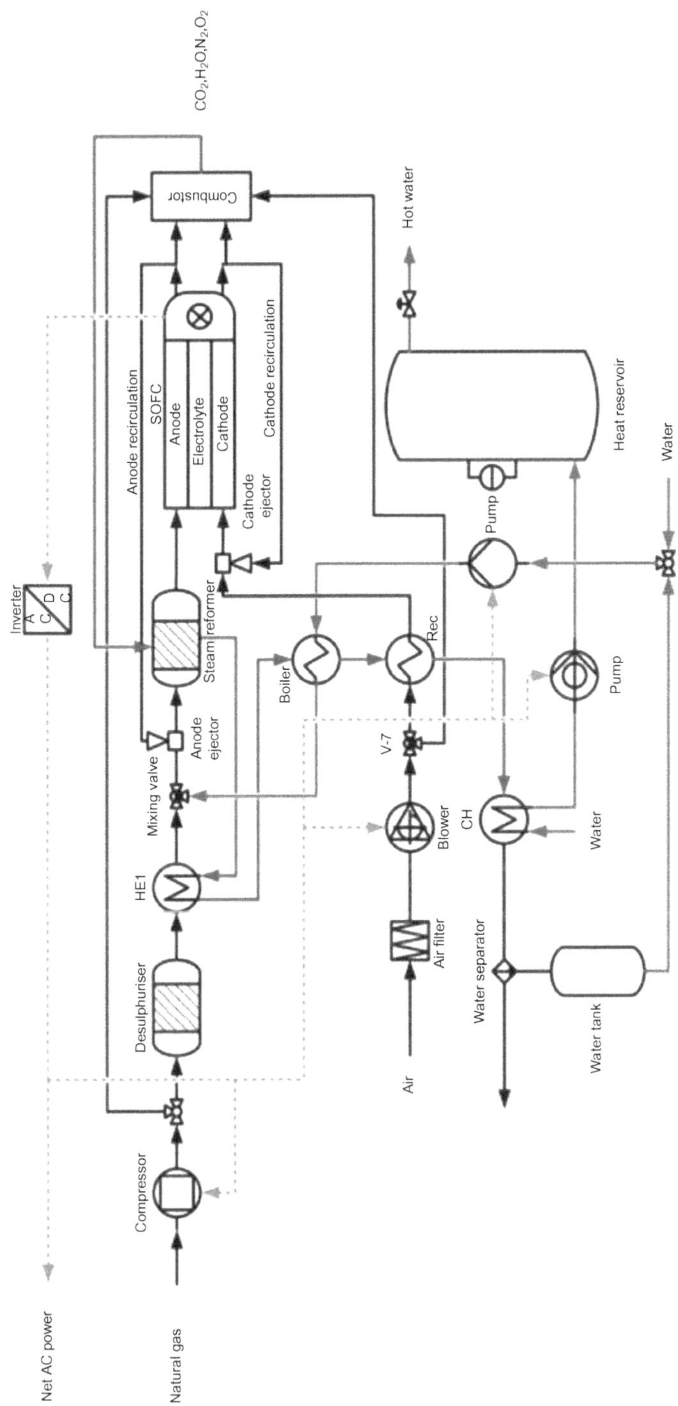

Figure 14.9 SR case with anode and cathode recirculation [90].

known that CO_2 can act as an oxidant for hydrocarbons in the presence of a suitable catalyst, so-called dry reforming (DR) as shown in Equations (14.10) and (14.11) for methane and a general higher hydrocarbon, respectively [57–61]:

$$CH_4 + CO_2 \rightarrow 2CO + 2H_2\ (\Delta H_{298} = 248\ \mathrm{kJ/mol}) \tag{14.10}$$

$$C_nH_m + nCO_2 \rightarrow 2nCO + (m/2)H_2 \tag{14.11}$$

Carbon deposition, described by Equations (14.8) and (14.9), is a particular problem with DR, especially with nickel-based catalysts [28,60,62,63]. Platinum- and rhodium-based catalysts show greater tolerance to carbon deposition [64–66]. Therefore, in addition to steam, CO_2 can also reform the methane, though it also represents a possible source of carbon deposition. Like SR, DR is also a strongly endothermic reaction.

14.3.3 Partial oxidation

In certain applications, especially in small-scale devices being developed for stand-alone or remote applications, oxygen, or simply air in many cases, is used as the oxidant rather than steam because of the cost and complexity associated with using large quantities of steam. POX occurs where a substoichiometric fuel–air (oxygen) mixture is partially combusted in a reformer, producing a hydrogen-rich syngas. Its chemical reactions are shown in Equations (14.12) and (14.13) for methane and a general higher hydrocarbon, respectively:

$$CH_4 + (1/2)O_2 \rightarrow CO + 2H_2\ (\Delta H_{298} = -35.4\ \mathrm{kJ/mol}) \tag{14.12}$$

$$C_nH_m + (n/2)O_2 \rightarrow nCO + (m/2)H_2 \tag{14.13}$$

There are two types of POX depending on whether a catalyst is used: thermal partial oxidation (TPOX) and catalytic partial oxidation (CPOX). Since the catalytic reaction rate is rapid, the CPOX reactor is extremely small, and in comparison to the non-catalytic partial oxidation (TPOX) process, there is no generation of soot or other unnecessary by-products. Some work has demonstrated the feasibility of using POX in SOFCs running on natural gas and bottled gas (propane/butane) [6,67–71].

Using oxygen, or air, is much simpler than steam, more convenient and less costly in terms of system configuration. Air is used as the source of oxygen, as a consequence of which the product gas is diluted with residual nitrogen, resulting in reduced cell voltage and electrical efficiency.

Being an exothermic reaction, in contrast to the strongly endothermic SR/DR reaction, POX can utilise the heat from the exothermic fuel cell reactions. In order to maximise the power output from the SOFC, it is necessary for the internal reforming catalyst or the anode to be selective for POX of the hydrocarbon to syngas.

The CPOX of hydrocarbons has been studied over many years [6,18,65,72–81]. A particular problem is to develop a catalyst and operating regime where high selectivity to POX at sufficiently high activity, whilst avoiding carbon deposition on the catalyst via Equations (14.8) and (14.9) [6,76,81–83]. For cost reasons, supported nickel catalysts are generally preferred although highly dispersed platinum and rhodium catalysts are also commonly used because of their greater activity and resistance to carbon deposition [75,78,79]. Clearly, carbon deposition is very undesirable both in heterogeneously catalysed hydrocarbon conversion processes and in internally reforming SOFCs. However, if an excess of oxygen is used, then there is a marked tendency for complete oxidation, that is, combustion, to CO_2 and H_2O to occur (Equations 14.14 and 14.15):

$$CH_4 + 2O_2 \rightarrow CO_2 + 2H_2O\,(\Delta H_{298} = -35.7\,\text{kJ/mol}) \tag{14.14}$$

$$C_nH_m + (n + m/4)O_2 \rightarrow nCO_2 + (m/2)H_2O \tag{14.15}$$

CO_2 and H_2O cannot be electrochemically oxidised at the anode of SOFCs. Hence there is a further loss of efficiency compared to using steam as the oxidant, if any complete oxidation products (as CO_2 and H_2O) are formed in the catalytic oxidation process.

Product gas composition will depend on the establishment of equilibrium between the accompanying reactions of Equations (14.3) and (14.5). In addition, the carbon limits are established by the potential of the carbon-forming reactions Equations (14.6)–(14.9). The knowledge of POX of hydrocarbons is primarily attributed to the decades of work devoted to POX of methane or natural gas [74,84–87], and not much attention has been given to POX of higher hydrocarbon [88,89]. Liquid hydrocarbons are activated at much lower temperatures than methane but are more challenging due to increased risk of carbon formation.

Schematic diagrams of a conceptual natural gas-fuelled SOFC system with external SMR (Steam Methane Reforming) and POX reactor are shown in Figures 14.9 and 14.10 [90]. By comparing the diagrams, it can be seen that heating of the POX reactor is not required during the fuel processing due to the exothermic reaction involved. Furthermore, hot air is mixed with the fuel flow before the POX as shown in Figure 14.10.

14.3.4 Autothermal reforming

ATR uses both oxygen (and sometimes carbon dioxide) and steam in a reaction with hydrocarbon to form syngas. The reaction takes place in a single chamber where the hydrocarbon is partially oxidised. The reaction is exothermic due to the oxidation. ATR is a combination of endothermic SR reactions (Equations 14.3 and 14.5) and exothermic POX reaction (Equations 14.12 and 14.13). The heat produced by the POX is used in the SR reaction to generate syngas. It is desirable to balance the reactions in

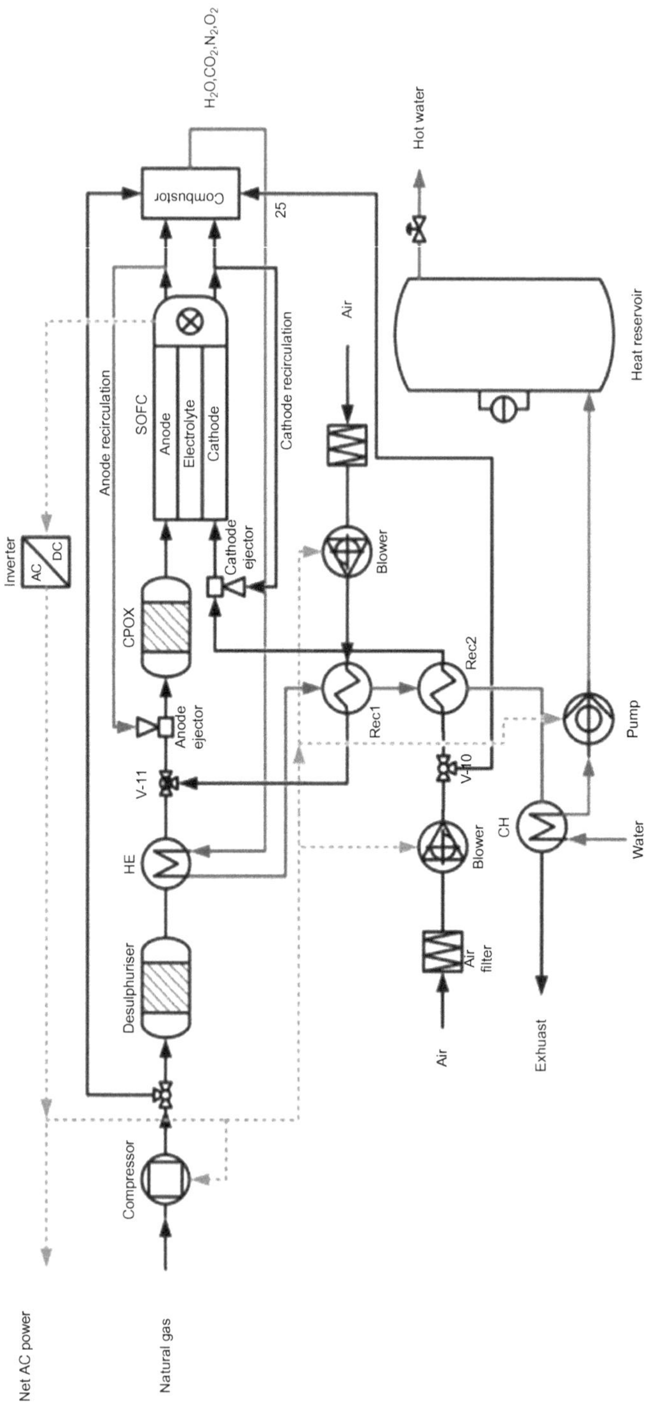

Figure 14.10 POX reforming case with anode and cathode recirculation [90].

such a way that it is essentially thermoneutral, requiring no net input of heat. The overall reaction of ATR can be described by the following Equations (14.16)–(14.19).

$$CH_4 + (1/4)O_2 + (1/2)H_2O \rightarrow CO + (5/2)H_2 \quad (14.16)$$

$$CH_4 + (1/2)O_2 + (1/2)CO_2 \rightarrow (3/2)H_2 + (3/2)CO + (1/2)H_2O \quad (14.17)$$

$$C_nH_m + (m/4)O_2 + 2(n - m/4)H_2O \rightarrow nCO + 2nH_2 \quad (14.18)$$

$$C_nH_m + ((n-1)/2)O_2 + CO_2 \rightarrow (n+1)CO + (m/2)H_2 \quad (14.19)$$

The thermoneutrality of the autothermal reactor depends on the values of n and m. In an ideal configuration, the first portion of the reactor is filled with a POX catalyst, followed by SR catalyst in the second part of the reactor. In this concept, the heat generated from the exothermic POX reaction is strategically coupled with the heat requirement of the endothermic SR reaction. With novel ATR catalysts developed in the past decade [91,92], only small amounts of the catalyst are required to initiate oxidation at low temperatures and to produce the heat required for the SR reaction. Two different classes of materials have been examined: Group VIII transition metals (e.g. platinum, rhodium or nickel) supported on an ion-conducting oxide (e.g. ceria, gadolinium-doped ceria), and mixed non-noble metal oxides with the ABO_3 stoichiometry and the perovskite structure.

Development of a low-cost and efficient fuel cell system hinges, to a large extent, on the development of small, compact and cost-effective reformers with reduced startup times. ATR is particularly attractive from these considerations [6]. The coupling of exothermic and endothermic reaction changes the dynamic behaviour of the reactor, which can be quite complex with respect to parametric sensitivities. Heat loss is minimised by integration of reaction and heat exchange, but making the reactor less sensitive to perturbations remains a challenge as these disturbances influence the catalytic activity in different zones along the axis of the reactor which, in turn, strongly affects reactor performance. Design must therefore involve optimisation of the operating conditions and reactor/catalyst structure [93].

14.3.5 Tri-reforming

Tri-reforming of methane involves three reforming reactions: endothermic SR (Equations 14.3 and 14.5), endothermic DR (Equation 14.10) and exothermic POX (Equation 14.11).

Tri-reforming of methane is a novel process of methane reforming by carbon dioxide, steam and oxygen simultaneously in one reactor [94,95]. The properties that make methane tri-reforming interesting are: a higher resistance against coke deactivation compared to DR due to the presence in the reaction environment of oxidants (H_2O and O_2) that could eliminate the coke previously generated (Equations 14.6–14.9); it is less endothermic than DR or SR as POX is exothermic and mitigates the endothermicity of the other two processes and finally, the H_2/CO molar ratio could be controlled by modifying the reagents ratio, yielding a syngas with a H_2/CO molar ratio around 2.

A biogas-fed SOFC operating in combination with an external tri-reforming system was investigated [96]. The tri-reforming process was carried out using a ceria supported Ni catalyst (1.75-wt% Ni/CeO_2). The results indicated that the process using biogas tri-reforming and SOFC was promising for application in small- and medium-sized stationary power systems.

14.3.6 Plasma reforming with/without catalyst

A plasma-driven process is also being developed for the reforming of hydrocarbon fuels into hydrogen-rich gases for use in fuel cells because DR of methane to syngas has attracted considerable interest [97–101]. The conversion process of CO_2 and CH_4 is, however, not straightforward. Indeed, CO_2 is a highly oxidised, thermodynamically stable molecule, and its use in redox reactions requires highly energetic reactants or electro-reducing processes. CH_4 is also chemically inert and slow to react. The high reaction temperature implied that this process is very costly. But recent plasma conversion of methane has gained especially remarkable progress [102–106].

High temperature non-catalytic reforming offers many advantages over thermal catalytic reforming. The high gas temperatures obtained via plasmas reduce the volume and weight of the reformer due to the fast kinetic rates. Plasma reformer-fuel cell systems could utilise a wide range of hydrocarbon fuels including diesel, gasoline, biomass, natural gas and alcohols [107,108].

14.3.7 Direct electrocatalytic oxidation

In principle, SOFCs can operate on natural gas or other hydrocarbon fuels without the addition of any oxidant to the fuel, instead directly oxidising the hydrocarbon fuel on the anode using the oxide ions which have passed through the solid electrolyte from the cathode [7,9,109–115]. The hydrocarbon can either be partially oxidised by the oxide ions to carbon monoxide and hydrogen (Equations 14.20 and 14.21), or fully oxidised to CO_2 and water (Equations 14.22 and 14.23), or undergo a mixture of partial and total oxidation:

$$CH_4 + O^{2-} \rightarrow CO + 2H_2 + 2e^- \tag{14.20}$$

$$C_nH_m + nO^{2-} \rightarrow nCO + (m/2)H_2 + 2ne^- \tag{14.21}$$

$$CH_4 + 4O^{2-} \rightarrow CO_2 + 2H_2O + 8e^- \tag{14.22}$$

$$C_nH_m + (2n + m/2)O^{2-} \rightarrow nCO_2 + (m/2)H_2O + (4n + m)e^- \tag{14.23}$$

This is especially true if the hydrocarbon is partially oxidised to CO and H_2 (Equations 14.20 and 14.21), rather than fully oxidised to CO_2 and H_2O (Equations 14.22 and 14.23), since this results in syngas from which other useful chemicals can be produced, in addition to electricity and heat. In this case, the SOFC can be viewed as an electrocatalytic reactor [116,117].

The major problem with DEO of the hydrocarbon fuel at the anode is the marked tendency towards carbon deposition via hydrocarbon decomposition (Equations 14.8 and 14.9). It is extremely difficult to avoid carbon deposition in the absence of a co-fed oxidant. However, some studies have reported anodes which show considerable promise for the DEO of hydrocarbons [97,114,115]. The conditions under which these anodes can be used may present problems for their widespread application, whilst their long-term durability with respect to carbon deposition must be established. Electrically conducting oxides have also been proposed as having potential for use as anodes for the DEO of hydrocarbons [111,112,118].

The concept of using an SOFC as an electrocatalytic reactor for chemical cogeneration has attracted much interest [116,117], offering the possibility of achieving higher product selectivity using electrochemically pumped oxide ions compared to gas-phase oxygen, whilst at the same time using pure oxygen rather than air as the oxidant, which would bring significant cost–benefits, since generation of pure oxygen from air constitutes a significant cost. In addition to syngas, it has also been shown that oxidative coupling of methane to ethene and ethane can be carried out in an SOFC electrocatalytic reactor [7,117] (Equations 14.24–14.26):

$$2CH_4 + O^{2-} \rightarrow C_2H_6 + H_2O + 2e^- \tag{14.24}$$

$$2CH_4 + 2O^{2-} \rightarrow C_2H_4 + 2H_2O + 4e^- \tag{14.25}$$

$$C_2H_6 \rightarrow C_2H_4 + H_2 \tag{14.26}$$

In some SOFC applications, especially small-scale and directly reforming systems, a lower steam to carbon ratio is desirable because of the costs and complexity associated with using large quantities of steam, and the problem of sintering of the nickel anode particles at high steam levels [11,119–122]. Consequently, there is considerable interest in developing catalyst and anode formulations which are resistant to carbon deposition at lower steam contents. Some work has shown that nickel-based anodes can have excellent selectivity for POX of methane with minimal carbon deposition [67,123].

Direct carbon fuel cells (DCFCs) are still at the very preliminary stage of development. Although a wide range of designs and concepts have been tested, there is currently no clear leader within the field with respect to optimal design or operating parameters. Figure 14.11 shows a schematic of the basic operating principle of DCFCs using molten tin at the anode to allow the carbon to react [11].

14.4 Carbon deposition and removal

Carbon formation is an important consideration in the conversion of hydrocarbon feedstocks either in the fuel processing reactor or in the anode chamber of the SOFC. The two main routes of carbon formation during SR, POX and ATR are

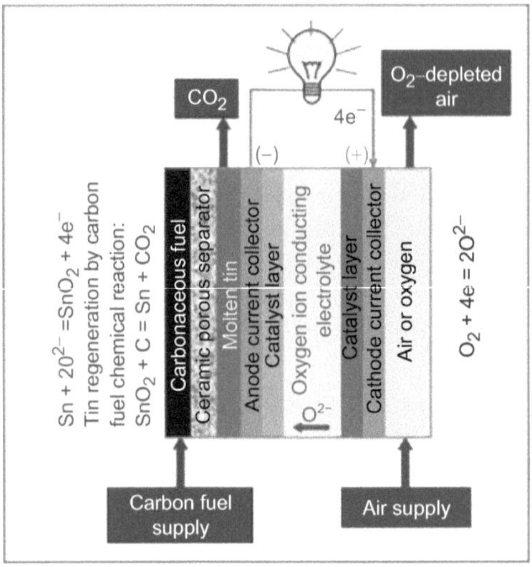

Figure 14.11 A schematic of the basic operating principle of DCFCs [11].

(i) hydrocarbon cracking or thermal decomposition (Equations 14.8 and 14.9) and (ii) the Boudouard reaction (Equation 14.7). Both reactions are catalysed by transition metals and can lead to the formation of large deposits of carbon on catalysts and metal surfaces. The Boudouard reaction, dissociating CO to carbon and CO_2, is thermodynamically favoured at temperatures below 700 °C and is not kinetically favoured at temperatures below 300 °C. Whilst methane decomposition is favoured at high temperatures (>650 °C), decomposition of higher hydrocarbons occurs at much lower temperatures: the higher the carbon number, the lower their activation temperature and hence higher the propensity for carbon formation [12,35,58,124]. Ethylene (C_2H_4) is reported to be the main reason for rapid carbon formation in the reforming process [125]. Olefins and aromatics have even higher thermodynamic potential for carbon formation. Carbon formation can also occur from gas-phase reactions, particularly with heavier hydrocarbon fuels. As it is not adsorbed on catalyst surfaces, gas-phase coke does not lead to catalyst deactivation. However, accumulation of gas-phase coke leads to an increase in pressure drop and may eventually lead to partial or complete blockage of the reactor [6,126–128].

Hydrocarbons can react homogeneously in the gas phase and heterogeneously on catalyst surfaces, possibly causing dehydrogenation and carbon deposition. These processes are complex and difficult to predict quantitatively. Detailed homogeneous chemistry including molecular-weight growth and the formation of polyaromatic hydrocarbons can involve several thousand elementary reaction steps. The heterogeneous chemistry is even more difficult. However, although not entirely quantitative, equilibrium predictions for stable graphite can provide a qualitative guide to where deposits may be problematic [9].

Coke is a high-molecular-weight polymer with low hydrogen content. Formation of coke on the catalyst surface is thought to occur in many instances by polymerisation of aromatic compounds originally present or formed in reactions. Olefin and aromatic contents in diesel and jet fuels are precursors of coke formation. Coke formation could cause degradation of the reformer catalyst and reduces its life time significantly. Nickel in particular is well known for its propensity to promote hydrocarbon pyrolysis and the build-up of carbon [6,35,49,52,53,57,58,124,129–131]. Carbon deposition can also occur on the reforming catalyst and anode in the SOFC via the disproportionation of CO (the Boudouard reaction) (Equation 14.7) and by reduction of CO by H_2 (Equation 14.6).

The build-up of carbon (coking) on either the internal reforming catalyst or on the anode, or indeed anywhere else in the fuel supply inlet manifold, is a critical problem to be avoided, or at least minimised, since over time this can lead to a loss of reforming activity and blocking of active sites on the reforming catalyst and the anode, and a loss of cell performance and poor durability. In extreme cases, growth of carbon filaments or whiskers can restrict the gas flow in the fuel supply system and result in actual physical blockages, whilst also pushing the anode particles apart and reducing conductivity [6,12,21,35,124,131].

The phenomenon of carbon deposition on SR catalysts has been extensively studied [21,25,28,57,58,124,130,132]. Carbon deposition through hydrocarbon decomposition is most prevalent at the inlet of the reforming catalyst or directly reforming anode where there is almost no hydrogen present, and the carbon-forming reactions proceed at a faster rate than the carbon-removal reactions.

To counter the problem of carbon deposition from higher hydrocarbons, a low-temperature pre-reforming stage, at 250–500 °C, is often employed to remove the higher hydrocarbons. At these temperatures, pre-reforming of higher hydrocarbons can be carried out with no resultant carbon deposition [12,125].

Much research has been devoted to studying and characterising carbon deposition and carbon build-up on reforming catalysts and SOFC anodes [15,50,52,53, 57,124,133,134], and in establishing oxidant/hydrocarbon operating envelopes which do not give rise to carbon formation on the reforming catalyst or on the anode surface at a given operating temperature. Activity has focused on developing catalyst and anode formulations which show greater resistance to carbon deposition [25,35, 51–53,123,133,134]. For nickel-based catalysts, this is generally achieved by the addition of small quantities of dopants. Dopants that have been added to nickel anodes include gold, copper, ruthenium and molybdenum [35,53,114,134,135]. It has been demonstrated that the incorporation of small amounts of gold or molybdenum into nickel/zirconia and nickel/ceria anodes can result in significantly improved resistance to carbon deposition when operating at low oxidant/carbon ratios [35,53,123, 133,134]. Incorporation of ceria into nickel/zirconia anodes has also been shown to give significantly improved resistance to carbon deposition [136–139]. Potassium is generally added to commercial nickel SR catalysts to minimise carbon deposition [12,13], and the addition of small quantities of gold to nickel catalysts has been shown to be particularly effective in preventing carbon deposition when using fuel mixtures with a low oxidant/carbon ratio [140].

Electrically conducting oxide materials have been investigated as potential anode materials. One of the main attractions of such materials is their resistance to carbon deposition compared to nickel cermet anodes under direct reforming conditions [21,25,28,111,112,118,130,141,142]. Rhodium, platinum and ruthenium have also been shown to exhibit very good activity for SR and POX of hydrocarbons. Unlike nickel, carbon filaments are not formed on these precious metals [12,58,124]. They are, however, still susceptible to deactivation by carbon deposition on the metal surface, which blocks active sites under conditions where hydrocarbon decomposition occurs. The major disadvantage of rhodium and ruthenium compared to nickel is their high cost which precludes their use as anodes, and essentially requires them to be present in a highly dispersed form. Despite this cost disadvantage, highly dispersed rhodium, platinum and ruthenium catalysts are used in some catalytic reformers, especially smaller-scale autothermal reformers [12].

Thermodynamically based on assuming that graphitic carbon is the only possible condensed phase, carbon deposit propensity decreases with the increase of S/C and O/C (non-internal-directly reforming) or the oxygen levels (internal-directly reforming). This also gives routes for the carbon removal from the catalyst bed of the fuel processing reactor or the SOFC anode chamber.

14.5 Impurity tolerance and purification

The performance and the durability of SOFCs are affected by the presence of the impurity species in fuel feeds. The impact of major impurities, such as sulphur and carbon on the Ni–YSZ anode, has been studied by researchers, whilst the short- and long-term effects of trace impurities (As and P) are still unclear. The presence of Cd and Se in the syngas impacts the SOFC anode performance to some extent. Impurity species such as Cl, Sb, As and P cause severe cell voltage degradation due to attack on the Ni–YSZ anode. Because they react with Ni to form secondary phases in the Ni–YSZ , which reduce the catalytic activity of the anode [66].

Hydrocarbon feedstocks such as natural gas and LPG contain a number of different kinds of sulphur compounds, which include inorganic sulphur species such as COS, H_2S and CS_2, and light organic sulphur compounds such as methyl mercaptan, ethyl mercaptan, dimethyl sulphide and diethyl sulphide. In addition, gas odorants such as mercaptans and tetrahydrothiophene are added by gas suppliers at concentrations that vary according to local regulations. Biogas produced by anaerobic digestion of biomass mainly consists of CH_4 and CO_2, but also small amounts of impurities, such as, for example, H_2S (>30 ppm), halogens and siloxanes. These sulphur compounds lead to deactivation of the fuel processing catalysts used in pre-reforming and of the fuel cell anode, and in most cases, the poisoning is irreversible. Even with reversible poisoning, as with H_2S, the catastrophic decline in activity in a matter of hours or days, at best, makes it impractical to operate real commercial systems by switching to S-free gas during operation to reverse the poisoning. All sulphur compounds must therefore be removed from the gas supply to the fuel cell system to parts per billion levels in order to achieve a commercially viable operating life of the pre-reforming catalyst and the fuel cell stack [6].

Even low levels of fuel impurities can significantly affect SOFC performance. Natural gas and liquid fuels often contain relatively high levels (100–1000 ppm) of sulphur compounds. The syngas produced from these fuels can contain significant H_2S and other trace species. Coal and biomass gasification can contain numerous impurities, including H_2S.

Conventional Ni/YSZ composite cermet anodes for SOFCs have poor tolerance to even low amounts of H_2S in the fuel, and more than 100 ppm H_2S in the fuel can cause considerable loss in performance of Ni/YSZ anodes. Therefore, H_2S-containing fuels used in SOFCs must be desulphurised, which increases both the complexity and cost of fuel systems and lower overall energy conversion efficiencies. Consequently, it is highly desirable to develop new anode materials with good sulphur tolerance as well as high catalytic activity [19,143,144].

Although at the elevated operating temperature of SOFCs, the nickel anode and any internal reforming catalyst show some tolerance to sulphur [12,145–152], generally the majority of the sulphur is removed from the natural gas prior to entering the SOFC to prevent poisoning of the anode and the reforming catalyst.

At low concentrations of sulphur-containing compounds, the adsorption of sulphur on nickel is reversible, and thus low concentrations of sulphur in the feed gas can be tolerated, especially at higher operating temperatures, since the tolerance of the anode and reforming catalyst to sulphur progressively increases with temperature [12,145,146]. Any adsorbed sulphur can be removed, and the activity restored to the original activity, by switching to a sulphur-free fuel feed or by a short exposure to steam. However, at higher sulphur concentrations, irreversible sulphidation of the catalyst or anode can occur [12,151,152].

Sulphur in hydrocarbon fuel or syngas can be removed by various methods. These include high temperature desulphurisation (hydrodesulphurisation) [121,153,154], whereby sulphur-containing organic compounds are reduced to hydrogen sulphide and the corresponding hydrocarbon by adding a small quantity of hydrogen to the fuel feed and passing this over molybdenum or cobalt catalyst supported on alumina [143]. This is shown in Equation (14.27) for diethyl sulphide. The hydrogen sulphide is then removed by adsorption on zinc oxide (Equation (14.28)). Hydrodesulphurisation is generally carried out in the temperature range 350–400 °C.

$$(C_2H_5)_2S + 2H_2 \rightarrow 2C_2H_6 + H_2S \tag{14.27}$$

$$H_2S + ZnO \rightarrow ZnS + H_2O \tag{14.28}$$

Low-temperature desulphurisation catalysts have also been developed which remove organic sulphur compounds and hydrogen sulphide at room temperature [105].

In smaller SOFC systems, sulphur can be removed by adsorption methods, most frequently using activated carbon or molecular sieves [105,143]. The ability of activated carbon to adsorb H_2S can be enhanced by chemically impregnating it. Such adsorbent systems can be reactivated by thermal treatment. However, their use becomes impractical for larger-scale SOFC systems because of the large quantities of adsorbent required and the subsequent problems associated with adsorbent reactivation and disposal of the desorbed sulphur [143,155].

14.6 Application of typical reforming processes for SOFCs

14.6.1 A 250 kW external fuel processor

There are five current commercially available SOFC power generation systems on the market, most of them are fuelled with natural gas [156]. A natural gas-fuelled distributed power generation system could be first to the marketplace, which would provide early manufacturing and operational experience on large commercial scale that would benefit SOFC power system developments with other fuels [157].

Birmingham et al. [158] at Rolls-Royce Fuel Cell Systems (US) Inc. reported the development of a 250 kW EFP for a SOFC under the DOE (Department of Energy) Award. The 250 kW EFP includes three subsystems: (1) desulphuriser subsystem to desulphurise the pipeline natural gas to a total sulphur content less than 80 ppbv; (2) synthesis gas subsystem to produce a hydrogen-rich synthesis gas during fuel cell start-up and low-load operation and (3) start gas subsystem to produce a non-flammable reducing gas during fuel cell start-up and shutdown.

The EFP is supplied with pressurised pipeline natural gas and pressurised air. The inlet pipeline natural gas feed total sulphur ranged between 2 and 5 ppmv and was made up of linear-chain mercaptan and sulphide odorants.

The desulphuriser subsystem consists of a selective catalytic sulphur oxidation reactor and sorbent beds.

In the synthesis gas subsystem, a CPOX reformer to partially oxidise the fuel with a substoichiometric quantity of air to produce a synthesis gas containing H_2, CO, CO_2 and water. The operating pressure up to 12 bar absolute (Bara) had a positive impact on the reformer performance by enabling higher throughput per unit volume of catalyst. They also confirmed that carbon will form at elevated pressures when proper operating conditions are not maintained. Test results indicated that the reactor operated at an average efficiency of 77%. This was near the expected value based on thermodynamic equilibrium. The efficiency, which was lowest at the lowest load levels, was slightly sensitive to operating load.

The start gas subsystem produces a relatively low-temperature reducing gas to the SOFC generator module during start-up and shutdown operations. A start gas reactor was developed to generate an oxygen-free, reducing gas from pipeline natural gas.

The intermediate-scale integrated system (ISIS) was installed and tested to verify operation of each subsystem over its various modes of operation. It included a full-scale synthesis gas generator, a near full-scale start gas generator, and a sub-scale (18 kW) desulphuriser system.

The ISIS was successfully run over all operating states including start-up, normal operations and shutdown. Overall results demonstrated important features in performance, hardware design and controls that were used for development of the 250 kW EFP.

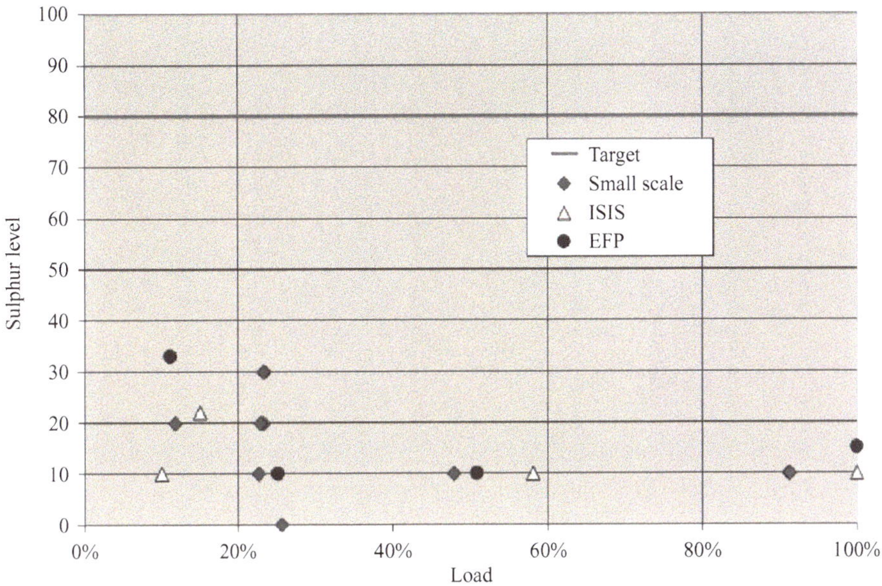

Figure 14.12 Desulphuriser performance comparison showing sulphur in ppb [158].

Figure 14.12 shows the sulphur removal levels as a function of load. The desulphuriser removed sulphur to below the target over the full load range. Also shown in the figure are test data from the small-scale developmental unit and the ISIS test.

Figure 14.13 shows a fairly constant hydrogen concentration in the synthesis gas as a function of load. Also shown in the figure are test data from the small-scale developmental unit and the intermediate-scale unit.

It was concluded that a 250 kW EFP was developed to supply all the needed gases for an SOFC fuelled by pipeline natural gas. The development tests provided scalable data for designing full-size components, and to develop an integrated system using these components. Factory testing of the 250 kW EFP demonstrated the functionality and controllability of the process overall operating states, including start-up, normal operations and shutdown. The test results demonstrated that this full-scale prototype unit is ready for integration with a 250 kW SOFC system.

14.6.2 On-board fuel processing for SOFC-APU

A gas turbine is usually used as an auxiliary power unit (APU) for an aircraft during ground time. It can release pollutant gases such as NO_x, hydrocarbons, CO, particulate matter and CO_2 in the combustion of jet fuel [159]. In recent years, to reduce emissions, SOFCs have been proposed to replace the gas turbine to provide power for aircraft and ships. Diesel engines are also highly polluting, especially when engines are

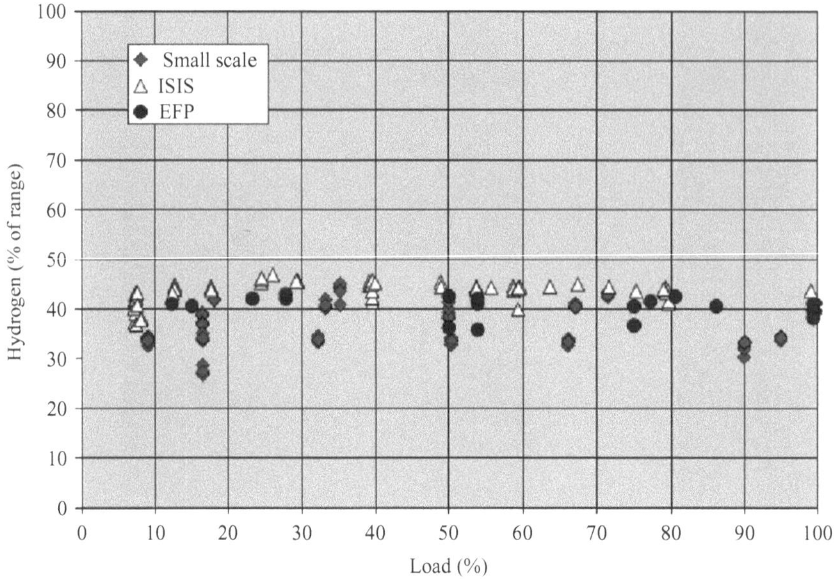

Figure 14.13 CPOX reformer performance comparison [158].

idling to provide living conditions for the driver. SOFC APUs could provide clean power and heat for these functions.

In 2012, Cummins Power Generation reported their work "Diesel-Fuelled SOFC for Class 7/Class 8 On-Highway Truck Auxiliary Power" under the support of the U.S. Department of Energy [160]. They successfully demonstrated SOFC APU operation on commercial ultra low sulphur diesel (ULSD) fuel and SOFC systems running on dry CPOX reformate. An integrated atomizer and catalyst CPOX reactor was developed as shown in Figure 14.14.

Diesel fuel is notoriously difficult to vapourise, as the final boiling point is close to the breakdown temperature of the fuel. Protonex's patent-pending MesoSpray™ atomizers can provide very small droplet sizes and uniform fuel–air mixing with very

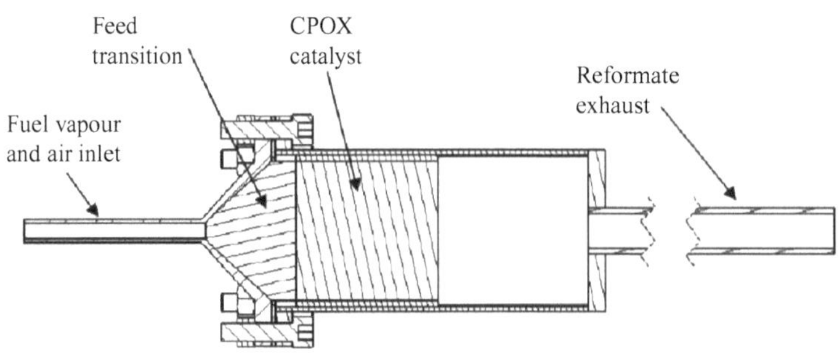

Figure 14.14 Integrated atomizer and catalyst CPOX reactor [160].

Table 14.1 **Performance at different ULSD feed rates [160]**

No.	Diesel (ml/ min)	O/C	Wet reformate composition (%)							H_2 yield	CO yield
			N_2	H_2	CO	CO_2	CH_4	Ar	H_2O		
1	6.0	1.1	50.6	21.9	23.9	0.50	0.24	0.61	2.2	0.93	0.97
2	5.0	1.1	50.8	21.5	23.8	0.49	0.40	0.61	2.5	0.91	0.96
3	1.0	1.5	57.0	15.9	16.9	4.2	0.05	0.68	5.4	0.82	0.83

low feed pressures of 6–15 kPa (1–2 psi). This atomizer produced sprays with mean diameters under 40 μm with a feed pressure of only 6 kPa. At a design pressure of 12 kPa, sprays with mean diameters under 25 μm are achievable. This leads to the formation of heavy tars that tend to clog vapourisers using small flow passages. The integrated reformer was ignited using an external torch, insulated and run at steady state for 24 h at a ULSD feed rate of 4.47 ml/min and 15.4 SLPM air without preheat. Hydrogen and carbon monoxide selectivity was each steady at 92% throughout the 24-h run. As shown in Table 14.1 [160], a product stream can be produced with about 20% H_2, 23% CO and less than 0.5% CH_4 at an O/C ratio of 1.1.

Catalyst performance was very steady for the lifetime test. After an initial decline in the first 40 h, hydrogen and carbon monoxide selectivity over the cumulative 500 h of testing (note: the catalyst sample was not changed or treated) remained constant at about 90%.

The PTX module demonstrated short-term operation on commercial ULSD without supplementary water or water recycling under controlled O/C ratio conditions. Their main results achieved were:

*Average power (net DC/AC) 1100/820 W
*CHP > 4 kW
*Start-up time (cold) 1 h.

In another DOE project, Delphi Automotive Systems teamed with PACCAR and Volvo Trucks North America. They defined system level requirements for truck auxiliary power, and subsequently designed and implemented an optimised system architecture using an SOFC APU to demonstrate and validate that the APU will meet system level goals [161]. Their primary focus was on APUs in the range of 3–5 kW for truck idling reduction. Fuels utilised were derived from low-sulphur diesel fuel. They believed that fuel consumption, pollutant levels and noise pollution can be significantly reduced if this technology is used. The SOFC APU (DPS3000D) was "fitted" to a vehicle (as shown in Figure 14.15) and initial vibration analysis/testing was performed.

14.6.3 Coal-based SOFC system

Siemens energy in the project of "SECA Coal-Based Systems" introduced a generation system plan with integrated configuration of a coal gasifier and SOFC [162]. The system design was to generate ~125 MWe net AC power at 50% efficiency, with a

Figure 14.15 The SOFC APU (DPS3000D) on vehicle [161]. *Source*: Delphi.

maximum power capability of ~170 MWe, and met CO_2 separation requirement. The coal gasification system plans to use the oxygen-blown KBR (Kellogg Brown & Root) TRIG (Transport Integrated Gasification) fluid-bed technology. At the gasifier exit, a WGS process is positioned to assure that maximum syngas carbon is available in the form of CO_2 to the next stage, a commercially available Selexol process. Syngas exiting the Selexol cleanup step will still contain sulphur at concentrations in the 20–50 ppmv range. A separate SOFC desulphurisation step could be used to reduce the sulphur concentration down to the ~0.1 ppmv level that is required for steady, long-term SOFC operation. Fuel compositions designed to enter the SOFC module for baseline system were H_2 92%, CH_4 3%, CO 2%, CO_2 2%, H_2O 0% and the others 1%. The residual fuel exiting the Selexol process was to be consumed in an oxy-combustor. Pressurised process air, serving both as the oxidant for the SOFC electrochemical process and the fuel cell coolant, was provided by a Siemens SGT-800 gas turbine. Oxygen for the coal gasification process and the oxy-combustor was provided by an ion transport membrane. The system plan was thermally integrated with a dual pressure heat recovery steam generator which satisfied the gasifier steam demand and also produced additional steam to be delivered to a steam turbine cycle. The power system cycle concept is illustrated schematically in Figure 14.16, where the magenta dashed frame labels the coal gasification and cleanup processes.

In additional, Fuel Cell Energy, Inc. (FCE) has been developing coal-fuelled, large-scale SOFC power plants [17]. The overall objective of this multiphase programme is to develop an economically viable high efficiency multi-MW (>100 MW) SOFC power system fuelled by coal-derived synthesis gas that features near-zero emissions.

14.6.4 Biomass

Biogas is an attractive fuel source that is indigenous, local, versatile and renewable. Digesters have been used on water treatment installations for many years to produce biomethane which has been used in gas engines to generate heat and power. This gas has been proved to work on molten carbonate fuel cells once desulphurised and reformed and could equally be demonstrated on SOFCs.

The alternative is to thermally gasify biomass and feed the resulting syngas into an SOFC. Such a biomass gasification process is thermally compatible with the SOFC

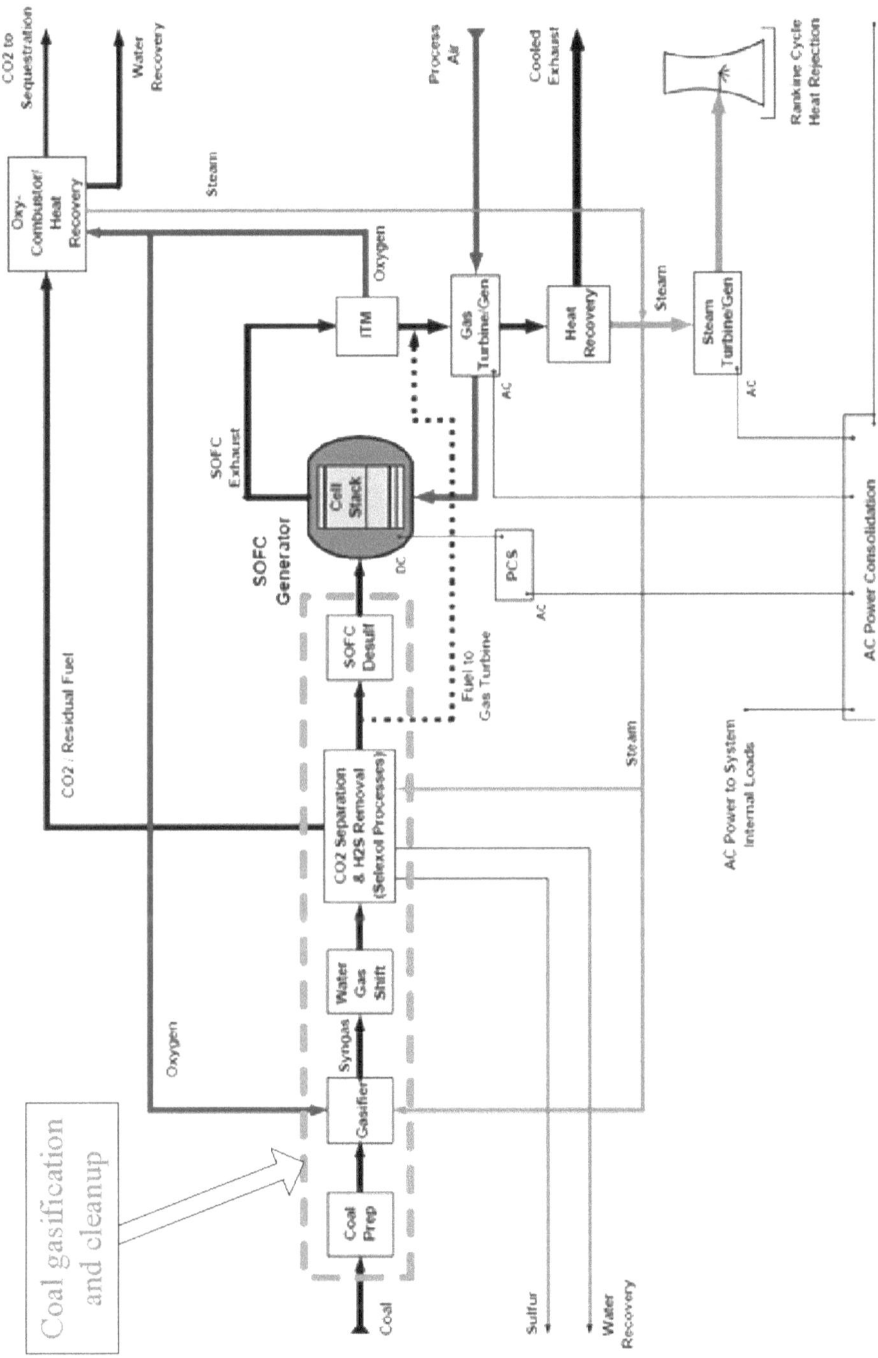

Figure 14.16 Integrated configuration of a coal gasifier and SOFC [162].

plant because they operate within the same temperature range. Different gasifying processes and biomass varieties will result in different gas compositions and qualities of the resulting biogas. Until now, the gasification technology has not been widely commercialised because of a number of technical/economic issues related to gasifier design and syngas cleanup.

Based on the apparent integration potential of SOFC and biomass gasification, some extensive modelling works based on the examination have been carried out [7] [163]. Figure 14.17 shows a flow chart of the combined SOFC–allothermal steam gasification CHPs [105]. The proposed CHP system flowchart consists of two fluidised bed reactors thermally coupled with heat pipes, a product gas cleaning train, a SOFC stack and its power conditioning, an air blower compressor, two gas-to-gas heat exchangers (HX1 and HX2), a heat recovery steam generator (HX3) and a hot-water boiler (HX4). The biomass is firstly dried and thermally decomposed to produce gases, tars and char. Then the raw product gas is cooled. In the clean steps, particulates are removed by a barrier type filter, and halogen and sulphide removal is

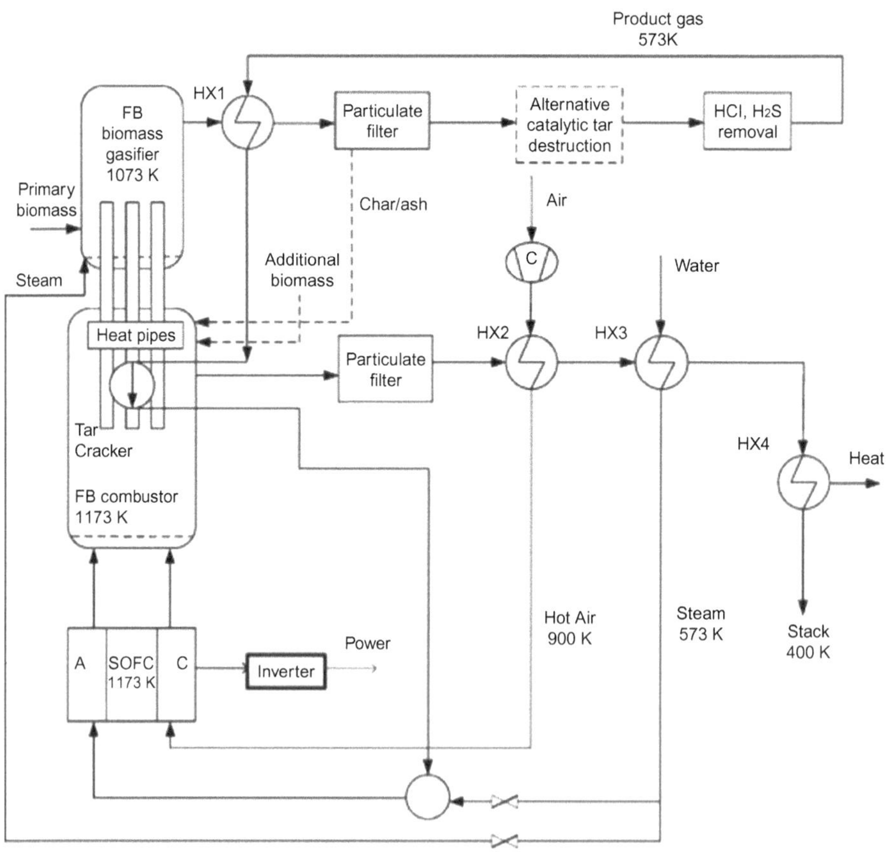

Figure 14.17 Flow chart of the combined SOFC/allothermal biomass gasification CHP [7].

accomplished in high temperature sorbent trap beds. Finally, the cleaned gas added with steam enters the anode of the SOFC stack. An energy analysis suggests that the electrical efficiency of the above-mentioned system is about 32%, whilst the thermal efficiency is 35% when olive kernel residues are used as biomass [163].

From spring 2011, a project was launched at Risø DTU in Denmark where the Biomass Gasification Group are conducting long-term operation of a 2 kW SOFC stack fuelled with gas from the Viking gasifier [164].

14.7 Brief consideration of present technology and future prospect

The choice of fuel and fuel reforming technology is strongly dependent on SOFC applications. In cases where fast heat-up, system simplicity and easy start-up are key product features, POX may be the optimum choice, but this choice significantly reduces the electrical efficiency of the fuel cell system. For optimum electrical efficiency, SR is the best reformer technology, and ATR lies in between POX and SR.

In fuel cells for stationary applications, SOFC is a strong future contender, where direct reforming of methane in the stack is desirable as it provides a mechanism for thermal management and the highest possible electrical efficiencies. This option requires a pre-reforming step with the objective of producing a methane-rich fuel stream. Similar to full reforming, POX, ATR and SR can be used for the pre-reforming process.

Direct reforming of the fuel on an SOFC anode, although offering the simplest and most cost-effective design for an SOFC system, remains a major challenge. Significant problems need to be overcome including the susceptibility of the nickel anode to carbon build-up and subsequent deactivation, and sintering of the nickel anode particles, together with the problem of steep temperature gradients across the cell caused by the strongly endothermic nature of the reforming reaction.

The development of DCFCs is very attractive even if it is in the preliminary stages of development. It is largely prohibited by the following key technical issues: (1) poor power densities, (2) high degradation rates and corrosion of cell components, (3) fuel feed system – getting the fuel to the anode, (4) up-scaling of the technology to kW and larger size systems and (5) establishing fuel processing requirements to produce low-cost fuel [11].

There is much interest at present in the development of anodes which are resistant to carbon deposition at lower oxidant/carbon ratios in the fuel feed. The development of such anodes is particularly important for SOFCs using co-fed oxygen as the oxidant or small quantities of steam added to the fuel. Furthermore, the possibility of developing SOFCs with such anodes which can run directly on practical hydrocarbon fuels, without any co-fed oxidant, something which has long been regarded as an ultimate goal for SOFCs, is now receiving renewed and vigorous effort, with some encouraging results [8,165].

An important consideration for fuel cell systems is the thermal cyclability of the fuel processing catalyst [6]. A fuel cell system is expected to undergo a number of thermal cycles during its operating life. The number of thermal cycles over the lifetime of the system is strongly applications-dependent, with up to 2000 cycles for APUs and portable systems and 20–50 cycles for stationary systems (the pre-reformer is expected to last 5 years). For the Ni-based SR catalyst, the challenge is to avoid deep oxidation of the catalyst during the regime of shutdown where fuel is turned off. A mild surface oxidation is beneficial to protect the catalyst by forming a passive surface layer which can be easily converted to the metal during heat-up to reactivate the catalyst.

Biomass processing technology will be highly relevant as biomass accounts for about 15% of the world's primary energy consumption and is renewable, clean and CO_2 neutral compared to fossil fuels. However, direct utilisation of biomass in SOFCs has yet to be fulfilled. For practical use, flexible biomass processing is needed to convert the feed into more favourable forms such as CO and H_2 as SOFC fuel inputs. Although local fuel resources are easily available and low cost, flexibility and efficiency in production are challenging for biomass conversion technologies at present.

It is more interesting to note the new development for direct, non-oxidative conversion of methane to ethylene, aromatics and hydrogen on the lattice-confined single iron sites [166]. Novel anode reactions will almost certainly prove fruitful in future.

References

[1] National Energy Policy Development Group. Reliable, Affordable, and Environmentally Sound Energy for America's Future: Report of the National Energy Policy Development Group, Washington, 2001.

[2] L. Carrette, K.A. Friedrich, U. Stimming, Fuel Cells 1 (2001) 5.

[3] World Bank Group, Environment Matters: Annual Review 2005, The International Bank for Reconstruction and Development, The World Bank, NW, 2005.

[4] M.F. Hordeski, Alternative Fuels – The Future of Hydrogen, second ed., CRC Press, Boca Raton, FL, 2008.

[5] J.A. Turner, Science 305 (2004) 972.

[6] K. Ahmed, K. Foger, Ind. Eng. Chem. Res. 49 (2010) 7239.

[7] X.W. Zhang, S.H. Chan, G.J. Li, H.K. Ho, J. Li, Z.P. Feng, J. Power Sources 195 (2010) 685.

[8] R.M. Ormerod, SOFC fuels and fuel processing, in: S.C. Singhal, K. Kendall (Eds.), High Temperature Solid Oxide Fuel Cells – Fundamentals, Design and Applications, Elsevier, UK, 2003.

[9] R.J. Kee, H.Y. Zhu, A.M. Sukeshini, G.S. Jackson, Combust. Sci. Technol. 180 (2008) 1207.

[10] K. Hemmes, A. Patil, N. Woudstra, J. Fuel Cell Sci. Technol. 5 (2008) 041010.

[11] S. Giddey, S.P.S. Badwal, A. Kulkarni, C. Munnings, Prog. Energy Combus. Sci. 38 (2012) 360.

[12] A.L. Dicks, J. Power Sources 61 (1996) 113.

[13] D. Hart, Chem. Ind. 9 (1998) 344.

[14] A.C. Lloyd, Sci. Am. 281 (1999) 64.
[15] K. Kendall, M. Palin, A small SOFC demonstrator for microelectronic applications, J. Power Sources 71 (1998) 268–270.
[16] K. Huang, J. Power Sources 196 (2011) 2763.
[17] H. Ghezel-Ayagh, S. Jolly, D. Patel, D. Stauffer, Ind. Eng. Chem. Res. 52 (2013) 3112.
[18] X.H. Xu, P.W. Li, Y.S. Shen, Appl. Energy 108 (2013) 202.
[19] C. Peng, J.L. Luo, A.R. Sanger, K.T. Chuang, Chem. Mater. 22 (2010) 1032.
[20] N. Siefert, D. Shekhawat, T. Kalapos, in: Proceeding of the ASME 2010 Eighth International Fuel Cell Science, Eng. and Tech. Conference, Fuel Cell 2010, June, Brooklyn, New York, USA, 2010.
[21] W.Y. Lee, J. Hanna, A.F. Ghoniem, J. Electrochem. Soc. 160 (2013) F94.
[22] G. Buchinger, P. Hinterreiter, T. Raab, S. Griesser, R. Claassen, D.P. Claassen, W. Sitte, D. Meissner, Trans. ASME 3 (2006) 280.
[23] F.N. Gayan, M.J. Zhi, S.R. Pakalapati, I. Celik, N.Q. Wu, R. Gemmen, J. Power Sources 185 (2008) 595.
[24] G. Krishnan, Effect of coal contaminants on solid oxide fuel system performance and service life, SRI Project No. P16935, 2008.
[25] H. Miao, W.G. Wang, T.S. Li, T. Chen, S.S. Sun, C. Xu, J. Power Sources 195 (2010) 2230.
[26] J.F.B. Rasmussen, A. Hagen, Fuel Cells 10 (2010) 1135.
[27] M.B. Pomfret, D.A. Steinhurst, J.C. Owrutsky, Energy Fuels 25 (2011) 2633.
[28] T.A. Trabold, J.S. Lylak, M.R. Walluk, J.F. Lin, D.R. Troiani, Int. J. Hydrogen Energy 37 (2012) 5190.
[29] A. Tatemi, S. Wang, T. Ishihara, H. Nishiguchi, Y. Takita, in: S.C. Singhal, M. Dokiya (Eds.), Solid Oxide Fuel Cells VIII, The Electrochemical Society Proceedings Series, PV 2003–07, 2003, p. 1260.
[30] B.C.H. Steele, A. Heinzel, Nature 414 (2001) 345.
[31] R. Doshi, V.L. Richards, J.D. Carter, X.P. Wang, M. Krumpelt, J. Electrochem. Soc. 146 (1999) 1273.
[32] J.M. Ralph, A.C. Schoeler, M. Krumpelt, J. Mater. Sci. 36 (2001) 1161.
[33] G.A. Nahar, Int. J Hydrogen Energy 35 (2010) 8891.
[34] G.A. Nahar, K. Kendall, Fuel Process. Technol. 92 (2011) 1345.
[35] C.L. Laycock, J.Z. Staniforth, R.M. Ormerod, Dalton Trans. 40 (2011) 5494.
[36] A. Abuadala, I. Dincer, Int. J. Energy Res. 36 (2012) 415.
[37] J. Huang, R.J. Crookes, Fuel 77 (1998) 1793.
[38] J. Staniforth, R.M. Ormerod, Green Chem. 3 (2001) G61.
[39] J. Staniforth, R.M. Ormerod, Catal. Lett. 81 (2002) 19.
[40] G. Campitelli, S. Cordiner, M. Gautam, A. Mariani, V. Mulone, Int. J. Hydrogen Energy 28 (2013) 320.
[41] J. Staniforth, K. Kendall, J. Power Sources 86 (2000) 401.
[42] J.F. Lin, T.A. Trabold, M.R. Walluk, D.F. Smith, Int. J. Hydrogen Energy 38 (2013) 12024.
[43] B. Grob, L. Blum, L.G.J. de Haart, A. Dengel, J. Power Sources 196 (2011) 5309.
[44] L.C. Brown, G.E. Besenbruch, R.D. Lentsch, K.R. Schultz, et al., High efficiency generation of hydrogen fuels using nuclear power, Final report to DOE, General Atomics report GA-A24285, June 2003.
[45] K.R. Schultz, in: World Nuclear Association Annual Symposium, London, September, 2003.

[46] D.J. Palmer, A combined gas cooled nuclear reactor and fuel cell cycle, PhD thesis, Polytechnic University, New York, USA, 2006.
[47] M.V. Twigg (Ed.), Catalyst Handbook, Wolfe, London, 1989.
[48] J.G. Xu, G.F. Froment, AIChE J. 35 (1989) 89.
[49] J.R. Rostrup-Nielson, L.J. Christiansen, Appl. Catal. Gen. 126 (1995) 381.
[50] R.H. Cunningham, C.M. Finnerty, R.M. Ormerod, in: U. Stimming, S.C. Singhal, H. Tagawa, W. Lehnert (Eds.), in: Solid Oxide Fuel Cells V, vol. 40, Electrochemical Society, Pennington, NJ, 1997, p. 973.
[51] C.M. Finnerty, N.J. Coe, R.H. Cunningham, R.M. Ormerod, Catal. Today 46 (1998) 137.
[52] C.M. Finnerty, R.H. Cunningham, K. Kendall, R.M. Ormerod, J. Chem. Soc. Chem. Commun. (1998) 915.
[53] R.M. Ormerod, Stud. Surf. Sci. Catal. 122 (1999) 35.
[54] G.H. Xiu, P. Li, A.E. Rodrigues, Chem. Eng. Sci. 57 (2002) 3893.
[55] S. Sircar, K.B. Lee (Eds.), Sorption Enhanced Reaction Concepts for Hydrogen Production: Materials & Processes, Research Signpost, India, 2010.
[56] F. Galluccia, A. Basile, Int. J. Hydrogen Energy 33 (2008) 1671.
[57] C.N. Satterfield, Heterogeneous Catalysis in Industrial Practice, McGraw-Hill, New York, 1991.
[58] J.R. Rostrup-Nielson, Catal. Today 18 (1993) 305.
[59] H.M. Swann, V.C.H. Kroll, G.A. Martin, C. Mirodatos, Catal. Today 21 (1994) 571.
[60] V.R. Choudhury, A.M. Rajput, Ind. Eng. Chem. Res. 35 (1996) 3934.
[61] A. Lebouvier, S.A. Iwarere, P. d'Argenlieu, D. Ramjugernath, L. Fulcheri, Energy Fuels 27 (2013) 2712.
[62] D. Duprez, M.C. Demichali, P. Marecot, J. Barbier, O.A. Ferretti, E.N. Ponzi, J. Catal. 124 (1990) 324.
[63] G. Xu, K. Shi, Y. Gao, H. Xu, Y. Wei, J. Mol. Catal. A 147 (1999) 47.
[64] A.M. Gadalla, M.E. Sommer, Chem. Eng. Sci. 44 (1989) 2825.
[65] A.T. Ashcroft, A.K. Cheetham, M.L.H. Green, P.D.F. Vernon, Nature 352 (1991) 225.
[66] F.N. Cayan, M.J. Zhi, S.R. Pakalapati, I. Celik, N.Q. Wu, R. Gemmen, J. Power Sources 185 (2008) 595.
[67] K. Kendall, M. Prica, Integrated SOFC tubular system for small-scale cogeneration, in: U. Bossel (Ed.), 1st European SOFC Forum, Luzern, (1994) 163–170 .
[68] K. Kendall, T. Alston, M. Palin, P. Windibank, M. Prica, A 1000 cell SOFC reactor for domestic cogeneration, J. Power Sources 71 (1998) 271–274.
[69] C.M. Finnerty, J. Staniforth, K. Kendall, in: A.J. McEvoy (Ed.), Proceedings of 4th European Solid Oxide Fuel Cell Forum, European Fuel Cell Forum, 2000, p. 151.
[70] E. Batawi, U. Weissen, A. Schuler, M. Keller, C. Voisard, in: H. Yokokawa, S.C. Singhal (Eds.), Solid Oxide Fuel Cells VII, vol. 16, Electrochemical Society, Pennington, NJ, 2001, p. 140.
[71] M. Pastula, J. Devitt, R. Boersma, D. Ghosh, in: H. Yokokawa, S.C. Singhal (Eds.), Solid Oxide Fuel Cells VII, vol. 16, Electrochemical Society, Pennington, NJ, 2001, p. 180.
[72] M. Prettre, C. Eichner, M. Perrin, Trans. Faraday Soc. 43 (1946) 335.
[73] D. Dissanayake, M.P. Rosynek, K.C.C. Kharas, J.H. Lunsford, J. Catal. 132 (1991) 117.
[74] D.A. Hickman, L.D. Schmidt, Science 259 (1993) 343.
[75] S.S. Bharadwaj, L.D. Schmidt, J. Catal. 146 (1994) 11.
[76] S.C. Tsang, J.B. Claridge, M.L.H. Green, Catal. Today 23 (1995) 3.
[77] C.T. Au, H.Y. Wang, Catal. Lett. 41 (1996) 159.
[78] Y. Boucouvalas, Z. Zhang, X.E. Verykios, Catal. Lett. 40 (1996) 189.
[79] K.H. Hofstad, O.A. Rokstad, A. Holmen, Catal. Lett. 36 (1996) 25.

[80] Y.H. Hu, E. Ruckenstein, J. Catal. 158 (1996) 260.
[81] V.A. Tsipouriari, Z. Zhang, X.E. Verykios, J. Catal. 179 (1998) 283.
[82] A.T. Ashcroft, A.K. Cheetham, J.S. Foord, M.L.H. Green, C.P. Grey, A.J. Murrell, P.D.F. Vernon, Nature 344 (1990) 319.
[83] A.M. Diskin, R.H. Cunningham, R.M. Ormerod, Stud. Surf. Sci. Catal. 122 (1999) 393.
[84] P.D.E. Vernon, M.L.H. Green, A.K. Cheetham, A.T. Ashcroft, Catal. Today 13 (1992) 417.
[85] A. Hickman, L.D. Schmidt, AIChE J. 39 (1993) 1164.
[86] V.R. Choudhury, V.H. Rane, A.M. Rajput, Appl. Catal. 162 (1997) 235.
[87] A. Slagtern, H.M. Swaan, U. Olsbye, I.M. Dahl, C. Microdatos, Catal. Today 46 (1998) 107.
[88] L. Ma, D.L. Trimm, C. Jiang, Appl. Catal. 138 (1996) 275.
[89] E. Newson, C.C.T.B. Truoug, Int. J. Hydrog Energy 28 (2003) 1379.
[90] V. Liso, A.C. Olesen, M.P. Nielsen, S.K. Ker, Energy 36 (2011) 4216.
[91] T. Takeguchi, S. Furukawa, M. Inoue, K. Eguchi, Appl. Catal. 240 (2003) 223.
[92] M. Ferrandon, T. Krause, Rh-based catalysts for autothermal reforming of liquid fuels into hydrogen, in: Presented at the 4th DOE National Laboratory Catalysis Conference, October 22–24, Oak Ridge, TN, 2004.
[93] G. Kolios, J. Frauhammer, G. Eigenberger, Chem. Eng. Sci. 55 (2004) 5945.
[94] C.S. Song, W. Pan, Catal. Today 98 (2004) 463.
[95] J.M. Garcc-Vargas, J.L. Valverde, J.D. Jv, P.S. Jverd, F. Dorado, Appl. Catal. B Environ. 148 (2014) 322.
[96] M.L. Faro, A.V.L. Pino, A.S. Aricò, Fuel Process. Technol. 115 (2013) 238.
[97] M.G. Poirier, J. Trudel, D. Guay, Catal. Lett. 21 (1993) 99.
[98] S. Wang, G.Q. Lu, G.J. Millar, Energy Fuels 10 (1996) 896.
[99] M. Tsui, T. Miyao, S. Naito, Catal. Lett. 69 (2000) 195.
[100] I. Omae, Catal. Today 115 (2006) 33.
[101] R. Snoeckx, R. Aerts, X. Tu, A. Bogaerts, J. Phys. Chem. C 117 (2013) 4957.
[102] L. Bromberg, D.R. Cohn, A. Rabinovich, N. Alexeev, Int. J. Hydrogen Energy 24 (1999) 1131.
[103] G. Petitpas, J.D. Rollier, A. Darmon, J. Gonzalez-Aguilar, R. Metkemeijer, L. Fulcheri, Int. J. Hydrogen Energy 32 (2007) 2848.
[104] H.J. Gallon, Dry reforming of methane using non-thermal plasma-catalysis, PhD dissertation, The University of Manchester, Manchester, 2010.
[105] J.W.C. Liberatori, R.U. Ribeiro, D. Zanchet, F.B. Noronha, J.M.C. Bueno, Appl. Catal. A Gen. 327 (2007) 197.
[106] X. Tu, H.J. Gallon, M.V. Twigg, P.A. Gorry, J.C. Whitehead, J. Phys. D Appl. Phys. 44 (2011) 274007.
[107] L. Bromberg, D.R. Cohn, A. Rabinovich, Int. J. Hydrogen Energy 22 (1997) 83.
[108] J.M. Cormier, I. Rusu, J. Phys. D Appl. Phys. 34 (2001) 2798.
[109] Y.L. Sandler, J. Electrochem. Soc. 118 (1971) 1378.
[110] K. Otsuka, S. Yokoyama, A. Morikawa, Bull. Chem. Soc. Jpn. 57 (1984) 3286.
[111] B.C.H. Steele, I. Kelly, H. Middleton, E. Rudkin, Solid State Ionics 28–30 (1988) 1547.
[112] E.S. Putna, J. Stubenrauch, J.M. Vohs, R.J. Gorte, Langmuir 11 (1995) 4832.
[113] J.T.S. Irvine, D.P. Fagg, J. Labrincha, F.M.B. Marques, Catal. Today 38 (1997) 467.
[114] E.P. Murray, T. Tsai, Nature 400 (1999) 649.
[115] S. Park, R. Craciun, V. Radu, R.J. Gorte, J. Electrochem. Soc. 146 (1999) 3603.
[116] V.V. Galvita, V.D. Belyaev, V.N. Parmon, V.A. Sobyanin, Catal. Lett. 39 (1996) 209.
[117] V.A. Sobyanin, V.D. Belyaev, V.V. Galvita, Catal. Today 42 (1998) 337.

[118] P. Han, W.L. Worrell, J. Electrochem. Soc. 142 (1995) 4235.
[119] T. Kawada, N. Sakai, H. Yokokawa, M. Dokiya, M. Mori, T. Iwata, Solid State Ionics 40/41 (1990) 402.
[120] G.J.M. Janssen, J.P. de Jong, J.P.P. Huijsmans, in: B. Thorstensen (Ed.), Proceedings of 2nd European Solid Oxide Fuel Cell Forum, European Fuel Cell Forum, 1996, p. 163.
[121] B. de Boer, M. Gonzales, H.J.M. Bouwmeester, H. Verweij, Solid State Ionics 127 (2000) 269.
[122] D. Simwonis, F. Tietz, D. Stöver, Solid State Ionics 132 (2000) 241.
[123] I.A. Proctor, A.L. Hopkin, R.M. Ormerod, Ionics 9 (2003) 242.
[124] C.H. Bartholomew, Catal. Rev. Sci. Eng. 24 (1982) 67.
[125] F. Joensen, J.R. Rostrup-Nielsen, J. Power Sources 105 (2002) 195.
[126] D.L. Trimm, Catal. Today 37 (1997) 233.
[127] D.L. Trimm, Catal. Rev. Sci. Eng. 16 (1977) 155.
[128] H.S. Bengarrd, J.K. Norskov, J.S. Sehested, B.S. Clausen, L.P. Nielsen, A. Molenbroek, J.R. Rostrup-Nielsen, J. Catal. 209 (2002) 365.
[129] S.H. Clarke, A.L. Dicks, K. Pointon, T.A. Smith, A. Swann, Catal. Today 38 (1997) 411.
[130] J.H. Koh, Y.S. Yoo, J.W. Park, H.C. Lim, Solid State Ionics 149 (2002) 157.
[131] T. Lide, M. Kawano, T. Matsui, R. Kikuchi, K. Eguchi, J. Electrochem. Soc. 154 (2007) B234.
[132] D.E. Ridler, M.V. Twigg, in: M.V. Twigg (Ed.), Catalyst Handbook, Manson, London, 1996, Chapter 5.
[133] C.M. Finnerty, R.H. Cunningham, R.M. Ormerod, in: S.C. Singhal, M. Dokiya (Eds.), Solid Oxide Fuel Cells VI, vol. 19, Electrochemical Society, Pennington, NJ, 1999, p. 568.
[134] C.M. Finnerty, R.M. Ormerod, in: S.C. Singhal, M. Dokiya (Eds.), Solid Oxide Fuel Cells VI, vol. 19, Electrochemical Society, Pennington, NJ, 1999, p. 583.
[135] N.J. Coe, R.H. Cunningham, R.M. Ormerod, in: P. Stevens (Ed.), Proceedings of 3rd European Solid Oxide Fuel Cell Forum, European Fuel Cell Forum, vol. 2, 1998, p. 39.
[136] V.D. Belyaev, T.I. Politova, O.A. Marina, V.A. Sobyanin, Appl. Catal. A Gen. 133 (1995) 47.
[137] T. Tsai, S.A. Barnett, J. Electrochem. Soc. 145 (1998) 1696.
[138] J. Staniforth, K. Kendall, in: S.C. Singhal, M. Dokiya (Eds.), Solid Oxide Fuel Cells VI, vol. 19, Electrochemical Society, Pennington, NJ, 1999, p. 603.
[139] M. Mogensen, N.M. Sammes, G.A. Tompsett, Solid State Ionics 129 (2000) 63.
[140] F. Besenbacher, I. Chorkendorf, B.S. Clausen, B. Hammer, A.M. Molenbrook, J.K. Norskov, I. Stensgaard, Science 279 (1998) 193.
[141] P. Tsiakaras, P.C. Athanasiou, G. Marnellos, M. Stoukides, J.E. ten Elshof, H.J.M. Bouwmeester, Appl. Catal. A Gen. 169 (1998) 249.
[142] O.A. Marina, C. Bagger, S. Primdahl, M. Mogensen, Solid State Ionics 123 (1999) 199.
[143] N.P. Brandon, S. Skinner, B.C.H. Steele, Annu. Rev. Mater. Res. 33 (2003) 183.
[144] Z. Cheng, J.H. Wang, Y.M. Choi, L. Yang, M.C. Lin, M.L. Liu, Energy Environ. Sci. 4 (2011) 4380.
[145] L.A. Shockling, P. Reichner, in: Proc. Symp. Fuel Cells, 14, Electrochemical Society, Pennington, NJ, 1989, p. 106.
[146] K. Kinoshita, F.R. McLarnon, E.J. Cairns, Fuel Cells, A Handbook, DOE/METC-88/6096, 1998.
[147] Y. Matsuzaki, I. Yasuda, Solid State Ionics 132 (2000) 261.
[148] S. Tao, J.T.S. Irvine, Nat. Mater. 2 (2003) 320.
[149] Y.H. Huang, R.I. Dass, Z.L. Xing, J.B. Goodenough, Science 312 (2006) 254.

[150] S.W. Zha, Z. Cheng, M.L. Liu, J. Electrochem. Soc. 154 (2007) B201.
[151] H. Yokokawa, H.Y. Tu, B. Iwanschita, A. Mai, J. Power Sources 182 (2008) 400.
[152] J.W. Yun, H.C. Ham, H.S. Kim, S.A. Song, S.W. Nam, S.P. Yoon, J. Electrochem. Soc. 160 (2013) F153.
[153] H. Topsoe, B.S. Clausen, F.E. Massoth, Hydrotreating Catalysis Science and Technology, Springer-Verlag, Berlin, 1996.
[154] K. Foger, B. Godfrey, in: A.J. McEvoy (Ed.), Proceedings of 4th European Solid Oxide Fuel Cell Forum, European Fuel Cell Forum, 2000, p. 167.
[155] W.S. Dong, H.S. Roh, K.W. Jun, S.E. Park, Y.S. Oh, Appl. Catal. A Gen. 226 (2002) 63.
[156] T.A. Adams, J. Nease, D. Tucker, P.I. Barton, Ind. Eng. Chem. Res. 52 (2013) 3089.
[157] S.D. Vora, ECS Trans. 57 (2013) 11.
[158] D.P. Birmingham, C.L. DeBellis, M.A. Perna, A.K. Upadhyayula, DOE final report (DE-FC26-06NT42809), 2008.
[159] A. Lindermeir, S. Kah, S. Kavurucu, M. Mühlner, Appl. Catal. Environ. 70 (2007) 488.
[160] Cummins Power Generation, DOE final report (DE-FC36-04GO14318), 2010.
[161] Delphi Automotive Systems, DOE final report (DE-FC36-04GO14319), 2010.
[162] Siemens Energy, DOE final report (DE-FC26-05NT42613), 2011.
[163] K.D. Panopoulos, L.E. Fryda, J. Karl, S. Poulou, E. Kakaras, J. Power Sources 159 (2006) 570.
[164] J. Ahrenfeldt, T.P. Thomsen, U. Henriksen, L.R. Clausen, Appl. Therm. Eng. 50 (2013) 1407.
[165] R.M. Ormerod, Chem. Soc. Rev. 32 (2003) 17.
[166] X.G. Guo, et al., Science 344 (2014) 616.

Index

Note: Page numbers followed by *f* indicate figures and *t* indicate tables.

F

G

H

I

K

L

FSC
www.fsc.org
MIX
Papier | Fördert gute Waldnutzung
FSC® C083411

Kolibri 360 GmbH
Ferdinand-Jühlke-Straße 7
99095 Erfurt, Deutschland
produktsicherheit@kolibri360.de